24 hours) or

Medicinal Inorganic Chemistry

ACS SYMPOSIUM SERIES **903**

Medicinal Inorganic Chemistry

Jonathan L. Sessler, Editor
The University of Texas at Austin

Susan R. Doctrow, Editor
Eukarion Pharmaceuticals, Inc.

Thomas J. McMurry, Editor
EPIX Medical, Inc.

Stephen J. Lippard, Editor
Massachusetts Institute of Technology

Sponsored by the
ACS Divisions of Inorganic Chemistry, Inc. and
Medicinal Chemistry

American Chemical Society, Washington, DC

Library of Congress Cataloging-in-Publication Data

Medicinal inorganic chemistry / Jonathan L. Sessler, editor ... [et al.].

p. cm.—(ACS symposium series ; 903)

"Sponsored by the ACS Divisions of Inorganic Chemistry, Inc. and Medicinal Chemistry."

Includes bibliographical references and index.

ISBN 0-8412-3899-5 (alk. paper)

1. Inorganic compounds—Therapeutic use—Congresses. 2. Minerals in pharmacology—Congresses. 3. Organometallic compounds—Therapeutic use—Congresses.

I. Sessler, Jonathan L. II. American Chemical Society. Division of Inorganic Chemistry, Inc. III. American Chemical Society. Division of Medicinal Chemistry IV. Series.

RS166.M43 2005
615′.2—dc22 2004062719

The paper used in this publication meets the minimum requirements of American National Standard for Information Sciences—Permanence of Paper for Printed Library Materials, ANSI Z39.48–1984.

Distributed by Oxford University Press

PRINTED IN THE UNITED STATES OF AMERICA

Foreword

The ACS Symposium Series was first published in 1974 to provide a mechanism for publishing symposia quickly in book form. The purpose of the series is to publish timely, comprehensive books developed from ACS sponsored symposia based on current scientific research. Occasionally, books are developed from symposia sponsored by other organizations when the topic is of keen interest to the chemistry audience.

Before agreeing to publish a book, the proposed table of contents is reviewed for appropriate and comprehensive coverage and for interest to the audience. Some papers may be excluded to better focus the book; others may be added to provide comprehensiveness. When appropriate, overview or introductory chapters are added. Drafts of chapters are peer-reviewed prior to final acceptance or rejection, and manuscripts are prepared in camera-ready format.

As a rule, only original research papers and original review papers are included in the volumes. Verbatim reproductions of previously published papers are not accepted.

ACS Books Department

Contents

Indexes

Chapter 1

Metal Ion Chemistry for Sustaining Life

Stephen J. Lippard

Department of Chemistry, Massachusetts Institute of Technology, Cambridge, MA 02139

The human cell is very smart. In the latter half of the 20th century, scientists made great progress in understanding how information is stored, transmitted, and translated into function in cells and organisms. Detailed mechanisms emerged revealing how enzymes were able to catalyze specific cellular transformations. Large sums of money have been spent both by government as well as private industry in an attempt to translate this wealth of information into the improvement of the human condition by the rational design of pharmaceuticals for the diagnosis and, especially, the treatment of both chronic and acute illnesses. Despite these efforts, however, serendipity still plays a more important role than intention in the invention process. For just when it might appear that a cellular pathway or enzyme might be targeted, based on mechanistic and structural biochemistry, the human cell devises a resistance mechanism that sends the medicinal chemist back to the drawing board.

Against this background it is clear that new approaches are needed. The articles in this book, which describe the activities of scientists in the fledgling area of "Medicinal Inorganic Chemistry," detail activities that have the potential to provide the requisite breakthrough science. In September of 2003 at the national meeting of The American Chemical Society in New York there was a symposium on Medicinal Inorganic Chemistry in which all of the present authors, together with a number of other chemists, were participants. The full program is attached as Table 1. In a prescient leadoff lecture, John Kozarich drew upon his academic and Merck and Activix industrial experiences to discuss how metal ions offer a form of diversity for drug discovery waiting to be exploited by the pharmaceutical sector. Historically, medicinal inorganic chemistry is rich in metal- or metalloid-based drugs, including Paul Erlich's organoarsenic compound for the treatment of syphilis, antiarthritic gold preparations, and diagnostic agents for magnetic resonance imaging (Gd, Mn, Fe) among others. But drug discovery and development is much more complex today than in the past and, to quote Kozarich, "Despite the explosion in medicinally-oriented chemical diversity, inorganic compounds have not captured a significant share of library space within the [drug industry]." The statistics speak for themselves. Of 89 new molecules granted 210 specific indications approved by the FDA from 1949 to 2003 for cancer therapy, only 6 were metal complexes or inorganic compounds. Three were platinum complexes, discussed in a later chapter by Reedijk. An attempt to redress the balance for both therapeutic and diagnostic applications has been pioneered by the National Institute of General Medical Sciences under the capable leadership of Peter Preusch, who reviews here the status of the Metals in Medicine program. Of particular interest in his chapter is the call for new research on metal metabolism and for more biomedical applications in the bioinorganic portfolio. According to Preusch, "the greatest barrier to further development is the perception that metals are toxic." Further discussion of the drug discovery process, including lead optimization, may be found in the chapters by Giandomenico describing dendritic gadolinium complexes now undergoing clinical trials for cardiac magnetic resonance imaging (MRI), by Farrell on polyplatinum compounds for cancer treatment, and by Schenck on detection of neurodegenerative disease by high field MRI studies of brain iron as an endogenous contrast agent.

Metal compounds offer special advantages as diagnostic agents. The field of magnetic resonance imaging relies upon the use of endogenous metal ions, or their exogenous introduction, to alter T_1 or T_2 values of water molecules in the vicinity of the paramagnetic center. Of particular interest in this respect is the use of Gd^{3+}, Mn^{2+}, and iron complexes. A session on metal complexes and MRI at the symposium highlighted recent challenges and advances in this area. A texaphrin complex of Gd^{3+} (Xcytrin®; motexafin gadolinium), an MRI-detectable compound with anticancer properties, is now in clinical trials and studies in vitro have defined models for how it might exert its antineoplastic effects through the generation of reactive oxygen species (ROS), as detailed in a chapter by Magda and colleagues. Additional work on bile acids as novel intravascular gadolinium MRI contrast agents, on controlling water exchange rates as a key parameter in the next generation of Gd^{3+} imaging compounds, and on targeted MRI imaging is described, respectively, by Anelli, Sherry, Caravan and co-workers, as well as by Platzek and Schmitt-Willich. In most of this research a stong foundation in the principles of coordination chemistry and ligand design is required for ultimate success in the biomedical application of interest.

The control of the proper concentration of metal ions in cells, a process known as homeostasis, is extremely important for maintaining good human health. Failure to perform this function adequately underlies many diseases, including those of aging such as Alzheimer's. In a session devoted to endogenous metals in disease, and their manipulation as approaches to therapy, Rogers presented evidence the the Alzheimer's precursor protein (APP) messenger RNA contains an iron regulatory element and thus must be involved in metal metabolism. Brewer discussed the ability of the simple inorganic ion tetrathiomolybdate, MoS_4^{2-}, to modulate copper concentrations and treat Wilson's disease patients. Malaria is another illness in the sights of medicinal inorganic chemists, as Wright relates, with peptide dendrimers targeting iron to block aggregation following heme release by the parasite that carries the disease. Details of how heme can be the trigger of malaria are described by Meunier and co-workers, who review mechanistic studies of artemisin, an organic peroxide with antimalarial properties.

The last session of the symposium covered redox- and chelation-based drugs. The dioxygen molecule is essential for human life, providing energy through the oxidation of carbohydrates and creating membrane gradients by a process coupled to oxidative phosphorylation. In many instances, metalloproteins are required for the transport and activation of O_2, but when these reaction are not fully controlled, toxic byproducts of dioxygen metabolism can escape the local environment and do damage to tissues. A chapter by Crow details the use of manganese porphyrin complexes as antioxidants to delay the onset of amyotrophic lateral sclerosis (ALS) in a mouse model for this devastating human disease; mutations in copper-zinc superoxide dismutase (SOD) are causative for the inherited form of ALS, producing oxidative stress in the spinal cord through unknown mechanisms. The manganese complexesare proposed as a potential therapy for ALS. Doctrow similarly describes Mn salen complexes as SOD and catalase mimics, the latter converting peroxide into dioxygen and water, and as potential therapeutics for neurodegenerative diseases. Hart and Valentine discuss how the mutations in the Cu-Zn SOD may cause protein misfolding, believed by many to be the cause of the ALS phenotype, and present crystallographic evidence to support the theory. Bergeron and colleagues describe treatments for iron overload using specific chelating agents, iron being one of the metal ions that can contribute to the generation of toxic ROS if its levels are not properly controlled. A long known link between vanadium and the enhancement of insulin release is exploited by maltolate complexes of the vanadyl ion (VO^{2+}) for delivery and human treatment of diabetes as related in a chapter by Orvig. Farmer and colleagues present a strategy for targeting melanoma by delivery of metal complexes to these cancer cells. The medicinal applications of N-heterocyclic carbene complexes of Ag^+ are reported in the final chapter by Youngs.

As can be seen from this brief overview, the area of medicinal inorganic chemistry is replete with creative approaches to the diagnosis and treatment of human disease. Many novel strategies have opened new frontiers that expand the purview of traditional medicinal chemistry. The work described intersects with many domains of importance for maintaining human health, and these areas are therefore of considerable commercial potential. This realization should encourage our colleagues in the pharmaceutical industry to reach out to bioinorganic chemsts and create the marriage required to bring a new focus on metal ions for sustaining life.

Chapter 2

Medicinal Inorganic Chemistry: Promises and Challenges

John W. Kozarich

ActivX Biosciences, Inc., 11025 North Torrey Pines Road, La Jolla, CA 92037

Medicinal inorganic chemistry remains a field of great promise with many challenges. The potential for a major expansion of chemical diversity into new structural and reactivity motifs of high therapeutic impact is unquestionable.

Introduction: Quest for Chemical Diversity

The search for new, effective medicines for human health and for the nearly $500 billion world-wide pharmaceutical industry invariably requires the ability to access new regions of chemical diversity. Chemical diversity for the purposes of this discussion refers to the arrangements of atoms within molecules that create a broad range of structural, spatial and reactivity combinations that can be interrogated against a biological or pharmacological response. We normally refer to these collections of chemically diverse compounds as libraries and the interrogated responses as assays. The sorting of chemical libraries against biological assays has been the linchpin of drug discovery for over one hundred years. The range of biological assays available today is unprecedented – from whole animal evaluation that was the mainstay of early discovery a few decades ago to miniaturized, high-throughput, multi-array analysis against individual molecular targets. The range of chemical diversity that can be accessed is stupendous.

The drug discovery industry has created the lion's share of this chemical diversity. During the past century, organic medicinal chemists have synthesized millions of new compounds – some purely synthetic creations; some variations on the natural products that have been identified along the way; some as single well-characterized compounds; some as mixtures of isomers or related compounds. In general, the diversity libraries created by medicinal chemists have largely been a historical record of the therapeutic targets their particular company has pursued. Thus, some libraries are rich in steroid-type structures and others are rich in antibiotic pharmacophores. The advent of combinatorial chemistry and chemoinformatics over the past 15 years has enabled drug discovery companies to quantify the scope of chemical diversity within their libraries, identify sparsely-represented regions, and rapidly fill those regions in with many millions of synthetic molecules as either single entities or as a cocktail of related compounds.

Despite the explosion in medicinally-oriented chemical diversity, inorganic compounds have not captured a significant share of library space within the pharmaceutical sector. Despite the impressive promise of medicinal bioinorganic chemistry clearly revealed in the subsequent chapters of this book, few inorganic compounds have reached the goal of FDA-approved drug. The reasons for this are at once simple and complex. I offer my own perspective on the promise and challenges of medicinal bioinorganic chemistry from the vantage point of a scientist who has functioned at the periphery of this discipline but believes that the field will play a crucial role in our understanding of human biology and in the development of innovative new medicines.

Promise of Medicinal Inorganic Chemistry

The use of metals in medicine is as old as recorded human history (*1*). Modern successes span from what was arguably the first medicinal chemistry screening campaign by Paul Erlich to the recent development of sophisticated bioimaging agents. The therapeutic applications of metal-based drugs span virtually every disease area: anticancer (Al, Ga, In, Ti, Ru, Pt, Au, Sn); antimicrobial (As, Cu, Zn, Ag, Hg, Bi); antiarthritic (Au); antipsychotic (Li); antihypertensive (Fe, Zn); antiviral (Li, Pt, Au, W, Cu); antiulcer (Bi); antacids (Al, Na, Mg, Ca); metalloenzyme mimetics (Mn, Cu, Fe); radiotherapy (e.g. Re,

Y, Pb); β-emitters (90Y, 212Bi); and metal chelators. Diagnostic applications are equally impressive and have received generally greater acceptance in mainstream medical practice: Radiosensitization (Pt, Ru); magnetic resonance imaging (e.g. Mn, Gd, Fe); X-ray imaging (e.g. Ba); radio-imaging (e.g. 99mTc, 111In). Recent reviews have nicely described the scope and potential of these applications (*2,3*).

Nearly one hundred years ago, modern medicinal chemistry was off to an impressive start in a decidedly inorganic direction. Paul Erlich developed the paradigm for medicinal chemistry and drug screening in his search for a new arsenic compound for the treatment of syphilis. He created a chemical library of organoarsenates designed to decrease the reactivity/toxicity of arsenic while retaining or increasing its therapeutic efficacy against the disease. Erlich screened a library of compounds and discovered that compound number 606 had the characteristics he wanted. The compound was arsphenamine (trade name, Salvarsan; Figure 1).

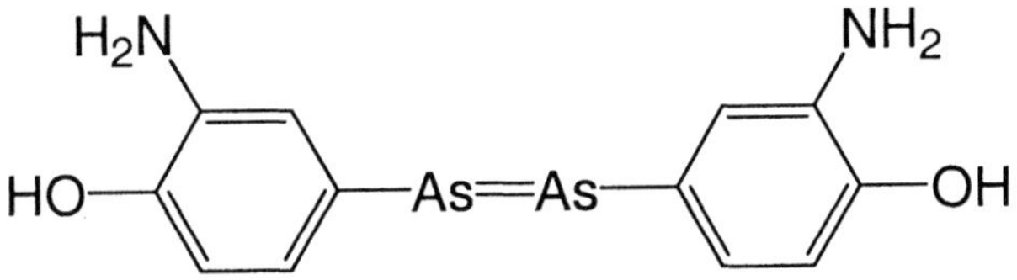

Figure 1. Arsphenamine (trade name Salvarsan), the result of the first modern medicinal chemistry program, discovered by Paul Erlich in 1909 for the treatment of syphilis.

This compound became the standard of treatment for syphilis for over thirty years until it was phased out by other arsenicals and, finally, penicillin. Erlich's approach – create chemical diversity and assay for improved therapeutic properties – has changed little in the last century with the exception of the vast expansion of chemical space and the sophistication of the biological assays.

Drug Discovery and Development Today

The process of drug discovery and development today is vastly more complex and expensive than a century ago. The Tufts Center for the Study of Drug Development estimated in 2001 that the average approval cost per new prescription drug is $802 million which was based on information from 10 companies; included were expenses of project failures and the impact that long development times have on investment costs (*4*). The development process, while somewhat formulaic once a development candidate is chosen, presents an often bewildering array of regulations.

The requirements for the filing of an IND (Investigational New Drug) application are focused on safety, chemical manufacturing and clinical protocols and are the same for metal therapeutics (Figure 2) (*5*). Animal pharmacology and toxicology studies are required to permit an assessment as to whether the product is reasonably safe for initial testing in humans. Animal studies to support the scientific hypothesis underlying drug efficacy are also important but not the primary focus of the safety review. Manufacturing information pertaining to the composition, manufacture, stability, and controls used for manufacturing the drug substance and the drug product is assessed to ensure the company can adequately produce and supply consistent batches of the drug. Detailed protocols for proposed clinical studies are assessed to determine whether the initial-phase trials will expose subjects to unnecessary risks. Information on the qualifications of clinical investigators--professionals (generally physicians) who oversee the administration of the experimental compound—is also reviewed to determine whether they are qualified to fulfill their clinical trial duties.

Once the FDA has determined that it is safe to proceed, the clinical trials and subsequent NDA (New Drug Application) must address three issues: whether the drug is safe and effective for its proposed use(s), and whether the benefits of the drug outweigh its risks; whether the drug's proposed labeling is appropriate, and, if not, what the drug's labeling should contain; whether the methods used in manufacturing the drug and the controls used to maintain the drug's quality are adequate to preserve the drug's identity, strength, quality, and purity (Figure 3). If these criteria are adequately addressed the FDA will approve the drug for the specific disease indications claimed (*5*).

Medicinal Inorganic Chemistry Therapeutics Scorecard

The number of metal-based drugs that have achieved FDA approval is remarkably few. Consider all oncology indications where the tolerance for drug side-effects and the demand for new treatments are relatively high. From 1949 to 2003, the FDA approved 89 new molecules that have been granted 210 specific claims for oncology treatments. Only 6 of these molecules are unambiguously defined as metal complexes or inorganics and these molecules have been granted a total of 9 claims for oncology. Thus, only 7% of the new molecules and 4% of specific claims approved by the FDA for oncology over the past 50+ years represent the fruits of medicinal inorganic chemistry. This is consistent with pharmaceutical sales; of the ~$16 billion world-wide oncology drug market in 2001, $1 billion was accounted for by the platinum(II) drugs (*6,7*).

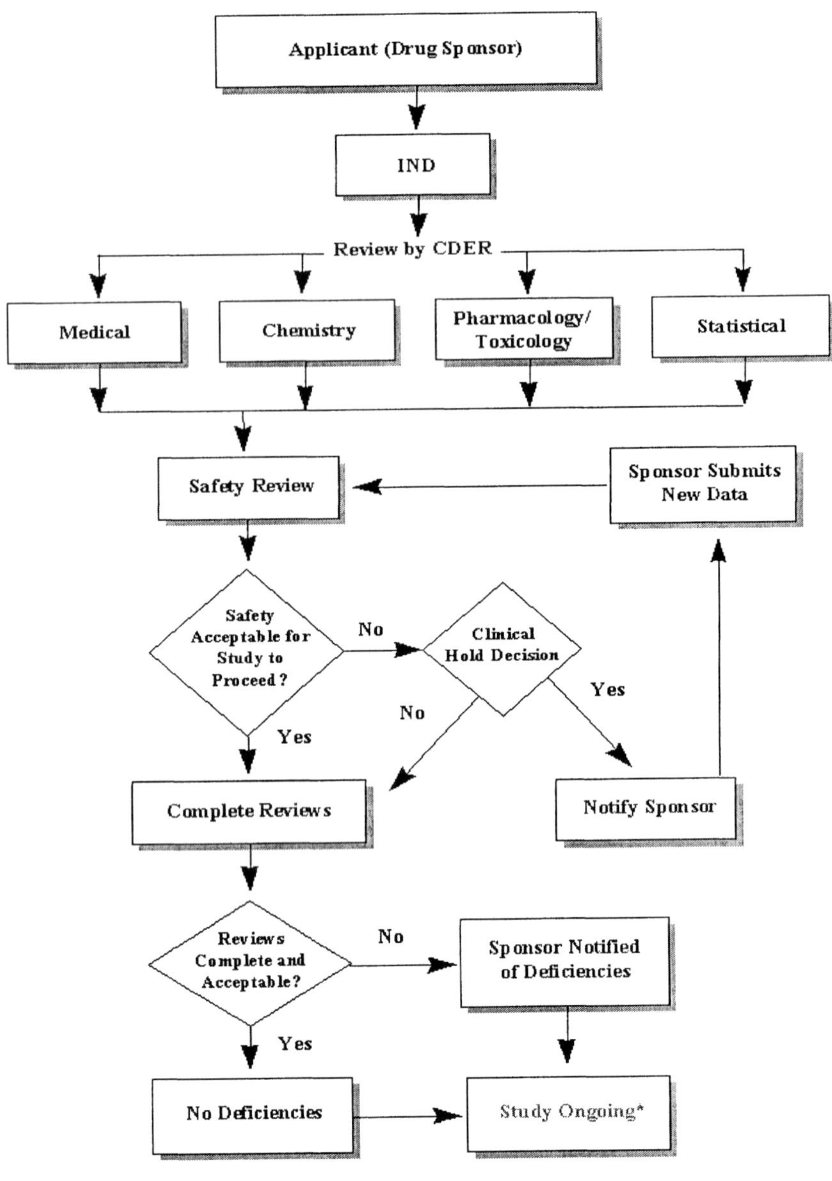

Figure 2. The Investigational New Drug review process chart describing the reviews and decision points leading to an acceptable IND application.

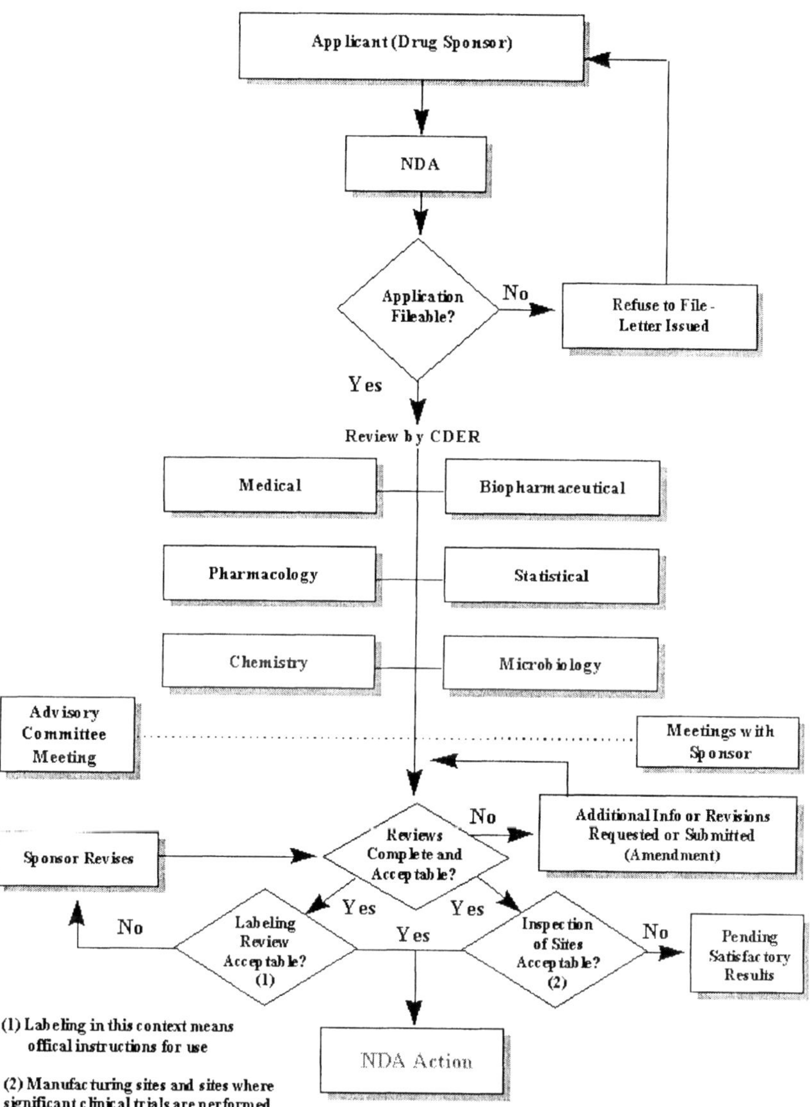

Figure 3. The New Drug Application review process chart describing the reviews and decision points leading to FDA approval.

The 6 metal complexes/inorganics that have made it across the FDA goal line are revealing. The 3 most significant are the covalent DNA-crosslinking platinum(II) complexes which are described in later chapters. Cisplatin (Platinol; Bristol Myers Squibb) received FDA-approval in 1978 for the treatment of ovarian and testicular cancers. Approval came 17 months after NDA filing. In 1993, an additional claim for transitional cell bladder cancer was approved 4 months after filing. Cisplatin has become a front-line cancer agent and has an impressive 90% cure rate for testicular cancer. Nephrotoxicity is the major adverse effect. In 1989, Bristol Myers Squibb received approval of carboplatin (Paraplatin) for the treatment of recurrent ovarian cancer a rapid 8 months after filing. A second claim for carboplatin was approved in 1991 for advanced ovarian cancer. The limiting toxicity for carboplatin is myelosuppression. Another platinum(II) variation on cisplatin, oxaliplatin (Eloxatin; Sanofi-Synthelabo) was approved for the treatment of colon or rectal cancer in combination with 5-fluorouracil/leukovorum (5-FU/LV). Accelerated approval in 2002 was granted 6 weeks from filing, a clear indication on the lack of treatment for this devastating cancer. Neuropathy appears to be the limiting toxicity for oxaliplatin. These 3 platinum(II) drugs and several others in the clinic represent a classic expansion of chemical diversity, reminiscent of Erlich's work, around a pharmaceutically useful motif to alter pharmacological properties. Unfortunately, the platinum(II) class currently remains the only example of medicinal inorganic chemistry in its fullest potential for drug discovery (*6*).

In 2002, the first radiopharmaceutical, ibritumomab tiuxetan (Zevalin; IDEC Pharmaceuticals), received accelerated FDA approval for the treatment of low-grade, B-cell, Non-Hodgkin's Lymphoma (NHL). Zevalin is an immunoconjugate of IDEC's successful antibody drug for NHL, Rituxan, and a high affinity chelation site for the inorganic radionuclides indium-111 or yttrium-90. Rituxan itself was the first monoclonal antibody approved for cancer therapy; it binds specifically to the CD20 antigen expressed on greater than 90% of B-cell NHLs. Thus, Rituxan targets and destroys only B cells. Zevalin is used for low-grade, B-cell NHLs that have not responded to chemotherapy or to Rituxan alone. The antibody-targeted chelation site for the radionuclide permits the delivery of high dose radiation (mCi's per dose) while reducing the amount of full body radiation. Zevalin received accelerated approval although long-term clinical efficacy remains to be established. This type of antibody-targeted delivery has considerable potential for metal-based therapeutic, as well as diagnostic, applications (*6*).

The two remaining FDA-approved oncology "drugs" are true inorganics. In 1997, an aerosol formulation of talc (Sclerosol; Bryan Pharmaceuticals) was approved for the treatment of malignant pleural effusion associated with lung cancers. The talc, which has a general structure of $Mg_3Si_4O_{10}(OH)_2$, is delivered directly into the pleural cavity and functions as a sealant to prevent re-accumulation of fluids into to lungs through cancer-associated lesions. Calling this an oncology drug is not entirely accurate; it is better viewed as supportive therapy. The final and, in many ways, the most astounding approval was for Trisenox (Cell Therapeutics) in 2000. Trisenox is an i.v. formulation of arsenic trioxide (As_4O_6), a fact that must have Paul Erlich scratching his head in bewilderment. But even more remarkable, was the speed of the FDA in approving one of the most toxic arsenic-containing compounds. Trisenox was approved 6 months after NDA filing for an orphan status indication as a second line treatment of relapsed or refractory acute promyelocytic leukemia (APL) following trans-retinoic acid (ATRA) plus an anthracycline. In one multicenter study, the remission rate was found to be 70%. Finally, it is sobering to note that Trisenox, a toxic, medicinal chemistry-unadorned arsenate, currently holds the FDA record for fastest time for any drug from initial IND filing to approval – 3 years (*6*)! The considerable human experience with this drug from uncontrolled studies conducted in China was a significant factor in the speed of the FDA approval.

The point of the above analysis is that there is both good news and bad news for medicinal bioinorganic chemistry. The good news is that the field has at least one clear-cut therapeutic success story, the platinum(II) complexes that has all the elements of a bona fide medicinal chemistry program with continued potential for new drugs. Another positive is that the FDA will move aggressively to approve new drugs, even less than optimal ones, if they address major unmet medical needs. Recent changes in the FDA appear to be moving them toward an even more proactive position. This is good news for the pharmaceutical industry in general. There is also bad news for medicinal inorganic chemistry to consider. Most obvious is the thin record of therapeutic successes over the past half-century in a complex, multi-etiological disease (cancer) with a high tolerance for side-effects in drugs and many niches for accelerated approval for unmet medical needs. Why is the record so thin? Why has progress been so slow and, in some cases (such as Trisenox) essentially retrograde to modern medicinal chemistry? The reasons lie in the fundamentals of how pharmaceutical drug discovery is conducted and who conducts it.

Diagnostics versus Therapeutics

The development of metal diagnostics has faired better than metal therapeutics. For diagnostics in general, a key requirement is that the compound must have minimal or no biological effect on the organism. Significant perturbation of the organism by a diagnostic undermines the validity of the information obtained. This requirement necessarily puts dosing of the compound below the no effect level for any therapeutic or potentially toxic effects. Diagnostics thus require high sensitivity. Here, metals are robust, due to their nuclear and/or electronic properties, so low doses and high sensitivity substantially reduce toxicity problems. Metal therapeutics, of course, must have a biological effect on a disease process, thereby requiring higher doses. With the possible exception of radiopharmaceuticals, the chemical properties of the metal come into play increasing the likelihood of toxic effects. Managing the chemical reactivity of metal therapeutics is a major challenge and in most cases the basis for efficacy.

Medicinal chemistry: organic versus inorganic

The vast majority of medicinal chemists in the pharmaceutical industry are trained as organic chemists. The philosophy that pervades medicinal chemistry programs is organocentric. Although the application of inorganic chemistry to drug discovery as we noted earlier began a century ago, the emergence of bioinorganic chemistry as a pharmaceutical discipline, as exemplified in this book, was much slower to develop. Thus, organic chemistry which is the primary language of natural products chemistry, the reference point for most drugs, gained a decisive upper hand during the emergence of the major pharmaceutical research labs like Merck and Pfizer in the '30s and 40s from their beginnings as fine chemical suppliers. For medicinal inorganic chemistry to succeed, the chemical culture of the industry must change. This will take time and a new generation of medicinal chemists trained and willing to integrate inorganic principles into drug design.

The chemical principles that are driving drug discovery in the industry also put inorganic approaches at a disadvantage. The medicinal chemist has grown to abhor drug leads with intrinsic chemical reactivities. Although the field grew out the development of reactive drugs such as the penicillins, chemical reactivity is intimately and, sometimes, inextricably associated with adverse drug effects. For example, the penicillins and cephalosporins rely on the intrinsic reactivity of the β-lactam moiety in the molecules to specifically and covalently acylate the active site serine residue in bacterial cell wall transpeptidases. This covalent enzyme inhibition is the basis on the drugs' antibacterial activity. This reaction is highly specific for the bacterial tranpeptidases but not absolutely so. The β-lactam will occasionally acylate other amino acids, such as lysine, in other proteins, some from the mammalian host. The resulting modified proteins can lead to adverse drug effects such as the known immunogenic effects of the β-lactam family. And so it goes with many drug leads; intrinsic reactivity while useful can be difficult to control in order to maximize the therapeutic window of a compound. The focus on increasingly safer drugs has driven the medicinal chemist toward the goal of identifying unreactive compounds that rely on potent, noncovalent interaction (tight binding) with their molecular target to modulate a biological process. Unreactive drugs without covalent modification and redox potential generally present fewer issues with respect to toxicities or so it is believed. The development of large combinatorial libraries to explore chemical diversity is also more easily industrialized when the structure to not require sophisticated and expensive methods to deal with intrinsically reactive compounds, such as, redox-active metal-containing compounds. In contrast to purely organic drug leads, most metal complexes that medicinal inorganic chemists are pursuing have intrinsic reactivities that are essential to their biological effect. The rich redox chemistries of the SOD mimics and vanadium complexes are critical to the biological efficacy but also present challenges with respect to the management of side effects. Thus, the medicinal inorganic chemist is engaged in a quest to design metal complexes that harness the metal reactivities to optimize a specific beneficial effect while the medicinal organic chemist is committed to the design of molecules that function as pure tight binders devoid of as much reactivity as possible. The pharmaceutical industry clearly favors the latter path and, until more successes can be marked on the medicinal inorganic score card, it will be difficult to change that trajectory. In the biotechnology arena, companies like Berlex which developed the first MRI agent in 1988 and Pharmcyclics which is developing metal-containing texaphyrins are two notable standard-bearers for the medicinal inorganic chemistry approach to diagnostics and therapeutics.

Conclusions

Medicinal inorganic chemistry remains a field of great promise with many challenges. The potential for a major expansion of chemical diversity into new structural and reactivity motifs of high therapeutic impact is unquestionable. As evidenced from the contributions to this book, the field is thriving largely in academic institutions, small biotech companies and some pharmaceutical companies with diagnostic units. A significant presence in the therapeutic arena of Big Pharma has yet to emerge. Creating some key therapeutic breakthroughs that capture the interest of the pharmaceutical industry is essential to ensure the growth of the field in the future.

References

(1) Orvig, C.; Abrams, M.J. *Chem. Rev.* **1999**, *99*, 2201-2203.
(2) Orvig, C.: Thompson, K.H. *Science* **2003**, *300*, 936-939.
(3) Zhang, C.X.; Lippard, S.J. *Curr. Op. Chem. Biol.* **2003**, *7*, 481-489 and references cited therein.

(4) The study can be accessed at http://csdd.tufts.edu/InfoServices/Publications.asp .
(5) The IND and NDA review processes can be accessed at http://www.fda.gov .
(6) Information on approved oncology drugs may be obtained from http://www.fda.gov/cder/regulatory/applications/ind_page_1.htm .
(7) In this analysis I am excluding drugs, such as bleomycin, which, while strong evidence exists for the role of metal ions in the mechanism of action, was not formulated nor developed as a specific metal complex.

Chapter 3

Metals in Medicine: Biomedical Significance and Inorganic Chemistry

Peter C. Preusch

National Institute of General Medical Sciences, National Institutes of Health, Department of Health and Human Services, Bethesda, MD 20892

This chapter reviews the origins and current status of the Metals in Medicine program of the National Institutes of Health. It provides a summary of research accomplishments and opportunities in this area, insight into public and congressional interests in metals, comments regarding peer review of grant proposals, an inside view of grant portfolio management and advisory committee activities, details of the Metals in Medicine Program Announcement, and a summary of responses it has elicited thus far.

Toward the Principles of Medicinal Inorganic Chemistry A Case Study in Program Development

The principles of drug design are well developed for organic molecules. Given a well characterized protein target, well established methods are available to the medicinal chemist for identification and optimization of compounds that affect the function of that protein. Reasonably well understood approaches are available to optimize the pharmacokinetic properties of lead compounds. The ability to predict drug metabolism and potential toxicity, and to design molecules taking these factors into account, is improving.

In contrast, the principles of drug design applicable to inorganic molecules are poorly understood. The process of discovery is largely serendipitous and optimization is mostly empirical. Understanding of the actions of metals, metal metabolism, and toxicity mechanisms is limited. On the other hand, the range of properties of the inorganic elements suggests that there may be opportunities for drug discovery that cannot be met by organic chemistry alone.

Research into the basic principles of inorganic medicinal chemistry will not likely be supported by industry; thus, there is a need for government supported research. The majority of government research is not conducted by government scientists, themselves, but rather through the development of competitive research grant programs designed to stimulate the needed research.

This review will trace the origins and current status of the Metals in Medicine program of the National Institutes of Health (NIH) led by the National Institute of General Medical Sciences (NIGMS). In brief, the primary objective of this program is to increase the biomedical impact of the inorganic chemistry research that is supported by the NIH. It was motivated by growing public interest in metals, observations of the NIH peer review process, analysis of the NIH portfolio of research grants, interactions with the bioinorganic research community, formal input from advisory group meetings, and a publicly held workshop. The program has generated significant community interest and resulted in a significant number of new applications and awards.

Public Interest: What do we know, what do we need to know, and who cares?

Setting the stage -- the following real life scenario occurred several years ago: The NIH received an inquiry suggesting a study into the hypothesis that manganese toxicity contributes to criminal behavior. The Office of the Director, NIH, contacted all of the component institutes to find out what was known and what related research NIH supported. A search of CRISP (the record of NIH funded grants) showed little research relevant to either the normal physiology of manganese or the toxicity of excess manganese (*1*). NIGMS supported more research on manganese than any other part of NIH; however, none was relevant to the question at hand. A significant amount of the NIGMS supported research on manganese was focused on the mechanism of oxygen evolution by the photosynthetic apparatus of green plants.

As part of her testimony before the House Appropriations Labor/HHS/Ed Sub-Committee for FY2003, then Acting NIH Director, Ruth L. Kirschstein was asked to answer the following questions posed by Rep. Ernest Ishtook, Jr. (*2*). Paraphrasing:

- How many grants have been issued on chromium and diabetes?
- How many grants have been issued on mercury? Mercury metabolism in brain? Diagnosis of mercury load in brain?
- How many grants have been issued on lead?
- How many grants on metals known to be nutrients: zinc, copper, selenium, molybdenum?

Notably absent from this list are any questions about basic mechanisms of electron transfer, metal site geometry, control of redox potentials, spin-spin coupling, interpretation of spectral properties, and so forth.

These two examples illustrate differences between what may be scientifically interesting, valid basic research questions and public interest in the biological chemistry of metals. There are important general scientific principles to be learned from the study of simple inorganic complexes and model enzymes, but given the role of NIH as a health research organization, it is important that this research lead to applications with biomedical impact. These examples also highlight the general perception that metals are toxic. This perception represents the single greatest obstacle to development of metallopharmaceuticals. On the other hand, the human body does contain several grams of transition metals (*3*), suggesting that a least some metals may be safe when they are controlled by appropriate mechanisms.

NIH Review Criteria – What is "Significance"?

The NIH mission statement includes the following: [The NIH] mission is science in pursuit of fundamental knowledge about the nature and behavior of living systems and the application of that knowledge to extend healthy life and reduce the burdens of illness and disability (*4*). The NIH carries out this mission mainly through the support of peer reviewed research grants. The review criteria are: Significance, Approach, Innovation, Investigators, and Environment (*5*). Significance is defined by the following questions: Does this study address an important problem? If the aims of the application are achieved, how will scientific knowledge be advanced? What will be the effect of these studies on the concepts or methods that drive the field? These instructions are intentionally broad and leave much to the judgment and interpretation of the reviewers. Furthermore, the weighting of Significance relative to the other review criteria is also intentionally left up to the reviewers. These review criteria were adopted by the NIH in 1997, to "focus review of grant applications on the quality of the science and the impact it might have on the field, rather than on details of technique and methodology" (*6*).

However, in the opinion of the author, Significance is still often not adequately and consistently addressed during review. For example, there are a number of well-accepted systems in bioinorganic chemistry for which intrinsic scientific interest, biological, or global environmental importance is accepted as the standard by many reviewers. For such cases, reviewers' comments on significance are often very brief. It is clear that these systems do present research challenges and do provide insights into the general workings of metalloenzymes, but the direct biomedical relevance of these systems appears to be limited. In other cases, where studies of less well characterized, but potentially more biomedically relevant systems are proposed, the Significance may be questioned because the medical importance of the problem is not yet fully established. There is a need for balance not only in the projects that are submitted to the NIH by bioinorganic chemists, but also for balance in the review of proposals, to be sure that both basic principles and public health impact are served by the portfolio of projects recommended for support.

NIGMS Portfolio Evaluation

In 1996, then NIGMS director, Marvin Cassman, had his staff carry out a thorough analysis of all research supported by the institute. NIGMS does not support all of the bioinorganic chemistry and metallobiochemistry funded by the NIH, but it does support the largest readily identifiable groups of projects. The portfolio evaluation showed that NIGMS is very heavily invested in studies of metalloenzyme active sites, including chemical model studies, of enzymes that are generally not of direct medical significance. Many of these projects had been reviewed by the Metallobiochemistry study section. A number of goals for the future were recommended by NIGMS staff: i) Increase the diversity of the NIGMS metallobiochemistry portfolio by capturing emerging areas of metabolism and the role of metals in regulation; ii) Increase the biomedical relevance of the inorganic chemistry under study; iii) Increase the numbers of funded female, minority, and new investigators in the field; iv) Increase interactions between researchers in inorganic chemistry and in biology. Progress toward these goals has been facilitated, in part, through the Metals in Medicine program.

Metals in Biology - Metals in Medicine

The Metals in Biology Gordon Research Conference (GRC) (*7*) is the primary annual meeting for scientists interested in biological inorganic chemistry. The International Conference on Biological Inorganic Chemistry (ICBIC) (*8*) is another valuable meeting. Conversations with scientists at these meetings had a major impact on the development of the Metals in Medicine program. In particular, comments made by Stephen J. Lippard after his talk on the mechanism of cisplatin action at the 1998 GRC meeting are informative. Paraphrasing: Why are there so few inorganic drugs on the market? Is it that metals have no promise? Is it that metals are too toxic? Is it that the pharmaceutical industry is dominated by organic chemists? Or conversely, is it because relatively few inorganic chemists have devoted themselves to drug discovery research?

Research into metal metabolism and metallopharmaceuticals, particularly platinum anticancer drugs, has been on-going for some time (*9-10*). A workshop, entitled, Metals in Medicine, was held February 3-4, 1984, organized by the Metallobiochemistry study section (Elizabeth C. Theil, Chair; Marjam G. Behar, Executive Secretary) and co-sponsored by the Division of Research Grants and the then National Institute of Arthritis, Diabetes, Digestive, and Kidney Diseases. NIH has issued a number of Program Announcements and Requests for Applications in various areas of bioinorganic chemistry over the years. David Badman of the National Institute of Diabetes and Digestive and Kidney Diseases (NIDDK) has been a notable NIH advocate in the metal metabolism area. However, metal metabolism and metallopharmaceuticals still represent modest subfields within the spectrum of bioinorganic research.

In 1998, as doubling the NIH budget over the next five years gained momentum in Congress, NIGMS director Marvin Cassman organized a series of NIGMS Advisory Council sub-group meetings to identify growth areas (*11*). Thomas V. O'Halloran was invited as a representative of the bioinorganic community. He emphasized the wealth of new discoveries to be made in the area of metal metabolism and the role of metals in biological regulation. In 1999, at the 9th ICBIC meeting in Minneapolis, conversations with Michael J. Clarke, Nicholas P. Farrell, C. Frank Shaw, and Thomas J. Meade, suggested the importance of new work on metals as therapeutic and diagnostic agents.

In 1999-2000, both Clarke (*12*) and Farrell (*13*) published books on inorganic medicinal chemistry and Mike Abrams and Chris Orvig edited a special issue of Chemical Reviews (*14*) compiling work by many prominent laboratories in the field.

NIH Metals in Medicine Meeting: Targets, Diagnostics, and Therapeutics

With the above background in mind, NIGMS organized a meeting held on the NIH campus in Bethesda, Maryland, June 28-29, 2000. The meeting was co-sponsored by the Center for Scientific Review, the National Cancer Institute (NCI), the National Institute of Allergy and Infectious Diseases (NIAID), the National Institute of Diabetes, Digestive, and Kidney Diseases (NIDDK), the National Institute of Environmental Health Sciences (NIEHS), and the Office of Dietary Supplements, Office of the Director, NIH.

The purpose of this meeting was to:

- Determine the role of metallobiochemistry in the pharmaceutical industry;
- Identify emerging areas of opportunity such as metal metabolism and metal regulation of cellular processes;
- Examine opportunities and obstacles to the development of metallo-pharmaceuticals.

The complete meeting agenda, abstracts and speaker biosketches, videocast of many of the talks, and an extensive meeting report are available on the Metals in Medicine page of NIGMS website at:
http://www.nigms.nih.gov/news/meetings/metals.html.

Session titles indicate the overall scope and thrust of the meeting:

- Molecular and Cellular Targets of Metal Action
- Metal-Containing Targets of Drug Action
- Imaging, Radiology, and Photodynamic Therapy
- Metal Metabolism as a Therapeutic Target
- Metallotherapeutics and Disease
- Medicinal Chemistry of Metallopharmaceuticals

Overall conclusions from the meeting were significant in guiding the further development of the NIH Metals in Medicine initiative. 1) Metalloproteins are important targets of current drug development programs. About one-third of all gene products are believed to be metal binding proteins and many are likely to become drug development targets in the future. However, current drug development programs only make modest use of NIH supported research in bioinorganic chemistry. 2) Metal metabolism is a growing area of research, the therapeutic opportunities of which remain to be determined. 3) Radioimaging, radiotherapy, magnetic resonance imaging, and photodynamic therapy are rapidly developing areas that will benefit from further basic inorganic research. 4) Medicinal inorganic chemistry is an underdeveloped area of science despite a few notable successes. Basic understanding of the principles necessary to design metal based drugs is lacking. Concerns about the potential toxicity of metals dominate. Metal-containing compounds have generally been dropped from the compound libraries screened in drug discovery programs.

A highlight of the meeting was the presentation by Jill Johnson on the experience of the National Cancer Institute (NCI) Developmental Therapeutics Program (DTP). See: http://dtp.nci.nih.gov. This program has conducted extensive screening of compounds for anticancer activity in animal models (until 1989) and against a panel of 60 cancer cell lines (1990 - present). The program has also screened compounds for anti-HIV activity in cell based assays. The results indicate that the record for identifying active metal-containing agents and bringing them to market is NOT worse, and is possibly better, than that for other classes of compounds.

Based on data reported at the meeting, 14,900 metal-containing compounds had been tested as of June, 2000, including a wide range of elements. Of those, 1,242 were selected as active in the animal model model screen, 191 were selected in the 60 cell line screen. Nine were investigated as clinical candidates. Five resulted in INDs. Two resulted in NDAs and are currently mainstream therapies (cisplatin and carboplatin). Reasons for dropping clinical candidates included: i) stability, formulation, or other pharmaceutical development issues; ii) renal or neurotoxicity; iii) insufficient advantage over current drugs. For comparison, of the 550,000 total compounds tested by the DTP, 14,475 were active in mouse screen models and 7,741 (out of 77,000) were active in the 60 cell-line assay; 59 INDs and 11 NDAs resulted.

Since 1997, the NCI has also operated a screen for compounds active against the cytotoxic effects of HIV in CEM cells. Of 80,000 compounds tested, 4,050 (or about 5%) were active. Of the compounds tested, 2,291 have included metals. Of those, 136 (about 6%) were active, two became clinical candidates. Both were dropped due to toxicity problems.

Nonetheless, NCI has posted a notice on its website that the program will not accept metal-containing compounds for further screening, unless a strong biological rationale is supplied (*15*). This statement also applies to several other compound classes that are likewise felt to have been adequately studied. The statement reflects in part a shift in the overall philosophy of the DTP from empirical drug discovery through screening large numbers of compounds toward a rational drug discovery approach. The new approach focuses on identifying the mechanistic foundations for activity and variations in activity across different cancer cell types. The Metals in Medicine program also stresses the importance of identifying mechanisms and rational design principles (see below).

Another highlight from the meeting was the presentation by Dennis Riley of research by Metaphore Pharmaceuticals exploiting the unique redox properties of metal complexes to achieve therapeutic activity that would be very difficult to accomplish with non-metal containing agents. In brief, Metaphore Pharmaceuticals successfully designed small molecule mimics of superoxide dismutase (SOD) with potential application to ischemia-reperfusion injury, stroke, shock, autoimmune and inflammatory disorders, and pain treatment. Manganese complexes were chosen because of their relatively low toxicity. An aza-crown lead compound was optimized to enhance complex stability, while also enhancing catalytic rate. These goals were accomplished by recognizing the rate determining step in the dismutation reaction ($Mn^{+2} + HO_2 \bullet \Rightarrow Mn^{+3} + H_2O_2$). Stability was improved by developing a rigid, pre-organized chelator through the application of molecular mechanics modeling. The rate of the Mn^{+2} to Mn^{+3} reaction was improved by minimizing ligand reorganization and enhancing outer sphere H^+-coupled electron transfer. This example illustrates the application of thoughtful drug design principles to metallopharmaceutical development. The resulting compounds are currently in preclinical and clinical development for several indications with the most advanced applications undergoing Phase II trials (*16*).

Other applications of metal complexes discussed at the meeting included:

- DNA and RNA as targets of Fe, Cu, Ni, Co, Zn, Eu, Ru, and Rh - containing complexes and metal complex conjugated oligonucleotides
- Protein targets of Zn, Fe, Cu, Co, and Re complexes as inhibitors and receptor ligands
- Radioimaging and MRI applications of Tc, Re, Y, In, Lu, Cu, and Gd complexes
- Cancer chemotherapy applications of Pt compounds and texaphyrins
- Antiviral activities (HIV and HSV) of TiO^+, M^{++}bicyclams, and polyoxometalates
- Antidiabetic potential of V and Cr compounds as drugs and dietary supplements
- Antioxidant activities of other Mn SOD and catalase mimics for inflammation, ischemia, stroke, myocardial infarction, and neurodegenerative diseases
- Applications of ^{99m}Tc-sestamibi to study drug transporters and of related Plasmodium falciparum multidrug resistance pump (pfMDR) inhibitors for malaria

Updates on many of these projects are presented in other chapters of this monograph. For example, motexafin gadolinium (Xcytrin), which was discussed by Richard A. Miller of Pharmacyclics, has now been shown to be of benefit as a potentiator of whole brain radiation therapy for brain metastases in Phase III clinical trials (*17*).

NIH Metals in Medicine Program Development

A report of the Metals in Medicine meeting was presented to the NIGMS Advisory Council at its meeting in January, 2001, along with a Concept Clearance document authorizing a program announcement (PA) in this area (*18*). The Council endorsed the emphasis on new research into metal metabolism and the basic mechanisms of metal ion interaction with cells, but was wary of NIGMS getting involved in drug development in general. Past negative industrial experience with metals was also cited as indicating a low probability of success; therefore, this area was deemphasized. NIDDK, NIEHS, and the Office of Dietary Supplements joined NIGMS in sponsoring the Metals in Medicine Program Announcement (PA01-071), which includes statements of interest by each NIH component. See: http://grants1.nih.gov/grants/guide/pa-files/PA-01-071.html. Although NIAID and NCI co-sponsored the meeting, these institutes chose not to participate in the PA. Because the areas of cancer chemotherapy, radiation therapy, and medical imaging fall outside the missions of the participating institutes these topics were not included in the PA.

Metals in Medicine Program Announcement

The objectives of the Metals in Medicine Program Announcement (PA-01-071; now reissued as PA-05-001) are: i) identification of new targets for drug discovery; ii) development of new in vitro and in vivo diagnostics; iii) elucidation of new basic concepts that will enable future therapeutics development; and iv) stimulation of collaborations between inorganic chemists and medical scientists; to ultimately increase the impact of bioinorganic chemistry research on health.

The program announcement solicits proposals for regular research (R01), program project (P01), and collaborative research (R24) grant awards. Proposals using other NIH grant mechanisms not specifically mentioned in the PA are also welcomed (e.g., Academic Research Enhancement Award (R15), high risk/high impact (R21), and small business (R41-R44) projects).

Areas of interest mentioned in the Metals in Medicine Program Announcement include:

- Role of metals in cell regulation and cell-cell signaling
- Metal ion homeostasis and the role of aberrant metal metabolism in disease
- Metabolic roles of essential trace element nutrients
- Interactions of metal complexes with living systems and their components
- Mechanisms of metal toxicity and for the amelioration of toxic metal accumulations

The following were highlighted as opportunities in metal metabolism: mechanisms of metalloenzyme assembly; metal trafficking proteins and pumps; metal sequestering vesicles and free/bound metal concentrations; mechanisms of metal ion homeostasis; mechanisms of metal mediated gene expression; mechanisms of metal accumulation; targets of metal toxicity; and chelation therapy to reduce the body burden of metals in genetic diseases and toxic exposures. These are areas of expected growth based on current trends in bioinorganic research.

The greater challenge is the development of basic principles to better utilize metals in the development of new in vivo diagnostics and therapeutics. The fact that such opportunities exist is reflected in the several billion dollar market for current metallopharmaceuticals. (See chapter by John Kozarich). The greatest barrier to further development is the perception that metals are toxic. Obviously some metals are toxic, depending on the dose and chemical form, and others are relatively non-toxic. According to an FDA spokesperson at the NIH Metals in Medicine meeting, there is no difference in the regulatory requirements for metal-containing and non-metal-containing drug candidates. Both must be proven safe and efficacious, under their conditions of use.

Key issues in the development of metallopharmaceuticals fall into two areas: i) development of rational design principles to exploit the unique properties of metals to accomplish diagnostic or therapeutic objectives that cannot be achieved in other ways; ii) development of rational design principles needed to avoid metal toxicities. The following lists give examples of some of the specific topics in need of study.

Key Issues in the Development of Metallopharmaceuticals I:
Opportunities to Exploit the Unique Properties of Metals

- Redox chemistry, substrate activation chemistry, catalytic activity
- Lewis acidity, electrophilicity, selective binding preferences, hydrolytic chemistry

- Access to anions, cations, and radicals not achievable without metal species
- Unique geometries, multivalency, reversible, and irreversible reactions
- Facile SAR, combinatorial synthesis, variation in metal and ligand structure
- Magnetic, spectroscopic, and radiochemical properties

Key Issues for Metallopharmaceuticals II
Obstacles include Perceptions, Reality, & Regulation

- Toxicity and accumulation of metals - limits long term applications
- Specificity, cellular uptake, sites of action - need to be known or improved
- Stability of complexes in vitro and in vivo - need to be documented
- Pharmacokinetics and pharmacodynamics - need to be predictable and determinable
- Metalation of chelators in vivo - needs to be documented
- Metal ion speciation in vivo - needs to be better understood

A great deal of attention has been paid to the direct interaction of metal complexes with DNA, somewhat less to interactions with RNA, and much, much less to other potential modes of affecting cell function. In contrast most existing therapeutic agents act on cell surface receptors, membrane transport proteins, intracellular signal transduction cascades, and nuclear receptor proteins, or as inhibitors of metabolic enzymes. Additional research in these latter areas is particularly needed.

As recommended by its Advisory Council, NIGMS interest in the above area is not in the development of particular drugs, but rather exploration of the basic principles of metal interactions with living systems that are relevant to these issues in metallopharmaceutical development.

Metals in Medicine - Responses

An NIH program is not likely to succeed without enthusiastic support of the scientific community. Response to the Metals in Medicine program has been very encouraging. Workshops and symposium sessions in this area have been organized by Rick Holtz (ACS meeting in Chicago, August, 2001); Tom O'Halloran (Metals in Biology Gordon Conference, January, 2002); Nick Farrell, Mike Clarke, and Frank Shaw (Metals in Medicine Gordon Conference, inaugural meeting June, 2002); David Westmoreland (Leermakers Symposium at Wesleyan University, May, 2003); John Sheats (ACS MidAtlantic Regional meeting at Princeton, June, 2003); and Jonathan Sessler, Steve Lippard, Susan

Doctrow, and Tom McMurry (ACS meeting in New York, September, 2003, reported further in this monograph).

The Metals in Medicine Program Announcement (PA01-071) was posted in the NIH Guide on March 19, 2001, for applications to be received on the regular NIH receipt dates beginning June, 2001. Approximately 10-20 proposals have been submitted in response to the PA on each of the subsequent receipt dates. Responses citing PA01-071 have generally been in the targeted areas. However, a greater number of applications directed toward understanding how to use metal complexes to specifically intervene in molecular and cellular function would be welcomed.

A total of 130 competing applications have been submitted in response to the PA through the Oct/Nov 2003 receipt date (*19*). Eleven were yet to be reviewed when this manuscript was prepared. Of the 119 reviewed proposals, 21 will have resulted in awards, once awards anticipated from the September 2003 and January 2004 Councils are completed. A few proposals have been revisions of previously submitted projects or continuations of previously funded projects. A number of responsive proposals have been resubmitted in revised form. Of the 88 unique, new projects submitted, 18 (20%) will have resulted in awards. This is comparable to the overall NIH success rate for New Applications (24.5%).
See: http://grants1.nih.gov/grants/award/success/rpgbyacttype7002.htm.

Competing awards totaling $2,211,780 and $3,084,741 (total costs) were made in FY02 and FY03, respectively. In total, over $5 million was awarded in FY03, including non-competing continuations and supplements to previous awards. Titles and investigators of projects funded in response to the PA can be retrieved through the Metals in Medicine web page.

PA01-071 will remain in effect through the June, 2004 receipt date. Plans are in progress to reissue the PA in 2004 to continue the program beyond its current expiration date. Responses starting with the October 1, 2004 receipt date should cite the new announcement.

Portfolio Evaluation Update

In 2002, NIGMS Acting Director, Judith Greenberg, asked NIGMS staff to repeat the portfolio evaluation conducted in 1996. Over that interval, the bioinorganic chemistry portfolio grew by a net 20% (*20*). Forty nine new projects were funded, including twelve in the area of metal metabolism, metalloenzyme assembly, and cell regulation. Nine new projects were added to explore basic chemistry with a view to eventual medical applications of metal complexes. New investigators received 65% of the new awards. Women scientists received 22% of the new awards. Minority scientists received 24% of the new awards, although none are considered to be under-represented minorities in science. Awards resulting from the Metals in Medicine program were only beginning to be reflected in the portfolio.

Where Do We Go From Here?

As shown in this review, the NIH Metals in Medicine program initiative was developed in partnership with the bioinorganic research community and it has elicited strong community participation. The goal has been to increase the biomedical significance of the bioinorganic chemistry supported by the NIH. If successful, the program should establish a foundation of basic knowledge that will facilitate the development of both metal targeted agents and metallopharmaceuticals by pharmaceutical companies.

What steps should be taken to further develop this area of research? What can be done to further advertise the program? What can be done to improve the success rate of applicants? What activities, meetings, books, and reviews are forthcoming in the field? What other areas of bioinorganic chemistry warrant special attention? Continued input from the community is welcomed by the NIH. Check the NIGMS Metals in Medicine web page periodically for updated information. See: http://www.nigms.nih.gov/metals/.

References

1. NIH Computer Retrieval of Information on Scientific Projects (CRISP) system. URL http://crisp.cit.nih.gov/.

2. Department of Labor, Health and Human Services, Education, and Related Agencies Appropriations for 2003, Hearings Before a Subcommittee of the Committee on Appropriations, House of Representatives, One Hundred Seventh Congress, Part 4A, National Institutes of Health, Document #80-980, U.S. Government Printing Office, Washington, D.C., 2002, p. 346-9.

3. Emsley, J.; *The Elements*, 3rd ed., Clarendon Press, Oxford, 1998. A very complete table from this text is reproduced at:
URL http://web2.iadfw.net/uthman/elements_of_body.html.

4. *The NIH Almanac*. URL http://www.nih.gov/about/almanac/index.html.

5. Guide For Assigned Reviewers' Preliminary Comments On Research Grant Applications (R01), Center for Scientific Review, National Institutes of Health. URL http://www.csr.nih.gov/CDG/CD%20Guidelines/r01.pdf

6. NIH Peer Review Notes, October, 1997. URL http://www.csr.nih.gov/prnotes/oct97.htm.

7. Metals in Biology Gordon Research Conference. URL http://www.grc.uri.edu/programs/2004/metalbio.htm

8. 12th International Conference on Biological Inorganic Chemistry. URL http://www.umich.edu/~icbic/

9. *Inorganic Chemistry in Biology and Medicine;* Martell, A.E., Ed.; ACS Symposium Series Vol 140, American Chemical Society: Washington, D.C., 1980.

10. *Platinum, Gold, and Other Metal Chemotherapeutic Agents;* Lippard, S.J., Ed.; ACS Symposium Series Vol 209, American Chemical Society: Washington, D.C., 1983.

11. News & Events: Planning and Priorities of the National Institute of General Medical Sciences, August 2, 2001. Available on request from Office for Communication and Public Liaison, NIGMS.

12. *Metallopharmaceuticals I & II;* Clarke, M.J.; Sadler, P., Eds.; Springer-Verlag, New York, Inc., 1999.

13. *Uses of Inorganic Chemistry in Medicine;* Farrell, N., Ed.; Springer-Verlag, New York, Inc., 1999.

14. *Medicinal Inorganic Chemistry;* Orvig, C.; Abrams, M.J., Eds.; *Chemical Reviews* Vol. 99, No. 9; American Chemical Society: Washington, D.C., 1999.

15. NCI Developmental Therapeutic Program. URL http://dtp.nci.nih.gov/screening.html. The comment about acceptance of compound classes is located on the page used by scientists registered with NCI to submit compounds for testing after logging in. Registration required.

16. Metaphore Pharmaceuticals. See press releases at: URL http://www.metaphore.com/

17. Survival and neurologic outcomes in a randomized trial of motexafin gadolinium and whole-brain radiation therapy in brain metastases. Mehta, M.P.; Rodrigus, P.; Terhaard, C.H.; Rao, A.; Suh, J.; Roa, W.; Souhami, L.; Bezjak, A.; Leibenhaut, M.; Komaki, R.; Schultz, C.; Timmerman, R.; Curran, W.; Smith J.; Phan, S.C.; Miller, R.A.; Renschler, M.F.; *J. Clin. Oncol.* **2003,** 21(13), 2529-36.

18. Minutes of the National Advisory General Medical Sciences Council meeting on September 13-15, 2000. URL http://www.nigms.nih.gov/about_nigms/council_sept00.html.

19. Data from NIH IMPACii QVR system on December 4, 2003.

20. NIH record of active grants on August 1, 1996 and July 1, 2002.

Chapter 4

Discovery and Development of Third-Generation Platinum Antitumor Agents with Oral Activity

Christen M. Giandomenico[1,2] and Ernest Wong[2]

[1]778 Elm Avenue, Blaine, WA 98230
[2]AnorMED Inc., 20353 64th Avenue, #200, Langley, British Columbia V2Y 1N5, Canada

The drug discovery process is illustrated by describing the collaborative process that was used to identify three potential platinum anti-tumor compounds (JM216 (satraplatin), AMD473 and JM335) and ultimately select two for clinical development. The selection cascade and medicinal chemistry program and selection criteria that lead to the identification of these preclinical candidate compounds is described. The additional preclinical studies and chemical development activities required to qualify these compounds as preclinical leads and ultimately as clinical candidates is described.

Cisplatin and carboplatin are widely used anti-cancer agents because they are effective against a broad range of cancers including testicular, ovarian, bladder, head and neck, and non-small cell lung cancer. A third platinum compound, oxaliplatin, was approved in 2002 for the treatment of colorectal cancer. Unfortunately, many initially responsive tumors develop resistance to these agents and each have serious toxicities such as nephrotoxicity, nausea and vomiting, ototoxicity, peripheral neuropathy, and myelosuppression that significantly degrade patient quality of life.

A new platinum agent that successfully addresses these unmet needs would find utility in chemotherapy. The question facing the drug discovery scientist is how to devise a program that has a reasonable probability of identifying a developable platinum compound. There are a variety of possible approaches to this problem. This article will illustrate the drug discovery process that was used to identify three potential platinum anti-tumor compounds and ultimately select two for clinical development. Some of the early stage development requirements will also be discussed.

AnorMED and its predecessor, the Johnson Matthey Biomedical group undertook a project to discover and develop new platinum antitumor agents, in collaboration with the Institute of Cancer Research. This program prepared and evaluated a large number of platinum compounds and ultimately identified three distinct classes of platinum compound having development potential. The final preclinical lead from each of these classes is depicted in *Figure 1*. The orally

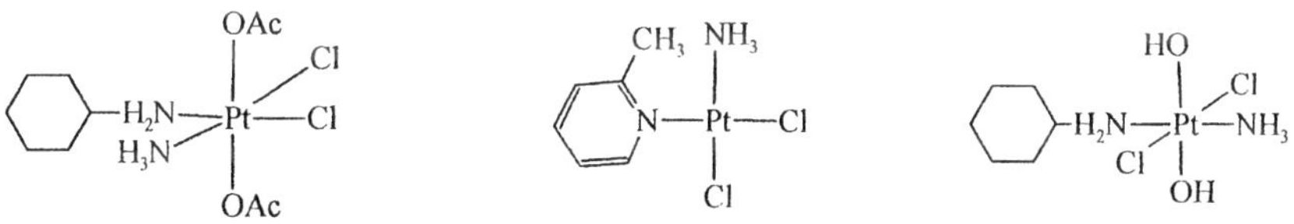

JM216 AMD473 JM335

Figure 1. Preclinical Leads

active platinum compound having comparable activity to cisplatin and carboplatin (JM-216, satraplatin) is being developed by GPC Biotech and entered Phase III trials for the treatment of hormone refractory prostate cancer in 2003. The third generation platinum (AMD473, NX 473) is being developed by NeoRx Inc. has completed several phase II clinical trials. The final lead (JM335) represented an unusual trans-platin compound that demonstrated *in-vivo* activity[1] but never progressed to clinical trials.

Lead Identification

In-vitro screening

Our collaborative platinum antitumor discovery strategy employed a selection cascade, depicted in Figure 2, to efficiently identify potent compounds

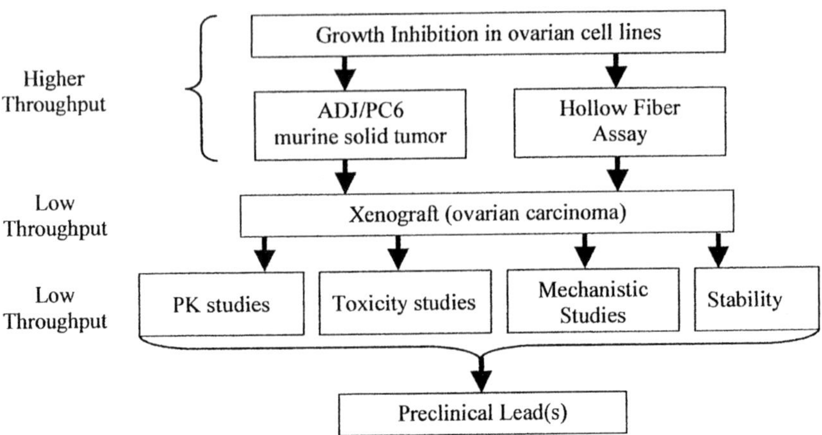

Figure 2. Selection Cascade

with appropriate pharmaceutical properties. In addition, the discovery group prepared a large number of potential compounds whose properties could be systematically varied to identify trends. This was accomplished employing the synthetic methodology depicted in *Figure 3* that is based on the laboratory scale preparation of the trichloro(ammine)platinate anion (TCAP)[2]. A wide variety of *cis*-mixed amine/ammine Pt(II) complexes that contained an amine ligand that would influence pharmacokinetics and DNA binding could be readily prepared from this intermediate. Oxidation of these compounds produced Pt(IV) dihydroxy compounds which were further functionalized to form carboxylates, carbonates and carbamates[3] which contained groups that would influence pharmacokinetics but would be lost prior to forming the active DNA lesion. The trans complexes were prepared by an alternate route[4].

The first element of the screening cascade was a disease based *in-vitro* cytotoxicity assay. *In-vitro* cytotoxicity testing is rapid, economic and an effective method of identifying and ranking compounds and probing the basic

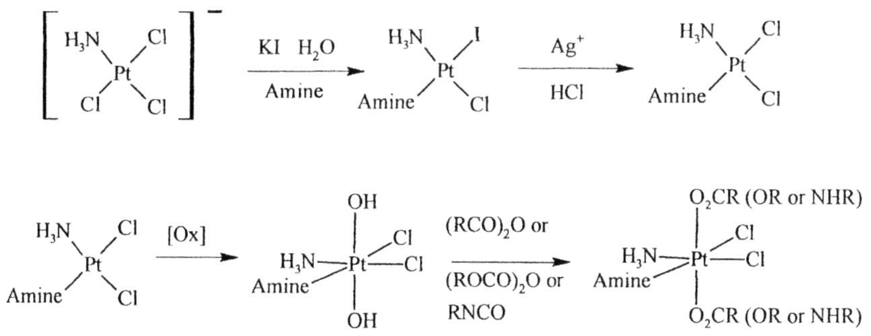

Adapted with permission from reference 3. Copyright 1995 American Chemical Society.

Figure 3. Synthesis of cis-Pt(II) and cis-Pt(IV) screening candidates

biochemical processes responsible for activity. The relative ability of compounds to overcome critical cellular resistance mechanisms such as decreased uptake or increased export of a drug, enhanced DNA repair, enhanced expression of protective agents such as such as glutathione, or enhanced ability to survive DNA damage can be evaluated by appropriate selection and characterization of the cell lines used in the screen. The ICR screen consisted of an *in-vitro* assay of ovarian cell lines[5,6] selected to exhibit a range of sensitivities to platinum and contained three pairs of cell lines with a acquired cisplatin resistant sublines. Ovarian cell lines were selected because this cancer is responsive to, but not cured by, platinum antitumor agents. The mechanism of resistance for each subline was charaterized[1,7,8] (41M/41McisR resistance due to reduced drug accumulation; CH1/CH1cisR resistance due to increased damage repair/tolerance; A2780/A2780cisR resistance due to elevated glutathione, reduced drug accumulation, and increased DNA damage repair/tolerance).

The cytotoxicity profile of the three leads[9] relative to the reference compounds cisplatin and carboplatin is shown the Table I. Cytotoxicity of the oral lead JM216 is greater across all cell lines tested compared to either cisplatin or carboplatin. However the relative activity is similar. Nonetheless this result met the selection criteria for cytotoxicity because the objective of discovery program was to identify a compound having a comparable activity to the current compounds but orally active. However, the program designed to identify a third generation platinum agent sought to identify potent compounds that had new activity profiles. As can be seen, the cytotoxicity profiles of JM335 and AMD473 relative to cisplatin and carboplatin met these criteria. In the case of JM335 the relative cytotoxicity of two cell lines, HX/62 and SKOV3 are

Table I ***In-vitro*** **cytotoxicity in selected ovarian cell lines**

Drug	*HX62*	*SKOV3*	*PXN/94*	*41M*	*CH1*	*A2780*
Cisplatin	13	4.4	3.0	0.26	0.11	0.33
Carboplatin	70	38	31	3.3	1.3	2.6
JM216	3.7	1.6	1.2	0.48	0.084	0.34
JM335	4.4	6.0	2.7	1.3	1.1	0.42
AMD473	41	24	5.4	3.5	1.5	3.8

Adapted with permission from reference 9. Copyright 1998 Cognizant Communication Corporation.

reversed compared to cisplatin and difference in cytotoxicity for the HX/62 and CH1 is dramatically reduced. A comparison of the data for AMD473 and cisplatin for the PXN/94 and 41M data exhibit similar trends. Consequently these compounds were selected for further testing.

In-vivo screening

The danger of relying solely on cytotoxicity data to predict or rank compounds is demonstrated in the data for two Pt(IV) carboxylates shown in Table II. The cytotoxicity of the Pt(IV) carboxylate[10] where R=propyl is an

Table II. ***In-vitro*** **cytotoxicity vs.** ***in-vivo*** **activity of** **$[PtCl_2(OCOR)_2NH_3(c\text{-}C_6H_{11}NH_2)$**

R	*In-vitro CH1* IC_{50} *(μM)*	*In-vivo ADJ/PC6* LD_{50}	ED_{90}	*TI*
Me	0.1	330	6	55
Propyl	0.0066	280	5	56

Data from reference 10 and 11

order of magnitude greater than R=Me (JM216). However, the antitumor activity in the ADJ/PC6 model as expressed by a Therapeutic Index (TI), the ratio of the LD_{50} to ED_{90} (the dose required to reduce tumor mass by 90%), are effectively identical[11]. The TI index is an indication of size of the therapeutic window for each candidate. Presumably, the increasing cytotoxicity trend, observed in the series of Pt(IV) carboxylates[10], is related to increased transport into cells as the lipophilicity of the complexes increase[12,13] and does not produce more active compounds *in-vivo* because the carboxylate groups are rapidly metabolized[14,15,16].

Animal models address a number of critical pharmacological determinants of clinical anti-tumor activity. For example, platinum compounds are administered at maximum tolerated doses. The difference between the dose in a patient that causes a response in cancerous tissue and the toxic dose (i.e. the Therapeutic Window) is of critical importance. While this characteristic of the compound can be definitively determined only in clinical trials, an *in-vivo* model can provide a means of estimating or ranking compounds based on a measured therapeutic window. A lead should also exhibit dose dependent activity and toxicity via the desired route of administration because if dose response is variable, a patient is at risk of receiving an ineffective or toxic dose. An appropriate animal screening model can distinguish potentially efficacious leads from toxic molecules due to the influence of route of administration and these pharmacology based determinants. Our screen utilized the ADJ/PC6 plastocytoma implanted into female mice[17,18]. A large comparative database of platinum activity data was generated in this screen over the course of screening many compounds. A high throughput hollow fiber assay[19] that permits *in-vivo* evaluation of compounds in human ovarian cell lines in an animal model was employed in later studies.

Candidate Selection

In-vivo efficacy Filter

Selected compounds that showed promise in the relatively high throughput screens, were tested in xenograft models in which the ovarian cell lines used for screening were grown as solid tumors. Xenografts provide the opportunity to evaluate the activity of drug against human tumors modified by animal pharmacology. The endpoint, growth delay (time it takes the solid tumor on the treated mouse to grow to a specified size compared to the untreated mouse) induced by an effective compound can exceed half a year, so these experiments are low throughput.

The *in-vivo* growth inhibition in selected xenografts for the lead compounds compared to cisplatin and carboplatin are shown in Table II. In these experiments dose dependence of activity is established (not shown) and the best results are at the MTD (maximum tolerated dose). JM216 administered orally has comparable activity to cisplatin and carboplatin[14,20]. This activity in an orally administered drug was considered sufficient to justify further development. The novel trans compound JM335 did exhibit *in-vivo* activity[1] which is extraordinary for this class of compound, however the activity was generally comparable or inferior to cisplatin and carboplatin. Since this represented the best in class data, development of this class of compounds was abandoned. AMD473 however exhibited activity[18,21] comparable or superior to cisplatin and carboplatin in several tumors and development of this compound continued.

Table III. *In-vivo* growth inhibition in selected xenografts

	Growth Delay (days)					
Drug	*PXN/65*	*HX/110*	*CH1*	*CH1R*	*SKOV3*	*HX/62*
Cisplatin (4mg/kg i.p.)	209	41	34	10	0	2.9
Carboplatin (80mg/kg i.p.)	ND	47	43	6.4	2	ND
JM216 (90mg/kg p.o.)	ND	36	43	3.5	6	ND
JM335 (4mg/kg i.p.)	ND	25	18	6	6.8	ND
AMD473 (30-40mg/kg i.p.)	>215	64	>139	34	6.8	0

ND Not determined

Adapted with permission from reference 18. Copyright 1997 Clinical Cancer Research.

Preclinical Pharmacology and Toxicity Filters

A successful drug will exhibit defined, predictable, and manageable toxicities. Preliminary studies to identify the key toxicities of the best compounds before undertaking major commitments of such as GLP toxicity studies help identify the best candidate from the few most efficacious. Selected data from the evaluation of AMD473[18] is presented in Table IV. In this case tissue accumulation, protein binding, organ toxicities, and dose limiting toxicities were similar to carboplatin. One particular advantage was the high oral bioavailability suggesting that this compound might be a candidate for oral administration. However some saturation of absorption was seen at higher doses that suggest optimization of an oral formulation may be required.

Table IV Selected PK and Toxicity data of AMD473

Tissue distribution	Liver & Kidney
Protein binding	75%
Oral bioavailability	40%
AUC, C_{max}	Non-linear above 100mg/kg
$T_{1/2}$	0.5-2hr
Organ Toxicity	No kidney, liver, cardiac or neurotoxicity. Mild gut
Dose Limiting Toxicity	Myelosuppression

Data from reference 9

Mechanistic Evaluation

Likewise, studies to distinguish the mechanism of action of a lead can help guide clinical development, provide a basis for progressing a new platinum compound, or lead to new classes of compounds that should be explored. For example AMD473 demonstrated that it overcomes acquired cisplatin resistance in tumor cells lines[22] exhibiting the major mechanisms of resistance to platinum drugs. *Figure 4* shows that AMD473 generally overcomes acquired resistance better than other platinum compounds. Preliminary studies may also suggest additional tests that should be incorporated into the screening cascade.

Stability and Physiochemical Property Filter

Physiochemical properties are also determinants of successful development of a platinum compound and will impact the design and selection of screening candidates. A potential candidate should have properties appropriate to the route of administration. A parenteral (i.v.) drug must be soluble enough to dissolve in a volume that can be handled and administered, whereas an oral candidate drug must be of an appropriate size, lipophilicity and hydrogen bonding characteristics to facilitate oral absorption[23]. The basic stability of the potential drug and a preliminary formulation should be evaluated. Degradation produces a loss of potency but can also produce toxic impurities. The biological safety of each impurity present at levels ≥0.15% must be established (qualification) by the time of product registration[24]. Desirable target stability for a formulation is 2 years at 25°C at 60% relative humidity. A compound or

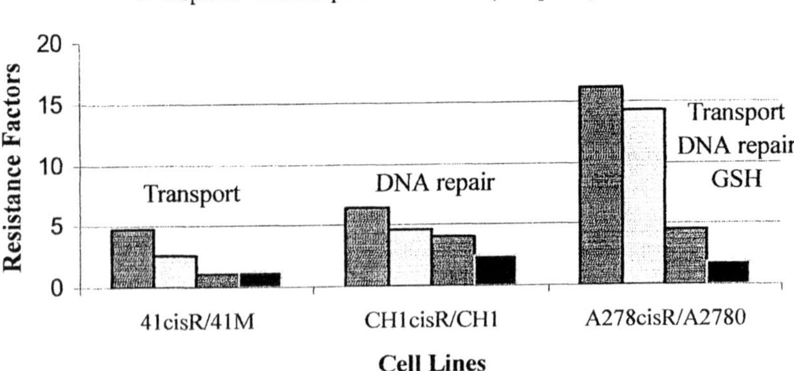

Data from reference 21

Figure 4 Circumvention of acquired cisplatin resistance

formulation that is not stable for extended periods in the dark at 5°C may present significant development challenges. If a lead does not intrinsically posses the desired physiochemical properties, additional studies to determine whether special formulations can overcome the limitation may be necessary at an early stage to avoid a costly development failure. Preliminary investigation of stability of AMD473 solutions revealed that aqueous AMD473 forms two aqua species[25] that are in equilibrium with AMD473 but that solutions are stable in the absence of light and heat. Exposure to light and heat does cause degradation therefore manufacturing and storage conditions must be optimized to minimize these factors.

Development

Regulatory Considerations

Prior to filing an IND, a number of additional studies and tasks must be completed. A formal characterization of the structure and properties of the compound must be completed, methods to analyze the strength, purity & stability of the pure compound and any formulation must be completed and bioanalytical methods to analyze drug in pharmacology and toxicology studies must be developed. Regulatory toxicity in two species, at minimum a single dose toxicity test and a repeat dose toxicity test that matches or exceeds the duration of the planned clinical trial, must be completed. Once a clinical plan is developed, the

IND package assembled submitted, and the FDA has reviewed and approved it, clinical trials may commence.

With respect to drug substances, the FDA recognizes that in early phases of development, information may not be complete. However it does require that sufficient information be submitted to ensure the proper identification, quality, purity, and strength of the investigational drug. Development of a reliable and scalable synthesis at an early stage is desirable to ensure that safety data generated in regulatory toxicity studies and early stage clinical trials is predictive of later batches of the drug. One common pitfall is to use a pure laboratory batch for initial regulatory toxicity studies. If the laboratory batch has a different (often cleaner) impurity profile than manufactured batches intended for initial clinical trials, the regulatory animal safety data may not be considered predictive of human toxicity. In that case, additional bridging toxicity studies are required, resulting in additional cost and delay. In the case of AMD473, the TCAP intermediate used in the medicinal chemistry synthesis was difficult to prepare reproducibly on a large scale. A more direct three step synthesis shown in *Figure 5* was developed based on the direct reaction of 2-methylpyridine with tetrachloroplatinate in a non-aqueous solvent to form the trichloro(2-picoline)platinate(II) (TCPP).

Business Considerations

Business considerations also significantly impact developability of a new platinum antitumor compound. Intellectual property protection must be adequate and the value of the therapeutic application must be large enough to meet the business objectives of the developer to justify the expenditure necessary to achieve registration. It is ideal if a platinum compound is covered by a composition of matter patent, though patented formulations or other protection strategies may be appropriate in certain circumstances. Large pharmaceutical companies generally seek to develop candidates with large market potentials (sales ~1 billion/year). Smaller pharmaceutical companies often have more modest business objectives.

Conclusion

A collaborative effort relying on a disease based screening cascade and having a source of systematically designed platinum candidates identified three potential lead compounds. By selecting compounds that had an appropriate efficacy and toxicity profile, and pharmaceutical properties, two platinum clinical candidates were identified. Methods for analysis and preparation of the candidates need to be developed at an early stage of development. Once regulatory toxicity and clinical plans were completed and approved the compounds, JM216 and AMD473, were progressed into clinical trials.

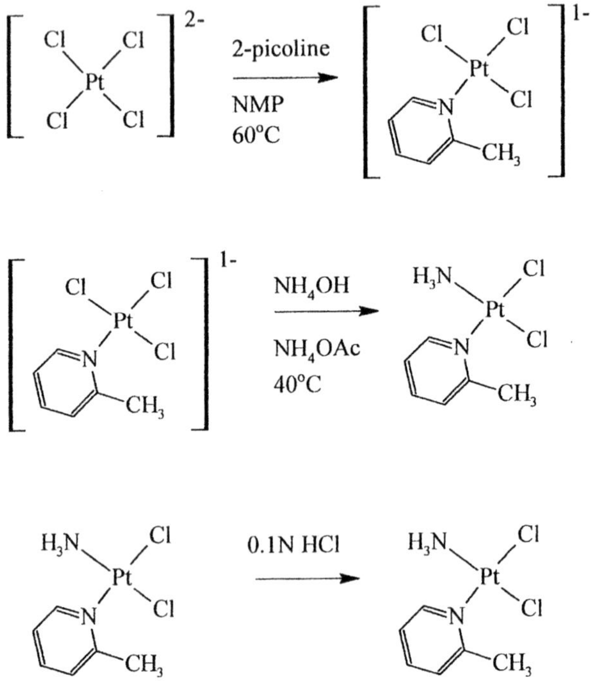

Figure 5. Manufacturing process for AMD473

Abbreviations

AUC — Area Under the Curve. The area under the plasma-vs-time curve is a measure of drug exposure and often correlates with toxicity and/or efficacy.

C_{max} — Maximum (peak) drug concentration in plasma. The C_{max} often correlates with toxicity and/or efficacy.

ED_{90} — The dose which causes a 90% reduction in tumor mass in test animals compared to untreated control animals.

FDA — Food and Drug Administration. The US regulatory body charged with promoting and protecting the public health by helping safe and effective products reach the market in a timely way, and monitoring products for continued safety after they are in use.

GLP — Good Laboratory Practice. The governmental regulations designed to assure that non-clinical laboratory (animal) studies, used by FDA to make public health decisions, are conducted according to scientifically sound protocols and with meticulous attention to quality. GLPs apply to non-clinical (animal) studies submitted in an IND.

IND — Investigational New Drug application. The document that requests authorization from the Food and Drug Administration (FDA) to administer an investigational drug to humans. Contains detailed information about the investigational drug including preclinical pharmacology and toxicity, the chemistry, manufacturing, and controls for the drug substance and product, and the clinical plan and previous human clinical data (if any).

LD_{50} — The dose which is lethal to half the test animals. This method of determining toxicity has been largely superceded by toxicity tests that are less lethal to animals.

MTD — Maximum Tolerated Dose is the maximum dose that does not cause unacceptability morbidity or is not lethal to the test animals.

$T_{1/2}$ The time required to eliminate half the administered dose for a drug with first-order elimination rates.

TI Therapeutic Index is a ratio of a measured toxicity vs measured potency of a compound. This calculated value can be used to estimate the size of the window between the efficacious dose and the toxic dose. In the studies described above TI= LD_{50}/ ED_{90}.

References

[1] Kelland L.R.; Barnard C.F.; Mellish K.J.; Jones M.; Goddard P.M.; Valenti M.; Bryant A.; Murrer B.A.; Harrap K.R.; Cancer Res. **1994,** Nov 1; Vol. 54(21), pp 5618-22.

[2] Abrams M.J.; Giandomenico C.M.; Vollano J.F.; Schwartz D.A.; Inorganica Chimica Acta; **1987,** Vol. 131, pp 3-4

[3] Giandomenico C.M.; Abrams M.J.; Murrer B.A.; Vollano J.F.; Rheinheimer M.I.; Wyer S.B.; Bossard G.E.; Higgins J.D. III; Inorg. Chem. **1995,** Vol. 34(5) pp 1015-1021

[4] Kelland L.R.; Barnard C.F.; Evans I.G.; Murrer B.A.; Theobald B.R.; Wyer S.B.; Goddard P.M.; Jones M.; Valenti M.; Bryant A.; et al.; J. Med. Chem. 1995 Aug 4;38(16):3016-24.

[5] Hills C.A.; Kelland L.R.; Abel G., Siracky J.; Wilson A.P.; Harrap K.R.; Br. J. Cancer. **1989,** Vol. 59(4), pp 527-34.

[6] Holford, J.; Sharp S.Y.; Murrer, B.A.; Abrams, M.; Kelland, L.R.; Br. J. Cancer. **1998,** Vol. 77(3), pp 366-73.

[7] Loh S.Y.; Mistry P.; Kelland L.R.; Abel G.: Harrap K.R.; Br. J. Cancer. **1992,** Vol. 66(6), pp 1109-15.

[8] Kelland L.R.; Mistry P.; Abel G.; Loh S.Y.; O'Neill C.F.; Murrer B.A.; Harrap K.R..; Cancer Res. **1992,** Vol. 52(14), pp 3857-64.

[9] Holford J.; Raynaud F.; Murrer B.A.; Grimaldi K.: Hartley J.A.; Abrams M.; Kelland L.R.; Anti-cancer Drug Des. **1998,** Vol.13(1), pp 1-18.

[10] Kelland L.R.; Murrer B.A.; Abel G.; Giandomenico C.M.; Mistry P.; Harrap K.R.; Cancer Res. **1992**, Vol. 52(4), pp 822-8.

[11] Harrap K.R.; Murrer B.A.; Giandomenico C.M.; Morgan S.E.; Kelland L.R.; Jones M.; Goddard P.M.; Schurig J.; In Platinum and other coordination compounds in Cancer Chemotherapy; Editor, Stephen Howell Ed.; Plenum Press NY New York; **1991**, pp 391-400

[12] Mistry P.; Kelland L.R.; Loh S.Y.; Abel G.; Murrer B.A.; Harrap K.R.; Cancer Res. **1992,** Vol. 52(22), pp 6188-93.

[13] Sharp S.Y.; Rogers P.M.; Kelland L.R.; Clin Cancer Res. **1995,** Vol. 1(9), pp 981-9.

[14] Kelland L.R.; Abel G.; McKeage M.J.; Jones M.; Goddard P.M.; Valenti M.; Murrer B.A.; Harrap K.R.; Cancer Res. **1993,** Vol. 53(11), pp 2581-6.
[15] Kelland L.R.; Expert Opin. Investig. Drugs. **2000,** Vol. 9(6), pp 373-82.
[16] Raynaud F.I.; Mistry P; Donaghue A.; Poon G.K.; Kelland L.R.; Barnard C.F.; Murrer B.A.; Harrap K.R.; Cancer Chemother. Pharmacol. **1996,** Vol 38(2), pp 155-62.
[17] Harrap K.R.; Kelland L.R.; Jones M.; Goddard P.M.; Orr R.M.; Morgan S.E.; Murrer B.A.; Abrams M.J.; Giandomenico C.M.; Cobbleigh T.; Adv. Enzyme Regul. **1991,** Vol. 31, pp 31-43.
[18] Raynaud F.I.; Boxall F.E.; Goddard P.M.; Valenti M.; Jones M.; Murrer B.A.; Abrams M.; Kelland L.R.; Clin. Cancer Res. **1997,** Vol. 3(11), pp 2063-74.
[19] Hollingshead M.G.; Alley M.C.; Camalier R.F.; Abbott B.J.; Mayo J.G.; Malspeis L.; Grever M.R.; Life Sci. **1995,** Vol. 57(2), pp 131-41.
[20] Kelland, L. R.; Jones, M.; Gwynne, J. J.; Valenti, M.; Murrer, B. A.; Barnard, C. F. J.; Vollano, J. F.; Giandomenico, C. M.; Abrams, M. J.; Harrap, K. R., Int. J. Oncol. **1993,** Vol. 2(6), pp 1043-8.
[21] Kelland R.; Barnard C.F.J; Drugs Future **1998,** Vol. 23(10), pp 1062-1065
[22] Holford J.; Sharp S.Y.; Murrer B.A.; Abrams M.; Kelland L.R.; Br. J. Cancer. **1998,** Vol. 77(3), pp 366-73.
[23] Lipinski, C. A.; Lombardo, F.; Dominy, B. W.; Feeney, P. J.; Adv. Drug Del. Rev. **1997**, Vol. 23, pp 3-25
[24] Guidance for Industry Q3A Impurities in New Drug Substances
U.S. Department of Health and Human Services Food and Drug Administration; February **2003**
[25] Chen, Y; Guo, Z.; Parsons, S.; Sadler, P.; Chem. Eur. J. **1998,** Vol. 4(4), pp 672-6

Chapter 5

Brain Iron as an Endogenous Contrast Agent in High-Field MRI

John F. Schenck

General Electric Global Research Center, 1 Research Circle, Schenectady, NY 12309

A new generation of clinical magnetic resonance imaging (MRI) systems operating at the high magnetic field strength of 3 tesla is becoming available. In contrast to conventional MRI scanners operating at fields of 1.5 tesla and below, these high field systems provide improved resolution and signal-to-noise levels. They are also markedly more sensitive to the presence of iron oxide deposits in the human brain. These iron deposits have long been known from postmortem studies to have a specific distribution in the normal brain. Changes in brain iron distribution, generally, but not always, involving increases above normal levels, have been associated with several neurodegenerative diseases including Alzheimer's and Parkinson's diseases. At the higher field strength these iron deposits are more intensely magnetized by the applied magnetic field and, under appropriate conditions (T2-weighted imaging), they can be studied at millimeter resolution in living subjects. Therefore, brain iron serves as an endogenous contrast mechanism for high field MRI. These new scanners may provide a major new technique for monitoring the progression of many neurodegenerative diseases.

Introduction to Magnetic Resonance Imaging

Since its introduction to clinical medicine in the early 1980s magnetic resonance imaging (MRI) has become fully accepted as an essential aspect of modern medicine. On a worldwide basis more than 15,000 scanners are now located in clinical settings and it is estimated that approximately 300,000,000 diagnostic studies have been performed to date. A panel of practicing internists recently selected MRI, along with computed tomography, as the most significant technical innovation in patient care over the last 25 years (*1*).

MRI was derived from the classical analytical chemistry technique of nuclear magnetic resonance (NMR) by scaling up the apparatus from one designed for test-tube sized samples to machines capable of encompassing the entire human body. In addition to the static field magnet, B_o, and the radiofrequency stimulating field, B_1, MRI utilizes three additional gradient field coils (one for each Cartesian direction, x, y and z) that encode position information into the MR signal (*2*). In common with the related imaging modalities computed tomography (CT) and positron-emission tomography (PET), MRI is used to produce cross-sectional images of the human anatomy. Among the advantages of MRI are its high quality of soft tissue discrimination, the ability to electronically adjust the scanning plane to any oblique orientation desired and the lack of a requirement for ionizing radiation or the use of radioisotopes. This last feature makes it possible to use MRI repeatedly on a given patient as often as medically necessary without concern for tissue injury from multiple imaging exposures.

The defining component of each MRI scanner is the static field magnet, which is specified in terms of its operating field strength measured in tesla (T). Modern high performance MRI devices almost always use a large superconducting magnet in a cylindrical cryostat. This permits the generation of highly uniform, intense magnetic fields without coil heating. A field uniformity on the order of 10 parts per million is required over the entire region of the body being imaged and temporal drift in field strength must be less than one part in 10^7/hr. MRI scanners designed for human use have been reported that utilize magnetic fields ranging from as low as 0.02T (*3*) to 8.0T (*4*). However, from the mid-1980s until the present time the majority of high performance MRI studies have been done using magnets operating at a field strength of 1.5T or lower. Recently, fully featured clinical scanners operating at 3T have become available (Figure 1). These higher field scanners have approximately twice the signal-to-noise ratio (SNR) of the 1.5T scanners. As discussed below, the 3T scanners have a particular advantage over lower field scanners in studying the distribution of iron deposits in the brain.

The signal utilized in conventional MRI is derived entirely from mobile protons within the human tissues. For the most part this signal originates in water

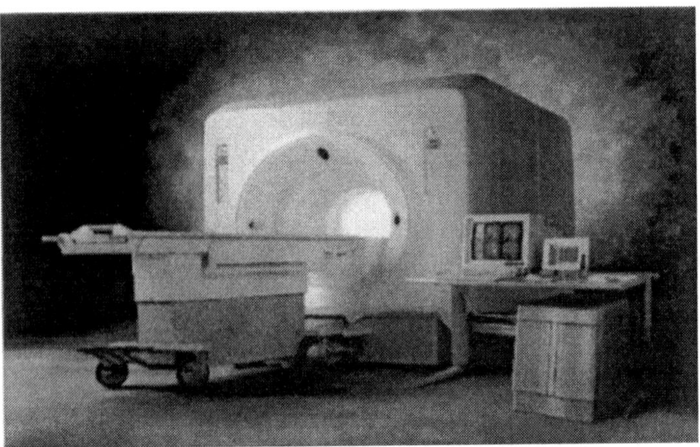

Figure 1. MRI scanner operating at a field strength of 3.0 tesla. (Photograph courtesy of General Electric Global Research.)

molecules, which comprise approximately 80% of most tissues. In some fatty tissues there are enough lipid-associated mobile protons to contribute a significant portion of the MR signal. However, in brain imaging, which is the subject of this article, the signal may be considered as arising entirely from water molecules.

In a typical brain imaging sequence one set of gradient coils is used to select spins in a planar thin slice with a chosen thickness to be imaged. The other two gradient coils are used to encode position information into the received MR signal. A Fourier transformation of this data is utilized to create a grid of picture elements (pixels), which span the electronically selected field of view (FOV). In 3T imaging of the whole brain it is typical to utilize a slice thickness of 2 mm and a FOV of 22 cm x 22 cm with 256 pixels in each direction. This produces picture elements of 220/256 = 0.86 mm on a side. Each pixel then corresponds to a volume element (voxel) 0.86 mm x 0.86 mm x 2 mm =1.48 mm^3 in the brain. If the head is scanned from front to back the images are referred to as coronal sections and approximately 96 2-mm slices will be required to cover the entire brain. The brain typically has a volume close to 1400 cc and approximately 950,000 pixel elements are required to cover the entire structure at the resolution just described. Depending on the level of image quality desired, this set of images can be acquired at 3T in an imaging time ranging from two to ten minutes. When scanning postmortem brains much longer scan times can be utilized and, using scan times of about 3 hours at 3T, SNR levels can be achieved that permit the voxel volume to be reduced to about 0.2 mm^3. By use of even longer scan times small bore systems operating at much higher field strengths (7-16T) can perform MRI microscopy on small animals and small human tissue samples by achieving voxel sizes on the order of 40 μm on a side (*5*). Whole-body MRI systems for human use operating at fields as high as 7 and 8T are currently available at a few research centers.

Patient safety is an important aspect of MR scanning (*6, 7*). In most cases exposure to MRI is inherently very safe and scans may be repeated as often as is medically necessary without tissue injury. This is an important advantage of MRI over techniques requiring ionizing radiation or the injection of radioisotopes. However, the presence of ferromagnetic materials either in the vicinity of the scanner or within the patient as a medical implant can be life threatening. Such materials must be rigorously excluded from the scanning environment.

MR Relaxation Times and Contrast in MR Images

Contrast between different tissue regions is essential to any medical imaging modality. In MRI the signal is derived largely from mobile protons on water molecules diffusing in the tissues. Image contrast arises from differences in the tissue water concentrations and from variations from one tissue region to another of the two spin relaxation times, T1 and T2 (*8, 9*). Here T1, the longitudinal relaxation time, is a measure of the time required for the spins to come into a statistically equilibrated alignment with the applied magnetic field. In pure water this time is relatively long, approximately 4200 ms, and is governed by the interaction of the spin of one water proton with that of the other proton on the same molecule and by somewhat smaller interactions with proton spins on neighboring water molecules. In living tissues T1 relaxation times are much shorter than in pure water, ranging approximately from 500 to 1500 ms.

The transverse relaxation time, T2, is a measure of the time required for the loss of MR signal after it has been excited. T2 is also long in pure water, approximately 2500 ms. In tissues T2 is shortened even more than T1 and is generally in the range 20 to 100 ms. The value of T1 in tissues is up to fifty times that of T2. Both parameters depend on subtle and hard to model magnetic interactions between water molecules and tissue macromolecules. As a result, understanding of the contrast between tissues based on differences in T1 and T2 depends on empirical knowledge of the variations of these parameters from tissue to tissue and, within a given tissue, of the variation between health and disease.

As discussed in detail below, it has been known since 1986 that there is strong evidence that tissue iron stores, presumably in the form of ferritin and the related compound hemosiderin, have a specific effect on MR relaxation parameters (particularly T2), resulting in a loss of signal (hypointensity) on T2-

weighted[1] images. In the liver, tissue iron stores are important markers of disease and contribute to the pathogenesis of hemochromatosis. In the brain, iron has been implicated in Parkinson's and Alzheimer's disease and in other neurodegenerative disorders. Therefore, the development of a full understanding and utilization of iron-dependent MR contrast would establish MRI as a quantitative measuring device of an important clinical parameter.

MR Contrast Agents

By adjusting the scanning parameters it is possible to produce T1-weighted images (tissues with short T1 appear bright or hyperintense) or T2-weighted images (tissues with short T2 appear dark or hypointense). The addition of magnetic ions to a tissue modifies T1 and T2 relaxation and thereby the tissue contrast. There has been a major effort to develop non-toxic magnetic compounds containing strongly magnetic ions (e.g., gadolinium and dysprosium) to serve as injectable, or exogenous contrast agents, and such agents are now widely used in clinical medicine (*10*). Most metallic ions present in brain tissues (sodium, potassium, calcium, magnesium, and zinc) are non-magnetic and do not contribute to MR image contrast. On the other hand, certain magnetic ions of the transition group (e.g., iron, manganese, and copper) are normally present in human tissues, including brain. However, under normal circumstances, only iron is present in concentrations sufficient to provide endogenous MRI contrast.

Brain Iron Deposition

Metals are of fundamental importance to biological tissues (*11-14*) and maintenance of iron homeostasis is a basic biological requirement (*15-18*). By use of Perls stain for non-heme iron and other histochemical techniques pathologists have long been able to demonstrate a consistent pattern of iron deposition (*19-23*) in the normal human brain. This iron concentration is most prominent in specific deep brain nuclei of the basal ganglia and associated structures. These nuclei include the globus pallidus, the putamen, the caudate nucleus, the red nucleus, the substantia nigra and the dentate nucleus of the cerebellum. The iron concentration is low at birth and increases with age until a

---

[1] A T2-wighted image is one taken with a long duration (tens of milliseconds) between signal excitation and detection. This provides a strong contrast between regions of differing T2 values.

relatively constant level is reached at an age of 20 - 30 years (*24, 25*). The histochemical techniques available to pathologists do not permit premortem studies of brain iron.

Clinical MRI became a reality in the early 1980s and it was soon noted that the high field systems available at that time (1.5T) could detect a pattern of shortened T2 in the brain regions classically known to have high iron concentrations (*26*) and when higher field research systems (4T) became available it was apparent that this T2 shortening was more pronounced as the magnetic field was increased (*27-29*).

In tissues, iron is present predominately in a storage form as mineralized iron oxide cores approximately 80 Å in diameter within the spherical shell of the protein ferritin and in the less structured material hemosiderin which is often described as a partially degraded aggregation of ferritin molecules (*15-17*). The mineralized iron cores of ferritin and hemosiderin are believed to be formed from the poorly ordered iron oxide, ferrihydrite. The chemical formula $Fe_5HO_8{\cdot}4H_2O$ has been suggested for this compound but the actual formula is not firmly established (*30*). Electron microscopy of brain shows iron to be in located in the cell bodies and processes of glial cells and in the inner and outer loops of myelin sheets (*31*). There is also evidence for brain iron associated with neuromelanin and liposfuscin (*32-35*). In brain regions characterized by high iron deposition the concentration of storage iron can reach 0.20 mg Fe/g wet tissue (*24*) that corresponds to 3.6 mM of iron in this tissue. In the cytoplasm and the extracellular spaces there is, in addition to the storage iron deposits, also a poorly characterized labile iron pool (LIP) involving iron in the μM range liganded to small molecules such citrate and ATP (*36*). Thus the iron distribution picture is that of a relatively high iron concentration (mM range) in a relatively chemically inert storage depot in exchange with a low concentration in a metabolically active LIP. The T2 shortening observed in MRI is attributed to the storage iron. Iron-associated tissue damage is presumably mediated by the LIP, which, because of its low total iron concentration, is not expected to be detected directly by MR imaging.

Mechanism of Iron-Induced T2 Shortening

A good deal of effort has gone into developing physical models to infer the tissue iron concentration and the configuration of the iron deposits from the measured MRI signals. Such models relate the signal loss in iron-rich regions to the applied field strength and to the imaging sequence employed. This makes it possible to use the scanner most effectively to extract the desired information from the subject. Development of effective models has turned out to be somewhat more difficult than originally anticipated and fully satisfactory models have not yet been achieved. In addition to obtaining optimum contrast, the choice of a practical imaging sequence must take patient comfort and the ability to tolerate the scan into consideration when determining the total scan duration and the total volume of brain to be imaged.

In a perfectly uniform magnetic field the spins of all water protons, once excited to a transverse orientation to the applied magnetic field, precess in phase with one another around the direction of this field. The rate of this precession (the Larmor frequency) is directly proportional to the applied field strength (*37*). For protons in a 3T field this precession completes one cycle every 7.8 ns (the Larmor period). The presence of magnetized iron-containing materials such as ferritin and hemosiderin produces magnetic field perturbations of microscopic extent in the cytoplasm and other tissue spaces surrounding the deposits. The spin precession of protons on water molecules diffusing through these microscopic fields will not remain in phase with one another. Instead, the phases of individual spins will come to vary by differing amounts that depend on the path taken by each water molecule as it diffuses in the vicinity of the magnetized particles. This perturbation increases relaxation rates and results in a reduction of both T1 and T2 from the values that would obtain in a perfectly uniform field. The fundamental theory of spin relaxation processes (37) explains that, because the process of water molecules diffusing through the microscopic magnetic field surrounding magnetized particles of macromolecular size or larger is on a much slower time scale than the Larmor precession, the effect of this diffusion on T2 will be greater than that on T1 (*38-40*). As a result the most prominent effect of tissue iron deposits on MRI is a reduction in the T2 relaxation time, which manifests itself as a loss of signal (hypointensity) in T2-weighted images. As mentioned above, efforts to model this T2 reduction and correlate it with measurements of tissue iron concentrations have not yet achieved quantitative precision (*41*). The most likely cause of discrepancies is that iron deposits in tissues are clumped or aggregated within neural cells in ways that are not well characterized and that these aggregated iron-containing clusters have stronger effects on T2 relaxation than are measured in the simple solutions of iron salts or ferritin on which much of the early modeling was based. More recent modeling work has taken a more complete account of the effects of water diffusion, magnetic particle size and field strength into consideration and has improved agreement with experiment (*42-45*).

It is possible, of course, that mechanisms other than regional iron deposition may be responsible for at least some of the hypointensities in specific brain regions found in high field MRI (*46*). In all tissues a variety of relaxation mechanisms are acting simultaneously and iron is never expected to be responsible for all the T2 relaxation rate in a given brain region. However, a number of factors point to an important role for iron deposits in contributing significantly to the observed relaxation of specific brain regions. These include

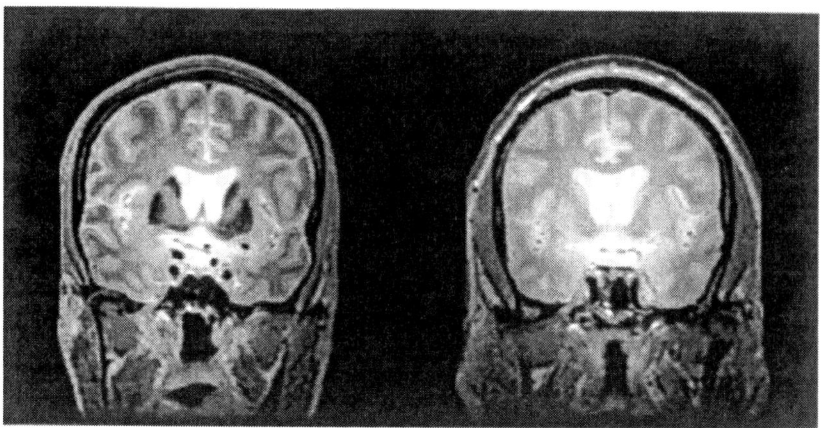

Figure 2. Brain iron deposition. The image on the left is from a 56-year old patient with aceruloplasminemia. This rare disease is known to lead to iron accumulation in the brain and other tissues. The image shows evidence for heavy accumulation of iron in the caudate nucleus, putamen and globus pallidus. The image on the right is from a 64-year old normal control subject. Both images were taken at 3 tesla using a spin-echo sequence with echo time of 18 ms, a 22 cm square field of view and a 2 mm slice thickness. (Images courtesy of J. F. Schenck and E. A. Zimmerman).

(i) the precise co-localization of short T2 regions with the regions of high iron concentration demonstrated with Perls staining of autopsy specimens, (ii) the strong increase of measured relaxation rates in these regions with magnetic field strength, (iii) the strong increase in the rate of signal loss when gradient echoes (T2*)[2] are used in place of radiofrequency-recalled echoes (T2) and (iv) the strong increase in T2 relaxation rate at high fields measured in diseases known to have large increases in brain iron deposition such as aceruloplasminemia (Figure 2) (*47-49*) and Hallervorden-Spatz syndrome (now known as neurodegeneration with brain iron accumulation or NBIA) (*50*).

Iron and Neurodegenerative Diseases

Iron is present to some degree in all tissues and is essential to normal cell metabolism (*51, 52*). It is important in numerous pathological states as well (*53, 54*). The relatively high concentration of iron and its unique pattern of

[2] T2* is the transverse relaxation time measured using gradient-recalled, rather radiofrequency (RF)-recalled echoes (*37*). The RF technique (T2) cancels some of the inhomogeneity-induced relaxation and, therefore, T2*<T2.

distribution in the brain suggest that iron plays an important role in brain function that is distinct from its role in other tissues. Iron involvement is suspected in many neurological diseases (*55-59*). For example, insufficient dietary iron during infancy leads to irreversible mental retardation (*60-64*).

The mechanism of iron-related tissue toxicity is generally held to be the enhancement of oxidative stress in tissue by catalysis of reactions leading to the generation of reactive oxygen species (ROS) and reactive nitrogen species (RNS) (*65*). A major pathway by which iron is proposed to bring about oxidative injury to tissues derives from its ability to catalyze the Haber-Weiss reaction and thereby to generate the highly reactive hydroxyl free radical, $OH^{\bullet}$, from the less reactive oxidants, H_2O_2 and the superoxide radical, $O_2^{\bullet-}$. The overall Haber-Weiss reaction is

$$H_2O_2 + O_2^{\bullet-} \rightarrow OH^{\bullet} + OH^{-} + O_2.$$

This reaction proceeds very slowly in aqueous solutions unless catalyzed by a metal such as iron (*65*, pp. *131-133*)

$$Fe^{+++} + O_2^{\bullet-} \rightarrow Fe^{++} + O_2$$

$$Fe^{++} + H_2O_2 \rightarrow OH^{\bullet} + OH^{-} + Fe^{+++}.$$

At the end of this process the iron atom is returned to its initial oxidation state and is prepared to take part in another round of the sequence. The hydroxyl radical is extremely reactive and is believed to diffuse only an average distance of about 50 Å before participating in a damaging oxidative reaction with cellular macromolecules. This radical produces cellular injury through a variety of processes including lipid peroxidation, DNA injury and protein degradation. Iron-enhanced oxidative tissue injury is suspected in a large number of neurological diseases (*66-78*).

In addition to the longstanding suspicion of iron involvement in the familiar neurological diseases in the above references, recent work has shown a link between abnormalities of iron regulatory proteins (*52*) and neurodegenerative diseases (*79, 80*). Genetic defects in ferritin synthesis have also recently been described as leading to neurodegenerative diseases (*81, 82*). A role for iron in Parkinson's disease has been suspected for many years (*83-90*).

As early as 1953 Goodman noted that iron deposits were frequently seen in association with amyloid plaques of Alzheimer's disease (AD) and that the disease "may be due to a disturbance in the cerebral metabolism of iron..." (*91*). The possibility of iron involvement in the pathogenesis of AD is consistent with the prevailing view that aggregations of amyloid A-beta are the proximate cause of this disease. There have been several reports of iron and iron-related proteins, such as ferritin, as components of senile plaques (*92-94*). A number of MRI and other studies have added credence to a role for iron in AD (*95-99*). A statistically significant reduction in T2 at 3T within the hippocampus of AD patients has been recently reported (*100*). Another recent MR microscopy report has provided evidence for small local regions of reduced T2, possibly the result of iron increases, in the vicinity of amyloid plaques (*101*). Of particular interest is the finding that the messenger RNA for amyloid precursor protein (APP) contains an iron-responsive element, indicating that iron may have a significant role in the metabolism of this protein, which is believed to be central to the pathogenesis of AD (*102*). There is evidence for iron involvement in several other neurological diseases, including multiple sclerosis (*103-105*), Huntington's disease (*106, 107*), and Friedreich's ataxia (*108-110*). For the most part, it is uncertain at present whether the association of iron with a specific neurodegenerative disease reflects a role for iron in the pathogenesis of the disease or is simply a consequence of the disease progression. It should be noted that in either case the ability to visualize the iron deposits by MRI has great potential in terms of monitoring disease progression and the response to therapy. In most diseases it is iron excess, not iron deficit, which is suspected of pathogenesis; however, the common neurological disorder, restless legs syndrome, may be associated with a deficit of brain iron (*111-113*).

Conclusions

The physiological purposes of the remarkable distribution of iron in the normal brain and the pathological correlates of alterations in this pattern are incompletely understood at present but advances are occurring rapidly. It is hypothesized that pathological changes in the LIP are reflected in changes (usually increases) in the storage iron pool. It is important to remember that MRI is likely sensitive only to the high concentrations of iron in the storage pool and not to the low concentrations of iron in the metabolically active and, in some situations presumably toxic, LIP. Therefore, the use of MRI alone would be insufficient to unravel the complete relationship between iron and brain physiology and pathology. However, MRI provides a crucial tool for this effort because these scanners, alone among the technologies available to neuroscience, permit direct, noninvasive observation of iron in the living human brain. It is anticipated that, through a combination of a fuller understanding of the genetics and molecular biology of brain iron metabolism and the availability of high field MRI, this phenomenon may soon provide a major clinical approach to the study and treatment of neurodegenerative diseases.

Acknowledgements

It is a pleasure to acknowledge technical discussions and other assistance provided by, among others, E. A. Zimmerman, Z. Li, K. M. Fish, A. H. Koeppen, S. J. Lippard, J. F. Graf, Z. L. Harris, S. J. Kalia, D. J. Brooks, M. C. Grubb, S. Adak, T. O'Keefe, R. Mullick, A. N. Ishaque, G. Kenwood, R. A. Frank, C. J. Belden and N. D. Chasteen

References

1. Fuchs, V. R.; Sox, H. C., Jr. Physicians' views of the relative importance of thirty medical innovations. *Health Aff. (Millwood)* **2001**, *20*, 30-42.
2. Schenck, J. F.; Bulkes, C. Instrumentation: magnets, coils and hardware. In *Magnetic Resonance Imaging of the Brain and Spine*; 3rd ed.; Atlas, S. W., Ed.; Lippincott, Williams and Wilkins: Philadelphia, 2002, pp 3-31.
3. Wahlund, L. O.; Agartz, I.; Almqvist, O.; et al. The brain in healthy aged individuals: MR imaging. *Radiology* **1990**, *174*, 675-679.
4. Robitaille, P. M.; Warner, R.; Jagadeesh, J.; et al. Design and assembly of an 8 tesla whole-body MR scanner. *J. Comp. Assist. Tomogr.* **1999**, *23*, 808-820.
5. Benveniste, H.; Einstein, G.; Kim, K. R.; et al. Detection of neuritic plaques in Alzheimer's disease by magnetic resonance microscopy. *Proc. Natl. Acad. Sci. USA* **1999**, *96*, 14079-14084.
6. Shellock, F. G.; Kanal, E. *Magnetic resonance: bioeffects, safety, and patient management*; 2nd ed.; Lippincott-Raven: Philadelphia, 1996.
7. Schenck, J. F. Safety of strong, static magnetic fields. *J. Magn. Reson. Imaging* **2000**, *12*, 2-19.
8. Bottomley, P. A.; Foster, T. H.; Argersinger, R. E.; Pfeifer, L. M. A review of normal tissue hydrogen NMR relaxation times and relaxation mechanisms from 1-100 MHz: dependence on tissue type, NMR frequency, temperature, species, excision, and age. *Med. Phys.* **1984**, *11*, 425-448.
9. Schenck, J. F. The role of magnetic susceptibility in magnetic resonance imaging: MRI magnetic compatibility of the first and second kinds. *Med. Phys.* **1996**, *23*, 815-850.
10. Merbach, A. E.; Toth, A. *The chemistry of contrast agents in medical magnetic resonance imaging*; Wiley: Chichester ; New York, 2001.
11. Lippard, S. J.; Berg, J. M. *Principles of bioinorganic chemistry*; University Science Books: Mill Valley, Calif., 1994.

12. Williams, R. J. P.; Silva, J. R. R. F. d. *The natural selection of the chemical elements: the environment and life's chemistry*; Clarendon Press; Oxford University Press: Oxford; New York, 1996.
13. Silva, J. J. R. F. d.; Williams, R. J. P. *The biological chemistry of the elements: the inorganic chemistry of life*; 2nd ed.; Oxford University Press: Oxford; New York, 2001.
14. Burdette, S. C.; Lippard, S. J. Meeting of the minds: metalloneurochemistry. *Proc. Natl. Acad. Sci. USA* **2003**, *100*, 3605-3610.
15. Frankel, R. B.; Blakemore, R. P. *Iron biominerals*; Plenum Press: New York, 1991.
16. Sigel, A.; Sigel, H. *Iron transport and storage in microorganisms, plants, and animals*; Marcel Dekker: New York, 1998.
17. Crichton, R. R; Boelaert, J. R. *Inorganic biochemistry of iron metabolism : from molecular mechanisms to clinical consequences*; 2nd ed.; Wiley: Chichester; New York, 2001.
18. Williams, R. J. P. The biological chemistry of the brain and its possible evolution. *Inorganica Chimica Acta* **2003**, *356*, 27-40.
19. Spatz, H. Über den Eisennachweis im Gehirn, besonders in Zentren des extrpyramidal-motorischen Systems. *Z. Ges. Neurol. Psychiat. (Zeitschrift fur die gesamte Neurologie und Psychiatrie)* **1922**, *77*, 261-390.
20. Hill, J. M. The distribution of iron in the brain. In *Brain Iron: Neurochemical and Behavioural Aspects.*; Youdim, M. B. H., Ed.; Taylor and Francis: London, 1988, pp 1-24.
21. Morris, C. M.; Candy, J. M.; Oakley, A. E.; et al. Histochemical distribution of non-haem iron in the human brain. *Acta Anat. (Basel)* **1992**, *144*, 235-257.
22. Koeppen, A. H. The history of iron in the brain. *J. Neurol. Sci.* **1995**, *134 Suppl*, 1-9.
23. Koeppen, A. H. A brief history of brain iron research. *J. Neurol. Sci.* **2003**, *207*, 95-97.
24. Hallgren, B.; Sourander, P. The effect of age on the non-haemin iron in the human brain. *J. Neurochem.* **1958**, *3*, 41-51.
25. Aoki, S.; Okada, Y.; Nishimura, K.; Barkovich, A. J.; et al. Normal deposition of brain iron in childhood and adolescence: MR imaging at 1.5 T. *Radiology* **1989**, *172*, 381-385.
26. Drayer, B.; Burger, P.; Darwin, R.; Riederer, S.; et al. Magnetic resonance imaging of brain iron. AJR Am J Roentgenol. *AJR Am. J. Roentgenol.* **1986**, *147*, 103-110.
27. Schenck, J. F.; Mueller, O. M.; Souza , S. P.; Dumoulin, C. L. Magnetic resonance imaging of brain iron using a 4 tesla whole-body scanner. In

Iron Biominerals; Frankel, R. B., Blakemore, R. P., Eds.; Plenum Press: New York, 1991, pp 373-385.

28. Schenck, J. F. Imaging of brain iron by magnetic resonance: T2 relaxation at different field strengths. *J. Neurol. Sci.* **1995**, *134 Suppl*, 10-18.
29. Schenck, J. F. Magnetic resonance imaging of brain iron. *J. Neurol. Sci.* **2003**, *207*, 99-102.
30. Cornell, R. M.; Schwertmann, U. *The iron oxides: structure, properties, reactions, occurrences, and uses*; 2nd ed.; Wiley-VCH: Weinheim, 2003. pp. 26-27.
31. Francois, C.; Nguyen-Legros, J.; Percheron, G. Topographical and cytological localization of iron in rat and monkey brains. *Brain Res.* **1981**, *215*, 317-322.
32. Castelnau, P. A.; Garrett, R. S.; Palinski, W.; et al. Abnormal iron deposition associated with lipid peroxidation in transgenic mice expressing interleukin-6 in the brain. *J. Neuropathol. Exp. Neurol.* **1998**, *57*, 268-282.
33. Double, K. L.; Gerlach, M.; Schunemann, V.; et al. Iron-binding characteristics of neuromelanin of the human substantia nigra. *Biochem. Pharmacol.* **2003**, *66*, 489-494.
34. Gerlach, M.; Double, K. L.; Ben-Shachar, D.; et al. Neuromelanin and its interaction with iron as a potential risk factor for dopaminergic neurodegeneration underlying Parkinson's disease. *Neurotox. Res.* **2003**, *5*, 35-44.
35. Zecca, L.; Zucca, F. A.; Costi, P.; et al. The neuromelanin of human substantia nigra: structure, synthesis and molecular behaviour. *J. Neural. Transm. Suppl.* **2003**, 145-155.
36. Cabantchik, Z. I.; Kakhlon, O.; Epsztejn, S.; et al. Intracellular and extracellular labile iron pools. *Adv. Exp. Med. Biol.* **2002**, *509*, 55-75.
37. Slichter, C. P. *Principles of magnetic resonance*; 3rd ed.; Springer: Berlin ; New York, 1996.
38. Gillis, P.; Koenig, S. H. Transverse relaxation of solvent protons induced by magnetized spheres: application to ferritin, erythrocytes, and magnetite. *Magn. Reson. Med.* **1987**, *5*, 323-345.
39. Vymazal, J.; Brooks, R. A.; Zak, O.; et al. T1 and T2 of ferritin at different field strengths: effect on MRI. *Magn. Reson. Med.* **1992**, *27*, 368-374.
40. Vymazal, J.; Brooks, R. A.; Bulte, J. W.; et al. Iron uptake by ferritin: NMR relaxometry studies at low iron loads. *J. Inorg. Biochem.* **1998**, *71*, 153-157.
41. Chen, J. C.; Hardy, P. A.; Clauberg, M.; et al. T2 values in the human brain: comparison with quantitative assays of iron and ferritin. *Radiology* **1989**, *173*, 521-526.

42. Yablonskiy, D. A.; Haacke, E. M. Theory of NMR signal behavior in magnetically inhomogeneous tissues: the static dephasing regime. *Magn. Reson. Med.* **1994**, *32*, 749-763.
43. Jensen, J. H.; Chandra, R. Strong field behavior of the NMR signal from magnetically heterogeneous tissues. *Magn. Reson. Med.* **2000**, *43*, 226-236.
44. Jensen, J. H.; Chandra, R. NMR relaxation in tissues with weak magnetic inhomogeneities. *Magn. Reson. Med.* **2000**, *44*, 144-156.
45. Jensen, J. H.; Chandra, R. Theory of nonexponential NMR signal decay in liver with iron overload or superparamagnetic iron oxide particles. *Magn. Reson. Med.* **2002**, *47*, 1131-1138.
46. Chen, J. C.; Hardy, P. A.; Kucharczyk, W.; et al. T2 values in the human brain: comparison with quantitative assays of iron and ferritin. *AJNR Am. J. Neuroradiol.* **1993**, *14*, 275-281.
47. Gitlin, J. D. Aceruloplasminemia. *Pediatr. Res.* **1998**, *44*, 271-276.
48. Miyajima, H.; Takahashi, Y.; Kono, S. Aceruloplasminemia, an inherited disorder of iron metabolism. *Biometals* **2003**, *16*, 205-213.
49. Harris, Z. L. Aceruloplasminemia. *J. Neurol. Sci.* **2003**, *207*, 108-109.
50. Hayflick, S. J.; Westaway, S. K.; Levinson, B.; et al. Genetic, clinical, and radiographic delineation of Hallervorden-Spatz syndrome. *N. Engl. J. Med.* **2003**, *348*, 33-40.
51. Aisen, P.; Wessling-Resnick, M.; Leibold, E. A. Iron metabolism. *Curr. Opin. Chem. Biol.* **1999**, *3*, 200-206.
52. Testa, U. *Proteins of iron metabolism*; CRC Press: Boca Raton, Fla., 2002.
53. Andrews, N. C. Disorders of iron metabolism. *N. Engl. J. Med.* **1999**, *341*, 1986-1995.
54. Brittenham, G. M.; Weiss, G.; Brissot, P.; et al. Clinical consequences of new insights in the pathophysiology of disorders of iron and heme metabolism. *Hematology (Am. Soc. Hematol. Educ. Program)* **2000**, 39-50.
55. Gelman, B. B. Iron in CNS disease. *J. Neuropathol. Exp. Neurol.* **1995**, *54*, 477-486.
56. Qian, Z. M.; Shen, X. Brain iron transport and neurodegeneration. *Trends Mol. Med.* **2001**, *7*, 103-108.
57. Rouault, T. A. Iron on the brain. *Nat. Genet.* **2001**, *28*, 299-300.
58. Thompson, K. J.; Shoham, S.; Connor, J. R. Iron and neurodegenerative disorders. *Brain Res. Bull.* **2001**, *55*, 155-164.
59. Connor, J. R. Iron transport proteins in the diseased brain. *J. Neurol. Sci.* **2003**, *207*, 112-113.
60. Lozoff, B.; Jimenez, E.; Wolf, A. W. Long-term developmental outcome of infants with iron deficiency. *N. Engl. J. Med.* **1991**, *325*, 687-694.

61. Lozoff, B.; Wolf, A. W.; Jimenez, E. Iron-deficiency anemia and infant development: effects of extended oral iron therapy. *J. Pediatr.* **1996,** *129*, 382-389.
62. Wolf, A. W.; Jimenez, E.; Lozoff, B. Effects of iron therapy on infant blood lead levels. *J. Pediatr.* **2003,** *143*, 789-795.
63. Beard, J.; Erikson, K. M.; Jones, B. C. Neonatal iron deficiency results in irreversible changes in dopamine function in rats. *J. Nutr.* **2003,** *133*, 1174-1179.
64. Beard, J. Iron deficiency alters brain development and functioning. *J. Nutr.* **2003,** *133*, 1468S-1472S.
65. Halliwell, B.; Gutteridge, J. M. C. *Free radicals in biology and medicine*; 3rd ed.; Clarendon Press; Oxford University Press: Oxford New York, 1999.
66. Riederer, P.; Sofic, E.; Rausch, W. D.; et al. Transition metals, ferritin, glutathione, and ascorbic acid in parkinsonian brains. *J. Neurochem.* **1989,** *52*, 515-520.
67. Halliwell, B. Oxidants and the central nervous system: some fundamental questions. Is oxidant damage relevant to Parkinson's disease, Alzheimer's disease, traumatic injury or stroke? *Acta Neurol. Scand. Suppl.* **1989,** *126*, 23-33.
68. Halliwell, B. Reactive oxygen species and the central nervous system. *J. Neurochem.* **1992,** *59*, 1609-1623.
69. Cohen, G. The brain on fire? *Ann. Neurol.* **1994,** *36*, 333-334.
70. Morris, C. M.; Edwardson, J. A. Iron histochemistry of the substantia nigra in Parkinson's disease. *Neurodegeneration* **1994,** *3*, 277-282.
71. Loeffler, D. A.; LeWitt, P. A.; Juneau, P. L.; et al. Increased regional brain concentrations of ceruloplasmin in neurodegenerative disorders. *Brain Res.* **1996,** *738*, 265-274.
72. Smith, M. A.; Harris, P. L.; Sayre, L. M.; Perry, G. Iron accumulation in Alzheimer disease is a source of redox-generated free radicals. *Proc. Natl. Acad. Sci. USA* **1997,** *94*, 9866-9868.
73. Castellani, R. J.; Smith, M. A.; Nunomura, A.; et al. Is increased redox-active iron in Alzheimer disease a failure of the copper-binding protein ceruloplasmin? *Free Radic. Biol. Med.* **1999,** *26*, 1508-1512.
74. Hirsch, E. C.; Hunot, S. Nitric oxide, glial cells and neuronal degeneration in parkinsonism. *Trends Pharmacol. Sci.* **2000,** *21*, 163-165.
75. Sayre, L. M.; Perry, G.; Harris, P. L.; et al. In situ oxidative catalysis by neurofibrillary tangles and senile plaques in Alzheimer's disease: a central role for bound transition metals. *J. Neurochem.* **2000,** *74*, 270-279.

76. Rottkamp, C. A.; Nunomura, A.; Raina, A. K.; et al. Oxidative stress, antioxidants, and Alzheimer disease. *Alzheimer Dis. Assoc. Disord.* **2000**, *14 Suppl 1*, S62-66.
77. Halliwell, B. Role of free radicals in the neurodegenerative diseases: therapeutic implications for antioxidant treatment. *Drugs Aging* **2001**, *18*, 685-716.
78. Lovell, M. A.; Xie, C.; Xiong, S.; Markesbery, W. R. Protection against amyloid beta peptide and iron/hydrogen peroxide toxicity by alpha lipoic acid. *J. Alzheimers Dis.* **2003**, *5*, 229-239.
79. Smith, M. A.; Wehr, K.; Harris, P. L.; et al. Abnormal localization of iron regulatory protein in Alzheimer's disease. *Brain Res.* **1998**, *788*, 232-236.
80. Grabill, C.; Silva, A. C.; Smith, S. S.; et al. MRI detection of ferritin iron overload and associated neuronal pathology in iron regulatory protein-2 knockout mice. *Brain Res.* **2003**, *971*, 95-106.
81. Curtis, A. R.; Fey, C.; Morris, C. M.; et al. Mutation in the gene encoding ferritin light polypeptide causes dominant adult-onset basal ganglia disease. *Nat. Genet.* **2001**, *28*, 350-354.
82. Crompton, D. E.; Chinnery, P. F.; Fey, C.; et al. Neuroferritinopathy: a window on the role of iron in neurodegeneration. *Blood Cells Mol. Dis.* **2002**, *29*, 522-531.
83. Youdim, M. B. Iron in the brain: implications for Parkinson's and Alzheimer's diseases. *Mt. Sinai J. Med.* **1988**, *55*, 97-101.
84. Dexter, D. T.; Carter, C. J.; Wells, F. R.; et al. Basal lipid peroxidation in substantia nigra is increased in Parkinson's disease. *J. Neurochem.* **1989**, *52*, 381-389.
85. Dexter, D. T.; Wells, F. R.; Lees, A. J.; et al. Increased nigral iron content and alterations in other metal ions occurring in brain in Parkinson's disease. *J. Neurochem.* **1989**, *52*, 1830-1836.
86. Hirsch, E. C.; Faucheux, B. A. Iron metabolism and Parkinson's disease. *Mov. Disord.* **1998**, *13 Suppl 1*, 39-45.
87. Vymazal, J.; Righini, A.; Brooks, R. A.; et al. T1 and T2 in the brain of healthy subjects, patients with Parkinson disease, and patients with multiple system atrophy: relation to iron content. *Radiology* **1999**, *211*, 489-495.
88. Castellani, R. J.; Siedlak, S. L.; Perry, G.; Smith, M. A. Sequestration of iron by Lewy bodies in Parkinson's disease. *Acta Neuropathol. (Berl.)* **2000**, *100*, 111-114.
89. Cole, G. M. Ironic fate: can a banned drug control metal heavies in neurodegenerative diseases? *Neuron* **2003**, *37*, 889-890.
90. Kaur, D.; Yantiri, F.; Rajagopalan, S.; et al. Genetic or

Pharmacological Iron Chelation Prevents MPTP-Induced Neurotoxicity In Vivo. A Novel Therapy for Parkinson's Disease. *Neuron* **2003**, *37*, 899-909.

91. Goodman, L. Alzheimer's disease; a clinico-pathologic analysis of twenty-three cases with a theory on pathogenesis. *J. Nerv. Ment. Dis.* **1953**, *118*, 97-130.
92. Grundke-Iqbal, I.; Fleming, J.; Tung, Y. C.; et al. Ferritin is a component of the neuritic (senile) plaque in Alzheimer dementia. *Acta Neuropathol. (Berl.)* **1990**, *81*, 105-110.
93. Morris, C. M.; Kerwin, J. M.; Edwardson, J. A. Non-haem iron histochemistry of the normal and Alzheimer's disease hippocampus. *Neurodegeneration* **1994**, *3*, 267-275.
94. Lovell, M. A.; Robertson, J. D.; Teesdale, W. J.; et al. Copper, iron and zinc in Alzheimer's disease senile plaques. *J. Neurol. Sci.* **1998**, *158*, 47-52.
95. Bartzokis, G.; Sultzer, D.; Cummings, J.; et al. In vivo evaluation of brain iron in Alzheimer disease using magnetic resonance imaging. *Arch. Gen. Psychiatry* **2000**, *57*, 47-53.
96. Bartzokis, G.; Tishler, T. A. MRI evaluation of basal ganglia ferritin iron and neurotoxicity in Alzheimer's and Huntingon's disease. *Cell. Mol. Biol. (Noisy-le-grand)* **2000**, *46*, 821-833.
97. Bishop, G. M.; Robinson, S. R.; Liu, Q. et al. Iron: a pathological mediator of Alzheimer disease? *Dev. Neurosci.* **2002**, *24*, 184-187.
98. Bishop, G. M.; Robinson, S. R. Human Abeta1-42 reduces iron-induced toxicity in rat cerebral cortex. *J. Neurosci. Res.* **2003**, *73*, 316-323.
99. Wadghiri, Y. Z.; Sigurdsson, E. M.; Sadowski, M.; et al. Detection of Alzheimer's amyloid in transgenic mice using magnetic resonance microimaging. *Magn. Reson. Med.* **2003**, *50*, 293-302.
100. Zimmerman, E. A.; Li, Z.; O'Keefe, T.; Schenck, J. F. High field strength (3T) magnetic resonance imaging in Alzheimer's disease. *Ann Neurol* **2003**, *54 (suppl 7)*, S68.
101. Helpern, J. A.; Lee, S. P.; Falangola, M. F.; et al. MRI assessment of neuropathology in a transgenic mouse model of Alzheimer's disease. *Magn. Reson. Med.* **2004**, *51*, 794-798.
102. Rogers, J. T.; Randall, J. D.; Cahill, C. M.; et al. An iron-responsive element type II in the 5'-untranslated region of the Alzheimer's amyloid precursor protein transcript. *J. Biol. Chem.* **2002**, *277*, 45518-45528.
103. Craelius, W.; Migdal, M. W.; Luessenhop, C. P.; et al. Iron deposits surrounding multiple sclerosis plaques. *Arch. Pathol. Lab. Med.* **1982**, *106*, 397-399.
104. Gerber, M. R.; Connor, J. R. Do oligodendrocytes mediate iron regulation in the human brain? *Ann. Neurol.* **1989**, *26*, 95-98.

105.Bakshi, R.; Dmochowski, J.; Shaikh, Z. A.; Jacobs, L. Gray matter T2 hypointensity is related to plaques and atrophy in the brains of multiple sclerosis patients. *J. Neurol. Sci.* **2001,** *185*, 19-26.
106.Klintworth, G. K. Cerebral iron deposition in Huntington's disease. In *Proceedings of the Second International Congress of Neuro-genetics and Neuro-ophthalmology of the World Federation of Neurology*; Barbeau, A., Brunette, J. R., Eds.; Excerpta Medica Foundation: Amsterdam, 1969; Vol. 1, pp. 589-596.
107.Bartzokis, G.; Cummings, J.; Perlman, S.; et al. Increased basal ganglia iron levels in Huntington disease. *Arch. Neurol.* **1999**, *56*, 569-574.
108.Hilditch-Maguire, P.; Trettel, F.; Passani, L. A.; et al. Huntingtin: an iron-regulated protein essential for normal nuclear and perinuclear organelles. *Hum. Mol. Genet.* **2000**, *9*, 2789-2797.
109.Cavadini, P.; Gellera, C.; Patel, P. I.; Isaya, G. Human frataxin maintains mitochondrial iron homeostasis in Saccharomyces cerevisiae. *Hum. Mol. Genet.* **2000**, *9*, 2523-2530.
110.Cavadini, P.; O'Neill, H. A.; Benada, O.; Isaya, G. Assembly and iron-binding properties of human frataxin, the protein deficient in Friedreich ataxia. *Hum. Mol. Genet.* **2002,** *11*, 217-227.
111.Earley, C. J.; Connor, J. R.; Beard, J. L.; et al. Abnormalities in CSF concentrations of ferritin and transferrin in restless legs syndrome. *Neurology* **2000,** *54*, 1698-1700.
112.Connor, J. R.; Boyer, P. J.; Menzies, S. L.; et al. Neuropathological examination suggests impaired brain iron acquisition in restless legs syndrome. *Neurology* **2003**, *61*, 304-309.
113.Earley, C. J. Clinical practice. Restless legs syndrome. *N. Engl. J. Med.* **2003**, *348*, 2103-2109.

Chapter 6

Platinum Anticancer Drugs: From Laboratory to Clinic

Nicholas Farrell

Department of Chemistry, Virginia Commonwealth University, 1001 West Main Street, Richmond, VA 23284–2006

This chapter reviews the current state of platinum anticancer drug development. The principal chemical and biological properties a new platinum drug must have to generate interest from pharmaceutical companies are outlined, using as an example the class of polynuclear platinum compounds. The chapter further summarizes some of the factors to be considered in academia-pharmaceutical industry relationships.

Introduction

Platinum drugs have made a significant contribution to cancer treatment since their introduction into the clinic in 1978. The use of cisplatin in testicular cancer, the role of cisplatin and carboplatin in ovarian cancer and the emerging use of oxaliplatin as front line treatment in colon cancer since its approval in 2002 are by now well documented (1,2,3,4). Thus the time span of platinum drug approval to date has lasted approximately 25 years. It is instructive to note that the first alkylating agent, mechlorethamine, was approved in 1949, and one of the latest versions on the alkylating agent theme, iphosphamide, was approved in 1988. The commercial importance of platinum drugs, with approximately $2bn sales in 2003, mirrors their medical importance.

Despite the success of platinum in the clinic, no major drug companies have in-house platinum drug development programs and the obstacles to cytotoxic drug development must not be underestimated. The current paradigm for anticancer drug discovery is targeted drug research, where proteins involved in signalling pathways, and which are overexpressed in cancer, are selectively targeted or biological pathways such as angiogenesis are attacked (5). Cancer drug development is nothing if not market-driven and the opportunity exists for small companies to advance novel cytotoxic drugs to at least Phase I and preferably Phase II clinical trials where the interest of the large companies will be greater. Nevertheless, the "burden of proof" for a new platinum agent may still be higher than for kinase inhibitors or histone deacetylase inhibitors, to name but two of the current intensively-studied targets in cancer research.

The need for new agents in cancer chemotherapy is apparent from the inability to predictably cure or induce remissions in common tumors such as breast, lung, or prostate cancer. The complementary goals for new platinum drug development have traditionally been activity in cisplatin-resistant cell lines and activity in tumors insensitive to cisplatin intervention. To gain commercial interest new compounds must be very clearly differentiated from cisplatin and oxaliplatin in both chemical structure and biological mechanism. It is axiomatic that anticancer drug development will require at some stage the investment of a major pharmaceutical company. Some of the considerations a company will make with respect to platinum (or indeed any) drugs are:

- Company Positioning. How the new drug fits the company portfolio.
- At the mechanistic level is there a novel mode of action ?
- Is the preclinical biological profile different to cisplatin ?
- New Targets – is the drug active in cisplatin-resistant cancers and can the drug expand the use of platinum drugs to solid tumors and other tumors historically insensitive to cisplatin intervention ? The rapid advance of oxaliplatin in treatment of colon cancer is a case in point here.
- New Targets – does the drug show synergistic or complementary effects in combination chemotherapy ? Interestingly, current evidence is that the newer agents such as angiogensis inhibitors appear to work best in combination with cytotoxics (6).

The prospect for cancer treatment in, say, twenty years, is very likely to be a mix of the newer successful targeted therapies and improved existing ones. New cytotoxic agents will continue to play an important role in the clinical management of cancer and innovative platinum agents are therefore still of clinical interest. This chapter reviews experiences in platinum drug development from an academic perspective.

New Platinum Analogs. Differentiation from Current Drugs and Structural Diversity.

In developing distinct new platinum analogs it is reasonable to situate new agents from a mechanistic point of view. The three principal parameters influencing cytotoxicity for platinum agents are

(a) Uptake and Efflux
(b) DNA Repair and Tolerance of DNA adducts
(c) Enhanced Glutathione binding, causing chemical repair through detoxification.

These are also the major factors involved in acquired resistance to cisplatin (7,8). The general pharmacokinetic principle is that "free" unbound drug is the active principle, allowing distribution to tissue. Thus an important parameter in general is tissue availability - stability in plasma and protein binding may be manipulated by appropriate formulation and liposomes and other methods of formulation have been used (3).

Currently, there are three drugs of interest in clinical trials and they illustrate nicely the differing mechanistic approaches. Their development represents examples of steric control of reactivity, control of oxidation state and ligand lipophilicity aimed at producing orally active agents and manipulation of new structures to produce structurally new DNA adducts (Figure 1).

Potential oral activity of JM-216 is achieved by a Pt(IV) oxidation state with axial carboxylate ligands as well as replacement of one NH_3 group by the more lipophilic cyclohexylamine. Insertion of steric bulk at the platinum center slows the kinetics of substitution in comparison to cisplatin. AMD473 was designed specifically to circumvent thiol-mediated drug resistance by sterically hindering its reaction with glutathione while retaining the ability to form cytotoxic adducts with DNA (9). It is of interest that clinical trials have found activity in prostate cancer for both these lipophilic agents. Although not specifically designed for such a use, the importance of bringing new agents to clinical trials is underlined by these findings, which have given new relevance to these prospective drugs beyond the chemical rationales outlined.

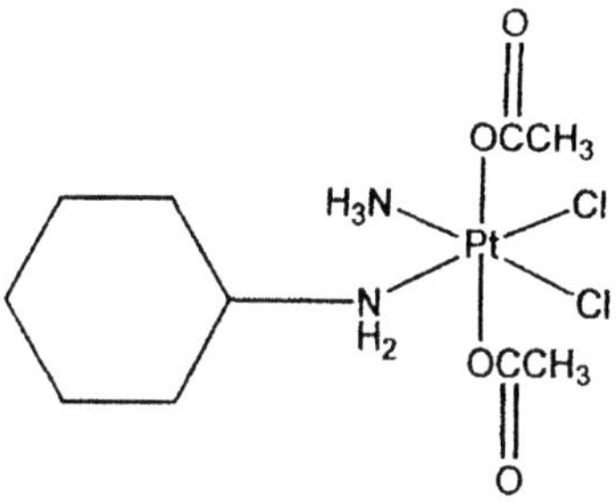

JM-216

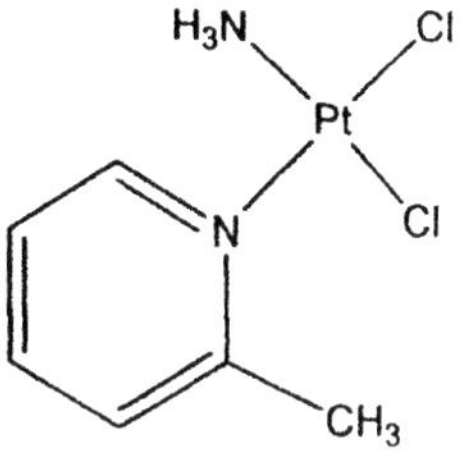

AMD-473

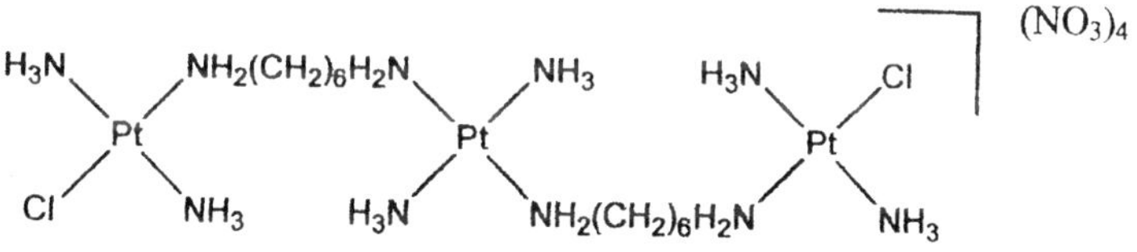

BBR-3464

Figure 1. Structures of platinum compounds currently in clinical trials

In the case of JM-216 and AMD473 the lesions induced on DNA are expected to be essentially similar to those of cisplatin and oxaliplatin. The trinuclear BBR3464 compound and its congeners were designed explicitly to produce structurally distinct DNA lesions – the chemistry arose from the hypothesis that development of platinum compounds structurally different to cisplatin may lead to a spectrum of clinical activity genuinely complementary to the parent drug (10,11,12). BBR3464 is the first polynuclear drug to enter clinical trials (in June 1998), and the first platinum drug not based on the cisplatin structure. The structure, derived from general structures of trinuclear systems (13), is notable for the presence of the central Pt which contributes to DNA affinity only through electrostatic and H-bonding interactions. The 4+ charge, the presence of at least two Pt coordination units capable of binding to DNA and the consequences of such DNA binding is a remarkable departure from the cisplatin structural paradigm.

DNA Binding of Polynuclear Platinum Complexes

The DNA binding modes of polynuclear platinum complexes are distinct from that of cisplatin and carboplatin. Further, the consequences of such modes of DNA binding in terms of downstream effects such as protein recognition and DNA repair are also clearly distinct. Long-range intrastrand and interstrand crosslinks (where the sites of platination are separated by up to 4 intervening base pairs) have been identified and characterized (12). The 1,4 interstrand crosslink has been studied in most detail and the major findings are:

(i) directional isomers have been observed where 3' → 3' and 5' → 5' crosslinks are formed. The relative proportion of these adducts may be dependent on hydrogen-bonding and charge effects within the linker (14).
(ii) the formation of the 1,4-interstrand crosslink produces a delocalized lesion where the conformational changes are transmitted beyond the actual site of platination. Especially the induction of the *syn* conformation in all purine bases in the crosslinked sequence is noteworthy (15,16).
(iii) Both 3' → 3' and 5' → 5' crosslinks bend DNA but not in the same rigid manner as that induced by cisplatin. The BBR3464-DNA interstrand crosslinks are not recognized by HMG group proteins which recognize cisplatin adducts. Using cell extracts the interstrand crosslinks were shown to be insensitive to DNA repair (14).

In the context of this tutorial session, these features clearly differentiate the class of compounds from cisplatin and analogs and, as mechanistic work proceeded in parallel with clincial trials, presented strong rationales for development, allowing polynuclear platinum in general to be situated as possibly repair-tolerant in cisplatin-resistant tumors.

Structural Diversity.

An important aspect is the structural diversity of any new class of potential drugs – this allows the selection of alternative candidates if and when problems arise even in the late stages of development. BBR3464 arose from studies of the dinuclear systems (Figure 2).

The polynuclear structure is extremely flexible and produces a wide series of compounds differing in functionality (bifunctional to tetrafunctional DNA-binding), geometry (leaving chloride groups *cis* or *trans* to diamine bridge). The findings leading to the choice of BBR3464 as the first clinical candidate and early structure-activity relationships have been summarized (*10,11*). The presence of charge and hydrogen-bonding capacity within the central linker (either in the form of a tetraamineplatinum moiety or a charged polyamine linker such as spermidine or spermine) produces very potent compounds significantly more cytotoxic (down to nanomolar levels in some cases) and antitumor active than the "simple" dinuclear species. The biological activity of dinuclear platinum-polyamine systems varies with chain length and charge, although the overall profile is similar (*17,18*). An example of clinical interest is shown in Figure 3. The polyamine system replicates nicely the charge and hydrogen-bonding effects and indeed in many ways is more attractive than the trinuclear system because of the ability to functionalize the central unit to produce pro drugs and targeted molecules (*19*). Again, the diversity of the structure is a strong positive for drug development – newer agents building on clinical experience (see below) are possible within the recognized structural classes.

Biological End Points

The principal parameters of biological activity remain *in vitro* cytotoxicity and *in vivo* activity in human tumor xenografts. Note that the term antitumor activity should be reserved for the *in vivo* situation (tumors in animals) - the tissue culture experiments measure only ability to kill cells. An example of a

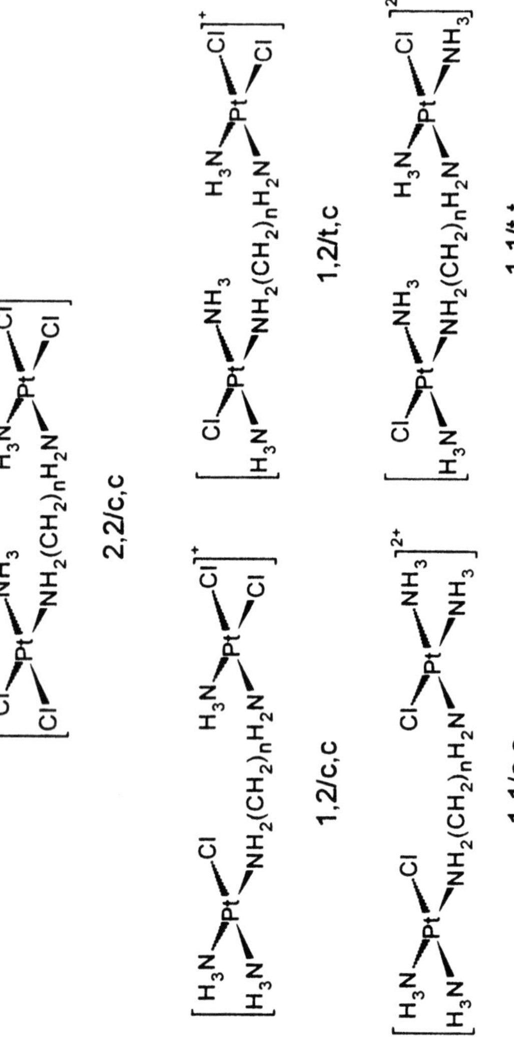
2,2/c,c
1,2/c,c
1,2/t,c
1,1/c,c
1,1/t,t

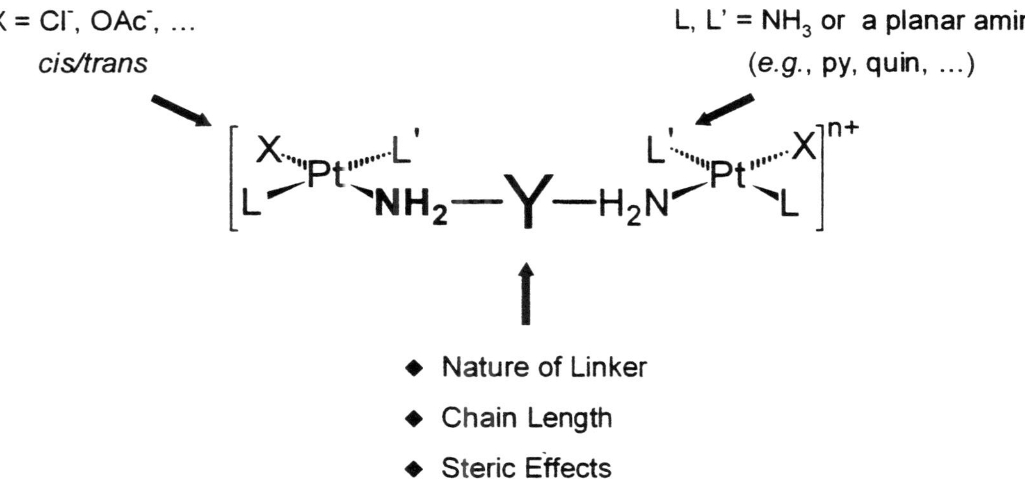

Figure 2. Structural diversity of the di(poly)nuclear structure. The bottom graph represents variations possible within the 1,1 series.

cytotoxicity profile for BBR 3464 is shown in Figure 4. In this case the trinuclear drug clearly circumvents the various resistance pathways produced by cisplatin.

Once an interesting profile is seen, *in vivo* studies most commonly using human tumor xenografts in mice will determine further potential. Newer, more productive and less intrusive assays (in terms of animal care and costs) to determine *in vivo* efficacy (such as the hollow-fibre assay used by NCI) are beyond the scope of this article. The *in vivo* results will be examined carefully not just in terms of increased life span (% ILS) or tumor weight inhibition (% TWI) but in terms of specific growth delay (SGD) or doubling time of tumors. An increased TWI or ILS, measured after, for example, a 30-day period may not be that significant if the tumor growth delay is simply slowed. The preclinical data for BBR3464 confirmed the high potency and low doses needed for antitumor efficacy (0.3-0.6 mg/kg versus 3.0 –6.0 mg/kg for cisplatin), again further strengthening the clinical potential (*20,21,22*).

Biological Assays.

The *in vitro* and *in vivo* assays are the classic parameters of drug potential. It is an increasingly important challenge to place the cytotoxic events in the context of cellular biology, especially signalling pathways, since much of cancer drug development lies at this juncture. Microarray data are now obtainable over a panel of tumors and gene expression profiles can be analyzed for meaningful differences to clinically approved agents. Further, when available the data from the NCI human tumor screen can also be analyzed to diferentiate biological activity (*23*). Note in this case enhanced activity for BBR3464 over cisplatin is seen in the panel of renal cancers (*20*). A good example of the use of the NCI panel in this manner has recently been given for a second series of platinum compounds based on the mononuclear *trans* geometry – in this case very obviously a different profile of anticancer activity can be expected (*24*).

It is desirable to show how new platinum structures may differ in signalling pathways from those of cisplatin and oxaliplatin. An important example is the role of p53, the tumor suppresor gene, in modulation of cytotoxicity of platinum drugs. BBR 3464 displays high activity in human tumor cell lines characterized by both wild type and mutant p53 gene (*21*). In contrast, on average, cells with mutant p53 are more resistant to the effect of cisplatin (*25*). It has been hypothesized that sensitivity or resistance of tumor cells to cisplatin might be also associated with the processes involving p53 (*26,27*). Transfer of functional p53 into p53-null SAOS osteosarcoma cells actually reduced cellular sensitivity

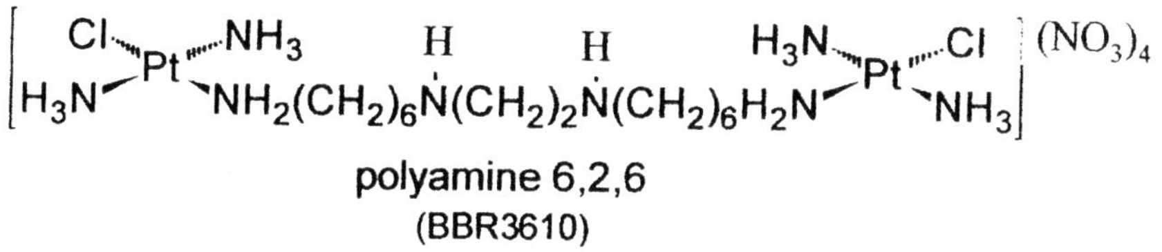

$$\left[\begin{array}{l} Cl \searrow Pt \swarrow NH_3 \quad\quad H \quad\quad H \quad\quad H_3N \searrow Pt \swarrow Cl \\ H_3N \nearrow \quad \nwarrow NH_2(CH_2)_6\dot{N}(CH_2)_2\dot{N}(CH_2)_6H_2N \nearrow \quad \nwarrow NH_3 \end{array}\right](NO_3)_4$$

polyamine 6,2,6
(BBR3610)

Figure 3. A dinuclear platinum-polyamine compound as analog of BBR3464.

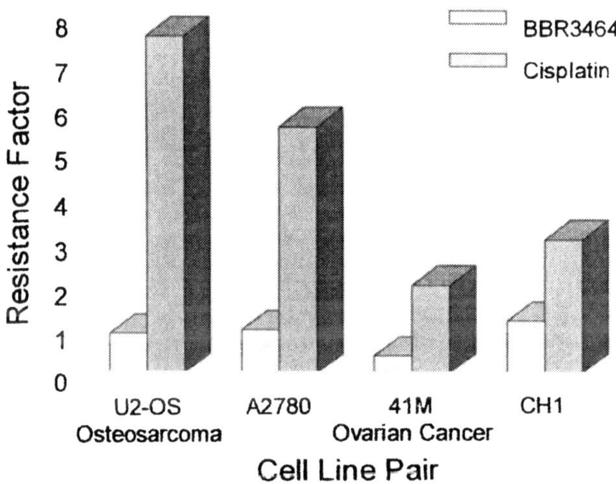

Figure 4. Cytotoxicity profile of BBR3464 compared to cisplatin. Profile is given as Resistance Factor (IC_{50} resistant/IC_{50} sensitive) where IC_{50} is dose required to kill 50% of cells. Adapted from 20 and 21.

White boxes are BBR3464 and lightly shaded boxes are cisplatin.

to BBR3464 compared to the parent p53-null line but caused moderate chemosensitization to cisplatin (*21*). Similarly, in a cisplatin-resistant ovarian cancer cell line (OAW42MER) the presence of a wild-type and functional p53 protein was considered an important determinant of cellular sensitivity to cisplatin but the presence of a non-functional p53 (as indicated by the lack of $p21^{waf1}$ induction for a p53-dependent apoptotic pathway) was not detrimental to the cytotoxicity of BBR3464 (*28*).

Clinical Status of BBR3464

A compound which is potent and has an interesting biological profile will be forwarded to toxicology studies before final decisions are made for advancement to trials or not. In some cases lack of toxicity is sufficient – for example carboplatin was developed because of its lack of nephrotoxicity. Approximately 28 direct structural analogs of cisplatin entered clinical trials but most have been abandoned through a combination of unacceptable toxicity profile and/or lack of improved or expanded anticancer efficacy (*29*). Some obstacles to successful execution of clincial trials are:

- Difficulty in supply and synthesis
- Inappropriate formulation
- Insufficient patent and licensing situation
- Excessive early toxicity
- Ineffective route or schedule of administration
- Long-term unpredictable toxicities
- Delays in execution of clinical trials
- Inappropriate compound choice
- Unexpected pharmacokinetics or metabolism

The clinical findings on BBR3464 may be briefly summarized (*12, 30,31,32*):
(i) Phase I studies fixed a dose of 0.9 – 1.1 mg/m^2 as maximum tolerated dose (MTD). Again the comparison with cisplatin at approx. 20 mg/m^2 is noteworthy and confirms extrapolation from preclinical studies.
(ii) Dose limiting toxicity is diarrheoa and limited nephrotoxicity was observed.
(iii) Partial response in pancreatic cancer and disease stabilization in colon cancer were observed in Phase I clinical trials.

(iv) In Phase II clinical trials 5/28 partial responses were seen with patients with cisplatin-relapsed ovarian cancer. These patients were p53wt. 1 partial response was observed in cisplatin-refractory ovarian cancer although notably the patient was p53 mutant.
(v) Some signs of activity were seen in non-small cell lung cancer.
(vi) Pharmacokinetic data indicated more rapid metabolism in human plasma than in mouse (*33*).

These data, while promising, may be improved if the pharmacokinetic issues of metabolism are resolved. The bar is high to go from Phase II to Phase III and indeed, while only 1 in 1000 molecules ever advance to human clinical trials, only 1 in 4 advance from Phase II trials to full use. The low dose of BBR3464 and the p53 status of patients is of considerable interest and possibly some extrapolation from the preclinical data is justified. Identification of the cause of rapid metabolism in humans can allow systematic study of analogs for an improved pharmacokinetic profile, as outlined above.

Advantages and Pitfalls of Academic-Industry Relationships.

It is axiomatic that the financial and physical resources of a large pharmaceutical company are necessary for a drug to advance to full clinical use. The type of data needed as described above may also be obtained in collaboration in the company. In the early stages it is probably essential to have an individual within the company who will champion the project over the other competing ones, all of which will surely have interesting science behind them. As long as this remains stable and the personal relationships are not obscured this is a generally positive situation. The academic scientist tends to see the excitement of the science, the company person may have other priorities and decisions are made not just on the science alone. A necessary consideration is to ensure that the scientists in close collaboration actually have the ear of management and can influence management decisions. Indeed, at times the rules for drug development may be summarized as the "Three Ps" – patents, politics and personalities ! Factors which may affect a successful conclusion to a project include (but are certainly not limited to) the following:

- *Company Positioning*. Most pharmaceutical companies will be willing to license a proven Phase II drug even if it is outside their usual expertise. Early stage investment will depend on the company strategy.
- *Patents and Papers*. The intellectual property situation must be absolutely clear before a company will move forward. While most universities now maintain technology transfer offices, the number of people actually

involved may be relatively small. The investment available for patenting depends very much on the size of the university and early partnering with companies is preferred in most cases. This has positives and negatives – the project moves along but the company naturally builds a bigger stake in the outcome. Another strategy favored by universities is the provisonal patent, which is cheaper initially but has a shorther activation time.

The balance between "publish or patent" is in fact nicely demonstrated in the cisplatin case – the patent was originally filed six days before the year anniversary of the publication (Figure 5). Under US law failure to file a patent within a year of the publication date places the intellectual property in the public domain where its value is greatly decreased. The relevant date on the patent is the first date of April 20, 1970. A secondary issue often problematic for academics is who should be on the patent. Books and guidelines from the US Patent Office website should be consulted (*34*). The overriding factor must be innovative contribution. Parenthetically, in drug development the people who obtain the biological data would not have automatic rights as the innovative step is usually the chemical entity – collegial understanding of these issues is best done beforehand to avoid problems later.

Technology Transfer. The feasibility of tech transfer and the ability to scale up are also criteria for evaluation before commitment.

- *Company unpredictability.* Drug development is a long-term process. Few, if any drugs, get to market in less than a 10-year time span. A current major problem is the lack of continuity in industrial relationships and the consolidation/merger trends of small companies. A good example is given by the platinums (Figures 6 and 7). Cisplatin development was relatively linear, despite the skepticism of drug companies at the time, and the project moved from an NCI/Johnson Matthey/Engelhard supported endeavor to that of one solid pharmaceutical company after Phase I clinical trials were found to be encouraging (*35*). Compare this to the situation for the three leading analogs today - AMD473, JM-216 and BBR3464 where mergers and company alliances have produced a distinctly non-linear development history. In the specific case of BBR3464 (Figure 7) the patent was first licensed to Boehringer Mannheim Italia which then was wholly acquired by Hoffman-LaRoche. This company then sublicensed the patent for development to a new company Novuspharma SpA which has since merged with CTI in Seattle (coincidentally the developers of Trisenox, Arsenic Trioxide, a novel success for inorganic chemistry in medicine

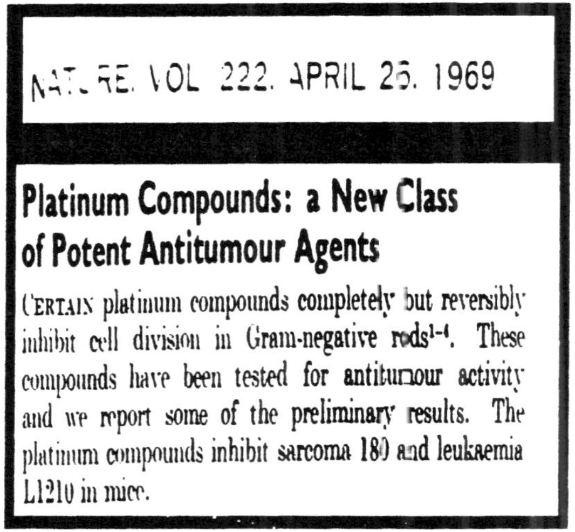

NATURE VOL 222 APRIL 26 1969

Platinum Compounds: a New Class of Potent Antitumour Agents

CERTAIN platinum compounds completely but reversibly inhibit cell division in Gram-negative rods[1-4]. These compounds have been tested for antitumour activity and we report some of the preliminary results. The platinum compounds inhibit sarcoma 180 and leukaemia L1210 in mice.

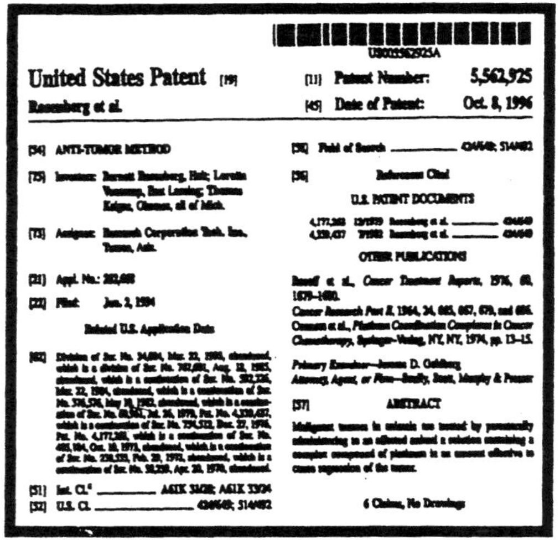

United States Patent [19]
Rosenberg et al.

[11] Patent Number: 5,562,925
[45] Date of Patent: Oct. 8, 1996

[54] ANTI-TUMOR METHOD

U.S. PATENT DOCUMENTS

OTHER PUBLICATIONS

[57] ABSTRACT

6 Claims, No Drawings

Figure 5. Patent or publish ?

(*36*). This situation may now well be the norm rather than the exception. At each stage delays due to reevaluation of the portfolio causes delays and loses precious patent protection as projects are reevaluated upon company consolidation.

Licensing Issues

Finally, once there is attention and interest on behalf of a pharmaceutical company a licensing agreement must be drawn up. Technology transfer offices are much more sophisticated than ten or twenty years ago in this arena but will always be in a disadvantageous position to the company professionals. The goals of a licensing agreement are sometimes not mutual – industry has different parameters than universities. There will inevitably be a time when any successful project is out of the hands of the academic researcher. At that time, the only recourse is the presence of the licensing agreement. Goals for inclusion in any agreement for the benefit of the academic researcher should be, at a minimum:

- Meaningful performance criteria for company to achieve goal.
- Timely progress reports from company to researcher.
- Unambiguous return of exclusive rights to the intellectual property if not developed in a timely manner.
- Technology transfer. Restriction of technology transfer to project at hand.
- Field of Use should be realistic and limited to interests of company.
- Milestone payments and royalties. Appropriate incentives should be agreed upon.

Summary

In this chapter I have used the example of BBR3464 to illustrate some of the considerations which go into the choice of new platinum analogs for cancer treatment. This is by no means exhaustive but is meant as an overview of the drug development process.

Acknowledgments

This chapter is a personal account of experiences at the academic-industry interface. It in no way represents the views of any individual working in the collaborating pharmaceutical companies. It is a pleasure to acknowledge the contribution and interest of the Boehringer Mannheim team in this project. The

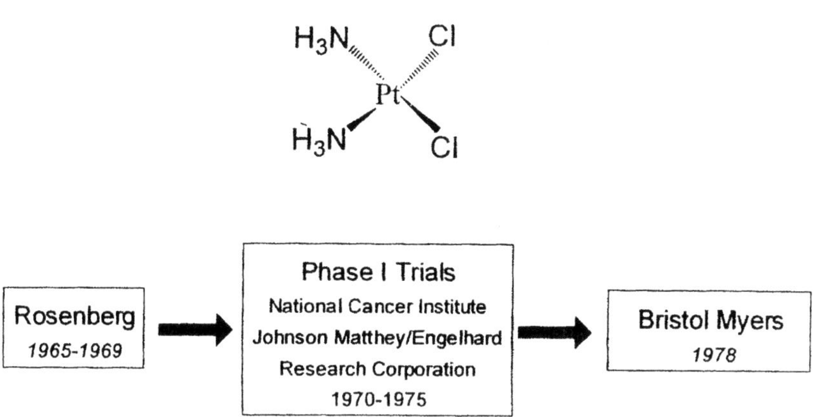

Figure 6. History of development of Cisplatin.

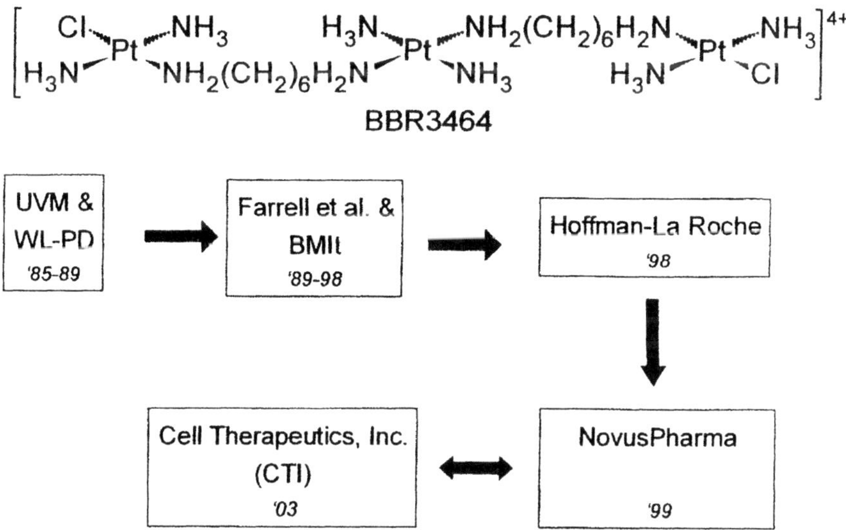

Figure 7. History of development of BBR3464.

work was supported by grants from the American Cancer Society, The National Institutes of Health and National Science Foundation.

References

1. Cisplatin: Chemistry and Biochemistry of a Leading Anticancer Drug; Lippert, B. Ed.; Wiley-VCH: New York, 1999.
2. *Platinum-Based Drugs in Cancer Therapy*; Kelland, L. R.; Farrell, N. P; Humana Press: New Haven, 2000.
3. Wong, E.; Giandomenico, C. M. *Chem Rev* **1999**, *99*, 2451.
4. Pelley, R. J., *Curr Oncol Rep* **2001**, *3*, 147.
5. Boyle, F. T.; Costello, G. F..*Chem Soc Rev* **1998**, *27*, 251.
6. Mousa, S.A.; Mousa, A.S. *Curr. Pharm. Des.*, **2004**, *10*, 1.
7. Kartalou, M.; Essigmann, J. M.*Mutat Res* **2001**, *478*, 23.
8. O'Dwyer, P. J.; Johnson, S. W.; Hamilton, T. C. *Cisplatin and Its Analogues*; 5th ed.; DeVita, V. T., Hellman, S. and Rosenberg, S. A., Eds.; Lippincott-Raven Publishers: Philadelphia, 1997; Vol. 2, pp 418-431.
9. Holford, J.; Raynaud, F.; Murrer, B. A.; Grimaldi, K.; Hartley, J. A.; Abrams, M.; Kelland, L. R., *Anticancer Drug Des* **1998**, *13*, 1.
10. Farrell, N.; Spinelli, S. *Dinuclear and Trinuclear Platinum Anticancer Agents*; Farrell, N., Ed.; The Royal Society of Chemistry: Cambridge, UK, 1999, pp 124-134.
11. Farrell, N.P. Polynuclear Charged Platinum Compounds as a New Class of Anticancer Agents: Toward a New Paradigm; Kelland, L. R. and Farrell, N. P., Ed.; Humana Press, 2000, pp 321-338.
12. Farrell, N. Metal Ions in Biol. Systems, **2004**, *41*, 252.
13. Qu, Y., Appleton, T.G., Hoeschele, J.D., Farrell, N.: *Inorg. Chem.* 1993, *32*, 2591.
14. Kasparkova, J., Zehnulova, J. Farrell, N., Brabec, V. *J. Biol. Chem.* **2002**, *277,* 48076..
15. Hegmans, A., Berners-Price, S.J., Davies, M.S., Thomas, D., Humphreys, A., Farrell, N. *J. Amer. Chem. Soc.* 2004, *12,*:2166.
16. Qu, Y., Scarsdale, N.J., Tran, M.-C., Farrell, N. *J. Biol. Inorg. Chem.* 2003, *8*, 19.
17. Rauter, H., Di Domenico R., Menta, E., Oliva, A., Qu, Y.,Farrell, N. *Inorg. Chem.* **1997,** *36,* :3919.
18. McGregor, T.D., Kasparkova, J., Neplechova, K., Novakova, O., Penazova, H., Vrana, O., Brabec, V., Farrell, N. *J. Biol. Inorg. Chem.* **2002,** *7,*:397.

19. Hegmans, A., Qu, Y., Kelland, L.R., Roberts, J.D., Farrell, N. *Inorg. Chem.* **2001**, *40,* 6108.
20. Manzotti, C.; Pratesi, G.; Menta, E.; Di Domenico, R.; Cavalletti, E.; Fiebig, H. H.; Kelland, L. R.; Farrell, N.; Polizzi, D.; Supino, R.; Pezzoni, G.; Zunino, F. *Clin Cancer Res* **2000**, *6*, 2626.
21. Pratesi, G.; Perego, P.; Polizzi, D.; Righetti, S. C.; Supino, R.; Caserini, C.; Manzotti, C.; Giuliani, F. C.; Pezzoni, G.; Tognella, S.; Spinelli, S.; Farrell, N.; Zunino, F. *Br J Cancer* **1999**, *80*, 1912.
22. Perego, P.; Caserini, C.; Gatti, L.; Carenini, N.; Romanelli, S.; Supino, R.; Colangelo, D.; Viano, I.; Leone, R.; Spinelli, S.; Pezzoni, G.; Manzotti, C.; Farrell, N.; Zunino, F. *Mol Pharmacol* **1999**, *55*, 528.
23. Weinstein, J. N.; Myers, T. G.; O'Connor, P. M.; Friend, S. H.; Fornace, A. J., Jr.; Kohn, K. W.; Fojo, T.; Bates, S. E.; Rubinstein, L. V.; Anderson, N. L.; Buolamwini, J. K.; van Osdol, W. W.; Monks, A. P.; Scudiero, D. A.; Sausville, E. A.; Zaharevitz, D. W.; Bunow, B.; Viswanadhan, V. N.; Johnson, G. S.; Wittes, R. E.; Paull, K. D., *Science* **1997**, *275*, 343.
24. Fojo, T., Farrell, N.; Myers, t.; Ortuzar, W.; Tanimura, H.; Weinstein, J. *Crit. Rev. Hematol. Oncol.* In Press.
25. P. M. O'Connor, J. Jackman, Bae, I., T. G. Myers, S. Fan, M. Mutoh, D. A. Scudiero, A. Monks, E. A. Sausville, J. N. Weinstein, S. Friend, A. J. Fornace, K. W. Kohn, *Cancer Res.* 1997, *57*, 4285.
26. Jordan, P.; Carmo-Fonseca, M. *Cell. Mol. Life Sci.* **2000,** *57,* 1229.
27. Riva, C.M. *Anticancer Res.* **2000**, *20*, 4463.
28. Orlandi, L.; Colella, G.; Bearzatto, A.; Abolafio, G.; Manzotti, C.; Daidone, M. G.; Zaffaroni, N. *Eur J Cancer* **2001**, *37*, 649.
29. Lebwohl, D.; Canetta, R. *Eur J Cancer* **1998**, *34*, 1522.
30. Sessa, C.; Capri, G.; Gianni, L.; Peccatori, F.; Grasselli, G.; Bauer, J.; Zucchetti, M.; Vigano, L.; Gatti, A.; Minoia, C.; Liati, P.; Van den Bosch, S.; Bernareggi, A.; Camboni, G.; Marsoni, S., *Ann Oncol* **2000**, *11*, 977.
31. Calvert, A.H., Thomas, H., Colombo, N.; Gore, M.; Earl, H.; Sena, L.; Camboni, G.; Liati, P.; Sessa, C. *European Journal of Cancer* **2001**, *37 (Supp6)* Poster Discussion 965.
32. Scagliotti, G., Crinò, L..; De Marinis, F.; Tonato, M.; Selvaggi, G.; Massoni, F.; Maestri, A.; Gatti, B.; Camboni, G. *European Journal of Cancer* **2001**, *37 (Supp6)* Poster 182.
33. See http://www.novuspharma.com
34. Pressman, D. "Patent It Yourself" Nolo Press, Berkely, **1998**.
35. Rosenberg, B. In "Lippert, B. Cisplatin: Chemistry and Biochemistry of a Leading Anticancer Drug"; Wiley-VCH: New York, 1999.
36. Novick, S. C.; Warrell, R. P., Jr., Arsenicals in hematologic cancers. *Semin Oncol* **2000**, *27*, 495-501.

Chapter 7

Mechanistic Studies of Pt and Ru Compounds with Antitumor Properties

Jan Reedijk

Leiden Institute of Chemistry, Leiden University, P.O. Box 9502, 2300 RA Leiden, The Netherlands (email: Reedijk@chem.leidenuniv.nl)

Key words: platinum, ruthenium, drugs; coordination chemistry, antitumor, ligand exchanges, DNA, cross-links, cisplatin, nami-A, tumor, cell killing

Abstract: Many metal compounds are known to be useful in medicinal applications, and from these the many anticancer Pt(II) and Ru(II)/Ru(III) compounds are a special class, as they show a particular metal-ligand exchange behavior. In fact these compounds often do have ligand exchange kinetics in the same order of magnitude as the division of (tumor) cells, which make them suitable candidates to interfere with this process. The present chapter discusses this class of compounds with a focus on the mechanism of action of cisplatin and related Pt and Ru compounds.

Even though we know that platinum antitumor drugs eventually end up on the DNA, it is not well understood how (fast) such compounds reach the DNA inside the cell nucleus, and how they are subsequently removed. The several types of Pt and Ru compounds that may reach and interact with DNA will be discussed, with an outlook for new drugs and other applications.

Introduction and History

Metal coordination compounds and free metal ions are well known to effect cellular processes [1, 2]. This metal effect not only deals with natural processes, such as cell division and gene expression where ions like Magnesium, Zinc and Manganese play a role, but also with non-natural processes such as toxicity, carcinogenicity, and anti-tumor chemistry [3, 4]. This chapter deals with a special aspect of medicinal metal biochemistry, namely the mechanism of Pt and Ru coordination complexes applied as antitumor drugs in humans [5].

In chemotherapy of tumors the critical issue is killing of the tumor cells, with as little as possible harm to healthy cells. It is generally accepted that most anticancer drugs act on DNA in one way or another, and this is also the case for Pt coordination compounds; the evidence for Ru compounds is less strong, but the very similar reactivity of both metals with nucleic acids, has been used as an important indication for this [6]. On the route to reach the cellular and nuclear DNA, such metal coordination compounds do have to pass many obstacles with as little as possible decomposition or side reactions. Some of our latest results in this field will be summarized below.

The success of cisplatin as an anticancer drug and its second generation derivatives in fact has started with *cis*-$[PtCl_2(NH_3)_2]$ (see Figure 1) [7]. Both cisplatin and carboplatin (Figure 1) are now widely used as a chemotherapeutic agents to treat urogenital tumors, whereas recently oxaliplatin was introduced to cure colon cancer [8]. Cisplatin was first described in 1845, but its possible anticancer properties appeared only after 1965, when Rosenberg [9] realized that the effects of an electric field on the growth of *E. coli* bacteria in a solution of NH_4Cl and Pt electrodes in sunlight (he found a strong filamentous growth and an arrest of cell division) might have something to do with dissolved Pt compounds - like *cis*-$[Pt(NH_3)_2Cl_2]$ and *cis*-$[Pt(NH_3)_2Cl_4]$ - from the electrodes, and with their DNA binding. He also realized that this finding might be important in tumor cell killing [10, 11].

Rather early it become clear from several studies on structure-activity relationships that virtually all Pt amine compounds with a *cis* geometry show significant activity, while the *trans* isomers did not show antitumor activity. In fact it has been recognized early that all such compound have at least a N-H group [12, 13]. Later, several other Pt compounds were reported by the groups of e.g. Farrell, Natile, Navarro-Ranninger, and recently also Gibson, that do not have the cis geometry and that do not have H-bond donor groups, but that nevertheless show significant antitumor activity [14-27]. A selection of 4 of such trans-Pt compounds is redrawn in Figure 2. In this case the steric effects and/or the slower ligand exchange kinetics might be important. This class of compounds, although likely to increase in importance, will not be discussed in great detail in this chapter. The interest reader is mainly referred a recent review [28]. Binding to DNA is prominent, and both interstrand and intrastrand

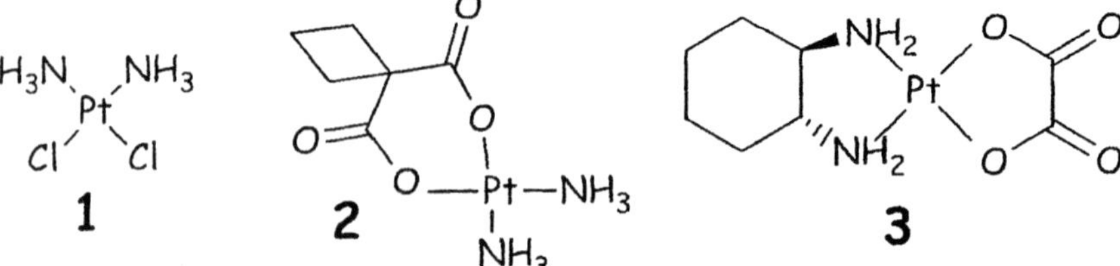

Figure 1: Structures of clinically used cisplatin (1) the first-generation drug: Carboplatin (2) and the recently introduced oxaliplatin (3).

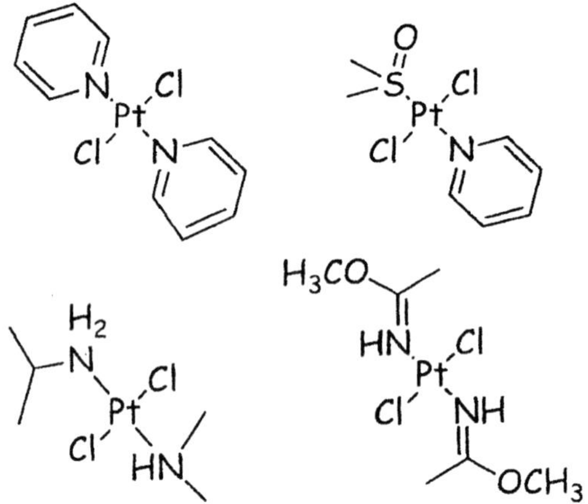

Figure 2: Selection of 4 highly active, non-classical trans Pt antitumor compounds.

crosslinking has been reported. An important property for the trans compounds appears to be their activity against tumor cell lines that are not sensitive for cisplatin.

After realizing that animal tumors could be cured by cisplatin, clinical trails on humans followed rapidly, and investigation of the efficacy of cisplatin against a variety of solid tumors was undertaken. Following Phase-I clinical trials of cisplatin [29] FDA approval was obtained in 1978 under the name Platinol. Carboplatin followed with FDA approval in 1989 initially under the name Paraplatin. In recent years a new compound, developed in Japan, oxaliplatin (Eloxatin), was added on the list for routine treatments of colon cancer (http://www.cancerbackup.org.uk/info). Cisplatin is particularly effective against solid tumor types, such as testicular, ovarian cancers and in fact the curing rates can be as high as 90%. More recently also curing of head and neck cancers, and even small-cell lung cancer curing has been reported [30]. The costs of the treatment are not extremely expensive, and typical drug prices are some 300 US $/gram. The sales for carboplatin have been reported to be 480 M$ in 2001 (http://www.fda.gov/bbs/topics/NEWS/2002/NEW00825.html).

The fact that the precise mechanism of cisplatin (and other derivatives) remains elusive, has resulted in a large interest to study metal DNA-binding in general and cisplatin – as well as several of its analogs - in particular. As a results of such questions, and the molecular details of it, this field has provided a flourishing area for challenging (bio)inorganic chemistry research [2, 31].

It should be clear from the beginning that - like all chemotherapeutic drugs - also cisplatin and related compounds show severe toxic side-effects (*e.g.* nausea, ear damage, vomiting).The toxic side effects of cisplatin limit the dose that can be administered to patients; typical doses are 100 mg/day for a period of 5 consecutive days. The nephrotoxicity can now be significantly reduced by saline (hydration) and administration of diuretic agents. Special drug-dosing protocols have been developed, making use of chemoprotecting agents, such as thiourea and sodium dithiocarbamate [32].

As a result of the toxic side effects, intense research to design new derivative Pt compounds has been developed [24]. The second-generation platinum drug carboplatin, $[Pt(C_6H_6O_4)(NH_3)_2]$, has less toxic side effects than cisplatin and is also more easily used in combination therapy. Its lower reactivity allows a higher dose to be administered (even up to 2000 mg/day). It appears that carboplatin is the reagent of first choice for ovarian cancer treatment, whereas oxaliplatin is known to be most effective in colon cancer treatment [24].

An important complication in Platinum chemotherapy appears to be development of spontaneous (intrinsic) drug resistance, and this is now one of the main limitations when treating patients. Fortunately (cross) resistance for new compounds is easily detected by using tumor cell lines, facilitating rapid screening. The non-classical Pt compounds mentioned above with different amines and not having the classical cis-diamine structure with two leaving groups is very important in this respect; they are sometimes considered as the third-generation Pt drugs [24-27].

The intermediate strength of the M-L bond in coordination compounds plays an important role in the explanation of the activity. The fact that the Pt-ligand coordination bond, which has a thermodynamic strength of a typical coordination bond (say 100 kJ/mole or even below), is much weaker than (covalent) C-C and C-N or C-O single and double bonds (which are between 250 and 500 kJ/mole, [33]), appears crucial, although this bond strength criterion might be valid for other metals as well. It has been realized, however, that isostructural compounds from other group-10 elements (Ni, Pd) do not yield antitumor-active compounds, and in fact the explanation easily follows from considerations of ligand-exchange kinetics. The kinetics for this type of ligands and platinum is of the order of a few hours half live, thereby preventing rapid equilibration reactions [34]. As a result ligand-exchange reactions vary from minutes to days, rather than from microseconds to seconds for many other coordination compounds. The same kinetic inertness holds for many ruthenium coordination compounds, and this kinetic behavior makes such compounds special indeed. Although several ruthenium compounds have been reported to show antitumor behavior in cell-line studies [35-38], only one of them as yet has entered clinical trials [39]. A selection of important active compound given in Figure 3. It should be noted here that Ru compounds are by definition octahedrally coordinated; even though a cis chelate is possible with some of them, the space that axial ligands require prevent octahedral Ruthenium compounds to form similar structures with DNA as Pt(II) compounds can. The compounds in figure 3 have 2 or more labile ligands, and compounds like Ru(terpy), may in theory even bind DNA trans (interstrand crosslink) or in a tridentate manner [38].

Platinum compounds are not only special in DNA binding and their kinetics. Also their preferred ligands are just not only N-donor atoms of the DNA. In fact Pt(II), and to a less extent Ru(II), has a strong thermodynamic preference for binding to S-donor ligands. For that reason one would expect that platinum compounds would perhaps never reach nuclear DNA, with so many cellular platinophiles (S-donor ligands, such as glutathione, methionine) as competing ligands in the cellular fluids; this appears not to be the case, as has been elucidated and reviewed elsewhere[40].

In the sections below more mechanistic details will be presented on Pt-DNA interactions, and also the as yet less-studied Ru-DNA interactions will be briefly discussed. The outcome of this analysis will be used in the subsequent discussion on new drugs.

Overview mechanistic insights for cisplatin and other Pt and Ru compounds

Is DNA is the target for all Pt and Ru species?

The activity of cisplatin is closely related to its binding to DNA [4]. Highly conclusive evidence for this target was the early observation that cells deficient in DNA repair are hypersensitive to cisplatin [41, 42]. In fact the local DNA kink resulting from the formation of the 1,2-intrastrand crosslink at d(GpG) site is considered to be most closely connected with the antitumor activity of cisplatin [2, 4, 31]. Cisplatin and several other platinum drugs have therefore been categorized as DNA-binding drugs. Early studies had made clear that from the 4 bases, guanine is strongly preferred and that the N7 site of G is by far the most prominent [34]. Two neighbouring G-N7 sites are even more favored [43], as shown schematically in figure 4. Many other DNA binding and cross-linking antitumor compounds are known, such as cyclophosphamides, nitrogen mustards, nitroso-ureas, epoxides, and anthracyclines [44]. A drug strongly bound to DNA may interfere with transcription, and/or DNA replication mechanisms [45], and such disruption processes may (eventually) trigger processes like apoptosis that will lead to cell death [46].

The main origins for the strong preference of the cisplatin (and related Pt compounds) for the G-N7 site are first of all the intrinsic strong basicity of that site, and the fact that many of the other potential binding sites are involved in ds DNA via Watson-Crick base pairing. A second important reason appears to be a combination of steric and hydrogen bonding effects. A schematic illustration of such hydrogen bonding and its effect on purine bases is given in Figure 5, where - as an example - the difference in approach of a cisPt unit to G and A is depicted. One can imagine that the binding of similar Pt compounds are preferred in the same way, but that trans-Pt compounds will not have such a beneficial effect for G-N7, and also the non-amine Ru compounds, when bound at DNA would not be expected to show this strong preference; and this is indeed observed and will be discussed below.

Uptake of cisplatin, carboplatin and NAMI-A in cells

Cisplatin and carboplatin are most commonly administrated to patients by intravenous injection. In human blood and in extracellular body fluids the physiological chloride concentration amounts to some 100 mM, and this is high enough to suppress hydrolysis, so that cisplatin can reach the outer surface of cells – perhaps recognized by receptor species in some cases – mainly as a neutral molecule. Carboplatin is even more hydrolytically stable. Early studies have shown that some 50% of the cisplatin may leave the body through the

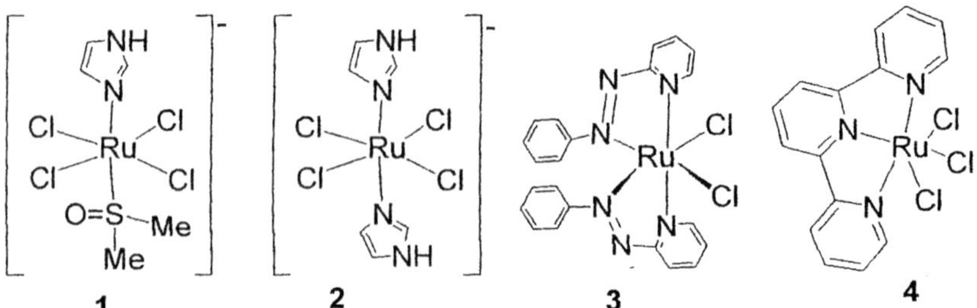

Figure 3: A selection of anticancer-active Ruthenium coordination compounds. The compounds NAMI-A (1) and NAMI (2), each have an imidazolium as a counter cation; a very active compound (3) of the α isomer of a Ru-azpy compound (azpy = 2-phenylazopyridine) and a Ru-terpy compound (4). From these NAMI-A has been in clinical trials since 2000.

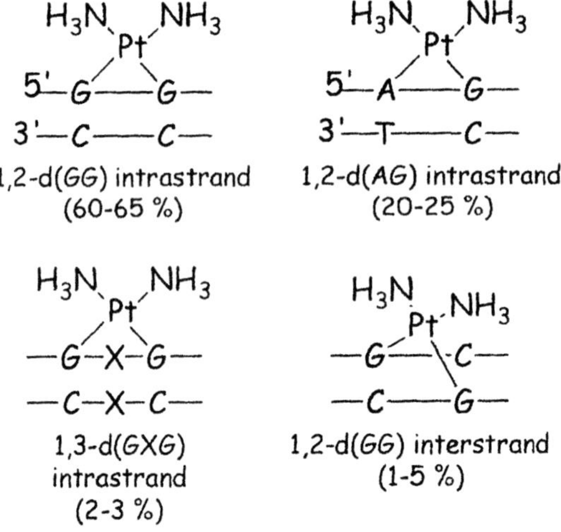

Figure 4: The 4 well-characterized bifunctional cisplatin-DNA adducts.

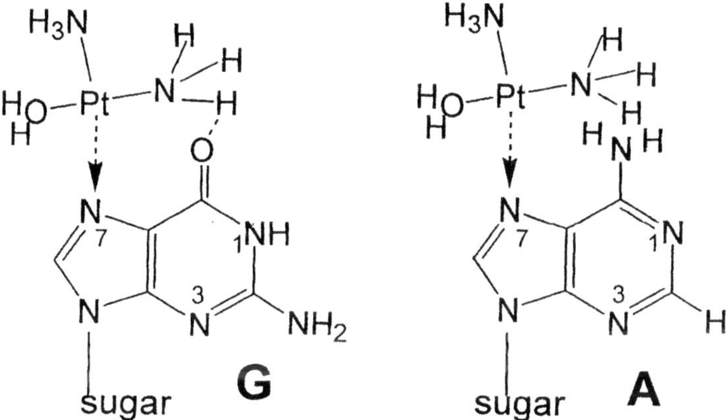

Figure 5. Amine N-H groups can have a strong effect on the binding of cis-amine-Pt compounds to purines, like G (Hydrogen bonding) and A (Hydrogen repulsion slows down the interaction). This is a major factor to explain the preference of cisplatin for G(N7) over A(N7).

kidneys within 48 h, and that the remaining 50% may will remain up to an additional 2 months in the body. The precise mechanism of cellular uptake of cisplatin has many unanswered questions, although we have published evidence that the presence of phosphatidyl serine in membranes plays an important role in the Pt uptake [47]. In fact a passive diffusion is believed the main mechanism, although some evidence has been reported that supports the involvement of active transport mechanisms [48]. More recently quite convincing evidence has been presented that the copper transport agent, Ctr1, is indeed able to mediate cisplatin uptake [49, 50].

The free chloride concentration inside cells is much lower (some 4 mM) and under these conditions hydrolysis of cisplatin can take place, albeit slowly. In fact the most important hydrolysis product is the $[PtCl(H_2O)(NH_3)_2]^+$ cation [51], which has a pK_a value of 6.5, and so above pH = 6 it starts to ionize to form $[PtCl(OH)(NH_3)_2]$ [52]. The cationic Pt species with a labile aqua ligand is much more reactive than cisplatin, so the monoaqua species is most likely to react with DNA and other molecules in the cell. Carboplatin hydrolyses much slower, and far less is known in this case. Studies of this type for oxaliplatin, are also still lacking. Transport of platinum compounds through the nuclear membrane is also hardly studied. Whether a special nuclear localizing signal peptides (NLS) may play a role remains uncertain [53].

For the Ru-antitumor compound NAMI-A the uptake and DNA-binding studies are still limited, but initial experiments have shown that many targets are available where - in theory - Ru can bind and in practice also has been found to bind [39, 54-57]. More recently, we have shown by a detailed NMR analysis that the hydrolysis of the complex NAMI-A significantly affects its DNA binding and also its antimetastatic activity. Hydrolysed NAMI-A species apparently are easier taken up by so-called metGM cells [54], showing intracellular ruthenium concentrations one order of magnitude greater than those of intact NAMI-A. So it has been proposed that the selective antimetastatic activity of NAMI-A during in vivo experiments can be attributed to its hydrolysed species [54].

The development of resistance against cisplatin and carboplatin

A disturbing clinical problem is the development of resistance of certain tumors against cisplatin. Some types of cancer are known to be intrinsically insensitive to cisplatin treatment, whereas other cancers only develop resistance during chemotherapy. Therefore, the applicability of cisplatin still is limited to a relatively narrow range of tumors. The mechanism of cisplatin-resistance seems to be multifactorial, and at least three factors have been identified as possible modulators of the observed cellular resistance [58-60]. These factors are now accepted to be:

(1) Decreased platinum uptake; indeed in many cisplatin-resistant cell lines reduced accumulation was indeed observed [58].

(2) Increased intracellular detoxification. Glutathione (GSH) reacts with platinum drugs to form deactivated adducts [59], and such adducts may be excreted by a glutathione S-conjugate export pump.

(3) Enhanced DNA repair; this phenomenon has been observed in some cisplatin-resistant cell lines [60]. Most recently even a link was proposed with drug-induced tumor cell death, such as decreased apoptosis [61]. More details on resistance development can be found elsewhere [62].

Metal-DNA complexes: the formation and structures of Pt and Ru adducts with DNA and DNA fragments

Over the last 2 decades many studies have been reported where Pt drugs and other platinum compounds were reacted with DNA and DNA fragments. Both the kinetics and the structures have been given attention. More recently also studies with several other metals, including the kinetically slow Ruthenium and Rhodium have been reported. In most cases the studies vary from simple nucleobases and nucleoside derivatives, to single-stranded and double-stranded oligonucletides, even up to plasmids and cellular DNA. This particular field has been regularly reviewed by others [4, 24, 63, 64], and us [2, 31, 34].

As addressed above, the guanine sites do have a very strong affinity for Pt amine compounds. Two neighbouring guanines are even more preferred, where the first binding step occurs most rapidly at the 3'G site in an oligonucleotide. Monofunctional Pt-G adducts at 5'G may live long, and these adducts may play a role in protein crosslinking [65]. The two most frequently occurring binding sites, *i.e.* at GG and AG, see figure 4 above, [43] appear to be found for very many sources of DNA [66].

The observation that Pt amine compounds do not all remain coordinated at S-donor ligands in the cell, but eventually - some of them - do reach purine N7 sites is still surprising. This process appears to be related to a migration of Pt from S ligand to purine N7 (where the above-mentioned O6 H-bond acceptor is likely to assist). Model reactions of S-guanosyl-L-homocysteine (SGH) indeed initially yield $[Pt(dien)(SGH\text{-}\underline{S})]^{2+}$ (starting from the model compound [PtCl(dien)]Cl at $2<pH<6.5$), but this species isomerizes intramolecularly into [Pt(dien)(SGH-$\underline{N7}$)] with Pt coordination at N7 of guanosine [67, 68]. Others found similar results for methionine-type ligands [69]. So one may suggest that also under *in vivo* conditions certain Pt species might migrate from S to N donor ligands [40]. This migration may not be the case for all types Pt-S compounds, as some (dinuclear) stable glutathione adducts have also been reported [59].

The coordination of G-N7 to other metals like Ni(II) and Cu(II) is also known, although the kinetics are fast, NMR spectroscopy allows proper detection [70]. When using metals that bear other ligands and that are kinetically slow, other types of interaction may arise, and in case of sterically crowded Ru(II) and dinuclear Rh(II) compounds, it appeared that certain adenine binding sites are at least competitive if not preferred, [71-76] }. In such case the role of the O6 H-bond acceptor in G is playing no role, but instead the H-donor function of the NH_2 in Adenine can become important, depending of course on whether the co-ligand is a H-bond acceptor.

Antitumor activity of other, less soft metals, like Ti(IV) en Sn(IV), appear to be related to their better binding at phosphate groups, and these will not be discussed here. As mentioned above, in all studies of metal binding to DNA and other ligands in the cell, one should discriminate between (reversible) kinetically fast, and kinetically slow binding metals. Through the early work of Taube [77] we know that ligand-exchange processes for the same ligands, but for a variation of metal ions, may differ some 14 orders of magnitude. For small ligands, like water, the ligand exchange for metals ions like Pt(II), Ru(II), Ru(III) and low-spin Co(III) appear to be in the order of magnitude of a few hours, *i.e.* they are within the same range as many cell-division processes. Co(III) is easy reduced to the fast-exchanging Co(II), leaving essentially Pt and Ru as good candidates for new drugs *(vide infra).* This statement is less clear for Pt(IV) compounds, as will be discussed below in detail, although there is now good evidence for oxidation of G residues by Pt(IV) [78]; whether or not other reagents in the cell will interfere with this reduction is not yet known.

In the case of Pt(II) the well-known *trans effect* further explains the stereochemically controlled kinetics of the ligand exchange reactions. Basically, with two cis-oriented amines at a Pt(II) ion, the other two ligands are more easily substituted, as a result of which the two amines remain (at least initially) coordinated when the Pt compound is bound to a biological target. If the target appears to be a soft ligand, the amine ligand may become labilized to undergo a subsequent reaction, especially when monodentate ligands, such as ammonia or primary amines are used. The kinetic *trans effect* is less pronounced for Ruthenium.

Structures of Pt and Ru compounds coordinated to DNA fragments are numerous now and especially for platinum many have been reviewed already [31]. A beautiful illustration of secondary binding effects in case of ruthenium, again showing that hydrogen bonding is of great importance, is presented in Figure 6 [79]

A great variety of structures of GG adducts in single-stranded and double-stranded DNA is now known, and for overviews of that work I refer to *e.g.* [65, 80] with cited references. The structure of the GG-Pt chelate species now appears to be quite general, but of course this is not the end point in the mechanistic discussions. A very important next mechanistic step must be recognition by proteins. In fact we know already that the Pt-GG adduct is recognized by proteins, followed by either stabilization of the distorted DNA structure, or by removal of the lesion through repair [4, 28]. Such Pt-GG chelates are basically the same [80] in many cases. An important study appears to be the case when a protein is bound simultaneously, and it has been shown that the kinked DNA is not only recognized by the so-called HMG (high-mobility gel) proteins, but even more importantly that the kinked structure is retained after binding [81]. The biological consequences of the protein binding at GG-platinated sites are the next step to investigate in the mechanism [82]. Crucial might be the time that the Pt stays on the DNA. In a preliminary study we have reported [83] that the bulk of the Pt compounds probably only stays coordinated on the DNA for a rather short time (less than 10 hours, as we found for a specially designed class of fluorophore-labeled Pt compounds). It appears that either or both the kinetic lability and thermodynamic instability may lead to a loss of the Pt from the nuclear DNA; the Pt species may eventually result in excretion via the Golgi organelles [83]. Whether or not all the Pt is released or not, is uncertain, as small fractions of Pt-DNA adducts could be detected in patients long after treatment [84, 85]. So from all properties related to antitumor activity of Pt and Ru compounds, and their DNA binding the kinetics of the reactions are crucial and structural studies are not enough. If a potentially highly active compound with the right geometric and electronic properties would not stay long enough coordinated to the DNA, no activity will be observed. So design and fine-tuning of new metal compounds with ligand substituents that have steric, electronic, and/or H-bonding implications for DNA binding, provides a great challenge for research towards new anticancer compounds. Recent examples of this multifunctional principle have appeared in the literature, [86, 87]. In the next section such examples other ways to new drugs, based on mechanistic studies will be presented.

Design and synthesis of new types of Pt and Ru Drugs

Making use of mechanistic findings, such as discussed in the previous section, coordination chemists have been searching from different perspectives to design and synthesize new metal-containing anticancer compounds. Generalising, a new DNA-binding metal compound with antitumor activity and high potential for clinical application would need to meet the following key requirements, all which can be controlled based on coordination chemistry principles and the current status of the mechanism:

A. The compound should have optimal intrinsic properties, *i.e.* saline solubility, stability in solution and survival decomposition before arriving at the main target.

B. The compound should be efficiently transported in the blood and through membranes.

C. The compound must have good DNA-binding properties, while at the same time reacting slowly with proteins.

D. The compound should have target specificity, if possible with a strong interaction at (the surface) of certain tumor cells.

E. It should display activity against tumors that are, or have become, resistant to cisplatin and derivatives. This latter requirement usually implies a different structure from cisplatin-type species.

Given the space limits of this review, I cannot discuss or describe the great variety of recently studied new compounds; so only a few will be given which are also based on our own work and work with collaborative partners in other laboratories. The first examples will be based on the classical cisplatin derivatives. The second group covers significantly different Pt and Ru drugs that may bind quite different to DNA.

Control of the kinetics and use of Pt(IV) compounds and prodrugs

A very recent and quite active compound has resulted from the approach to control the kinetics by steric effects, as shown in Figure 7,[88-91]. This quite simple classical Pt compound, has a structure where the axial attack of incoming ligands at one side of Pt-ligand plane is significantly slowed down by the bulk of the methyl group on the ligand. As a consequence of this geometry S-donor ligands bind much slower. This kinetic effect in Pt-DNA binding and with competing ligands has now also been studied theoretically [92].

For a number of years the compound JM-216, which is a Pt(IV) compound (Figure 7) [93, 94] has been in routine clinical application as an oral drug. Mechanistically, it is of great interest whether or not such compounds are reduced before entering the cell, or perhaps inside the cell before binding to the DNA, or maybe not at all [95], although recent evidence point towards oxidation of the DNA [78].

This last finding is in harmony with an early study by Brabec [96], which has shown that in the case of binding of *cis,cis,trans*-$PtCl_2(NH_3)_2(OH)_2$ the isolated DNA adducts differ from those obtained with cisplatin under identical conditions; they also showed that no external (added) reducing agent is needed for their formation. Whether or not the formed adducts contain Pt(II) or Pt(IV) is not completely clear, although most evidence points towards slowly formed Pt(IV) adducts [96]. Our own work using DNA model bases [97] has made clear that a reduction to Pt(II) can occur, albeit slowly. Recent work with a variety of different amines showed that the Pt redox potential is largely influenced by the coordinated amine [95], and also that a proper choice of spectroscopic and analytical methods can indeed prove that Pt(IV) compounds can oxidize GMP [78]

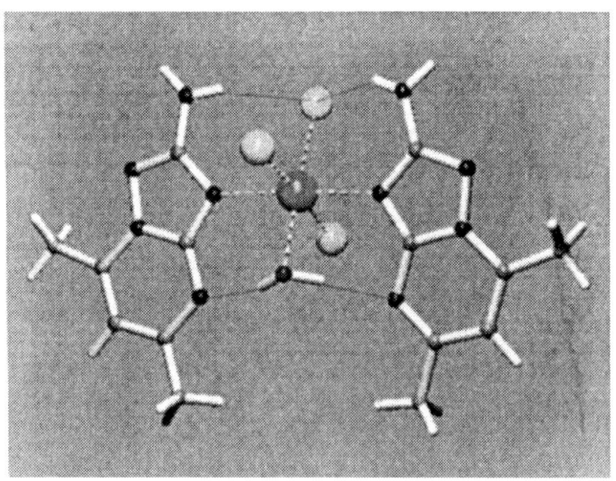

Figure 6: Molecular structure of a NAMI derivative-Ru(III) adduct (XRD) of formula $RuCl_3(H_2O)(L_2)$ containing a model purine nucleobase; the important of intramolecular hydrogen bonding stabilization is indicated [79]*.*

(See page 1 of color insert.)

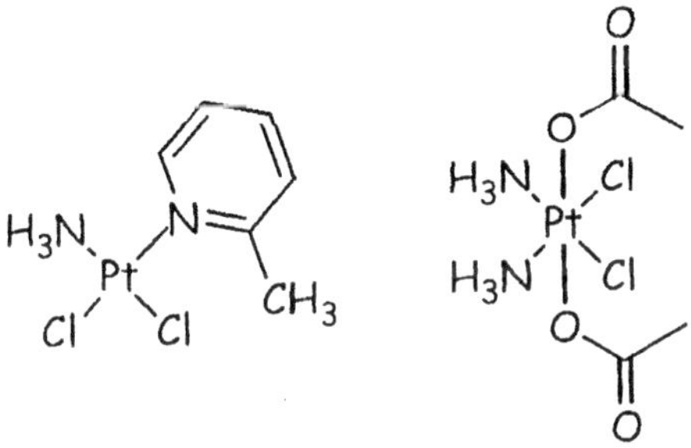

Figure 7. The new drug AMD-473 (left) with significantly slower reaction kinetics than cisplatin and the structure of a classical Pt(IV) drug, JM-216 (right), which needs reduction in order to become active.

Given the toxicity and side effects of cisplatin, much activity has been generated on slow-release drugs, such as generation from a polymer. Such biodegradable polymers can bind cisplatin analogs that are released gradually [98-100]. Examples of such structures are given in Figure 8. A different approach towards pro-drugs has been the formation of membrane-encapsulated cisplatin, which can be formed only by repeated freeze-thaw cycles [101]. The resulting nano-products are highly soluble and apparently result in slow Pt releases and high activity [102].

Pt species with G-N7-binding capability, also bearing a second DNA-binding function

Realizing that Pt(II) amine species have a strong tendency to go for G-N7 sites (single and double), one can now think on the design of compounds that carry a second DNA binding group, as a tail to the first one. So, in the amine part of the cisplatin derivatives, substitutions allow the introduction of DNA-binding, or repelling side arms, so that the kinetics of the DNA binding can be influenced by charge, hydrogen bonding and/or steric effects. Also the stabilization after Pt binding at guanines can be influenced in this way. A simple set of such compounds is given in Figure 9. In the case of the peptide ligands, application of solid-phase synthesis allows combinatorial variations [87, 103, 104]. Also other amino-acid derivatives can be used in such processes, as recently shown [105].

Cisplatin derivatives can also be linked to the DNA minor groove with the aim to alter the interaction of the platinum moiety with the DNA; examples have been given by Brabec [106] for netropsin and distamycin, two oligopeptides with potent antibacterial, antiviral and antineoplastic properties, both known to bind to DNA and related to their pharmacological activity. Such non-intercalative compounds do form non-covalent complexes with double-helical DNA in the minor groove and often exhibit a preference for AT-rich domains. Regretfully, the antitumor activities found were much lower than those of cisplatin [106]. Of course one can also use carrier groups that bind to targets other than DNA, such as estrogen analogs, targeting platinum to some breast and prostate cancer cells that are known to overexpress estrogen receptors, or even simpler variations, such as the drugs developed by Keppler and others [107, 108] with ligands containing amino phosphonic acid groups aimed for production of platinum drugs with selective activity in bone tumors.

In a next step of drug design one can bind a second metal to the tail, and we have done this recently for Platinum [109] and for Ruthenium, with several variations of the Ru group and of the spacer, type and length. A representative sample for a cationic Pt-Ru species [110] for which we have XRD and NMR evidence, has been included in Figure 9, item 3.

One can of course use the same principle for adding a second Pt species, and this area has been extensively explored by Farrell and co-workers. In fact extending studies of mononuclear Pt compounds to dinuclear (and oligonuclear) Pt compounds is quite logical. However, it took some time before the dinuclear compounds of the type given in Figure 10 were introduced [111, 112]. Especially the dinuclear compounds of general formula: $[ClPt(NH_3)_2(H_2N\text{-}(CH_2)_nNH_2Pt(NH_3)_2Cl](anion)_2$ were indeed found to be active, and they even appeared to chelate with two guanines at N7 [113]. From such dinuclear species, also other structures may arise, as we found an interesting hairpin structure for a certain DNA sequence only[25] after binding to this ds DNA. As yet the dinuclear compounds did not make it to clinical trials, mainly because of their toxicity. However, a trinuclear compound, also introduced by Farrell and apparently inspired by naturally occurring polyamines that bind DNA, was shown to have unique DNA binding properties [27, 114, 115]. The most successful of a group of such trinuclear compounds, *i.e.* BBR3464 (see Figure 10) is active against melanoma, lung cancer, pancreatic cancers, and retains active in cisplatin resistant cell lines. The compound shows activity at 10x lower concentrations than cisplatin, and makes typical long-range DNA crosslinks (both interstrand and intrastrand; however the interstrand crosslinks cannot be repaired).

A quite different approach within this area, and also to circumvent the cross-resistance has been followed by Komeda [116, 117]. The aim has been to design a Pt(II) complex which would react with DNA in a different way from that of cisplatin. We have looked for an intrastrand crosslink, and at the same time a quite small distortion of the double helix, to reduce recognition chances by repair enzymes. Simple modeling already made clear that a rigid azole bridge (pyrazole, triazole) would generate a base-to-base contact of some 350 pm. The compound used as prototype is redrawn in Figure 10 [116].

To check whether or not the design principle was correct, the compounds was first reacted with the model DNA base 9-ethylguanine (9-Etgua), and the X-ray structure of the bis adduct (see Figure 11) beautifully shows the parallel orientation of the two bases. This structure already suggests that also in ds DNA such parallel orientation would be possible, with as a further consequence a hardly distorted ds DNA, difficult to recognize by repair systems in the cells.

We did already report recently on the very high activity for two of these rigidly-bridged compounds against a variety of different tumor cell lines [118] and detailed binding studies of these compounds with DNA fragments are under further study, the first results of which have been reported. In one case an unprecedented ligand isomerisation was found upon DNA binding [117, 119]. A detailed NMR study of the reaction product with double-stranded oligonucleotides having GG sequences showed that at most a small distortion occurs [120-122]. A projection of an NMR structure is given in Figure 12, which illustrates the structure as hardly distorted from ds DNA [118, 119].

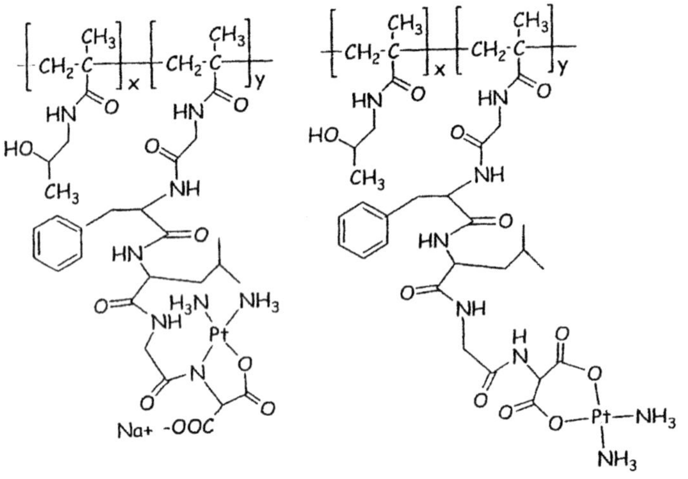

Figure 8: Two examples of polymer attached anticancer Pt species; left so-called AP-5280, and right the related AP-5279, which both slowly release cisplatin species in tissue.

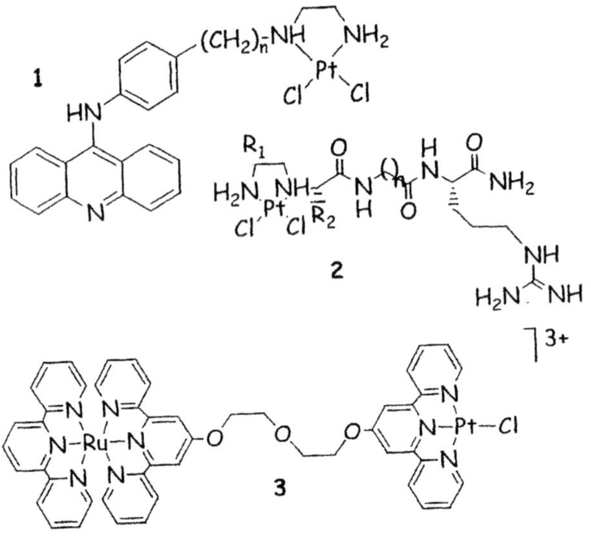

Figure 9: Examples of Pt(II) species, in which substituents at the amine part may (1) intercalate with ds DNA; (2) form complicated minor groove binding; or (3) may have a second metal containing group.

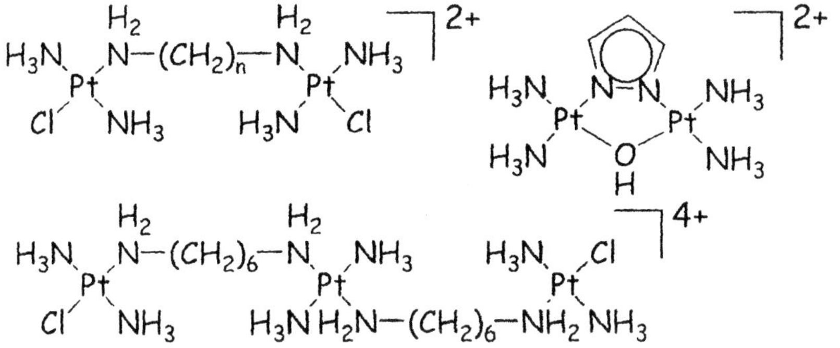

Figure 10: Dinuclear and Trinuclear cationic Pt species with high antitumor activity.

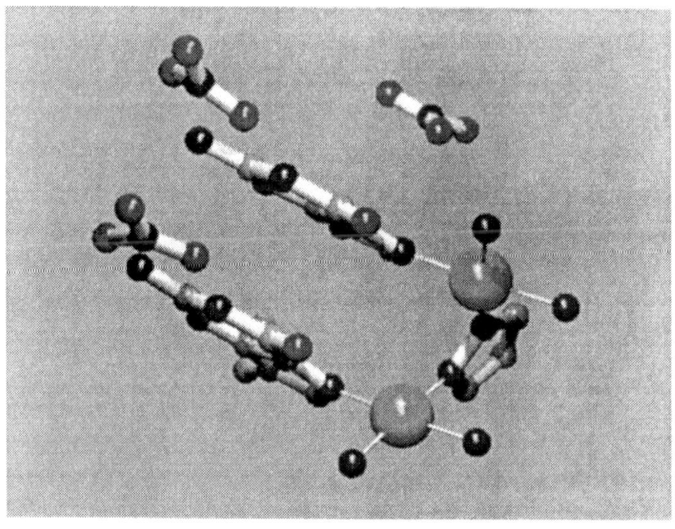

Figure 11: Two model bases (9-Etgua) can remain parallel when coordinated to the azole-bridged dinuclear Pt unit [116].

(See page 1 of color insert.)

Ruthenium drugs and new trans-amine Pt compounds

Although ruthenium antitumor chemistry is more recent than cisplatin and derivatives, the compounds presented in Figure 3 have generated a tremendous synthetic activity. Recently, we have modified the azapyridine compounds and made them water soluble by changing the ligands and the counter anions [6]; this has resulted in compounds with a quite high antitumor activity. Sadler [123] has shown that even organometallic half-sandwich Ru compounds show antitumor activity and quite interesting DNA binding properties. In further steps of developments one can think of monofunctional Pt compounds, and bifunctional trans-amine Pt compounds linked to other metals and thereby generating new classes of compounds. Of few of such cases, recently published or in progress, are given in Figure 13 [124, 125].

Metal-containing Drugs: Quo Vadis

The control of metal binding to DNA, by simultaneous coordination **and** other properties, such as stacking and hydrogen bonding has been a crucial item in our research, and we plan to use in the design of new drugs. We have introduced the view that in the applications of M-DNA binding the kinetics of the M-L binding are usually more important than the thermodynamics of binding. We also realize that secondary binding is very important, such as stacking and H-bonding interactions, both in the kinetics of the process and in the stabilization of the formed metal-adduct structure.

Predictions for the future are always risky, but we have little doubt that the above-presented mechanistic highlights and challenges for the future shall provide fascinating new possibilities for research in the coming decade and for new generations of chemists. Further development of new techniques, like microscopy, MS and NMR specialties, such as the powerful [^{1}H,^{15}N] method (HSQC) and fluorescent labeling applications, will allow following the reaction of Pt amines and nucleic acids and proteins in space and time, thereby allowing the detection and localization of otherwise invisible intermediate products [83, 126].

Application of such knowledge in drug design is nearby and in the next decade we are likely to see improved antitumor (but also antiviral, and other) drugs that will have been developed based on the knowledge of the molecular details of the M-DNA interactions (and their repair) and on the kinetics of binding of Pt and Ru compounds to proteins and DNA. At this point it is also of interest to refer to critical questions about whether ot not the intrinsically weak metal-ligand coordination bond will ever lead to new drug applications (http://www.callerio.org/forum); I believe that the kinetic control of drug stability is likely to overcome this.

Figure 12: A double-stranded DNA remains almost unkinked when the dinuclear azole-bridged Pt amine species is chelating to two neighbouring guanines [122].

(See page 2 of color insert.)

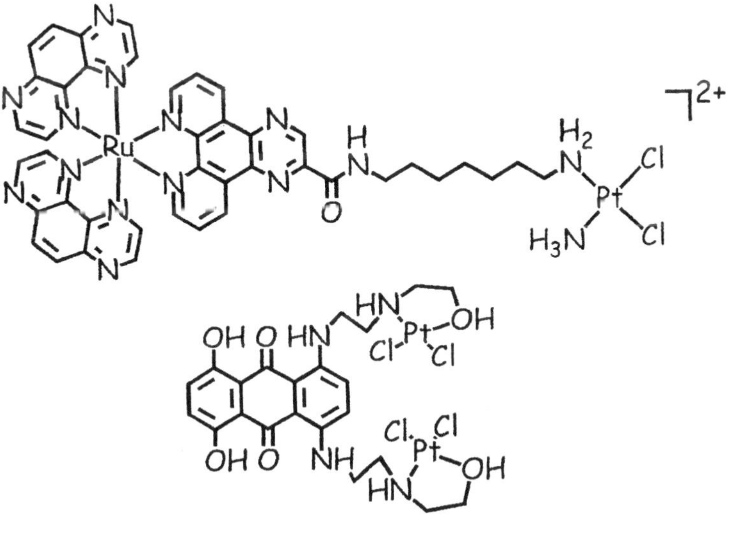

Figure 13: Selection of new Pt and Ru compounds under recent investigation as potential new drugs.

I want to close by mentioning a special application which is already in commercial use. The kinetically controlled reactions of certain cis- and trans-Pt compounds with nucleic acids, have led to a promising, easy applicable, high-potential labeling technique [127]. By starting from the basic unit Pt(diaminoethane), attached to a (selected) oligonucleotide and carrying a fluorescent label - after combination with hybridization to a complementary strand to form ds DNA - allows detection of specific sequences of DNA in biological samples. The critical issue here is the fact that the Pt should remain long enough coordinated to the DNA, *i.e.* the fluorescing label should only be present at the desired, recognized DNA sequence. A schematic structure of such a system is redrawn in Fig 14. In fact the intermediate stability of the coordination bond between platinum and the guanine-N7, allows the coupling with appropriately slow ligand-exchange kinetics. High-potential applications of this promising methodology have recently been reported [128, 129].

Acknowledgements

All work mentioned above and obtained in Leiden was only possible to obtained with the dedication, inspiration, quality and hard work of undergraduates, graduate students, postdocs, visiting scientist and many collaborators. Their names appear as co-authors in the references from our laboratory. Support from Leiden University and the Councils of Chemical Research and Applied Sciences of the Netherlands Organisation for Research

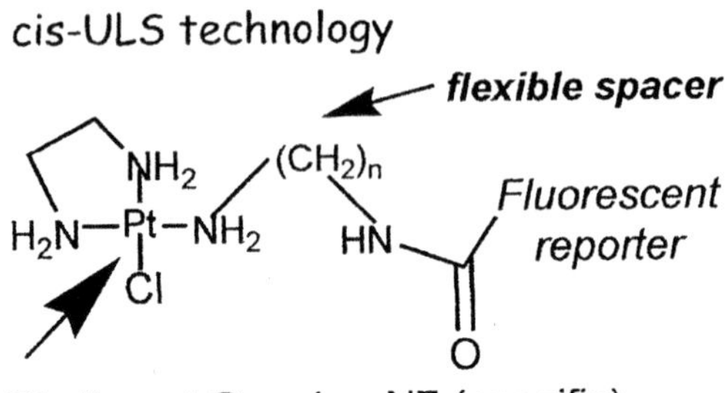

Figure 14: A new and high-potential application of kinetically controlled metal-DNA binding, in which a fluorescent marker (with a variable spacer shape, size and length) and a certain biological target are cross-linked by the Pt(en) unit.

(NWO) is gratefully acknowledged, as are the highly appreciated loan schemes for K_2PtCl_4 and $RuCl_3$ from Johnson & Matthey (Reading UK).

Finally help from the EU for postdocs and research trainees from COST and Framework Programmes 3, 4 and 5 since 1992 is thankfully mentioned, allowing regular research exchanges, including short visits of students, to partner laboratories inside EU countries.

References:

1. Reedijk, J.; Bouwman, E., *Bioinorganic Catalysis.* second, revised and expanded ed.; Marcel Dekker, Inc.: New York, 1999; 'Vol.' p 606.

2. Reedijk, J., *Proc. Natl. Acad. Sc. (USA)* **2003,** 100, 3611-3616.

3. Lippard, S. J., *Science* **1993,** 261, 699-700.

4. Jamieson, E. R.; Lippard, S. J., *Chem. Rev.* **1999,** 99, 2467-2498.

5. Reedijk, J., *Curr. Op. Chem. Biol.* **1999,** 3, 236-240.

6. Hotze, A. C. G.; Bacac, M.; Velders, A. H.; Jansen, B. A. J.; Kooijman, H.; Spek, A. L.; Haasnoot, J. G.; Reedijk, J., *J. Med. Chem.* **2003,** 46, 1743-1750.

7. Peyrone, M., *Ann. Chemie Pharm.* **1845,** 51, 1-29.

8. Desoize, B.; Madoulet, C., *Crit. Rev. Oncol./Hematol.* **2002,** 42, 317-325.

9. Rosenberg, B.; Van Camp, L.; Krigas, T., *Nature* **1965,** 205, 698.

10. Rosenberg, B.; Van Camp, L.; Trosko, J. E.; Mansour, V. H., *Nature* **1969,** 222, 385.

11. Lippert, B., *Cisplatin, Chemistry and Biochemistry of a Leading Anticancer Drug.* ed.; Wiley-VCH: Weinheim, 1999; 'Vol.' p.

12. van Kralingen, C. G.; Reedijk, J.; Spek, A. L., *Inorg. Chem.* **1980,** 19, 1481.

13. van Kralingen, C. G.; Reedijk, J., *Ciencia Biologica* **1980,** 5, 159.

14. Farrell, N., *Cancer Invest.* **1993,** 11, 578-589.

15. Malina, J.; Hofr, C.; Maresca, L.; Natile, G.; Brabec, V., *Biophys. J.* **2000,** 78, 2008-2021.

16. Delalande, O.; Malina, J.; Hofr, C.; Novakova, O.; Natile, G.; Kozelka, J.; Brabec, V., *J. Inorg. Biochem.* **2001,** 86, 201-201.

17. Janovska, E.; Novakova, O.; Natile, G.; Brabec, V., *J. Inorg. Biochem.* **2002,** 90, 155-158.

18. Novakova, O.; Kasparkova, J.; Malina, J.; Natile, G.; Brabec, V., *Nucleic Acids Res.* **2003,** 31, 6450-6460.

19. Pantoja, E.; Alvarez-Valdes, A.; Perez, J. M.; Navarro-Ranninger, C.; Reedijk, J., *Inorg. Chim. Acta* **2002,** 339, 525-531.

20. Gibson, D.; Najajreh, Y.; Kasparkova, J.; Brabec, V.; Perez, J. M.; Navarro-Ranniger, C., *J. Inorg. Biochem.* **2003,** 96, 42-42.

21. Farrell, N.; Ha, T. T. B.; Souchard, J. P.; Wimmer, F. L.; Cros, S.; Johnson, N. P., *J. Med. Chem.* **1989,** 32, 2240-2241.

22. Qu, Y.; Farrell, N., *Inorg. Chem.* **1992,** 31, 930-932.

23. Zakovska, A.; Novakova, O.; Balcarova, Z.; Bierbach, U.; Farrell, N.; Brabec, V., *Eur. J. Biochem.* **1998,** 254, 547-557.

24. Wong, E.; Giandomenico, C. M., *Chem. Rev.* **1999,** 99, 2451-2466.

25. Yang, D.; van Boom, S. S. G. E.; Reedijk, J.; van Boom, J. H.; Farrell, N.; Wang, A. H.-J., *Nature Struct. Biol.* **1995,** 2, 577-586.

26. Farrell, N.; Kelland, L. R.; Roberts, J. D.; Vanbeusichem, M., *Cancer Res.* **1992,** 52, 5065-5072.

27. Brabec, V.; Kasparkova, J.; Vrana, O.; Novakova, O.; Cox, J. W.; Qu, Y.; Farrell, N., *Biochemistry* **1999,** 38, 6781-6790.

28. Brabec, V., DNA modifications by antitumor platinum and ruthenium compounds: Their recognition and repair. In *Progress in Nucleic Acid Research and Molecular Biology, Vol 71*, ed.; 'Ed.'^'Eds.' 2002; 'Vol.' 71, p^pp 1-68.

29. Higby, D. J.; Wallace, H. J.; Albert, D. J.; Holland, J. F., *Cancer* **1974,** 33, 1219-1228.

30. Giaccone, G., *Drugs* **2000,** 59, supp. 4,9.

31. Reedijk, J., *Chem. Commun.* **1996**, 801-806.

32. Reedijk, J.; Teuben, J. M., Platinum-Sulfur Interactions Involved in Antitumor Drugs, Rescue Agents and Biomolecules. In *Cisplatin, Chemistry and Biochemistry of a Leading Anticancer Drug*, ed.; Lippert, B., 'Ed.'^'Eds.' Wiley-VCH: Weinheim, 1999; 'Vol.' p^pp 339-362.

33. Housecroft, C. E.; Sharpe, A. G., *Inorganic Chemistry*. ed.; Prentice Hall: Harlow, 2001; 'Vol.' p.

34. Reedijk, J., *Inorg. Chim. Acta* **1992,** 198-200, 873-881.

35. Clarke, M. J.; Zhu, F.; Frasca, D. R., *Chem. Rev.* **1999,** 99, 2511-2533.

36. Chen, H. M.; Parkinson, J. A.; Parsons, S.; Coxall, R. A.; Gould, R. O.; Sadler, P. J., *J. Am. Chem. Soc.* **2002,** 124, 3064-3082.

37. Velders, A. H.; Kooijman, H.; Spek, A. L.; Haasnoot, J. G.; de Vos, D.; Reedijk, J., *Inorg. Chem.* **2000,** 39, 2966-2967.

38. van Vliet, P. M.; Toekimin, S. M. S.; Haasnoot, J. G.; Reedijk, J.; Novakova, O.; Vrana, O.; Brabec, V., *Inorg. Chim. Acta* **1995,** 231, 57-64.

39. Sava, G.; Bergamo, A.; Zorzet, S.; Gava, B.; Casarsa, C.; Cocchietto, M.; Furlani, A.; Scarcia, V.; Serli, B.; Iengo, E.; Alessio, E.; Mestroni, G., *Eur. J. Cancer* **2002,** 38, 427-435.

40. Reedijk, J., *Chem. Rev.* **1999,** 99, 2499-2510.

41. Brouwer, J.; van de Putte, P.; Fichtinger-Schepman, A. M. J.; Reedijk, J., *Proc. Natl. Acad. Sci. U.S.A.* **1981,** 78, 7010.

42. Popoff, S.; Beck, D. J.; Rupp, W. D., *Mut. Res.* **1987,** 1987, 129-145.

43. Fichtinger-Schepman, A. M. J.; van der Veer, J. L.; den Hartog, J. H. J.; Lohman, P. H. M.; Reedijk, J., *Biochem.* **1985,** 24, 707-713.

44. Rajski, S. R.; Williams, R. M., *Chem. Rev.* **1998,** 98, 2723-2740.

45. Mello, J. A.; Lippard, S. J.; Essigmann, J. M., *Biochem.* **1995,** 34, 14783-14792.

46. Barry, M. A.; Behnke, C. A.; Eastman, A., *Biochem. Pharmacol.* **1990,** 40, 2353-2360.

47. Speelmans, G.; Staffhorst, R. W. H. M.; Versluis, K.; Reedijk, J.; de Kruijff, B., *Biochem.* **1997,** 36, 10545-10550.

48. Gately, D. P.; Howell, S. B., *Br. J. Cancer* **1993,** 67, 1171-1177.

49. Ishida, S.; Lee, J.; Thiele, D. J.; Herskowitz, I., *Proc. Natl. Acad. Sci. U. S. A.* **2002,** 99, 14298-14302.

50. Lin, X. J.; Okuda, T.; Holzer, A.; Howell, S. B., *Mol. Pharmacol.* **2002,** 62, 1154-1159.

51. Miller, S. E.; House, D. A., *Inorg. Chim. Act.* **1989,** 166, 189-192.

52. Berners-Price, S. J.; Frenkiel, T. A.; Frey, U.; Ranford, J. D.; Sadler, P. J., *J. Chem. Soc.-Chem. Commun.* **1992,** 789-791.

53. Nitiss, J. L., *Proc. Natl. Acad. Sci. U. S. A.* **2002,** 99, 13963-13965.

54. Bacac, M.; Hotze, A. C. G.; van der Schilden, K.; Haasnoot, J. G.; Pacor, S.; Alessio, E.; Sava, G.; Reedijk, J., *J. Inorg. Biochem.* **2004,** 98, 402-412.

55. Piccioli, F.; Messori, L.; Orioli, P.; Keppler, B.; Alessio, E.; Sava, G., *J. Inorg. Biochem.* **2001,** 86, 379-379.

56. Bergamo, A.; Zorzet, S.; Gava, B.; Sorc, A.; Alessio, E.; Iengo, E.; Sava, G., *Anti-Cancer Drugs* **2000,** 11, 665-672.

57. Bergamo, A.; Gagliardi, R.; Scarcia, V.; Furlani, A.; Alessio, E.; Mestroni, G.; Sava, G., *J. Pharmacol. Exp. Ther.* **1999,** 289, 559-564.

58. Hospers, G. A. P.; Mulder, N. H.; de Vries, E. G. E., *Med. Oncol. Tumor Pharmacother* **1988,** 5, 145-150.

59. Murdoch, P. D.; Kratochwil, N. A.; Parkinson, J. A.; Patriarca, M.; Sadler, P. J., *Angew. Chem.-Int. Edit.* **1999,** 38, 2949-2951.

60. Chaney, S. G.; Sancar, A., *J. Natl. Cancer Inst.* **1996,** 88, 1346-1350.

61. Borst, P.; Borst, J.; Smets, L. A., *Drug Resist. Updates* **2001,** 4, 129-131.

62. Brabec, V.; Kasparkova, J., *Drug Resist. Updates* **2002,** 5, 147-161.

63. Lippert, B., *Coord. Chem. Rev.* **1999,** 182, 263-295.

64. Hall, M. D.; Hambley, T. W., *Coord. Chem. Rev.* **2002,** 232, 49-67.

65. Reeder, F.; Guo, Z. J.; Murdoch, P. D.; Corazza, A.; Hambley, T. W.; BernersPrice, S. J.; Chottard, J. C.; Sadler, P. J., *Eur. J. Biochem.* **1997,** 249, 370-382.

66. Eastman, A.; Schulte, N., *Biochemistry* **1988,** 27, 4730.

67. van Boom, S. S. G. E.; Reedijk, J., *J. Chem. Soc. Chem. Commun.* **1993,** 1993, 1397-1398.

68. van Boom, S. S. G. E.; Chen, B. W.; Teuben, J. M.; Reedijk, J., *Inorg. Chem.* **1999,** 38, 1450-1455.

69. Chen, Y.; Guo, Z. J.; Murdoch, P. D.; Zang, E. L.; Sadler, P. J., *J. Chem. Soc.-Dalton Trans.* **1998**, 1503-1508.

70. Song, B.; Zhao, J.; Griesser, R.; Meiser, C.; Sigel, H.; Lippert, B., *Chem.-Eur. J.* **1999,** 5, 2374-2387.

71. Dunbar, K. R., **1997,** 36, 9999-10001.

72. Velders, A. H.; van der Geest, B.; Kooijman, H.; Spek, A. L.; Haasnoot, J. G.; Reedijk, J., *Eur. J. Inorg. Chem.* **2001**, 369-372.

73. Hotze, A. C. G.; Broekhuisen, M. E. T.; Velders, A. H.; van der Schilden, K.; Haasnoot, J. G.; Reedijk, J., *Eur. J. Inorg. Chem.* **2002**, 369-376.

74. Asara, J. M.; Hess, J. S.; Lozada, E.; Dunbar, K. R.; Allison, J., *J. Am. Chem. Soc.* **2000,** 122, 8-13.

75. Chifotides, H. T.; Koshlap, K. M.; Perez, L. M.; Dunbar, K. R., *J. Am. Chem. Soc.* **2003,** 125, 10703-10713.

76. Chifotides, H. T.; Koshlap, K. M.; Perez, L. M.; Dunbar, K. R., *J. Am. Chem. Soc.* **2003,** 125, 10714-10724.

77. Taube, H., *Chem. Rev.* **1952,** 50, 69-78.

78. Choi, S.; Cooley, R. B.; Hakemian, A. S.; Larrabee, Y. C.; Bunt, R. C.; Maupas, S. D.; Muller, J. G.; Burrows, C. J., *J. Am. Chem. Soc.* **2004,** 126, 591-598.

79. Velders, A. H.; Ugozzoli, F.; Biagini-Cingi, M.; Manotti-Lanfredi, A. M.; Haasnoot, J. G.; Reedijk, J., *Eur. J. Inorg. Chem.* **1999**, 213-215.

80. Marzilli, L. G.; Saad, J. S.; Kuklenyik, Z.; Keating, K. A.; Xu, Y. H., *J. Am. Chem. Soc.* **2001,** 123, 2764-2770.

81. Ohndorf, U. M.; Rould, M. A.; He, Q.; Pabo, C. O.; Lippard, S. J., *Nature* **1999,** 399, 708-712.

82. Zamble, D. B.; Mikata, Y.; Eng, C. H.; Sandman, K. E.; Lippard, S. J., *J. Inorg. Biochem.* **2002,** 91, 451-462.

83. Molenaar, C.; Teuben, J.-M.; Heetebrij, R. J.; Tanke, H. J.; Reedijk, J., *J. Biol. Inorg. Chem.* **2000,** 5, 655-665.

84. Reed, E., *Cytotechnology* **1998,** 27, 187-201.

85. Reed, E., *Cancer Treat. Rev.* **1998,** 24, 331-344.

86. Zenker, A.; Galanski, M.; Bereuter, T. L.; Keppler, B. K.; Lindner, W., *J. Chromatogr. B* **2000,** 745, 211-219.

87. Robillard, M. S.; Galanski, M.; Zimmermann, W.; Keppler, B. K.; Reedijk, J., *J. Inorg. Biochem.* **2002,** 88, 254-259.

88. Chen, Y.; Parkinson, J. A.; Guo, Z. J.; Brown, T.; Sadler, P. J., *Angew. Chem.-Int. Edit.* **1999,** 38, 2060-2063.

89. Holford, J.; Beale, P. J.; Boxall, F. E.; Sharp, S. Y.; Kelland, L. R., *Eur. J. Cancer* **2000,** 36, 1984-1990.

90. Hotze, A. C. G.; Chen, Y.; Hambley, T. W.; Parsons, S.; Kratochwil, N. A.; Parkinson, J. A.; Munk, V. P.; Sadler, P. J., *Eur. J. Inorg. Chem.* **2002,** 1035-1039.

91. Zhang, J. Y.; Liu, Q.; Duan, C. Y.; Shao, Y.; Ding, J.; Miao, Z. H.; You, X. Z.; Guo, Z. J., *J. Chem. Soc.-Dalton Trans.* **2002**, 591-597.

92. Deubel, D. V., *J. Am. Chem. Soc.* **2002,** 124, 58-34-5842.

93. Neidle, S.; Snook, C. F.; Murrer, B. A.; Barnard, C. F. J., *Acta Cryst.* **1995,** C51, 822-824.

94. BernersPrice, S. J.; Sadler, P. J., *Coord. Chem. Rev.* **1996,** 151, 1-40.

95. Choi, S.; Delaney, S.; Orbai, L.; Padgett, E. J.; Hakemian, A. S., *Inorg. Chem.* **2001,** 40, 5481-5486.

96. Novakova, O.; Vrana, O.; Kiseleva, V.; Brabec, V., *Eur. J. Biochem.* **1995,** 228, 616-622.

97. Talman, E. G.; Brüning, W.; Reedijk, J., *J. Inorg. Biochem.* **1995,** 59, 138.

98. Bouma, M.; Nuijen, B.; Stewart, D. R.; Shannon, K. F.; St John, J. V.; Rice, J. R.; Harms, R.; Jansen, B. A. J.; van Zutphen, S.; Reedijk, J.; Bult, A.; Beijnen, J. H., *PDA J. Pharm. Sci. Technol.* **2003,** 57, 198-207.

99. Bouma, M.; Nuijen, B.; Harms, R.; Rice, J. R.; Nowotnik, D. P.; Stewart, D. R.; Jansen, B. A. J.; van Zutphen, S.; Reedijk, J.; van Steenbergen, M. J.; Talsma, H.; Bult, A.; Beijnen, J. H., *Drug Dev. Ind. Pharm.* **2003,** 29, 981-995.

100. Bouma, M.; Nuijen, B.; Stewart, D. R.; Rice, J. R.; Jansen, B. A. J.; Reedijk, J.; Bult, A.; Beijen, J. H., *Anti-cancer Drugs* **2002,** 13, 915-924.

101. De Kruijff, B.; Speelmans, G.; Staffhorst, R.; Willibrordus, H. M.; Reedijk, J. Antitumor cisplatinum prodrugs,. 1998.

102. Burger, K. N. J.; Staffhorst, R. W. H. M.; de Vijlder, H. C.; Velinova, M. J.; Bomans, P. H.; Frederik, P. M.; de Kruijff, B., *Nature Medicine* **2002,** 8, 81-84.

103. Robillard, M. S.; Valentijn, A. P. M.; Meeuwenoord, N. J.; van der Marel, G. A.; van Boom, J. H.; Reedijk, J., *Angew. Chem. Int. Ed.* **2000,** 39, 3096-3099.

104. Robillard, M. S.; Jansen, B. A. J.; Lochner, M.; Geneste, H.; Brouwer, J.; Hesse, M.; Reedijk, J., *Helv. Chim. Acta* **2001,** 84, 3023-3030.

105. Moradell, S.; Lorenzo, J.; Rovira, A.; Robillard, M. S.; Aviles, F. X.; Moreno, V.; de Llorens, R.; Martinez, M. A.; Reedijk, J.; Llobet, A., *J. Inorg. Biochem.* **2003,** 96, 493-502.

106. Kostrhunova, H.; Brabec, V., *Biochemistry* **2000,** 39, 12639-12649.

107. Bloemink, M. J.; Diederen, J. J. H.; Dorenbos, J. P.; Heetebrij, R. J.; Keppler, B. K.; Reedijk, J., *Eur. J. Inorg. Chem.* **1999**, 1655-1657.

108. Bloemink, M. J.; Dorenbos, J. P.; Heetebrij, R. J.; Keppler, B. K.; Reedijk, J.; Zahn, H., *Inorg. Chem.* **1994,** 33, 1127-1132.

109. van Zutphen, S.; Robillard, M. S.; van der Marel, G. A.; Overkleeft, H. S.; den Dulk, H.; Brouwer, J.; Reedijk, J., *Chem. Commun.* **2003**, 634-635.

110. van der Schilden, K., *angew. chem.* **2004**.

111. Stetsenko, A. I.; Yakovlev, K. I.; Rozhkova, N. D.; Pogareva, V. G.; Kazakov, S. A., *Russ. J. Coord. Chem.* **1990,** 16, 560-565.

112. Farrell, N., *Comm. Inorg. Chem.* **1995,** 16, 373-389.

113. Bloemink, M. J.; Reedijk, J.; Farrell, N.; Qu, Y.; Stetsenko, A. I., *J. Chem. Soc. Chem. Commun.* **1992,** 1992, 1002-1003.

114. McGregor, T. D.; Balcarova, Z.; Qu, Y.; Tran, M. C.; Zaludova, R.; Brabec, V.; Farrell, N., *J. Inorg. Biochem.* **1999,** 77, 43-46.

115. Zehnulova, J.; Kasparkova, J.; Farrell, N.; Brabec, V., *J. Biol. Chem.* **2001,** 276, 22191-22199.

116. Komeda, S.; Ohishi, H.; Yamane, H.; Harikawa, M.; Sakaguchi, K.; Chikuma, M., *J. Chem. Soc., Dalton Trans* **1999**, 2959-2962.

117. Komeda, S.; Bombard, S.; Perrier, S.; Reedijk, J.; Kozelka, J. F., *J. Inorg. Biochem.* **2003,** 96, 357-366.

118. Komeda, S.; Lutz, M.; Spek, A. L.; Chikuma, M.; Reedijk, J., *Inorg. Chem.* **2000,** 39, 4230-4236.

119. Komeda, S.; Lutz, M.; Spek, A. L.; Yamanaka, Y.; Sato, T.; Chikuma, M.; Reedijk, J., *J. Am. Chem. Soc.* **2002,** 124, 4738-4746.

120. Komeda, S. Ph.D. thesis, Ph.D. Thesis, Leiden University), 2002.

121. Teuben, J. M. Ph.D. Thesis, Leiden University), 2000.

122. Komeda, S.; Kozelka, J.; Reedijk, J., *J. Am. Chem. Soc.* **2003**, to be submitted.

123. Morris, R. E.; Aird, R. E.; Murdoch, P. D.; Chen, H. M.; Cummings, J.; Hughes, N. D.; Parsons, S.; Parkin, A.; Boyd, G.; Jodrell, D. I.; Sadler, P. J., *J. Med. Chem.* **2001,** 44, 3616-3621.

124. Kalayda, G. V.; Jansen, B. A. J.; Reedijk, J., *J. Biol. Inorg. Chem.* **2004,** in press.

125. Meistermann, I.; Hotze, A. C. G.; Kalayda, G. V.; Reedijk, J., *Tet. Lett.* **2004,** 45, 2593-2596.

126. Barton, S. J.; Barnham, K. J.; Habtemariam, A.; Sue, R. E.; Sadler, P. J., *Inorg. Chim. Acta* **1998,** 273, 8-13.

127. Houthoff, H. J.; Reedijk, J.; Jelsma, T.; Heetebrij, R. J.; Volkers, H. H. Methods for labeling nucleotides, labeled nucleotides and useful intermediates. Application Information WO 97?NL559 19971008
Priority Application Information EP 96?202792 19961008, 1998.

128. van Gijlswijk, R. P. M.; Talman, E. G.; Peekel, I.; Bloem, J.; van Velzen, M. A.; Heetebrij, R. J.; Tanke, H. J., *Clin. Chem.* **2002,** 48, 1352-1359.

129. Heetebrij, R. J.; Talman, E. G.; Van Velzen, M. A.; Van Gijlswijk, R. P. M.; Snoeijers, S. S.; Schalk, M.; Wiegant, J.; Van der Rijke, F.; Kerkhoven, R. M.; Raap, A. K.; Tanke, H. J.; Reedijk, J.; Houthoff, H. J., *Chembiochem* **2003,** 4, 573-583.

Chapter 8

Mechanistic Studies of Motexafin Gadolinium (Xcytrin®): A Redox Active Agent That Reacts with Electron-Rich Biological Substrates

Darren J. Magda[1], Nikolay Gerasimchuk[1], Zhong Wang[1], Jonathan L. Sessler[2], and Richard A. Miller[1]

[1]**Pharmacyclics, Inc., 995 East Arques Avenue, Sunnyvale, CA 94085**
[2]**Department of Chemistry and Biochemistry, 1 University Station, The University of Texas at Austin, Austin, TX 78712–0165**

Motexafin gadolinium (MGd, Xcytrin®) is a water soluble gadolinium(III) texaphyrin complex that is currently undergoing a Phase III clinical trial as an anticancer agent used in combination with radiation therapy of metastatic non-small cell lung cancer. It is also in clinical trials for use in a number of other oncological applications. Its mechanism of action is thought to rely, at least in part, on its ability to function as a redox cycling agent, capturing electrons from electron-rich endogenous species, so-called reducing metabolites, and transferring them to oxygen to produce various reactive oxygen species that are known to trigger apoptosis. In this chapter, the reaction of MGd and various other congeneric metallotexaphyrin complexes with the prototypical reducing species NADPH, ascorbate, and dithiothreitol, is summarized.

Introduction

The texaphyrins are pentaaza, Schiff base-derived expanded porphyrins [1] that are known to form stable 1:1 complexes with many metal cations, including those of the trivalent lanthanide series. The gadolinium(III) complex of one particular texaphyrin derivative, motexafin gadolinium (MGd; Xcytrin®; **1**, Chart 1), has been extensively studied as an MRI detectable anti-cancer agent [2-9]. Clinical studies have confirmed that this agent is taken up into and retained in tumors with high specificity [7-9]. Currently, MGd is the focus of numerous clinical trials, including ones in advanced head and neck cancer, multiple myeloma, chronic lymphocytic leukemia, non-Hodgkin's lymphoma, and recurrent glioma.

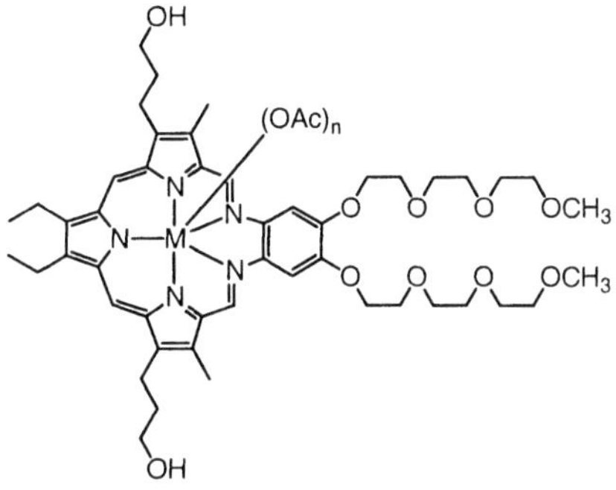

(1) M = Gd, n = 2 (MGd)
(2) M = Lu, n = 2 (MLu)
(3) M = Y, n = 2 (Y-Tex)
(4) M = Dy, n = 2 (Dy-Tex)
(5) M = Er, n = 2 (Er-Tex)
(6) M = Eu, n = 2 (Eu-Tex)
(7) M = Sm, n = 2 (Sm-Tex)
(8) M = Nd, n = 2 (Nd-Tex)
(9) M = Cd, n = 1 (Cd-Tex)
(10) M = Mn, n = 1 (Mn-Tex)
(11) M = Co, n = 1 (Co-Tex)
(12) M = $Fe(O)_{1/2}$, n = 1 (Tex-Fe-O-Fe-Tex)

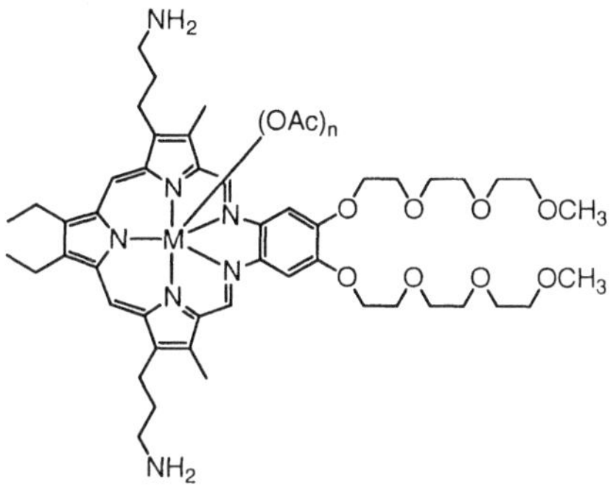

(13) M = Gd, n = 2 (DAGd-Tex)
(14) M = Lu, n = 2 (DALu-Tex)
(15) M = Mn, n = 1 (DAMn-Tex)

*Chart 1. Structure of motexafin gadolinium **1** and its congeners.*

Because of its advanced clinical status, efforts are ongoing to study in detail the chemical and biological properties of MGd. Initial mechanistic studies revealed a clearance pathway that was largely hepatic [3] and provided support for a very novel mechanism of action [10,11]. This latter mechanism, manifest in terms of increased apoptosis, is believed to involve redox modulation processes [12] that take place outside of the nucleus [13]. In this chapter we review the chemical attributes of motexafin gadolinium (MGd; **1**), focusing on those that make it attractive as a potential drug candidate. Particular attention will be paid to reactions of MGd and related metallotexaphyrin systems with electron-rich species, so-called reducing metabolites, since such reactions are thought critical to its biological activity. Here, key questions we are seeking to answer include specifically *i*) whether similar reactivity patterns are observed for different classes of reducing metabolites and *ii*) whether other metallotexaphyrins show greater or lesser levels of reactivity.

The most defining feature of motexafin gadolinium and, indeed, all texaphyrins, is their greater size relative to the porphyrins. Not only do they contain five, rather than four, inward-pointing nitrogen atoms, they are characterized by a central coordinating core that is roughly 20% larger than that of the porphyrins. As a consequence, they form stable, nonlabile 1:1 complexes with most trivalent cations of the lanthanide series, including gadolinium(III) [5, 14, 15]. Another salient feature of the lanthanide(III) texaphyrins is exhibited in their redox behavior. While not particularly easy to oxidize, these complexes are easy to reduce [15-17]. This stands in marked contrast to what is true for free-base and metalated porphyrins and is thought to have important mechanistic implications (as described in more detail later on in this chapter). At low proton concentrations, the first reduction wave, corresponding to a single electron reduction process, is found to be quite reversible. In the specific case of MGd in DMSO the first reduction wave is seen at about -294 mV vs. Ag/AgCl or about -72 mV vs. NHE (cf. Figure 1) [17]. The $E_{1/2}$ value associated with this reduction wave is seen to vary from complex to complex, with a greater ease of reduction being seen in the case of the smaller lanthanide cations; this is as would be expected based on considerations of size, charge density, and the fact that these smaller cations are better accommodated within the texaphyrin core [14]. In all cases, however, the reduction process remains facile and, in analogy to what is seen for metalloporphyrin complexes containing redox inactive metal centers, is thought to be ligand based.

Biolocalization Studies

Studies of tissue and sub-cellular localization have been carried out with MGd and other texaphyrins [13,18]. These have provided support for the notion that texaphyrins are taken up into cells via an active transport process and largely co-localize with endosomes, lysosomes, and, eventually, mitochondria. By contrast, no evidence of significant incorporation into the nucleus was obtained over the full multi-hour time scale associated with these studies. Such findings support the notion that MGd does not mediate its primary biological effect via a mechanism that involves the direct modification of nuclear DNA. Rather, alternative modes, involving perturbation of cytoplasmic processes, are thought to be operative (see below) [10].

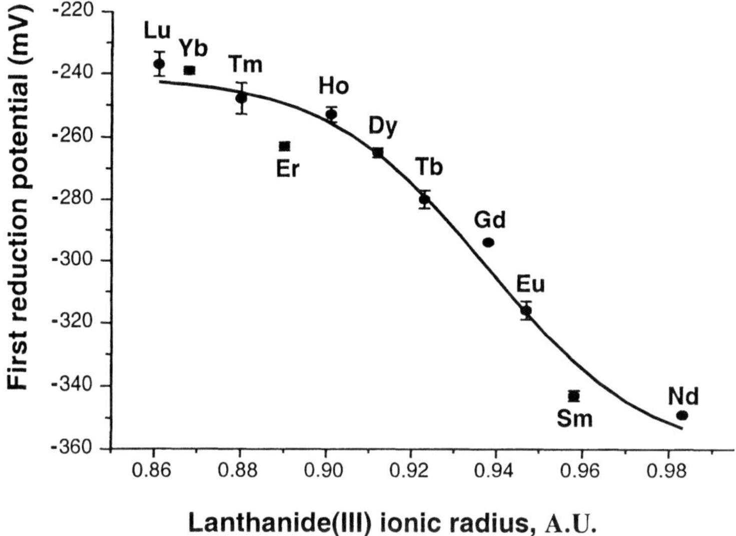

Figure 1. Plot of first reduction potential vs. Ag/AgCl in DMSO for various lanthanide(III) texaphyrin complexes vs. ionic radius. This figure was constructed using data first published in ref 17.

Early Mechanistic Studies

Prior to the completion of the sub-cellular localization studies, it was thought that MGd could be mediating its radiation enhancing effect by acting as an "electron sponge" [19]. Specifically, it was considered likely that it would serve to "trap" the solvated electrons produced in conjunction with hydroxyl radicals when water is subject to ionizing radiation. To the extent such "trapping" occurs within the nucleus, it would serve to increase the effective concentration of hydroxyl radicals, highly reactive species that are known to engender both reparable and irreparable modifications of DNA [20]. While such a direct mode of action now seems unlikely on the basis of the sub-cellular localization studies, ease of MGd reduction was identified as a key chemical property in the course of this early mechanistic work. In particular, it was found that solvated electrons, generated by pulse radiolysis, would serve to effect the one-electron reduction of

MGd. Furthermore, it was found that this reduced species would react with oxygen in a fast, reversible process to produce superoxide anion (cf. Scheme 1). This species is rapidly converted to hydrogen peroxide under biological conditions. The rate constant for non-enzymatic disproportionation of superoxide is estimated, for instance, to be 1 x 10^5 M^{-1} s^{-1} at pH 7.4 [21].

$$\mathrm{MGd} + O_2^{\bullet-} \rightleftharpoons \mathrm{MGd}^{\bullet-} + O_2 \qquad (1)$$

$$\mathrm{NADPH} + O_2 \xrightarrow{\mathrm{MGd}} \mathrm{NADP}^{+} + H_2O_2 \qquad (2)$$

$$\mathrm{Ascorbate} + O_2 \xrightarrow{\mathrm{MGd}} \mathrm{Dehydroascorbate} + H_2O_2 \qquad (3)$$

Scheme 1. Key redox reactions of MGd and reducing metabolites.

Independently, it was found that thiols, ascorbate, and NADPH could serve as a source of reducing equivalents, producing hydrogen peroxide and the oxidized forms of these key metabolites under conditions designed to mimic those present in typical cellular assays [10,11]. Since peroxide and the oxidizing metabolites formed in this process (disulfides, dehydroascorbate, etc.) are known to induce apoptosis and potentiate the effects of ionizing radiation in vitro [22,23], it was thought that these redox processes could play a key role in mediating the biological activity of MGd. This led to a working proposal, embodied in the Scheme 2, wherein this redox active agent acts in the presence of oxygen to generate the oxidized forms of these key metabolites and peroxide site-specifically within cancerous lesions in vivo. Initial evidence for this mechanism included *i*) the loss of MGd spectral features under hypoxic conditions in the presence of NADH and *ii*) the formation of NAD^+ in the presence of catalytic amounts of MGd in the presence of oxygen. Subsequent chemical findings supporting this hypothesis involved quantifying the amount of hydrogen peroxide produced under the assay conditions and measuring the oxygen consumption in the presence and absence of the antioxidant enzymes superoxide dismutase and catalase. This work was complemented by biological studies demonstrating: *i*) Cytotoxicity of MGd in the presence of NADPH or ascorbate; *ii*) formation of intracellular peroxide by oxidation of dichlorofluorecin diacetate (DCFA) to fluorescent dichlorofluorescein (DCF); *iii*) potentiation of the effects of ionizing radiation by the combination of MGd and ascorbate; and *iv*) synergistic biological effects with agents known to be involved in thiol metabolism, namely buthionein sulfoximine (BSO, an inhibitor of glutathione synthesis), diamide (a thiol oxidant), and antimycin A (a superoxide generator).

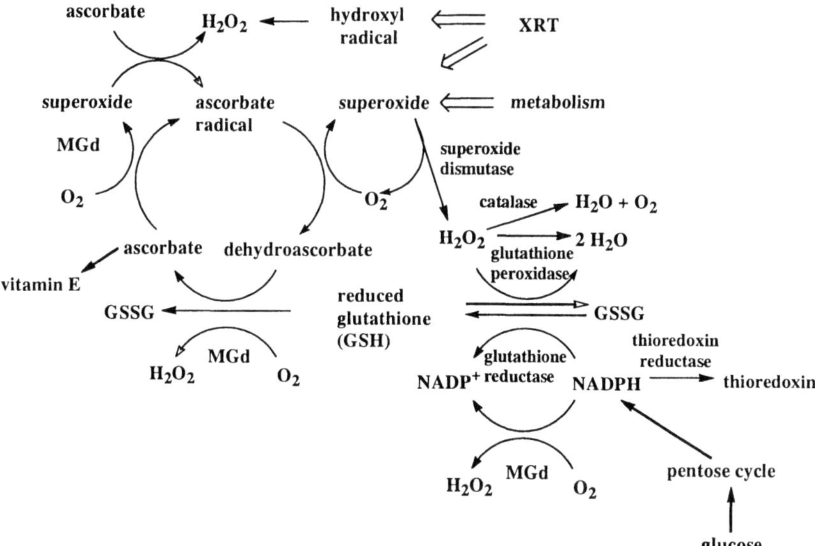

Scheme 2. Abridged description of the cellular antioxidant system and the effect that ionizing radiation and the presence of Xcytrin (MGd, ***1****) can have on this system. This scheme is modified from one that originally appeared in ref. 10.*

In this chapter we present further evidence in favor of the generalized mechanism shown in Scheme 2 while reviewing briefly some of the previously published findings. It is important to note that Scheme 2 does not distinguish between global oxidative stress as opposed to any specific biological "target" as the basis for MGd activity. The in-depth studies to be described herein were undertaken in an effort to identify the more important biological targets of MGd. The focus will be on three classes of reducing metabolites of critical biological importance, namely NADPH, ascorbate, and thiols. In the course of this discussion, we will compare MGd to other related metallotexaphyrins (e.g., **2**-**12**). All of these compounds share an identical equatorial ligand, and differ only in the identity of the coordinated metal cation. As might be anticipated, several of these analogues, e.g., those derived from transition metal cations, display properties markedly different from those of MGd. Other congeners, such as those drawn from the lanthanide series, differ in more subtle ways. In this regard, it is useful to note that the atomic radius of the lanthanide cation contracts, as the series is traversed from left to right. As a result, the corresponding texaphyrin complexes exhibit a graduated increase in stability, ease of reduction, and lipophilicity. It would thus be predicted that any structure activity relationships that are developed in the course of comparing lanthanide derivatives, however revealing, may be unable to distinguish the relative contributions to activity made by each of these individual properties. Additional data, e.g., derived from texaphyrin complexes bearing altered solubilizing substituents on the ligand, are invaluable in making this latter determination. Nonetheless, this review will focus primarily on compounds **1-12**, and make mention of other derivatives (i.e., **13-15**) only where most germane.

Reaction with Reducing Metabolites

NADH/NADPH

NADH and NADPH are important sources of reducing equivalents within cellular milieus. These chemically related species both have redox potentials of –315 mV vs. NHE ($E^{o'}$, i.e., at pH 7) [24]. As such, they represent some of the most reducing species present in cells. NADH is formed via the Krebs cycle and is principally involved in ATP generation through oxidative phosphorylation which occurs in the mitochondria. By contrast, NADPH is for the most part formed via the pentose cycle and is an important cofactor in a number of cellular enzymatic processes, including those involved in maintaining the cell in a reduced state as shown in Scheme 2 above. These two cofactors can be interconverted, albeit indirectly, via their oxidized forms through the action of NAD kinase [25]. Both species possess a negatively charged phosphoric anhydride linkage as part of their chemical structure and thus might be anticipated to coordinate to MGd in an axial position. It was thus considered important to understand how MGd and related metallotexaphyrin complexes would interact with these cofactors.

Reactions of MGd with NADH/NADPH

In initial studies, the oxidation of NADPH by MGd was quantified using HPLC methods [10]. We now favor a method based on oxygen consumption. In this approach, the decrease in dissolved oxygen concentration in a sealed vessel is measured as a function of time at a chosen concentration of texaphyrin and reducing metabolite using a commercially available ruthenium bipyridyl-based oxygen sensor (Ocean Optics) [11]. This approach has the important benefit of enabling comparisons involving a variety of redox cycling substrates using a general assay. It also allows the concentration of MGd (or other absorbing species) to be simultaneously monitored by UV-vis spectroscopy.

As illustrated by the time-dependent oxygen consumption curve reproduced in Figure 2, aqueous, air-saturated, pH 7.5 HEPES buffered saline solutions of NADPH are to a first approximation kinetically stable, as judged both by lack of oxygen consumption (cf. initial time points in Figure 2) and by independent HPLC-based analysis of $NADP^+$ as reported in ref. 10. However, in sealed vessels, where the dissolved oxygen is not replenished, addition of MGd induces a decrease in the overall oxygen concentration. Provided both the NADPH and oxygen are initially in substantial excess of the metallotexaphyrin complex, the $[O_2]$ vs. time plot is linear at early times (ca. an hour for the experiment shown in Figure 2). This allows an initial rate, V_o, of 0.70 μM/min to be calculated. Gratifyingly, this value is coincident with that obtained by monitoring the $NADP^+$ concentration increase as a function of time by HPLC and by the decrease in NADPH absorbance at 338 nm. Furthermore, this initial rate was found to be first order in [MGd], provided both oxygen and NADPH were present in excess.

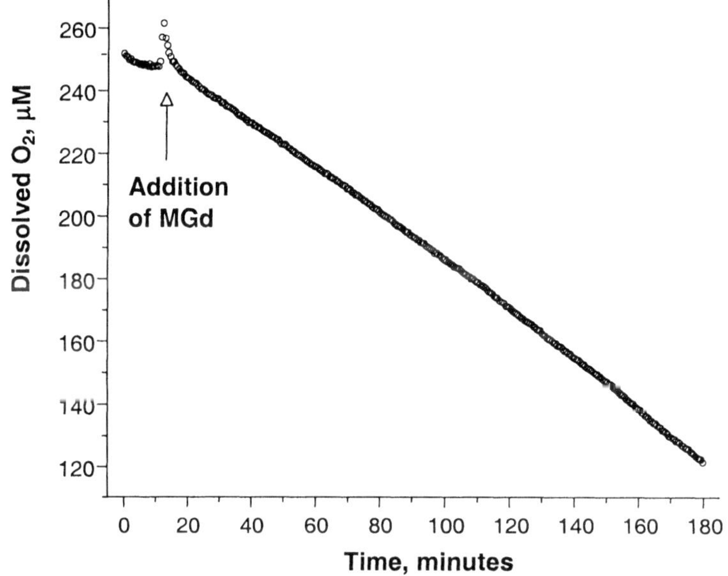

Figure 2. Plot of dissolved $[O_2]$ vs. time for a sealed vial containing NADPH (initial concentration: 250 μM) in HEPES/NaCl buffer at pH 7.5 showing the effects of MGd addition (12.5 μM initial concentration after dilution). From the linear portion of the decay profile, an initial reaction rate, V_o, of 0.70 μM/min was calculated.

Separate studies, carried out using UV-vis spectroscopy, served to demonstrate that the MGd-mediated oxidation of NADPH in the presence of air produces hydrogen peroxide. Thus, consistent with the pulse radiolysis experiments described above, we propose that electron transfer from NADPH to MGd serves to produce a reduced texaphyrin species that then reacts with oxygen to regenerate the starting MGd complex and to produce superoxide (not detectable under the reaction conditions), which disproportionates to form peroxide. The net result is shown in eqn. 2 of Scheme 1.

Effect of Oxygen

When the absorption spectrum of MGd, rather than the concentration of dissolved O_2, is monitored as a function of time, it is found that little change in the texaphyrin spectral features is observed during the linear phase of the $[O_2]$ vs. time curve. At longer times, or under conditions where oxygen is deliberately excluded from the reaction vessel, significant bleaching in the texaphyrin absorption bands is observed. This decrease in texaphyrin absorbance is accompanied by the formation of new absorbances typical of texaphyrin synthetic precursors. In contrast to what is observed when ascorbate or dithiothreitol (DTT) was used as the reductant (vide infra), addition of air did not serve to reverse the bleaching and restore the initial MGd spectrum. Thus, the decrease in texaphyrin spectral intensity observed in the absence of oxygen (which, presumably, functions as an electron acceptor) is thought to reflect the irreversible formation of multiply reduced species, including perhaps those wherein the basic texaphyrin framework is ruptured. While not addressed by experiment, it is considered that the formation of these latter species is facilitated in the case of NADPH because it can, inter alia, act as a more effective hydride donor than the other reductants considered in this study (see below).

Reactions of other metallotexaphyrins with NADH/NADPH

Monitoring both the oxygen concentration and that of $NADP^+$ as a function of time allowed the efficacy of MGd as an NADPH oxidation catalyst to be compared with that of related metallotexaphyrins. As illustrated in Figure 3, such a comparison leads to the conclusion, that for this particular reductant, MGd is substantially *less* effective than its Lu(III)-, Y(III)-, and Dy(III)-containing congeners, **2**-**4**. One interpretation of this finding is that these other metallotexaphyrins could have a role to play as therapeutic agents, although other factors, such as bioavailability may override the benefits of greater catalytic prowess. Another possibility is that the cofactor oxidation process as studied is not directly germane to biological activity. This latter interpretation is supported by the relatively low rates of reaction involved both in an absolute sense and as compared to those obtained using other reducing species (see below).

In preliminary comparative tests using intact cells, the human ovarian cancer cell line MES-SA was used and the extent of proliferation was monitored by

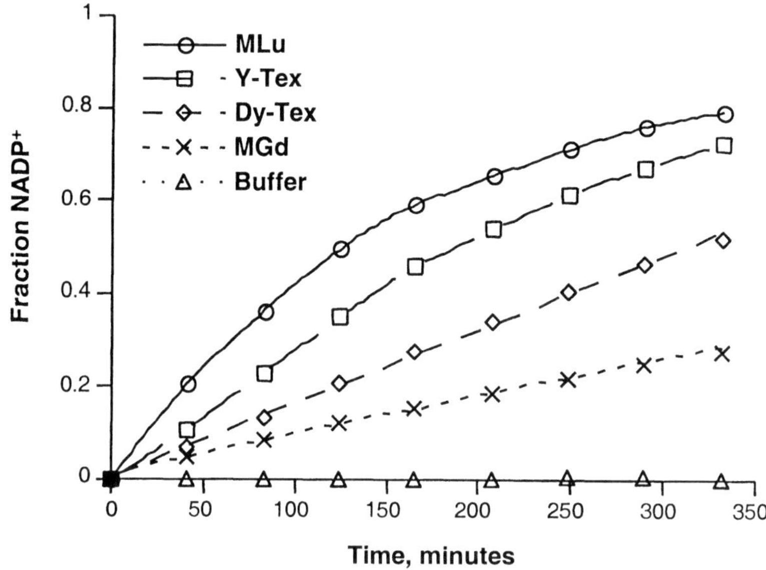

*Figure 3. Plot of the [NADP$^+$]/[NADP$^+$ + NADPH] ratio ("fraction NADP$^+$") vs. time for a sealed vial containing NADPH (initial concentration: 250 μM) in HEPES/NaCl buffer at pH 7.5 showing the effects of different metallotexaphyrins (12.5 μM initial concentration after dilution of MGd (**1**), MLu (**2**), Y-Tex (**3**), and Dy-Tex (**4**)) on the rate of NADPH reduction.*

tetrazolium salt (MTT) reduction [10]. RPMI 1640 with dialyzed serum was chosen as the medium in these experiments to assure the absence of adventitious ascorbate (vide infra), while NADPH was added at a concentration of 333 μM. Under these conditions, the presence of MGd at the 50 μM level caused no decrease in cell viability within error, whereas a similar concentration of MLu caused a 40 ± 10% reduction in cell survival. By contrast, replacement of NADPH by ascorbate reversed the order of complex activity. For instance, at an ascorbate concentration of 111 μM, 50 μM of MGd, **1**, led to a 65 ± 5% reduction in cell survival vs. no apparent reduction in the case of MLu, **2**. In all cases, the effects, where observed, were significantly (or even completely) obviated by the addition of catalase and superoxide dismutase. Such findings are consistent with the extracellular production of reactive oxygen species (i.e., outside of the cells themselves), under conditions that may or may not mimic those operative in vivo. They also serve to demonstrate that the redox cycling activity observed in buffered saline, outlined above, is recapitulated in tissue culture medium, i.e., in the presence of serum proteins.

Effect of thioredoxin reductase and other enzymes

During the course of the studies described above, it was recognized that other components present in the cellular milieu would need to be considered prior to any attempt to extrapolate these measurements to biological systems. In particular, it was of interest to examine the rate of NADPH oxidation by MGd in the presence of enzymes for which NADPH is a cofactor, such as thioredoxin reductase (EC 1.6.4.5). NADPH was therefore incubated with MGd as described above at concentrations providing a low rate of chemical reaction. When thioredoxin reductase was added to this system, a dramatic enhancement of oxygen consumption was observed. As might be expected, oxygen consumption was accompanied by a corresponding bleaching of the NADPH absorbance at 338 nm and, after most of the oxygen present in the cuvette was consumed, texaphyrin absorbances at 470 and 742 nm. Similar findings were obtained where the order of addition was reversed, i.e., a slow background reaction between NADPH and the enzyme, followed by rapid reaction upon addition of catalytic amounts of MGd (cf. Figure 4). Other enzymes, such as cytochrome c reductase and cytochrome P450 reductase, could be effectively substituted for thioredoxin reductase. By contrast, glutathione reductase and other cofactor-binding enzymes investigated did not display this effect. Experiments such as these lead us to suggest that MGd can oxidize cofactors indirectly, perhaps by accepting an electron from the flavin subunit of their enzyme complexes. Although the details of MGd interaction with enzymatic processes such as these remain the subject of ongoing investigation, it is reasonable to conclude from the data available at this time that such interactions are likely to contribute to the biological activity of the drug.

Ascorbate

Reactions of MGd and congeners with ascorbate

Ascorbate, with a one-electron reduction potential, E_0' of 282 mV vs. NHE [26], is another critical source of cellular reducing equivalents. Regenerated in vivo from the reduction of dehydroascorbate by NADPH either directly or via the intermediacy of glutathione, this species serves not only to maintain cells in a reduced state, it also plays a role in protecting against the deleterious effects of superoxide and other radicals. It thus complements enzymatic pathways (e.g., superoxide dismutase) that afford an alternative means of protection. Ascorbate is also of interest because it is not the product of biosynthesis in humans but rather must be obtained from the diet. This is not true, however, for other species, such as mice. Surprisingly, ascorbate is often absent in standard cell culture media. We, and others, have thus proposed that understanding the effects of MGd on ascorbate could provide important insights into the behavior of this agent both in vitro and in vivo [10,27].

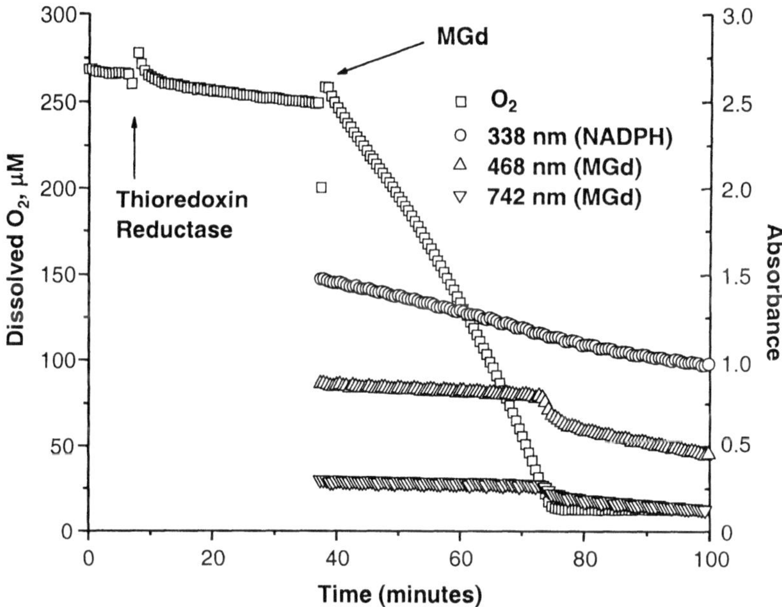

Figure 4. Change in dissolved oxygen concentration as a function of time observed when first thioredoxin reductase (0.6 unit) and then subsequently MGd (62.5 µM) is added to a sealed cuvette containing NADPH (initial concentration: 1.25 mM) in HEPES/NaCl buffer at pH 7.5. Also monitored in this experiment were the change in absorbance at wavelengths characteristic of NADPH (338 nm) and MGd (468 and 742 nm).

As proved true for NADPH, addition of a catalytic amount of MGd to aerated solutions of ascorbate led to rapid, catalytic oxidation and concurrent production of hydrogen peroxide in accord with Eqn. 3, Scheme 1. However, in contrast to what was observed with NADPH, in the case of ascorbate, MGd was found to be a substantially more effective oxidation catalyst than MLu or two other lanthanide(III) texaphyrin complexes containing smaller coordinated cations, namely Dy-Tex (**4**) and Er-Tex (**5**). Other lanthanide(III) texaphyrins containing larger cations, namely the Eu(III), Sm(III), and Nd(III) complexes (**6**-**8**, respectively) showed rates that were similar to those of MGd. Initial rates of reaction therefore appear to fall into two groups, that correlate with ionic radius, as can be seen for the seven texaphyrin lanthanide congeners shown in Figure 5.

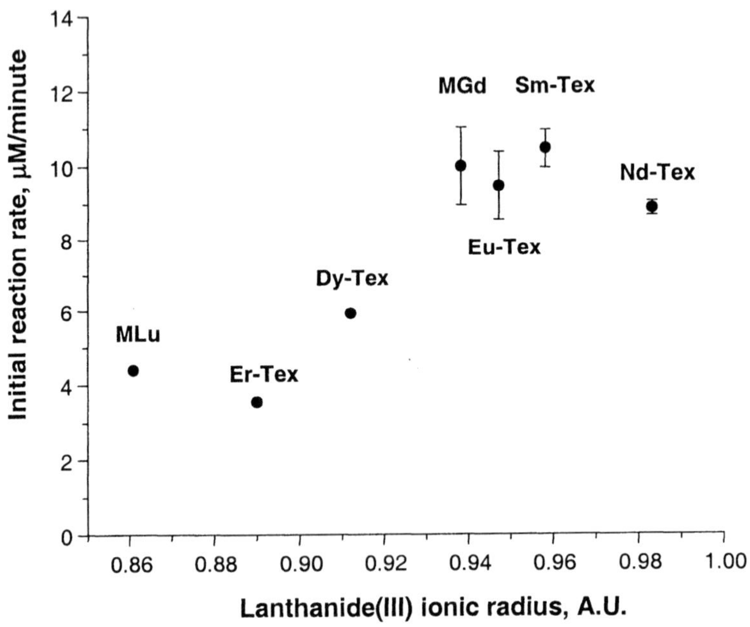

Figure 5. Plot of initial rates of ascorbate oxidation vs. lanthanoid(III) ionic radius, as determined from the change in absorbance at 266 nm. The texaphyrin complexes (0.05 molar equivalent) in question were added to solutions of ascorbate (1.25 mM) in buffered saline (100 mM NaCl, 50 mM HEPES, pH = 7.5). All reactions were conducted in 1 mm cuvettes.

As mentioned earlier, we observe changes in reduction potential, stability, and solubility as the lanthanide series is traversed. Presumably, this reflects the fact that the metal is out of the plane in complexes with the lighter cations (as inferred from previous X-ray diffraction analyses) [14], whereas in complexes with heavier cations, such as MLu, the cation is more centrally coordinated with the plane of the texaphyrin core. In the reaction with NADPH, it is plausible to rationalize the observed order of reactivity based on reduction potential. In the reaction with ascorbate, other factors, such as solubility, could be playing a role. In fact, a variety of experiments, including extinction coefficient studies, have led us to consider that MLu, but not MGd, is likely to be aggregated under conditions analogous to those employed in the ascorbate oxidation studies.

In order to assess a possible effect of aggregation, we therefore compared the reactivity of MLu with that of its (doubly protonated) diamine congener, DALu-Tex (**14**). As seen in Figure 6, the hydrophilic DALu-Tex catalyzed ascorbate oxidation with an initial rate of reaction faster than that of MLu and comparable to that of MGd. On the other hand, the rate of ascorbate oxidation by MGd was essentially the same as that of its diamine derivative, DAGd-Tex (**13,** data not shown). This, along with the observation that essentially first order rate behavior was seen as the MGd concentration was varied over a 1.5×10^{-6} to 6×10^{-6} M concentration range (the initial rate was found to increase by a factor of 4.8 as the ascorbate:MGd ratio was varied from 80:1 to 20:1), is completely consistent with the previous findings that MGd is not aggregated under normal aqueous buffer conditions, at least at concentrations relevant to the present experiments.

Perhaps not surprisingly, the initial rate of catalysis was found to increase as the concentration of oxygen was increased. For instance, when the normal air atmosphere in the cuvettes used to study the reaction dynamics was replaced by pure oxygen, the initial rate was increased by a factor of ca. 3 and the degree of completion (as judged by the loss of ascorbate) was also found to be enhanced. Conversely, when oxygen was excluded from the vessel, no reaction occurred, in contrast to the situation seen above using NADPH as the reductant. These observations are consistent with the texaphyrin undergoing a reversible 1-electron reduction with the driving force for the reaction being thermodynamically unfavorable in the absence of oxygen. In the presence of oxygen, peroxide is formed via the disproportionation of superoxide with the effect that the overall reaction becomes energetically favorable.

When the concentrations of oxygen and ascorbate are high relative to those of MGd, the rate of ascorbate oxidation is catalytic in MGd, as observed with NADPH above. Under these conditions, the MGd complex is stable, with no evidence of decomposition being observed. However, as the oxidation process is allowed to proceed, a new species, characterized by spectral features at 510 and 780 nm, is seen to grow in. This species undergoes precipitation when allowed to stand in saline, and reverts to MGd in organic solvent. On the basis of these observations and specific chemical analysis, it was assigned as being a coordination polymer with oxalate [11]. Oxalate is a known decomposition product of dehydroascorbate. It is also known to be unstable at physiological pH and in the presences of superoxide. Carboxylate anions coordinate well to gadolinium(III) and other lanthanide(III) cations. Thus, the formation of oxalate and its subsequent reaction with MGd over time is considered reasonable. The fact that it can bridge two Gd(III) centers in side-on fashion, allowing the planar aromatic texaphyrin ligands to stack, explains the ease of formation of the polymer and its spectral characteristics. Indeed, viscosimetry measurements indicate an average molecular weight of about 100,000 dalton for the species [28]. While the importance of such polymer formation under in vivo conditions is subject to current analysis, it is noteworthy that it is taken up into A549 lung cancer cells more effectively than MGd [11].

The potential importance of the reactions with ascorbate as a possible indicator of biological function propelled us to study whether other metallotexaphyrins besides those containing lanthanide(III) cations could

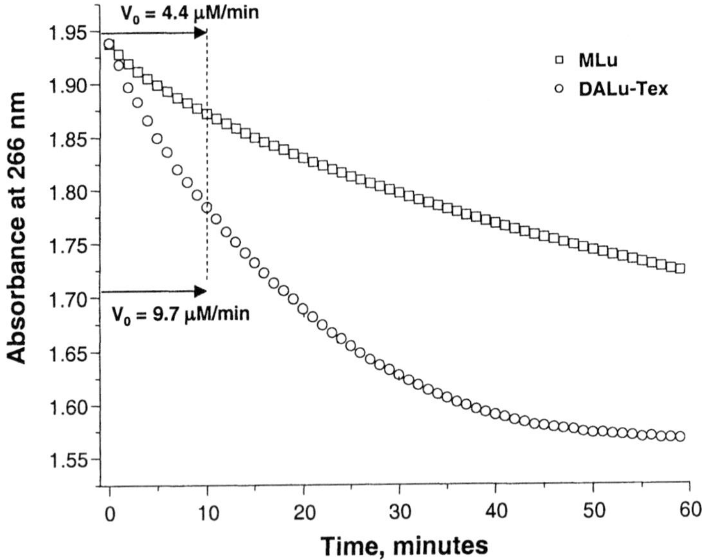

Figure 6. Time profiles for ascorbate oxidation observed in the presence of two lutetium (III) texaphyrin complexes, as determined from the change in absorbance at 266 nm Experiments were conducted as described in the caption to Figure 5.

function as oxidation catalysts for this key reducing metabolite. Recently, we have prepared several new transition metal complexes of texaphyrin, including the Mn(II), Co(II), and μ-oxo Fe(III) dimer derivatives (**10**-**12**), as well as a water soluble version of a Cd(II) complex (**9**) reported early on [14,29].

As can be seen from an inspection of Figure 7, the choice of central coordinated metal has a very substantial effect on the initial rate of ascorbate oxidation when both it and oxygen are present in excess relative to the metallotexaphyrin. While the Mn(II) and Cd(II) complexes displayed initial rates that were 2-3 times slower than MGd under identical experimental conditions (V_o = 3.0 and 3.4 μM/min vs. 8.7 μM/min, respectively), the Co(II) and μ-oxo Fe(III) dimer complexes **11** and **12** gave initial rate values (V_o = 23.8 and 30.6 μM, respectively) that were substantially larger. (In other experiments, Mn-Tex **10** and its diamine analogue, DAMn-Tex **15**, displayed similar activity,

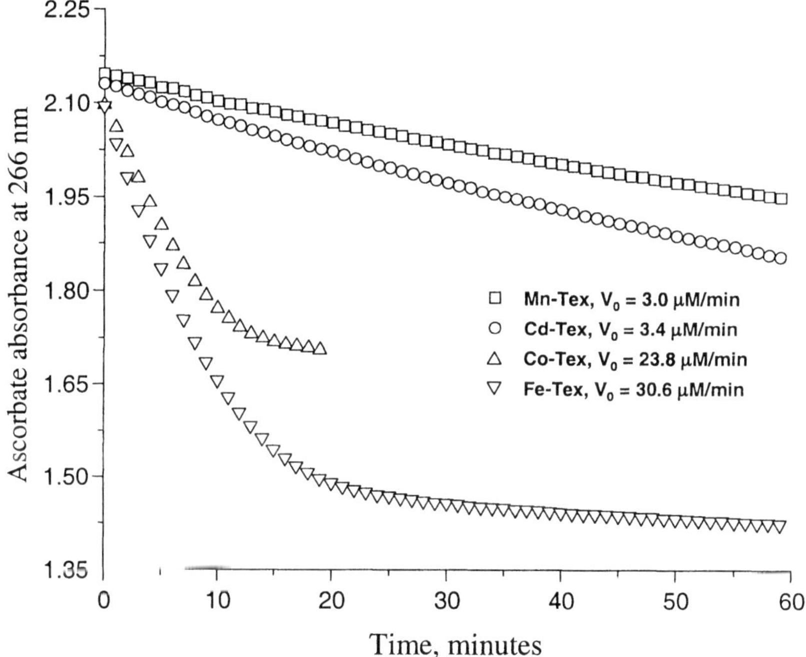

Figure 7. Comparison of the catalytic activity of transition metal texaphyrins in effecting the air-based oxidation of ascorbic acid in buffered aqueous media. Conditions: [Ascorbate] = 1.23 mM; substrate:catalyst ratio = 20:1; pH 7.5 HEPES/NaCl buffer; solution pretreated with Chelex 100 resin; ambient temperature.

presumably ruling out aggregation as the source of lowered activity in this instance.) Unfortunately, the Co(II) complex proved less stable than MGd under biological conditions, although independent control experiments carried out with simple cobalt salts served to alleviate concerns that the catalytic oxidation effects were due to free cobalt(II) ions. Likewise, the μ-oxo Fe(III) dimer forms a mixture of monomeric and dimeric species at neutral pH. Thus, at the present time neither complex looks attractive in terms of further drug development, at least for the uses for which MGd is being tested.

Effect of antioxidant enzymes

The effect of two antioxidant enzymes was investigated in the case of MGd and ascorbate. The first of these, catalase, was seen to have relatively little effect on the reaction, as judged both by the rate of oxygen consumption and ascorbate oxidation (cf. Figure 8) [11]. On the other hand, it reduced by a factor of roughly two, the rate at which the presumed MGd-oxalate polymer was

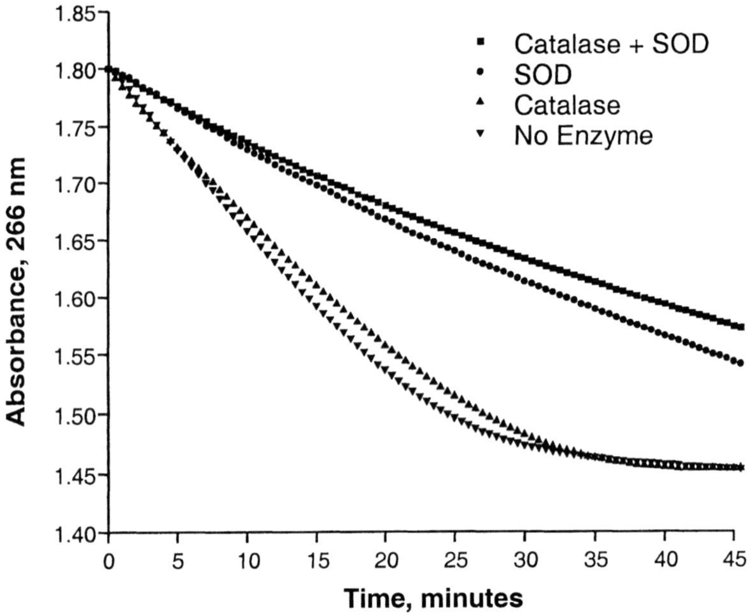

Figure 8. Effect of catalase and superoxide dismutase (SOD) on the MGd-promoted air-based oxidation of ascorbic acid in buffered aqueous media. Conditions: [Ascorbate] = 1.23 mM; ascorbate:MGd = 20:1; pH 7.5 HEPE/NaCl buffer; ambient temperature. The catalase (EC 1.11.1.6; 2600 units mL^{-1}) and SOD (EC 1.15.1.1; 100 units mL^{-1}) were added prior to the addition of MGd.

formed under the standard reaction conditions (cf. Figure 9). Catalase serves to remove hydrogen peroxide via the production of oxygen and water. Thus, initially it was considered likely that its addition would slow the apparent rate of oxygen loss. However, the HEPES buffer present in this system appears to react with the peroxide formed under the conditions of the reaction.* The net result is that the concentration of oxygen (and hence reaction rate) is not greatly effected by the addition of catalase. On the other hand, its addition does serve to reduce the rate of follow-up dehydroascorbate oxidation to form oxalate. As a consequence, the rate of MGd-oxalate polymer formation is reduced.

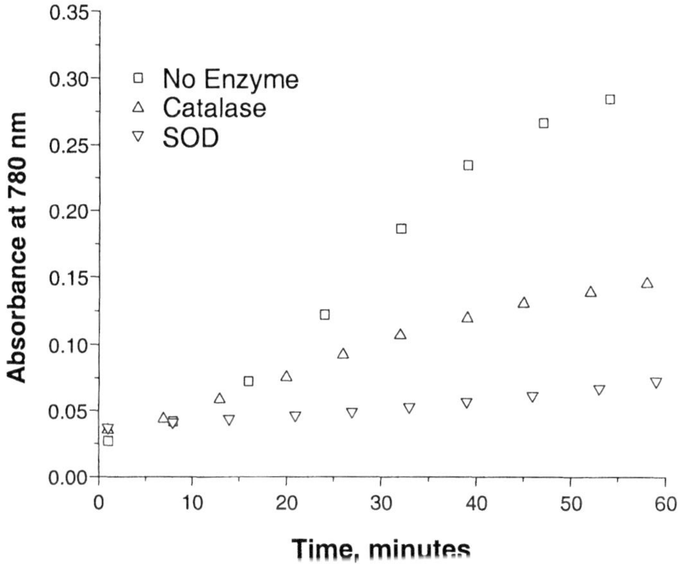

Figure 9. Effect of enzymes on MGd-oxalate polymer formation. Conditions as in Figure 8.

Very different behavior was seen when the second antioxidant enzyme, superoxide dismutase, was studied. In this case, the rates of both oxygen consumption and ascorbate oxidation were slowed by a factor of two. Moreover, the rate of MGd-oxalate polymer formation was reduced by a factor of at least 10. This is consistent with superoxide being critically involved in a rate determining process, presumably the reduction of superoxide to peroxide by

* In other, phosphate buffered systems, the expected two-fold decrease in the rate of oxygen consumption was observed [30].

ascorbate. It is also consistent, inter alia, with the proposal that ascorbate serves to produce a one electron reduced texaphyrin species that then reacts with oxygen to produce, as its initial product, superoxide, while regenerating the starting MGd complex.

Thiols

Peptides and proteins containing cysteine as one or more of their composite amino acids are redox active species. The sulfur substituents are found either in a reduced, sulfhydryl form, or as part of a disulfide linkage. The latter is formed by the oxidation of two sulfhydryl groups, in a reaction that is energetically favorable (E_o' = -240 mV for the GSSG / 2GSH couple, where GSH = glutathione [24]). Due to their high concentration, thiol-containing proteins and peptides constitute the principal redox buffer in biological systems [24]. It therefore seemed appropriate to consider whether thiols could serve as electron donors in redox reactions catalyzed by MGd.

Reactions of MGd with dithiothreitol

Although thiol oxidation is an energetically favorable process, some thiol-containing compounds are more reactive than others. In particular, since thiol oxidation to form a disulfide requires the participation of two sulfhydryl groups, those compounds which contain two groups that are spatially located in the same vicinity (so-called "vicinal thiols") undergo oxidation at much faster rate. Due to its availability and water solubility, dithiothreitol (DTT) was chosen as a prototypic vicinal thiol with which to study the interaction of MGd with this functionality [30]. When tested under our standard conditions (i.e., as a 1.25 mM solution in 50 mM HEPES, pH 7.5, 100 mM NaCl, Chelex 100®-treated buffer, at ambient temperature), DTT was found to undergo little oxidation in the presence of air, as judged by the reduction of dissolved oxygen in experiments involving sealed cuvettes. However, when MGd (ca. 62 μM) was added into the sealed cuvettes containing DTT and carefully mixed by repetitive pipetting, the amount of dissolved oxygen in solution was seen to decrease dramatically. Concurrent with monitoring of the dissolved oxygen, the changes in the texaphyrin spectrum were recorded every 30 sec (cf. Figure 10).

Figure 10 combines UV-vis data (at 470 nm) from the diode array spectrophotometer with that from the oxygen probe measurement. The reaction profile may be divided into three phases, which correspond to the following time periods: Zero to 8 minutes (where half of the oxygen is gone), 9-17 minutes (where all of the oxygen is gone), and 18-160 minutes. The complex is essentially stable in the first part, there is a point of inflection after the second period, and fairly continuous, albeit slower bleaching thereafter. This bleaching can be fit to a biexponential decay curve, with k_1 = 2.8 μM/min and k_2 = 0.26 μM/min. Here, it is important to note that even though bleaching of MGd occurs under these conditions, it takes place slowly. In fact, even after 3 hours, well past the time when all appreciable quantities of oxygen are removed, residual quantities of MGd are still present, as judged from the UV-vis spectrum recorded at this point.

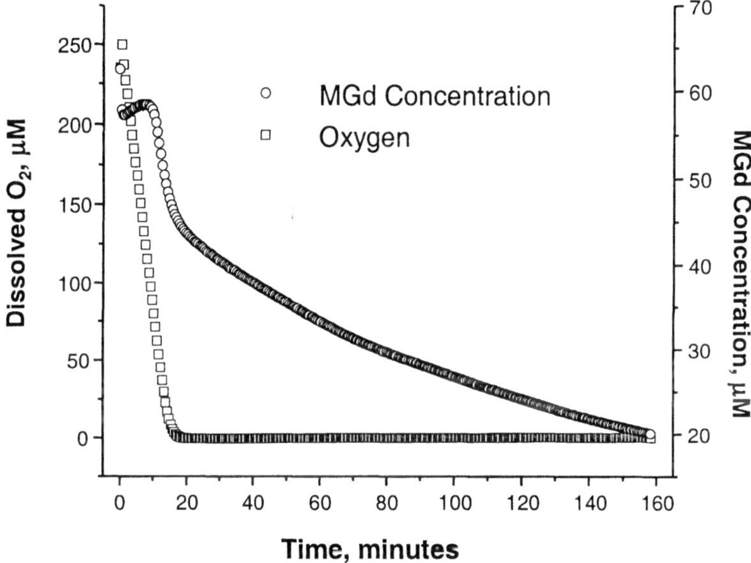

*Figure 10. Combined plot showing the decrease in dissolved oxygen and changes in MGd concentration based on change in texaphyrin-based absorption observed at 470 nm as a function of time. MGd (**1**) was added to a sealed cuvette containing a 1.25 mM aqueous solution of dithiothreitol (DTT); pH 7.5 HEPES/NaCl buffer; DTT:MGd = 20:1.*

Effect of oxygen

Addition of oxygen to the above solutions at a point where most of the original MGd spectral intensity is lost (e.g., at the 3 hour time point) served to restore it to its original intensity. When catalase was added to the original reaction mixture (to prevent peroxide-based chromophore bleaching) an essentially complete reversal, highlighted by isosbestic behavior, was observed (cf. Figure 11). Such observations are consistent with a mechanism wherein the MGd is reduced by the reductant (a vicinal thiol in this instance), in analogy to what occurs with NADPH or ascorbate (cf. eqns. 2 and 3 in Scheme 1) to generate a reduced texaphyrin species. This reduced texaphyrin species could exist free in solution or be stabilized as an adduct with the oxidized thiol products under the reaction conditions (cf. eqns. 4 and 5 in Scheme 3). The reduced texaphyrin species, free or bound, is then subsequently reoxidized to MGd by oxygen with the concomitant production of superoxide as shown in eqn. 1 of Scheme 1. As in the case of the other reductants studied, it is considered likely that hydrogen peroxide is formed from this superoxide "intermediate" via

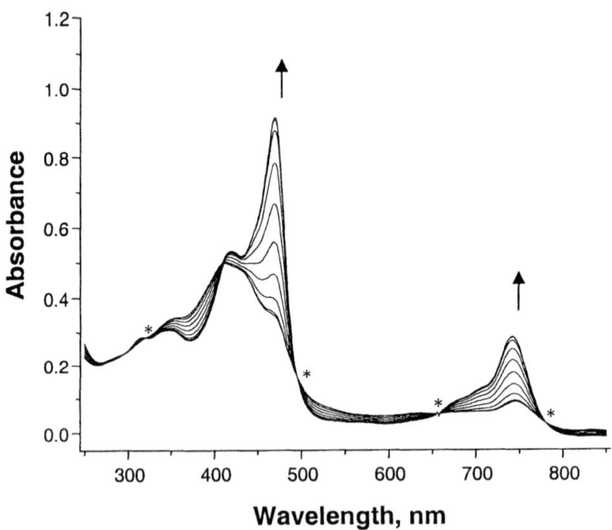

Figure 11. Overlay plot showing the restoration in the original vis absorption spectral features observed when air is added to bleached solution of MGd produced after letting sealed cuvettes containing MGd and DTT stand 180 min. Conditions are as in Figure 10, with the exception that catalase 2600 units mL^{-1} was added to the solution. The time elapsed between the first and last recorded spectrum is 55 min.

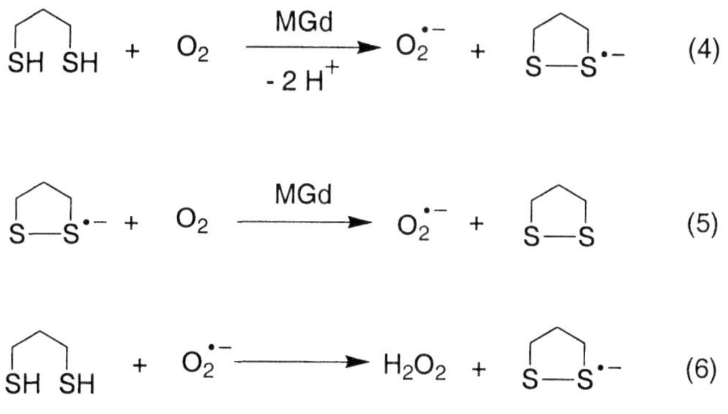

$$\text{SH SH} + O_2 \xrightarrow[-2\,H^+]{\text{MGd}} O_2^{\bullet-} + \text{S—S}^{\bullet-} \qquad (4)$$

$$\text{S—S}^{\bullet-} + O_2 \xrightarrow{\text{MGd}} O_2^{\bullet-} + \text{S—S} \qquad (5)$$

$$\text{SH SH} + O_2^{\bullet-} \longrightarrow H_2O_2 + \text{S—S}^{\bullet-} \qquad (6)$$

Scheme 3. Redox reactions involving a representative vicinal thiol and MGd.

a combination of disproportionation and direct reaction with the substrate. This latter chemistry is specifically illustrated in eqn. 6 of Scheme 3.

Effect of antioxidant enzymes

Experiments were also carried out in the presence of superoxide dismutase and catalase. As in the case of the corresponding experiments with ascorbate, the addition of catalase had relatively little effect on the reaction rate, presumably reflecting the propensity of hydrogen peroxide to react with the HEPES buffer. By contrast, the effect of the superoxide dismutase was even more substantial, reducing the initial rate by at least an order of magnitude, rather than the ca. two-fold seen in the case of ascorbate (cf. Figure 12).

Presumably, this very substantial difference reflects the greater contribution of an autocatalytic component of the reaction. This could arise from the fact that superoxide reacts effectively with DTT as shown in Eqn. 6 (Scheme 3) and that the one-electron oxidized product (comparable to the ascorbate radical formed in Scheme 2) produced in this way can, in turn, react with MGd (eqn. 5; Scheme 3). Apart from illustrating the effectiveness of SOD in inhibiting free radical processes, these findings lead us to suggest that SOD might serve to attenuate the oxidative stress caused by MGd redox cycling in cells.

Other thiol containing substrates

Several other thiol-containing substrates were also studied, in order to understand better the structural requirements inherent in thiol oxidation by MGd. These include the three DTT analogues shown in Figure 13, as well as glutathione (Figure 14). As can be seen from inspection of these two figures, the change from a vicinal dithiol to a simple thiol is dramatic. In fact, of the various reducing substrates studied in conjunction with MGd, glutathione was found to be the most slowly oxidized. By contrast, the fastest rate of initial reaction was seen with DTT. Consistent with this finding, it was observed that of the three thiols compared in Figure 13, the vicinal thiol (1,3-propanedithiol) is oxidized

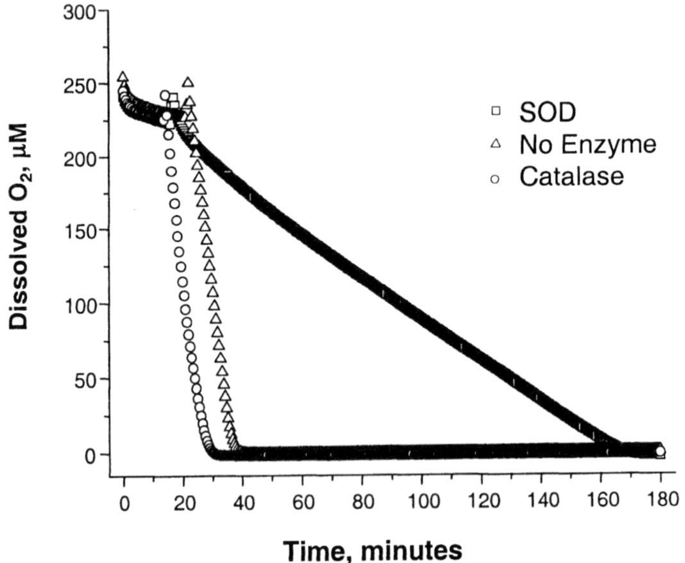

Figure 12. Comparison of added enzymes on the MGd-catalyzed, air-based oxidation of DTT in pH 7.5 NaCl/HEPES buffered aqueous media monitored by recording the concentration of dissolved O_2 in 2 mm sealed cuvettes as a function of time. The initial DTT concentration was 1.25 mM and the DTT:MGd ratio was 20:1.

more quickly than the two monothiol substrates, namely 3-mercaptopropionic acid and 3-thiapropan-1-ol. Thus, vicinal thiols forming 5-member ring products, 6-member ring products (DTT), or 7-member ring products (e.g., lipoic acid, data not shown [30]) upon oxidation all react with MGd. While not specifically addressed by experiment, these data provide support for the proposal that vicinal thiols present in larger molecules, i.e., proteins, might similarly become oxidized in the presence of MGd.

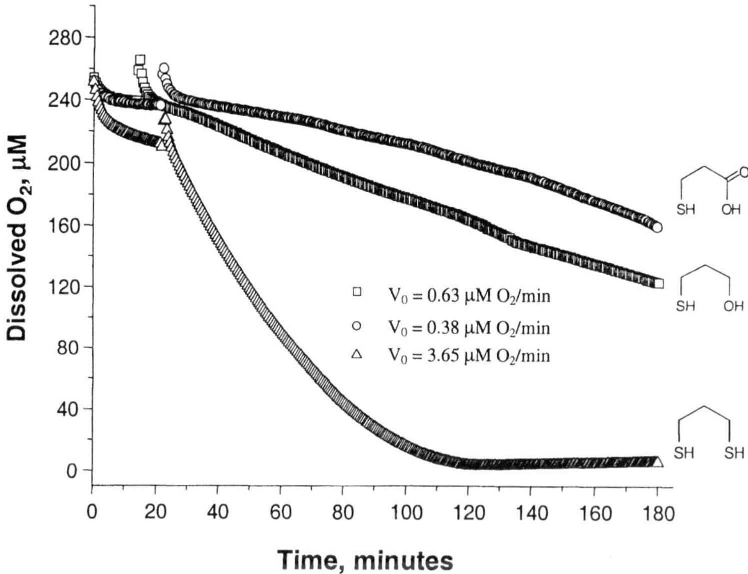

Figure 13. Combined data for MGd-catalyzed, air-based oxidation of thiols in pH 7.5 NaCl/HEPES buffered aqueous media. The concentration of dissolved O_2 in 2 mm sealed cuvettes was monitored as a function of time for each sample and the initial rates were calculated from initial linear region of the respective [O_2] vs. time plots. The initial thiol concentration was 1.25 mM and the thiol:MGd ratio was 80:1 in all cases.

Conclusions

In this study, the rates of various MGd-catalyzed, air-based oxidation processes have been compared. As highlighted by Figure 14, several common, representative electron-rich species of biological significance were found to be oxidized in the presence of MGd at different rates. Glutathione was found to react most slowly, and, despite the high concentration of this and similar species present in cells, its oxidation likely does not occur at a significant rate. Similarly, direct oxidation of NADPH seems unlikely, considering the slow rate of this reaction and the relatively low (ca. 200 μM) intracellular concentration of this cofactor. Ascorbate, on the other hand, was seen to undergo relatively facile oxidation in the presence of MGd, and is present at as high as 1 mM concentration in some tissues. We therefore consider that ascorbate oxidation may contribute to the biological activity of MGd. Given the central role ascorbate plays in protecting cells from radical species, oxidation of this species could render cells more vulnerable to free radical generating agents such as ionizing radiation. Alternatively, vicinal thiols reacted the fastest of the various biological reductants tested. This observation, coupled with the relatively high concentration of vicinal thiols present in cells (ca. 1 mM), makes vicinal thiols attractive candidates for oxidation by MGd. However, the fastest reactions observed to date are those observed in the presence of reductases, as shown in Figure 4. This serves as a reminder that biological systems are complex, and

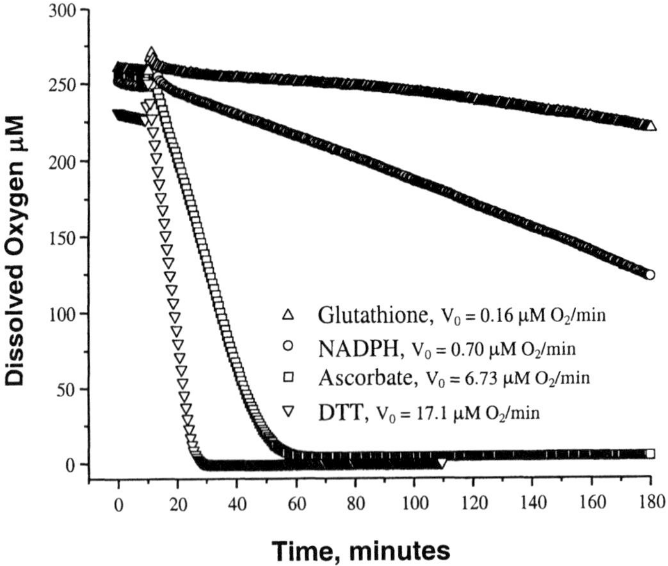

Figure 14. Combined data for MGd-catalyzed, air-based oxidation of reducing metabolites in pH 7.5 NaCl/HEPES buffered aqueous media. The concentration of dissolved O_2 in 2 mm sealed cuvettes was monitored as a function of time for each sample and the initial rates were calculated from initial linear region of the respective $[O_2]$ vs. time plots . The initial reductant concentration was 1.25 mM and the substrate:MGd ratio was 20:1 in all cases.

underscores the likelihood of a cellular mechanism of action based on a particular molecular target(s), as opposed to generic oxidative stress. DNA microarray analysis of cell cultures treated with MGd, indicating the elevated expression of RNA transcripts of a particular oxidative stress/metal response pathway, seems consistent with this latter assertion [31].

Several other metallotexaphyrin species were also tested, with MLu being subject to the greatest level of experimental scrutiny. This particular agent suffers from aggregation in aqueous media but nonetheless was found to be more effective than MGd in terms of catalyzing the air-based oxidation of NADPH. This greater activity was not seen in the case of ascorbate, where the advantages of MGd as an oxidation catalyst were clearly manifest. Several other metallotexaphyrins were also found to be active as catalysts in the case of ascorbate with the cobalt and iron species proving to be substantially more effective than the gadolinium complex. As such, these species or related congeners are potentially attractive for study as potential drug leads. However, the fact that MGd is more soluble in aqueous media and allows for facile MRI imaging in vivo, provides it with an important competitive advantage which justifies its choice as the texaphyrin complex now in late stage clinical testing as an anticancer agent.

Acknowledgment

This work was supported in part by the National Institutes of Health (grant no. CA 68682 to J. L. S.).

References

1. Sessler, J. L.; Seidel, D. *Angew. Chem. Int. Ed. Engl.* **2003**, *42*, 5134-5175.
2. Young, S. W.; Quing, F.; Harriman, A.; Sessler, J. L.; Dow, W. C.; Mody, T. D.; Hemmi, G.; Hao, Y.; Miller, R. A. *Proc. Natl. Acad. Sci. USA* **1996**, *93*, 6610-6615. Correction: *Proc. Natl. Acad. Sci. USA* 1999, *96*, 2569.
3. Miller, R. A.; Woodburn, K.; Fan, Q.; Renschler, M.; Sessler, J. L.; Koutcher, J. A. *Int. J. Biol. Radiat Oncol.* **1999**, *45*, 981-989.
4. Sessler, J. L.; Miller, R. A. *Biochem. Pharmacol.* **2000**, *59*, 733-739.
5. Mody, T. D.; Fu, L.; Sessler, J. L. *Progr. Inorg. Chem.* **2001**, *49*, 551-598.
6. Rosenthal, D. I.; Nurenberg, P.; Becerra, C. R.; Frenkel, E. P.; Carbone, D. P.; Lum, B. L.; Miller, R.; Engel, J.; Young, S.; Miles, D.; Renschler, M. F. *Clinical Cancer Res.* **1999**, *5*, 739-745.
7. Carde, P.; Timmerman, R.; Mehta, M. P.; Koprowski, C. D.; Ford, J.; Tishler, R. B.; Miles, D.; Miller, R. A.; Renschler, M. F. *J. Clin. Oncol.* **2000**, *19*, 2074-2083.
8. Mehta, M. P.; Shapiro, W. R.; Glantz, M. J.; Patchell, R. A.; Weitzner, M. A.; Meyers, C. A.; Schultz, C. J.; Roa, W. H.; Leibenhaut, M.; Ford, J.; Curran, W.; Phan, S.-C.; Smith, J. A.; Miller, R. A.; Renschler, M. F. *J. Clin. Oncol.* **2002**, *20*, 3445-3453.

9. Mehta, M. P.; Rodrigus, P.; Terhaard, C. H. J.; Rao, A.; Suh, J.; Roa, W.; Souhami, L.; Bezjak, A.; Leibenhaut, M.; Komaki, R.; Schultz, C.; Timmerman, R.; Curran, W.; Smith, J.; Phan, S.-C.; Miller, R. A.; Renschler, M. F. *J. Clin. Oncol.* **2003**, *21*, 2529-2536.
10. Magda, D.; Lepp, C.; Gerasimchuk, N.; Lee, I.; Sessler, J. L.; Lin, A.; Biaglow, J.; Miller. R. A. *Int. J. Radiat. Biol. Oncol. Phys.* **2001**, *51,* 1025-1036.
11. Magda, D.; Lepp, C.; Gerasimchuk, N.; Lecane, P.; Miller, R. A.; Biaglow, J. E.; Sessler, J. L. *Chem. Commun.* **2002**, 2730-2731.
12. Sessler, J. L.; Tvermoes, N. A.; Guldi, D. M.; Hug, G. L.; Mody, T. D.; Magda, D. *J. Phys. Chem.* **2001**, *105*, 1452-1457.
13. Woodburn, K *J. Pharmacol. Expt. Therapeutics* **2001**, *297*, 888-894.
14. Sessler, J.L.; Mody, T. D.; Hemmi, G. W.; Lynch, V. *Inorg. Chem.* **1993**, *32*, 3175-3187.
15. Sessler, J. L.; Hemmi, G.; Mody, T. D.; Murai, T.; Burrell, A.; Young, S. W. *Acc. Chem. Res.* **1994**, *27*, 43-50.
16. Sessler, J. L.; Tvermoes, N. A.; Guldi, D. M.; Mody, T. D. Allen, W. A. *J. Phys. Chem. A* **1999**, *103*, 787-794.
17. Guldi, D. M.; Mody, T. D.; Gerasimchuk, N. N.; Magda, D.; Sessler, J. L. *J. Am. Chem. Soc.* **2000**, *122*, 8289-8298.
18. Byrne, A. T.; Magda, D.; Nguyan, H.; Miles, D.; Boswell, G.; Miller, R. A. *Proc. Am. Assoc. Cancer Res.* **2003**, *44*, 393.
19. Adams, G. E.; Flockhart, I. R.; Smithen, C. E.; Stratford, I. J.; Wardman, P.; Watts, M. E. *Radiat. Res.* **1976**, *67*, 9-20.
20. Bump, E. A.; Brown, J. M. *Pharmac. Ther.* **1990**, *47*, 117-136.
21. Buettner, G. R. *Arch. Biochem. Biophys.* **1993**, *300*, 535-543.
22. Koch, C. J.; Biaglow, J. E. *J. Cell. Physiol.* **94**, 299-306.
23. Biaglow, J. E.; Mitchell, J. B.; Held, K. *Int. J. Radiat. Oncol. Biol. Phys.* **1992**, *22*, 665-669.
24. Schafer, F. Q.; Buettner, G. R. *Free Radical Biol. Med.* **2001**, *30*, 1191-1212.
25. Stubberfield, C. R,; Cohen, G. M. *Biochem. Pharmacol.* **1989**, *38*, 2631-2637.
26. Buettner, G. R.; Jurkiewicz, B. A. *Radiat. Res.* **1996**, *145*, 532-541.
27. Rockwell, S.; Donnelly. E. T.; Liu, Y.; Tang, L.-Q. *Int. J. Radiat. Oncol. Biol. Phys.* **2002**, *54*, 536-541.
28. Magda, J. J. personal communication.
29. Hannah, S.; Lynch, V.; Guldi, D. M.; Gerasimchuk, N.; MacDonald, C. L. B.; Magda, D.; Sessler, J. L. *J. Am. Chem. Soc.* **2002**, 124, 8416-8427.
30. Biaglow, J. E.; Miller, R. A.; Magda, D.; Lee, I.; Tuttle, S. Abstract of the 49th Ann. Mtg. Rad. Res. Soc., p. 107.
31. Hacia, J.; Magda, D.; Lepp, C.; Lee, I.; Miller, R. A. *Proc. Am. Assoc. Cancer Res.* **2002**, *43*, 647.

Chapter 9

Bile Acids at Work: Development of a New Intravascular MRI Contrast Agent

Pier Lucio Anelli[1], Marino Brocchetta[1], Vito Lorusso[1], Giuseppe Manfredi[1,2], Alberto Morisetti[1], Pierfrancesco Morosini[1], Marcella Murru[1], Daniela Palano[1], and Massimo Visigalli[1]

[1]Milano Research Centre, Bracco Imaging SpA, via E. Folli 50, 20134 Milan, Italy
[2]Current address: GSK Italy, via Asolana 68, 43056 S. Polo di Torrile, Parma, Italy

Complexes of Gd(III) featuring strong binding to human serum albumin (HSA) can be potentially used as intravascular MRI contrast agents. Accordingly, a series of Gd(III) complexes containing a bile acid moiety as the HSA binding subunit were synthesised. Diethylenetriaminepentaacetic acid was used as the chelating moiety of Gd(III) ion for all the conjugates. Derivatives of cholic, chenodeoxycholic, deoxycholic, and lithocholic acids, as well as bile acids in which the 12-α OH group was oxidized to a ketone, were prepared. Preliminary screening, which was based on the evaluation of binding to HSA and in vivo tolerability (in mice), led to the identification of the conjugate containing a deoxycholic subunit (*i.e.* **1c**) as the clinical candidate. During development, the route to **1c** used for the research phase was largely modified, taking advantage of a different synthetic approach.

Gadolinium complexes of polyaminopolycarboxylic ligands are by and large the most preferred species used as contrast agents for Magnetic Resonance Imaging (MRI) (*1*). All the complexes so far approved for administration in humans are so-called extracellular fluid (ECF) contrast agents (*2*). Such contrast agents are able to cross blood vessel walls. This means that, immediately after intravenous administration, they diffuse from plasma into the interstitial space until equilibrium between the two compartments is reached. These complexes usually do not enter cells.

Although the currently approved complexes can be successfully used for some angiographic procedures, they appear not suitable when visualization of coronary arteries is required (*3*). Indeed, imaging of coronary arteries, which are tiny and highly tortuous vessels, is furtherly complicated by cardiac and respiratory motion effects.

In order to fill this gap, contrast agents, which after administration are confined in the vascular space, have been developed by several research groups. Two main classes of compounds have been proposed to avoid, or at least to largely reduce, extravasation:

- *Species which, by virtue of their molecular size, are not able to cross blood vessel walls*. Accordingly, supramolecular systems (like micelles and liposomes) (*4*), proteins (*e.g.* albumin) (*5*) and polymers (*6*) labelled with a large number of gadolinium complexes have been proposed. Among these species, dendrimers (*e.g.* Gadomer-17) (*7*) have proved noticeably promising. Interestingly, also a gadolinium complex (*i.e.* P792) featuring four large hydrophilic side chains, due to its peculiar structure and size (6473 Da), behaves like these systems (*8*).
- *Small Gd(III) complexes (in the 800-1200 Da range) which feature suitable binding to Human Serum Albumin (HSA), the main component of human plasma proteins*. It is important to stress that this binding needs to be tuned in such a way to prevent leakage in the interstitial space but at the same time to allow elimination from the body in a reasonable timeframe. MS-325 (*i.e.* a Gd(III) complex of a diethylenetriaminepentaacetic acid, DTPA, ligand featuring a diphenylcyclohexyl residue and a phosphonate group) represented a breakthrough in this context (*9*). The compound is presently under late clinical development

We also focused our efforts on the latter class of compounds, the so-called "protein binders". Initially, variously mono-, di- and tri-substituted derivatives of Gd-DTPA^{2-} were investigated (*10, 11*). More recently, we took advantage of the experience gained in the use of bile acids as carriers for the development of hepatocyte-directed Gd(III) complexes (*12*). At that time, with the aim of addressing specific transporters of the hepatocyte trafficking, a number of conjugates of Gd(III) complexes with bile acids was prepared. In depth investigations involved several structural features of the conjugates like: *i*) nature of the bile acid; *ii*) site of conjugation on the bile acid skeleton; *iii*) nature of the polyaminopolycarboxylic ligand; *iv*) global charge of the conjugate.

Indeed, during such studies we had found that complex **1a** *N*-methylglucammonium salt (see Table I), was endowed with promising features in terms of binding to HSA (72% binding extent) and tolerability (LD_{50} 7.6 mmol/kg, in mice). Accordingly, this compound became our lead candidate and led us to explore structural modifications aimed at improving the HSA binding extent. It is worth noting that **1a** incorporates a subunit of cholic acid in which the 3α-OH residue has been converted into a 3β-NH_2 residue for conjugation purposes. Binding of simple bile acids to HSA had been studied more than twenty years ago by Roda *et al.* (*13*).

Table I. Chemical Structures of Compounds 1a-g

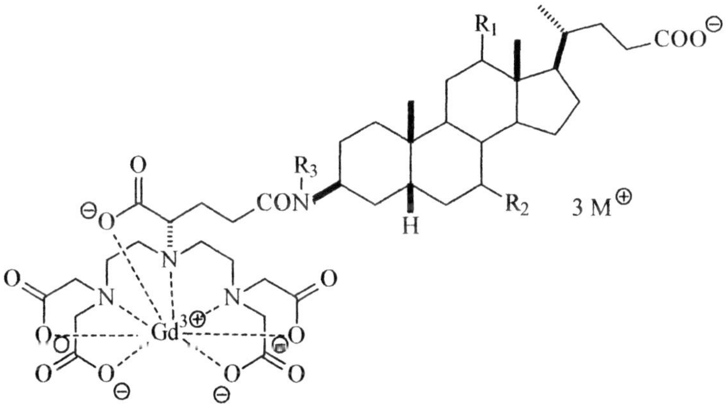

R1 COO⊖ R3 CON H R2 3 M⊕ Gd3⊕ N N N O O O O O O O O ⊖ ⊖ ⊖ ⊖

Compound[a]	R_1	R_2	R_3
1a	α-OH	α-OH	H
1b	H	H	H
1c	α-OH	H	H
1d	H	α-OH	H
1e	=O	H	H
1f	=O	α-OH	H
1g	H	H	$CH_2COO^-NMGH^+$

[a] M is either sodium or *N*-methylglucammonium (NMGH).

They had shown that the binding association constant of bile acids roughly increases 60 times on going from cholic acid to lithocholic acid, with intermediate values for deoxycholic and chenodeoxycholic acids. This clearly reflects the different lipophilicity of the bile acids, due to the different number and position of the hydroxy groups on the steroidic moiety. Accordingly, we decided to investigate how modifications to the basic structure of **1a** could affect binding to HSA and tolerability. To minimize the number of variables, the chelating moiety (*i.e.* the DTPA subunit derived from L-glutamic acid, *vide infra*) was kept unmodified in all the compounds which were synthesised. Conversely, trying to take advantage of the indications of Roda *et al.* (*13*), all structural variations were aimed at understanding the effect of modifications in the hydrophilicity/lipophilicity balance of the bile acid moiety.

Table I lists the conjugates which were prepared. In all compounds the cholanoic moiety stems from a natural bile acid in which the 3α-OH has been converted into a 3β-NH_2 group. Apart from taking advantage of the different number and position of OH groups on the steroidic moiety, hydrophilicity/lipophilicity of the structures has been tuned either oxidizing one of the OH residues to a ketone (like in conjugates **1e,f**) or introducing an auxiliary charged group (like in **1g**).

Chemistry of Conjugates 1b-g

The general synthetic route for the preparation of compounds **1b-g**, with the exception of **1f**, is the same previously used to achieve **1a** and reported in Figure 1 (*14*). The key step of such route is the assembly of the skeleton of ligand **2a** which occurs by diethoxyphosphoryl cyanide (DEPC) mediated condensation of the versatile intermediate **6** and the 3β-aminocholate derivative **8**. On the one hand, the monoacid pentaester **6** is obtained, using a Rapoport-like methodology (*15*), by double alkylation of L-glutamic *t*-butyl benzyl diester **3** with bis(*t*-butyl) *N*-(2-bromoethyl)iminodiacetate **4**, in its turn obtained by alkylation of ethanolamine with *t*-butyl bromoacetate followed by bromination of the hydroxyl group with *N*-bromosuccinimide/triphenylphosphine (NBS/PPh_3). Hydrogenolysis of the benzyl ester affords derivative **6**. On the other hand, 3β-aminocholate derivative **8** is obtained from methyl cholate and diphenylphosphoryl azide (DPPA) under Mitsunobu conditions followed by reduction of the azide group with triphenylphosphine/H_2O (Staudinger conditions) (*16*).

After condensation of **6** with **8**, all ester functions of intermediate **9** are deprotected in two steps and ligand **2a** is complexed with Gd_2O_3.

Complexes **1b-d** were synthesised according to the same path starting from

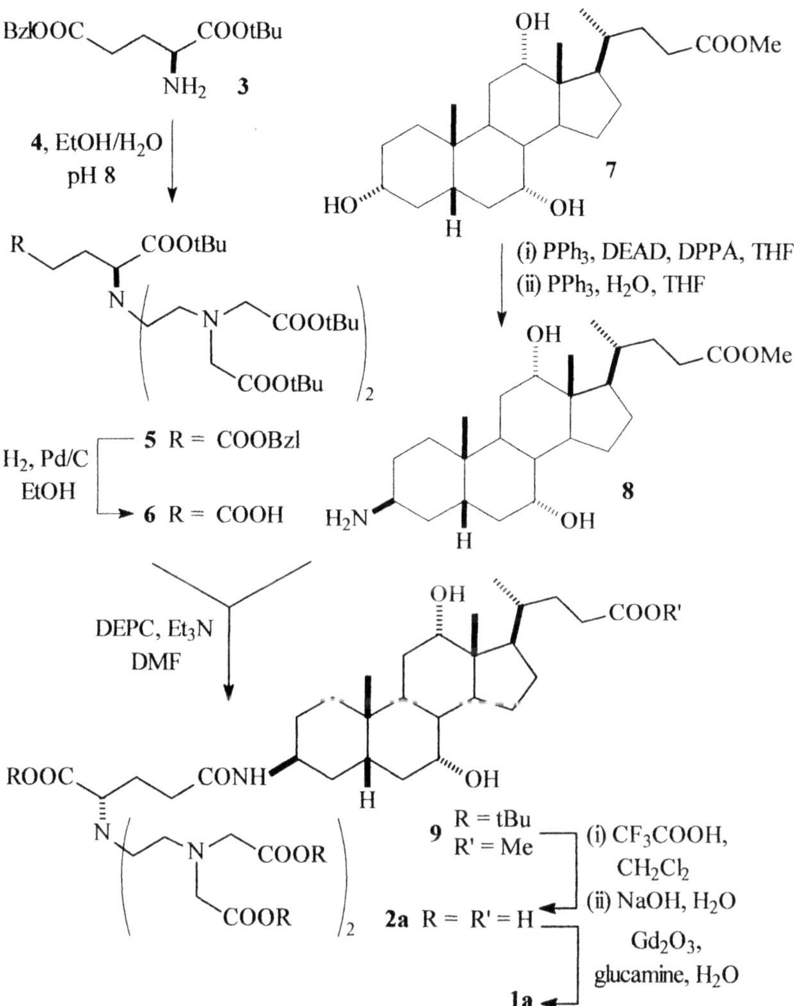

*Figure 1. Synthetic Route to **1a***

the methyl esters of lithocholic, deoxycholic and chenodeoxycholic acid, respectively. Noticeably, it was found that: *i*) the key condensation can be advantageously carried out replacing diethoxyphosphoryl cyanide with dicyclohexylcarbodiimide/hydroxybenzotriazole; *ii*) the final hydrolysis of the five *t*-butyl and of the methyl ester functions can be performed in a single step with a large excess of LiOH in dioxane/H_2O.

For the synthesis of **1e**, the intermediate **11** was prepared from the 3β-N_3 derivative **10** of methyl deoxycholate by oxidation of the hydroxy group with CrO_3 in H_2SO_4/acetone followed by Pd/C catalyzed hydrogenation of the azide moiety (Figure 2).

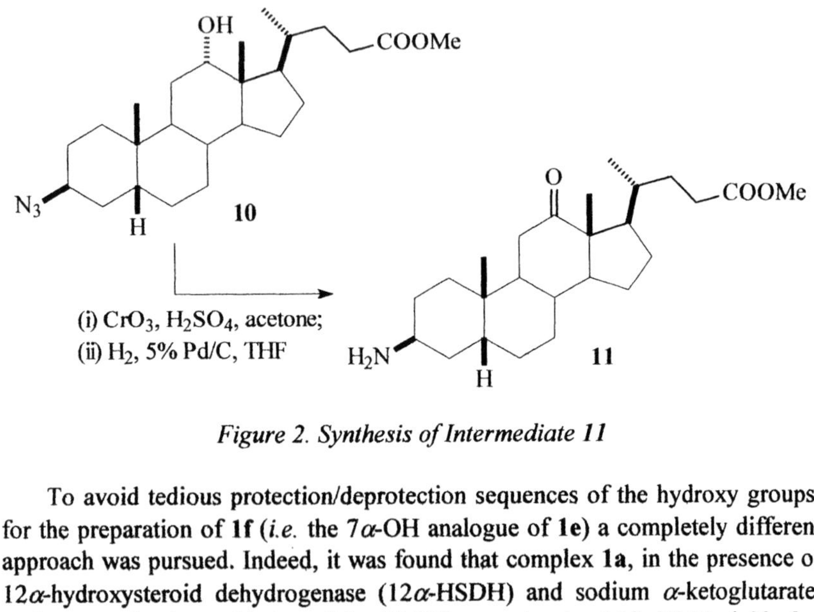

OH COOMe N3 H 10

(i) CrO_3, H_2SO_4, acetone;
(ii) H_2, 5% Pd/C, THF

O COOMe H2N H 11

Figure 2. Synthesis of Intermediate 11

To avoid tedious protection/deprotection sequences of the hydroxy groups, for the preparation of **1f** (*i.e.* the 7α-OH analogue of **1e**) a completely different approach was pursued. Indeed, it was found that complex **1a**, in the presence of 12α-hydroxysteroid dehydrogenase (12α-HSDH) and sodium α-ketoglutarate, undergoes selective oxidation of the 12-OH group to give **1f** in 72 % yield after purification (Figure 3) (*17*). In the light of this result, the enzymatic oxidation was successfully applied for an alternative synthesis of **1e** directly from **1c** in 75 % yield. The synthesis of **1g** was performed starting from **12** which was an intermediate already used for the preparation of **1b**.

The amino group of **12** was monoalkylated with *t*-butyl bromoacetate in dimethylformamide in the presence of diisopropylethylamine (DIEA) to afford **13** (Figure 4). The latter was coupled to **6** using (1-hydroxy-1*H*-benzotriazolato-*O*)tris(*N*-methylmethanaminato)phosphorous hexafluorophosphate (BOP) in dimethylformamide in the presence of diisopropylethylamine. Subsequently, the reaction path followed that of complexes **1b-d.**

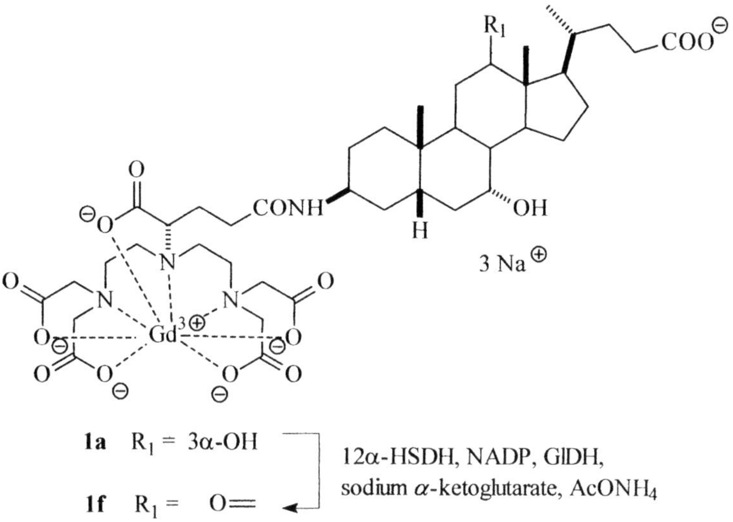
R1
COO⊖
O
⊖O
CONH
OH
H
3 Na⊕
N
O
N
N
O
Gd3⊕
O⊖
⊖O
O
O
O
O
⊖
⊖
1a R1 = 3α-OH
1f R1 = O=
12α-HSDH, NADP, GlDH,
sodium α-ketoglutarate, AcONH4

*Figure 3. Enzymatic Oxidation of **1a** to **1f***

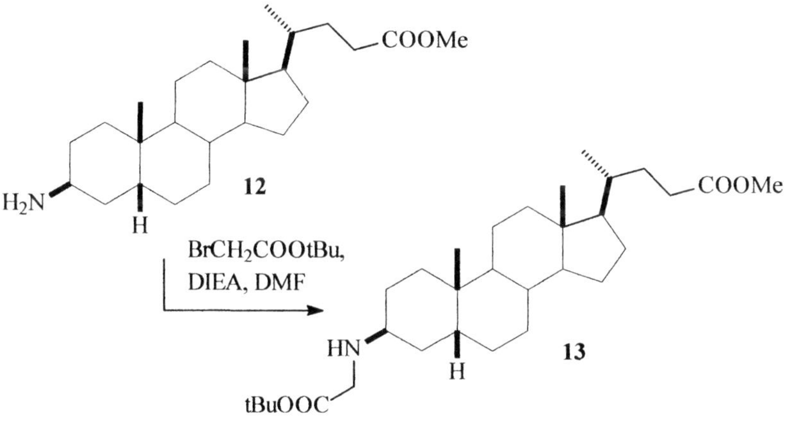
COOMe
12
H2N
H
BrCH2COOtBu,
DIEA, DMF
COOMe
HN
H
13
tBuOOC

Figure 4. Syntesis of Intermediate 13

Selection of the Clinical Candidate (*i.e.* 1c)

Preliminary screening of all complexes involved: *i*) determination of the HSA binding by means of ultrafiltration; *ii*) rough assessment of the *in-vivo* tolerability. Data for **1b-g** are reported in Table II in comparison with those of **1a** (*i.e.* our initial lead candidate). As far as tolerability concerns, for some of the complexes LD_{50} values are reported, whereas for other complexes mortality at a given dose level of administration is indicated. For the sake of comparison, in terms of safety index, for Gd-DTPA-dimeg (*i.e.* the most used ECF contrast agent currently approved for MRI procedures), which is usually administered to patients at 0.1 mmol/kg dose level, the LD_{50} value is 6 mmol/kg, in mice (*18*).

Table II. Preliminary Evaluation of Compounds 1a-g

Compound	*HSA Binding (%)*[a]	*LD_{50} (mmol/kg)*[b]
1a	72	7.6
1b	98	0.66
1c	94.5[c]	3.6[c]
1d	> 90[d]	3.0 (2/5)[e]
1e	91	3.0 (0/10)[e]
1f	57	5.0 (0/5)[e]
1g	> 90[d]	1.5 (8/10)[e]

[a] Determined by means of ultrafiltration. [b] In mice, after intravenous administration. [c] From ref. 20. [d] Binding was not more precisely assessed in the light of the tolerability data. [e] Single dose toxicity data reported in terms of dose level (dead animals/treated animals).

As anticipated, our aim was to increase the lipophilicity of complex **1a** to achieve a stronger binding to HSA. Indeed, with the lithocholic derivative **1b** binding to HSA remarkably increased to 98%. However, likely due to such strong binding, the tolerability (LD_{50} 0.6 mmol/kg, in mice) and the elimination of the complex proved definitively unacceptable.

Complexes characterized by an intermediate lipophilicity, like **1c** and **1d**, which derive from deoxycholic and chenodeoxycholic acid, respectively, showed HSA binding around 95% and very encouraging tolerability data.

Attempts to reduce the HSA binding extent of the lithocholic derivative **1b** introducing an additional charged residue in the structure led to complex **1g** which, however, proved only slightly more tolerable than **1b**.

Complex **1e** in which the 12-OH residue of the steroidic moiety has been converted into a ketone is characterized by a good tolerability but a lower HSA binding in comparison with the hydroxylated precursor **1c**.

A remarkable weakening of the binding to HSA as a consequence of the conversion of the 12-OH residue of **1a** into a ketone can also be noticed for complex **1f** in comparison with its hydroxylated precursor **1a**.

On the basis of the promising HSA binding and tolerability results, the properties of complex **1c** were furtherly investigated. The strong binding is also responsible for the relatively high relaxivity in human serum (~ 27 $mM^{-1}s^{-1}$). Under the experimental conditions used for compounds **1a-g**, MS-325 showed a relaxivity of ~ 35 $mM^{-1}s^{-1}$ but binding to HSA was reduced to 80% (*19*). As expected, studies carried out in rats and monkeys showed that, due to the high HSA binding, **1c** features a long plasma half-life if compared to that of routine ECF agents. Nonetheless, the elimination in rats of **1c**, which mainly occurs through the biliary route, accounted for more than 82% within 24 h and for 94% after 7 days. In the same study, biodistribution confirmed that there was no accumulation of gadolinium in any organs (*20*). Since size and anatomy of the coronary arteries of pigs are very similar to those of humans, imaging studies were carried out in micropigs. Such studies showed that administration of **1c** at 0.1 mmol kg^{-1} dose level could guarantee a sufficient signal enhancement of blood in the coronary arteries for a suitable time (*19, 20, 21*).

On these grounds, **1c** was selected as the candidate for clinical development as a contrast agent for MRI coronary angiography. Preliminary results on the first administration to humans have been reported (*22*).

Process Research and Development of 1c

Once complex **1c** was selected for clinical development, a synthetic route amenable to scale up was required. The research route which, as already discussed, roughly follows that of complex **1a** (see Figure 1) was critically analyzed. A not so appealing scenario appeared: *i*) the overall yield was very low (*i.e.* 6% from L-glutamic acid); *ii*) a large number of chemical steps (*i.e.* 12) and several isolated intermediates (*i.e.* 10), most of which were purified by silica gel chromatography, were involved. Although the latter issues are ordinary life in process research and development, the low overall yield deserves some further comments.

The monoacid pentaester **6** is a versatile intermediate, nearly perfect for the synthesis of a series of compounds like those above described. However, its synthesis (*14*) largely accounts for the low yield observed in the preparation of **1c**. Furthermore, the conversion of methyl deoxycholate into the corresponding 3β-NH_2 derivative **14** (see Figure 5; Research route) involves the intermediate production of the 3β-N_3 derivative **15.** Although the latter proved a perfectly safe compound, we must underline that: *i*) the reagent (*i.e.* diphenylphosphoryl azide) which is used in the Mitsunobu reaction is extremely expensive; *ii*) alternative approaches to generate the azide by phase transfer catalysed nucleophilic substitution on the 3-mesyloxy derivative had to rely upon the hazardous NaN_3. Therefore, a new route to derivative **14** was devised.

Again taking advantage of a Mitsunobu reaction, methyl deoxycholate was converted into the corresponding 3*β*-phthalimido derivative **16**. Subsequently, the phthalic residue was removed in two steps using $NaBH_4$ in dimethylacetamide/MeOH at first and then HCl in MeOH to give **14** (Figure 5; Development route). Additionally, during development the synthesis of the ligand precursor of **1c** starting from **14** was largely changed.

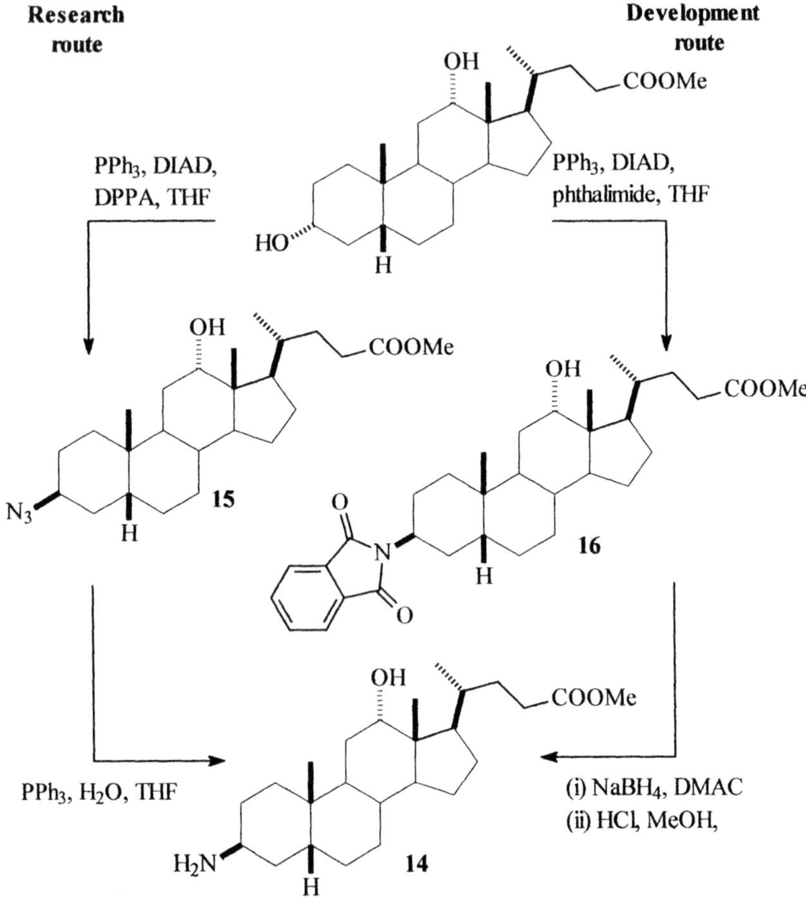

*Figure 5. Synthesis of Intermediate **14** Research vs Development Route*

Looking at Figure 6, it is easy to recognize that in the research route the disconnection approach involves first formation of bonds **a** and then bond **b**. To circumvent the drawbacks linked to the use of **6**, the disconnection was turned the other way around. Accordingly, the development route entails first the formation of bond **b** and then bonds **a**. The success of this approach was heavily dependent on the choice of a suitably protected L-glutamic acid derivative of some sort. Methyl *N*-Boc-L-pyroglutamate, **17** (see Figure 7), proved the ideal intermediate in this respect. Indeed, this choice came out of a massive work during which Cbz protected derivatives as well as different esters of L-pyroglutamic acid were screened. The final route to **1c**, which led to a successful pilot plant campaign, is depicted in Figure 7. The key step is the condensation between derivatives **14** and **17** to afford **18** which occurs in good yield in the absence of any catalyst. The Boc protecting group of **18** is then removed under acidic conditions to give **19**.

At this stage, according to a Rapoport-like approach, **19** could have been alkylated with the bromo derivative **4** (see Figure 1). However, we found much more convenient to *in-situ* generate the mesyl derivative of hydroxyethyl derivative **20** and directly proceed to the double alkylation of **19** in *n*-butyl acetate. The thus obtained hexaester **21** is deprotected to ligand **2c** in the presence of a large excess of NaOH in *i*-PrOH/H_2O. Eventually, the ligand is complexed with Gd_2O_3 to give the gadolinium complex **1c**. It is noteworthy that hexaester **21** differs from the analogous intermediate according to the research route, for the presence of a methyl instead of a *t*-butyl ester on the central acetic residue of the DTPA moiety. This proved a noticeable advantage in order to minimize racemization during basic hydrolysis of **21** to **2c**.

According to the synthetic route optimized during scale up, the number of isolated intermediates dropped to 6 and all silica gel chromatographies were eliminated. The overall yield for the new process is roughly 30% (from L-glutamic acid).

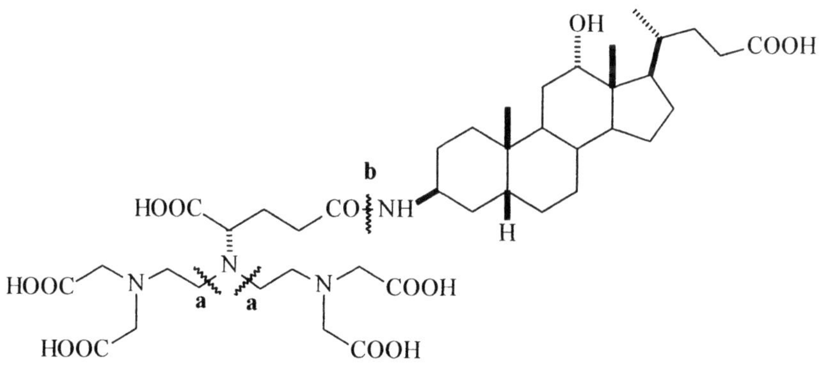

Figure 6. Disconnection Approach to 2c

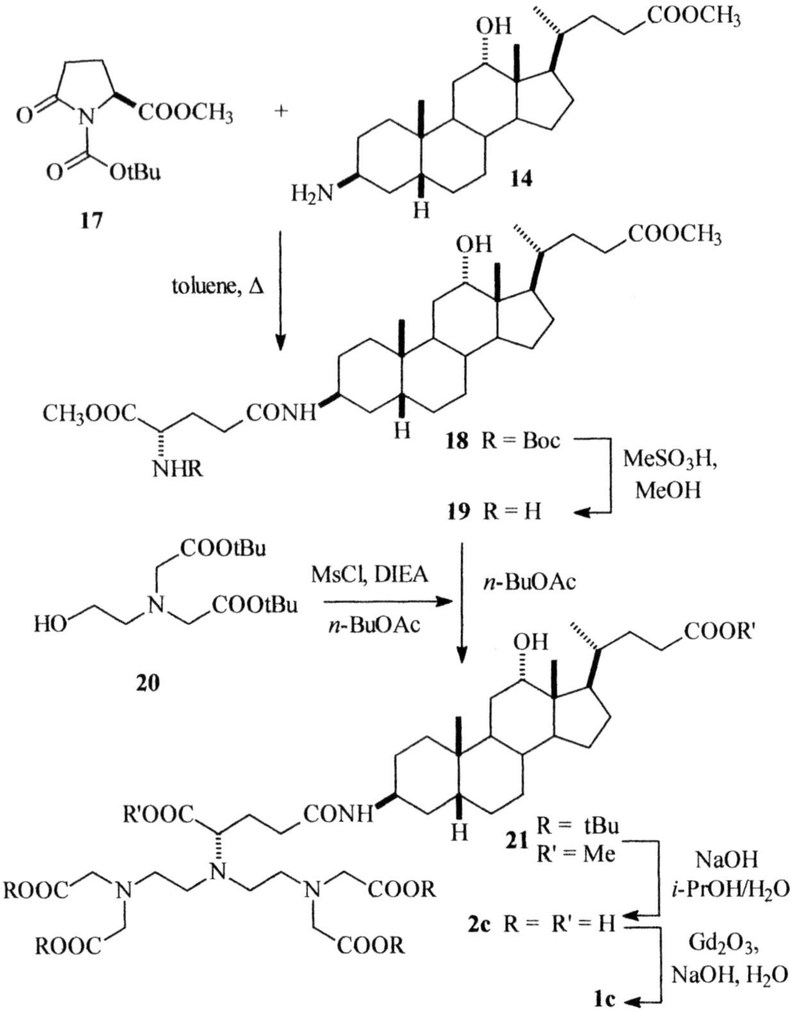

$COOCH_3$ OH $COOCH_3$

O N $COOCH_3$ +

O OtBu H_2N H **14**

17

OH $COOCH_3$

toluene, Δ

CH_3OOC CONH H **18** R = Boc $MeSO_3H$, MeOH

NHR

19 R = H

COOtBu

MsCl, DIEA *n*-BuOAc

HO N COOtBu *n*-BuOAc

20 OH COOR'

R'OOC CONH H **21** R = tBu R' = Me NaOH *i*-PrOH/H_2O

ROOC N N N COOR

ROOC COOR **2c** R = R' = H Gd_2O_3, NaOH, H_2O

1c

Figure 7. Syntesis of 1c: Final Route

Conclusions

A series of conjugates of Gd(III) complexes with bile acids were synthetised and investigated as MRI intravascular contrast agents. Complex **1c**, containing a subunit of deoxycholic acid, was selected as a clinical candidate for development. The synthetic route to **1c,** which had been used during the research phase, was completely changed during scale up to pilot plant in order to achieve an industrially appealing route. The compound is presently undergoing Phase IIa clinical trials.

References

1. *The Chemistry of Contrast Agents in Medical Magnetic Resonance Imaging.* Merbach, A.E.; Toth, E. Eds.; John Wiley & sons: Chichester, England, 2001.
2. Dawson, P. In *Textbook of Contrast Media,* Dawson, P.; Crosgrove, D. O.; Grainger, R. G. Eds; Martin Dunitz: London, 1999; pp 319-321.
3. Li, D.; Zheng, J.; Bae, K. T.; Woodward, P. K.; Haacke E. M. *Invest. Radiol.* **1998,** *33*, 578-586.
4. Anelli, P. L.; Lattuada, L.; Lorusso, V.; Schneider, M.; Tournier, H.; Uggeri, F. *MAGMA* **2001,** *12,* 114-120.
5. Brasch, R. *Magn. Reson. Med.* **1991,** *22*, 282-287.
6. Vexler, V.; Clement, O.; Schmitt-Willich, H.; Brasch, R. *J. Magn. Reson. Imaging* **1993,** *4,* 381-388.
7. Misselwitz, B.; Schmitt-Willich, H.; Ebert, W.; Frenzel, T.; Weinmann H.-J. *MAGMA* **2001**, *12,* 128-131.
8. Port, M.; Corot, C.; Rousseaux, O.; Devoldere, L.; Idee, J. M.; Dencausse, A.; Le Graneur, S.; Simonot, C.; Meyer, D. *MAGMA* **2001,** *12,* 121-127.
9. Caravan, P.; Cloutier, N. J.; Greenfield, M. T.; McDermid, S. A.; Dunham, S. U.; Bulte, J. W. M.; Amedio, Jr., J. C.; Looby, R. J.; Supkowski, R. M.; Horrocks, Jr., W. DeW.; McMurry, T. J.; Lauffer, R. B. *J. Am Chem. Soc.* **2002,** *124,* 3152-3162.
10. Anelli, P. L.; Lolli, M.; Fedeli, F.; Virtuani, M. U.S. Patent 6,458,337, 2002.
11. Calabi, L.; Maiocchi, A.; Lolli, M.; Rebasti, F. U.S. Patent 6,403,055, 2002.
12. Anelli, P. L.; Calabi, L.; de Haën, C.; Lattuada, L. ; Lorusso, V.; Maiocchi, A.; Morosini, P.; Uggeri, F. *Acta Radiol.* **1997,** *S142*, 125-133.
13. Roda, A.; Cappelleri, G.; Aldini, R.; Roda, E.; Barbara, L. *J. Lipid Res.* **1982,** *23,* 490-495.
14. Anelli, P. L.; Fedeli, F.; Gazzotti, O.; Lattuada, L.; Lux, G.; Rebasti F. *Bioconjugate Chem.* **1999,** *10,* 137-140.
15. Williams, M. A.; Rapoport, H. *J. Org. Chem.* **1993,** *58,* 1151-1158.
16. Anelli, P. L.; Lattuada, L.; Uggeri, F. *Synth. Commun.* **1998,** *28,* 109-117.

17. Anelli, P. L.; Brocchetta, M.; Morosini, P.; Palano, D.; Carrea, G.; Falcone, L.; Pasta, P.; Sartore, D. *Biocatal. Biotransform.* **2002,** *20,* 29-34.
18. Weinmann, H.-J.; Press, W.-R.; Gries, H. *Invest. Radiol.* **1990,** *25,* S49-S50.
19. Cavagna, F. M.; Anelli, P. L.; Lorusso, V.; Maggioni, F.; Zheng, J.; Li, D.; Abendschein, D. R.; Finn, P. J. *Proceedings,* ISMRM-ESMRMB Joint Annual Meeting, Glasgow, Scotland, 2001; p. 519.
20. de Haën, C.; Anelli, P. L.; Lorusso, V.; Morisetti, A.; Maggioni, F.; Uggeri, F.; Cavagna, F. M. *Invest. Radiol.* submitted.
21. Cavagna, F. M.; Lorusso, V.; Anelli, P. L.; Maggioni, F.; de Haën, C. *Acad. Radiol.* **2002,** *9 (Suppl. 2),* S491-S494.
22. La Noce, A.; Stoelben, S.; Scheffler, K.; Hennig, J.; Lenz H. M.; La Ferla R.; Lorusso, V.; Maggioni, F.; Cavagna, F. M. *Acad. Radiol.* **2002,** *9 (Suppl. 2),* S404-S406.

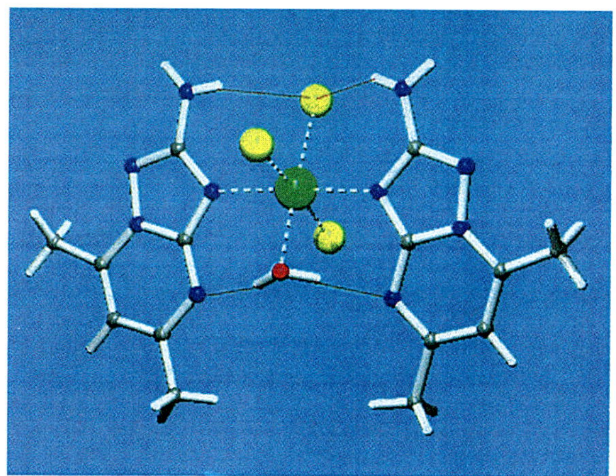

Figure 6: Molecular structure of a NAMI derivative-Ru(III) adduct (XRD) of formula $RuCl_3(H_2O)(L_2)$ containing a model purine nucleobase; the important of intramolecular hydrogen bonding stabilization is indicated [79].

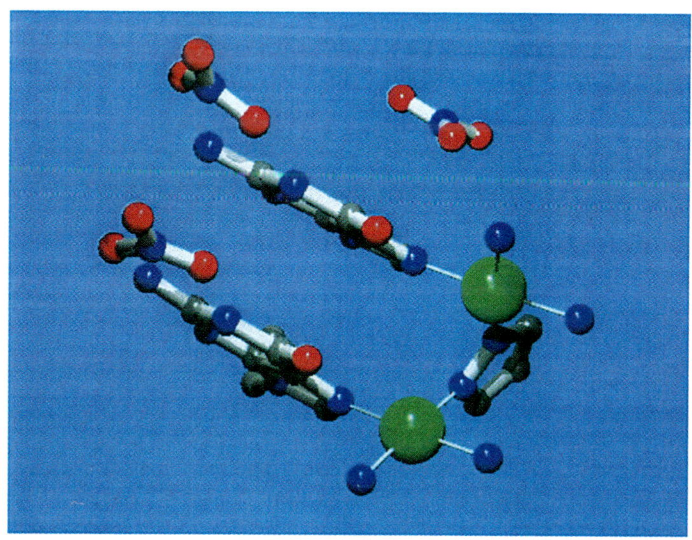

Figure 11: Two model bases (9-Etgua) can remain parallel when coordinated to the azole-bridged dinuclear Pt unit [116].

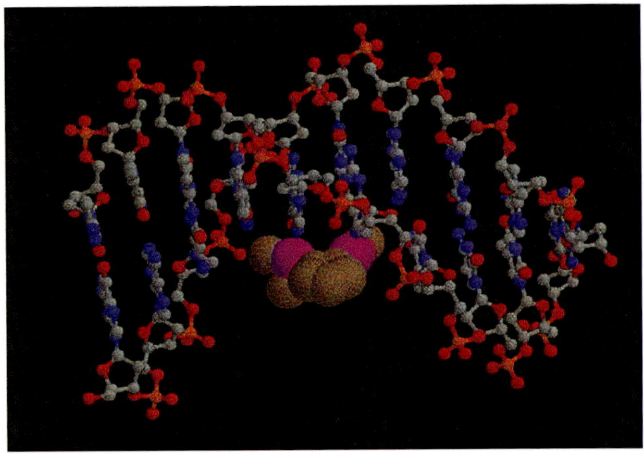

Figure 12: A double-stranded DNA remains almost unkinked when the dinuclear azole-bridged Pt amine species is chelating to two neighbouring guanines [122].

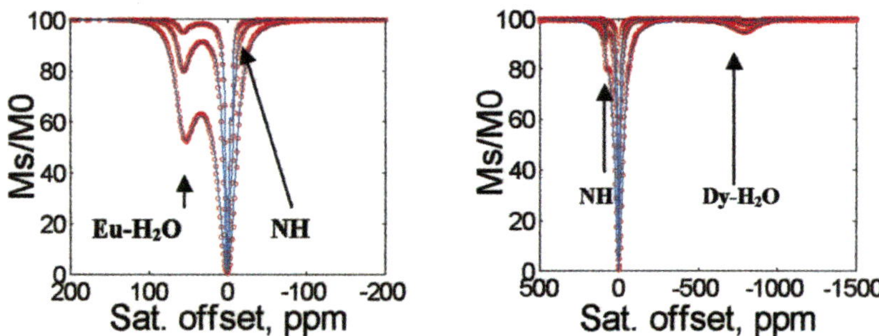

Figure 2. *Z-spectral data (individual points) and best fit simulations to Bloch theory (solid lines) for* ***10 mM*** *Eu(**10**)$^{3+}$ at B_1 values of 229, 505 and 1020 Hz (left, in descending order in the graph) and* ***1 mM*** *Dy(**10**)$^{3+}$ at B_1 values of 505, 1020 and 2000 Hz (right (see ligand structure in Figure 4).* [36]

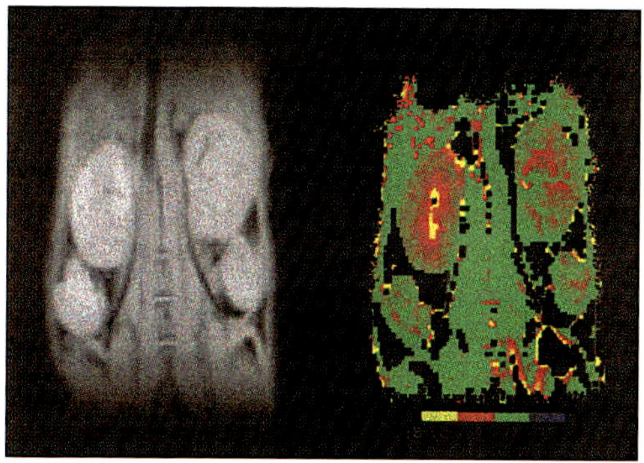

Figure 5. *Standard T_1 weighted MRI of a mouse (left) and its pH map (right) derived by comparing clearance of GdDOTA-4AmP^{5-} versus GdDOTP^{5-}. The color code shows the measured pH gradient across the kidney.*[45]

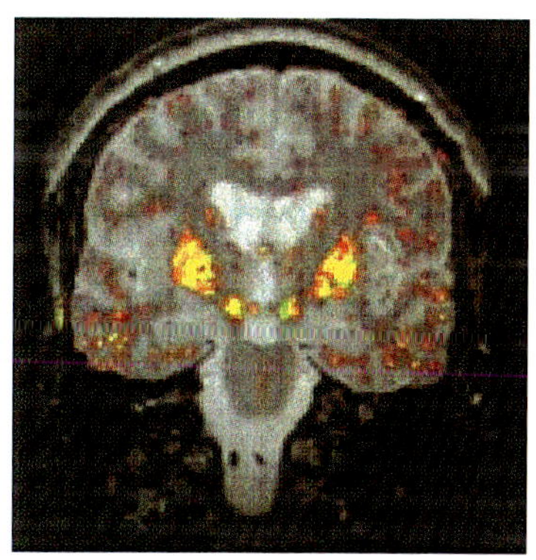

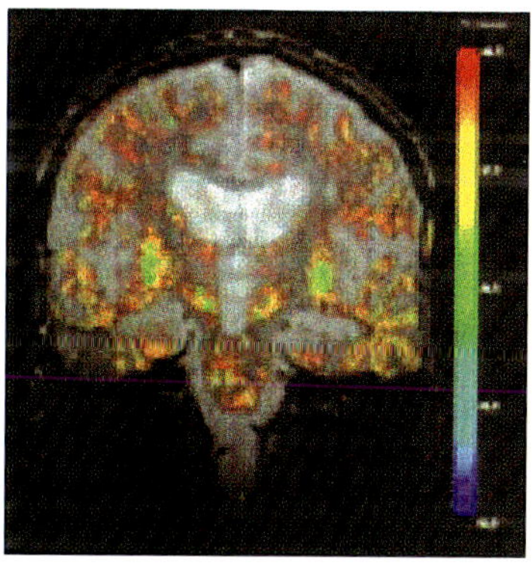

Figure 3. *Non-targeted molecular imaging with MRI: brain iron distribution as a potential biomarker for neurodegenerative disease. Thin-slice, T2-weighted coronal images demonstrate increased distribution of pixels with short effective T2 (Fe content) in brain of dementia patient as compared to normal volunteer. Work of Dr. D. Alsop, Beth Israel Deaconess Medical Center and Drs. J. Schenck and A. Alyassin, GE Global Research Center; used with permission.*

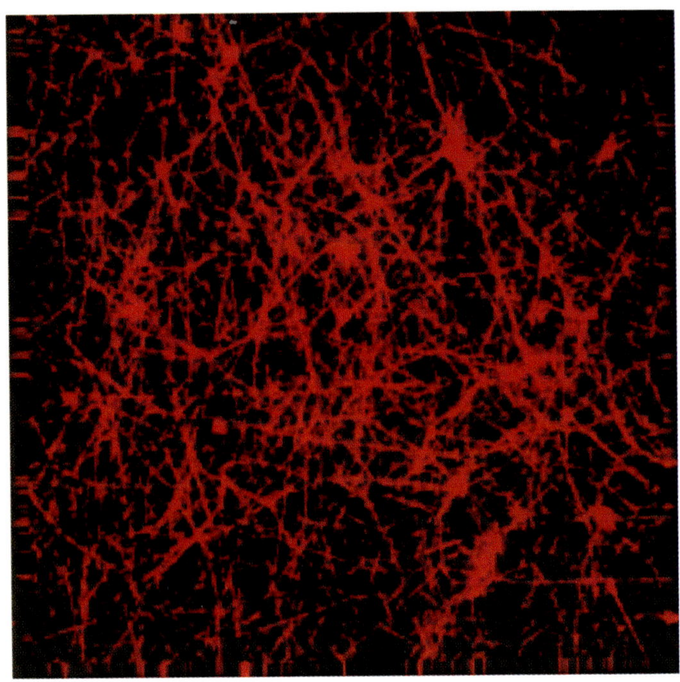

Figure 10. *Confocal fluorescence microscopy image of a human platelet rich plasma diluted 1:5 and clotted with 10 μg/mL of thrombin. After 2.5 minutes, the TMR-labeled petide is added and the clot is imaged. The fibrin network is apparent in the image. Image courtesy of Prof. P. Janmey, University of Pennsylvania.*

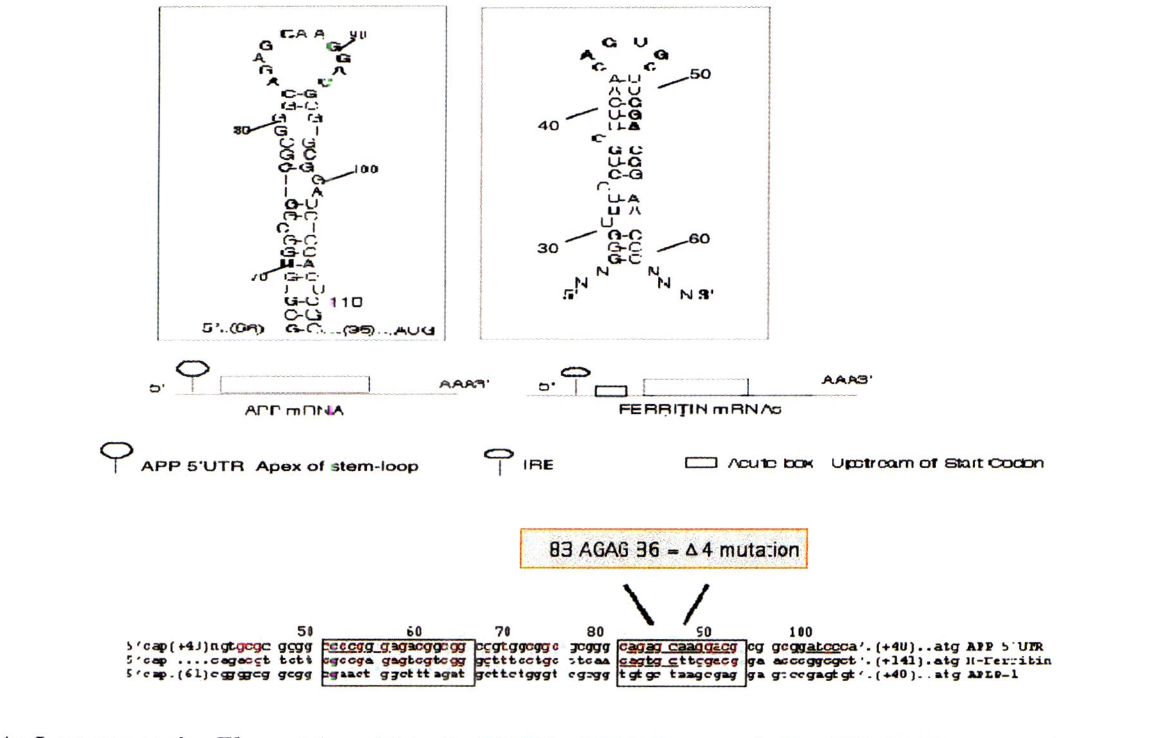

Figure 13.3. ***An Iron-responsive Element (type II) in the 5′UTR mRNA. Top panel:*** *The APR 5′UTR was predicted to fold into the stable RNA stemloop similar to the 5′UTR-specific IRE in H-ferritin transcript (′′APR 5′UTR′′ ΔG = –54 kCal/mol) (Zuker et al., 1989).* ***Bottom panel:*** *The APP mRNA IRE (type II) was identified after alignment with the ferritin Iron-responsive Element (shown in two clusters of > 70% sequence similarity). (Adapted from Rogers et al., 2002.)*

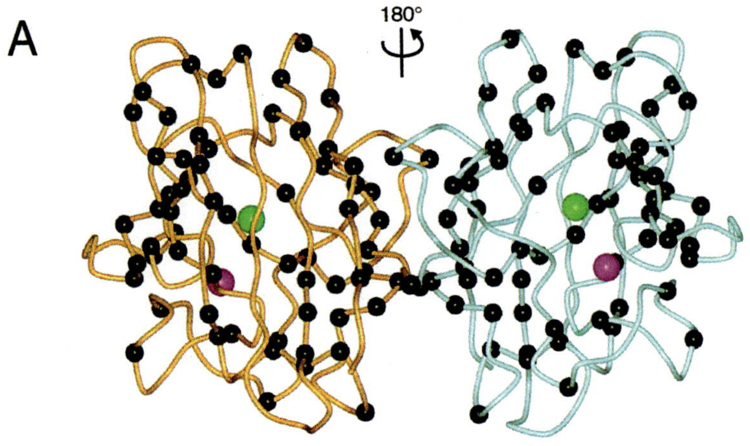

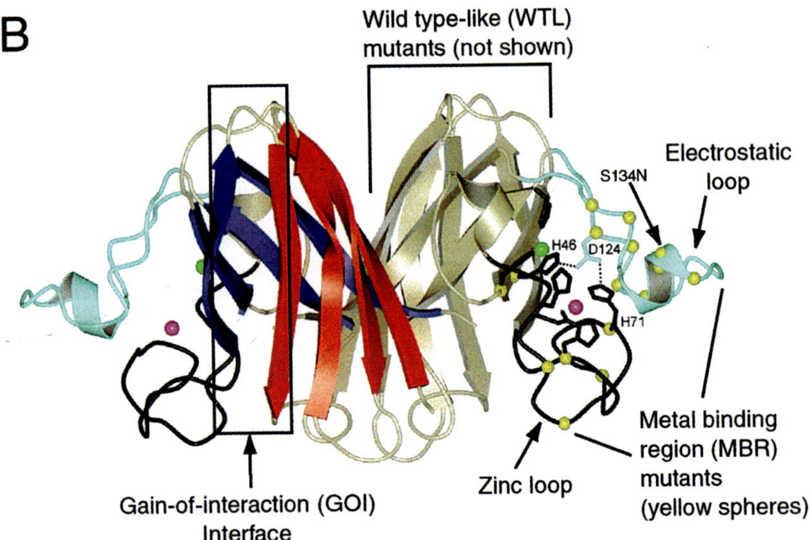

Plate 1. A. Human SOD1 showing the locations of the FALS mutations.
(See page 350 for full figure caption.)

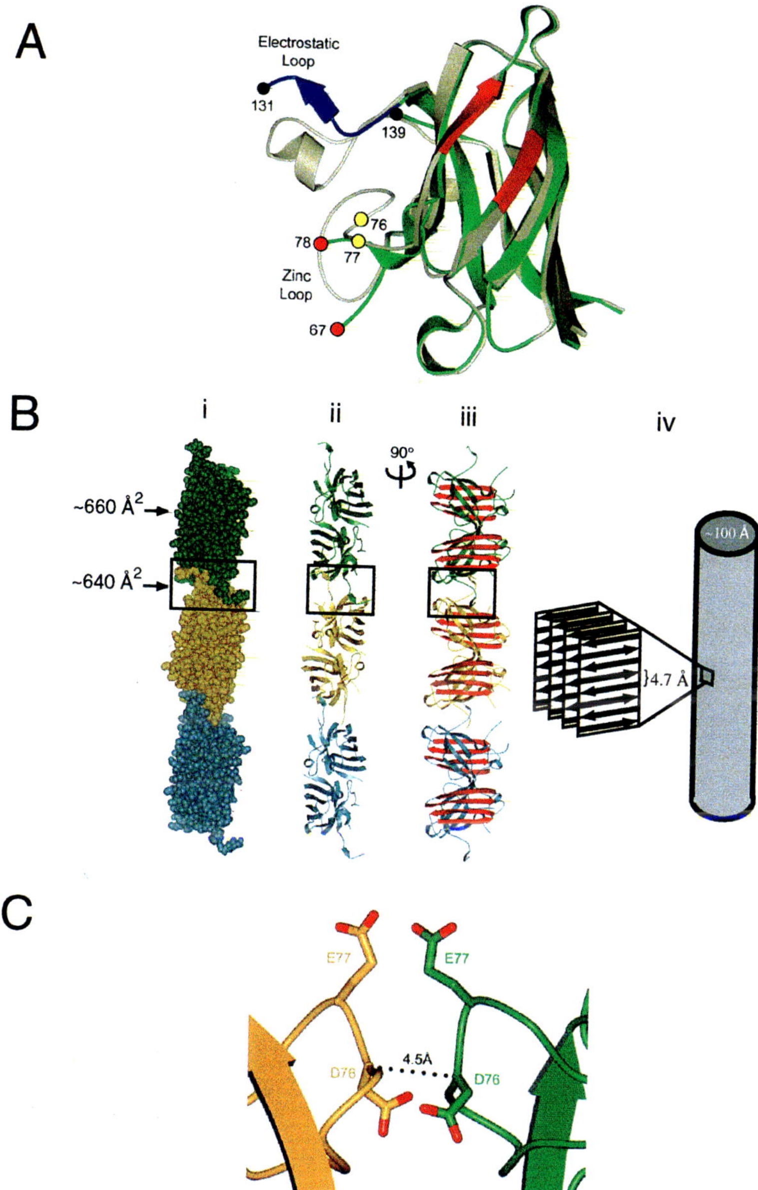

Plate 2. (See page 358 for full figure caption.)

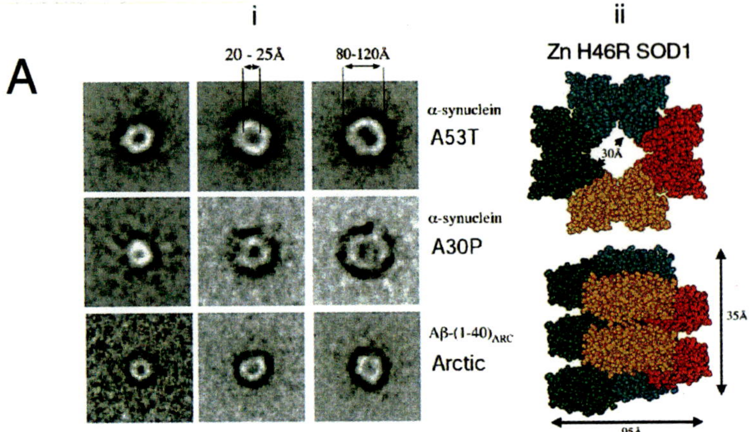

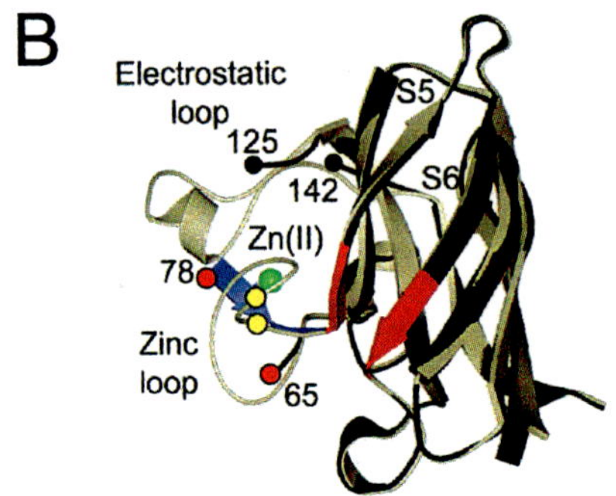

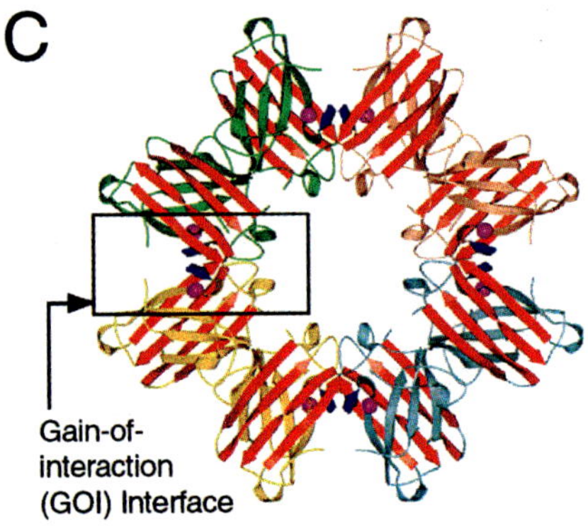

Plate 3. (See page 360 for full figure caption.)

Chapter 10

Water Exchange Is the Key Parameter in the Design of Next-Generation MRI Agents

A. Dean Sherry[1,2], Shanrong Zhang[2], and Mark Woods[1]

[1]Department of Chemistry, University of Texas at Dallas, Richardson, TX 75083–0688
[2]Department of Radiology, University of Texas Southwestern Medical Center, 5323 Harry Hines Boulevard, Dallas, TX 75390–9085

Abstract: Water exchange in lanthanide complexes is the key parameter one must consider in the design of new MRI agents. Fast water exchange is necessary for targeted Gd^{3+}-based agents while slow water exchange is important for novel agents that provide MR contrast through a chemical exchange saturation transfer mechanism (CEST). The paramagnetic CEST agents platform is shown to be particularly versatile for imaging physiologic or metabolic parameters.

Introduction

Scientific interest in lanthanide ion coordination chemistry has grown substantially over the past 25 years partially driven by the emergence of magnetic resonance imaging (MRI) as a preeminent tool in clinical medicine.[1] Complexes of Gd^{3+} are most widely used as "contrast agents" for MRI because they are most efficient at relaxing bulk water protons due to the high spin state, long electronic relaxation time, and fast water exchange characteristics of Gd^{3+}.[2, 3] The efficiency of a paramagnetic relaxation agent is usually reported in terms of relaxivity ($mM^{-1}s^{-1}$), the water proton relaxation rate ($1/T_1$)

catalyzed by the agent normalized to 1 mM concentration. Currently, all agents in clinical use have relatively low relaxivities (~4 $mM^{-1}s^{-1}$) compared to that of theoretical prediction (~140 $mM^{-1}s^{-1}$).[4] Optimizing the parameters necessary to achieve this theoretical limit has been the goal of much research in recent years.

The other paramagnetic lanthanide ions are less efficient at relaxing water protons but have other useful magnetic characteristics such as the ability to induce sizable NMR hyperfine shifts.[5] Hyperfine shifts can provide useful quantitative structural information or may simply be used to "resolve" resonances of NMR active nuclei in biological compartments. This feature has been particularly attractive for separating the NMR signals of intra- and extracellular $^{23}Na^+$ in isolated cells, organs and tissues *in vivo*.[6, 7] The hyperfine shifting ions have not been applied to MRI because, unlike Gd^{3+}-based T_1 agents where rapid water exchange is considered imperative, fast water exchange effectively reduces water proton hyperfine shifts to near zero (weighted average of a small shifted pool and a large bulk pool). Recently, lanthanide complexes with surprising slow water exchange characteristics were reported[8-16] and in some of these systems highly hyperfine shifted Ln^{3+}-bound water resonances can be detected by high resolution ^1H NMR (Table 1). As we shall see, this feature offers a novel way to introduce contrast into an image.[17] In this Chapter, we will summarize 1) ways to increase water exchange in Gd^{3+} complexes and 2) a new type of contrast agent that takes advantage of hyperfine shifted water resonances in slow water exchange complexes.

Table 1. Bound water ^1H chemical shifts (ppm) of paramagnetic LnDOTA-4AmCE^{3+} complexes in acetonitrile at room temperature. [18]

Ln^{3+}	Pr^{3+}	Nd^{3+}	Sm^{3+}	Eu^{3+}	Tb^{3+}	Dy^{3+}	Ho^{3+}	Er^{3+}	Tm^{3+}	Yb^{3+}
$\delta(^1H)$	-60	-32	-4	50	-600	-720	-360	200	500	200

Optimizing water exchange in Gd^{3+} complexes

As the aqua gadolinium ion is quite toxic, clinical MRI agents must be administered in the form of a kinetically and thermodynamically robust chelate. Octadentate polyaminocarboxylate ligands such as DTPA and DOTA have proven particularly effective for this purpose. DOTA effectively sandwiches the central gadolinium ion between the four coplanar nitrogen atoms of the cyclic amine and the four coplanar oxygen atoms of the pendant arms. A water molecule occupying the apical position completes the coordination sphere. The torsion angle between the square defined by the nitrogen atoms and the square defined by coordinating oxygen atoms defines the coordination geometry of the complex. A larger torsion angle, ~39°, defines a capped square antiprismatic (SAP) geometry whereas a smaller torsion angle, ~29°, and an opposite twist angle defines a capped twisted square antiprismatic (TSAP) geometry.[19] In solution, these two coordination isomers are in dynamic equilibrium, interconverting either *via* inversion of the macrocyclic ring conformation or rotation of the oxygen pendant arms relative to the square defining the nitrogen atoms.[20] Since the conformation of each ethylene bridge in the macrocycle maybe defined as λ or δ (according to the sign of the torsion angle) the conformation of the ring in $LnDOTA^-$ complexes may be designated either (δδδδ) or (λλλλ). The orientation of the pendant arms (*i.e.* the O-Gd-N torsion angle) may also be designated as Δ or Λ so the four stereoisomeric coordination geometries related as two enantiomeric pairs are possible: Δ(λλλλ), Λ(δδδδ) (SAP structures) and Δ(δδδδ), Λ(λλλλ) (TSAP structures). Sequential arm rotation and ring flip interconverts enantiomers. Recent studies have shown that the rate of water exchange in these two coordination isomers may differ by as much as two orders of magnitude. The bound water lifetime (τ_M) as a weighted average of the two coordination isomers of $GdDOTA^-$ has been estimated at 244 ns[21] by NMR. This is sufficiently short for $GdDOTA^-$ to be used as an extracellular MRI agent but too long for use in the design of higher relaxivity targeted agents based upon the DOTA structure.

Consequently, there has been considerable effort made toward understanding the physical parameters that determine the water exchange rate in this and other Gd^{3+} complexes in hopes that new complexes with optimal exchange rates can be designed that will yield much higher relaxivity agents when bound to a biological target. A novel approach to this problem has been implemented by Raymond and coworkers. In contrast to the usual octadentate polyamino carboxylate ligands, they reported that TREN-3,2-Me-HOPO, a hexadentate ligand, forms stable complexes with Gd^{3+} and the resulting complexes have two inner sphere exchanging water molecules.[22] Unfortunately the complex has extremely poor solubility in water which has made it difficult to evaluate as a contrast agent. However, addition of a teraphthalamide (TAM) group with an appended polyethylene glycol chain imparts sufficient water solubility to allow further characterization. Longer polyethylene glycol chains containing 44 or 123 oxygen atoms displace one of the inner sphere coordinated water molecules. Of most interest however was the finding that the single water molecule in these $q = 1$ complexes have inner-sphere bound lifetimes précising in the range (τ_M = 19 - 31 ns) for achieving optimal relaxivity upon slowing rotation.[22] It was also reported that a shorter PEG chain having 3 oxygens does not displace either of the two inner sphere water molecules and that these also have an optimal lifetime (τ_M = 19 ns) and exchange *via* an associative mechanism.[22]

A second approach to achieving faster water exchange Gd^{3+} complexes became evident after it was learned that an α-substitution on the acetate arms of DOTA could lock the conformation of the pendant arms in complexes[23] and that the orientation of the arms is determined by the configuration at the α-carbon, an *R* configuration confers a Λ orientation while *S* gives rise to a Δ orientation. It was also known that substitution on one ethylene position of a cyclen ring slows interconversion of the macrocyclic conformers.[24, 25] Thus, by adding appropriate substitutions on both the pendant arms and the macrocycle it should be possible to selectively lock the complex into one or other coordination geometry. A TSAP geometry may be selected by employing the same configuration at the macrocycle and the pendant arm whereas a SAP geometry may be obtained if the configurations are opposing.

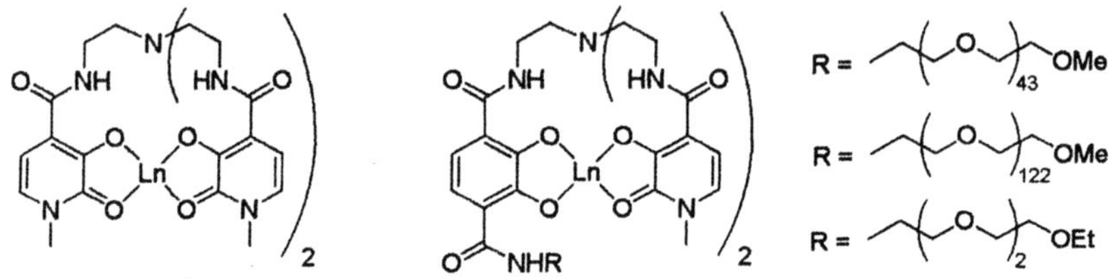

GdTREN-3,2-HOPO

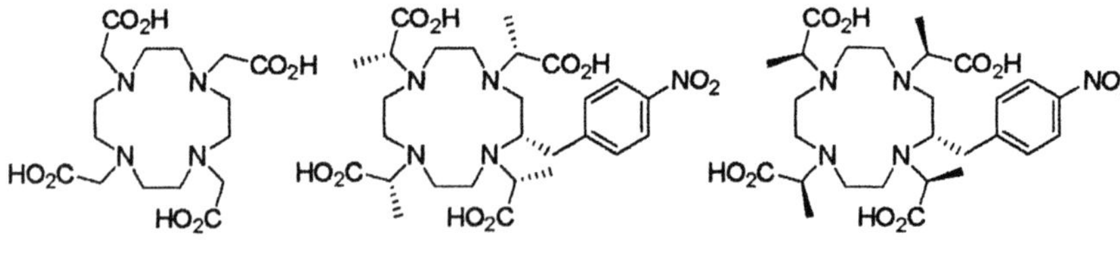

DOTA   *S-RRRR*-1   *S-SSSS*-1

Hence, the two stereoisomers of 2-(nitrobenzyl)-DOTMA shown above were synthesized and studied.[26] High resolution proton NMR spectra of the europium complexes showed quite clearly that Eu(*S-RRRR*-**1**)$^-$ exists in solution as a single SAP coordination isomer while Eu(*S-SSSS*-**1**)$^-$ is in the form of a TSAP coordination isomer. Both isomers were found to have a single inner-sphere water molecule by fluorescence spectroscopy and the corresponding Gd(*S-RRRR*-**1**)$^-$ and Gd(*S-SSSS*-**1**)$^-$ complexes had τ_M values that differed by about one order of magnitude. The TSAP isomer had a bound water lifetime of 15 ns, a value that is again close to optimal for producing high relaxivity complexes once bound to a biological target.

Using hyperfine shifting Ln^{3+} cations as contrast agents

An alternative way to produce contrast in an image would be to alter the total water signal detected in the experiment. This was demonstrated by Balaban and coworkers[27] when they used a presaturation pulse to saturate the broad water signal that lies beneath the sharper bulk water signal in many tissues. This broad resonance is thought to reflect water tightly associated with tissue proteins, lipids, and/or subcellular structures and appears to change with progressive diseases such as multiple sclerosis, ALS, *etc.* Upon saturation, exchange occurs between the bound water pool and the larger bulk water pool resulting in a decrease in bulk water signal intensity. This technique is now commonly known as magnetization transfer (MT) imaging.[28] More recently, this same group demonstrated that low molecular weight compounds with slowly exchanging –NH or –OH protons may also be used to alter tissue contrast *via* chemical exchange saturation transfer (CEST) of pre-saturated spins to bulk water.[29] Van Zijl and coworkers have demonstrated that the CEST effect can be amplified considerably by using macromolecules with a large number of chemically equivalent, exchangeable NH groups[30] and that the amide protons of intracellular proteins may be used in a CEST experiment to image tissue pH.[31] In all early reports (endogenous CEST agents or intrinsic protein exchanging sites), the molecules first used to demonstrate the CEST effect were diamagnetic with proton chemical shifts close to that of bulk water (-NH and –OH groups are typically within 5 ppm of bulk water). Paramagnetic analogs of such compounds were not known until the discovery of lanthanide complexes that have unusually slow water exchange kinetics.[17, 32]

In the simplest CEST experiment, one applies frequency-selective RF irradiation on the bound pool until all nuclear magnetizations at that frequency are at steady-state and then samples any remaining Z magnetization of the bulk water protons by applying a non-selective sampling RF pulse. Experimentally, this can be accessed from a plot of residual bulk water proton Z magnetization *versus* the frequency offset of the saturation pulse varied over a range frequencies that includes the Larmor frequencies of both pools of protons. The resulting plot is referred to as a Z-spectrum[33] or a CEST spectrum.[29] The physical features of paramagnetic lanthanide complexes that yield an optimal CEST effect (concentration, bound water lifetimes, hyperfine shifts, T_1, and T_2, offset frequency, strength of RF irradiation) have been described by the Bloch equations modified for two site chemical exchange.[34] Such complexes are now referred to as PARACEST agents.[34]

Figure 1 illustrates the large influence of the relaxation times of pool A (bulk water) on theoretical Z-spectra.[35] The top panel shows that increasing

T_{1a} has a large influence on the CEST efficiency (the residual intensity at the bound water pool, B) of a PARACEST agent with a hyperfine shifted water peak at 50 ppm. The bottom panel illustrates that changes in T_{2a} of bulk water have little effect on the CEST efficiency of this same agent in the region of 50 ppm, but certainly affects the region between the two water resonances. This illustrates the importance of PARACEST agents that induce large hyperfine shifts (large $\Delta\omega$) so that off-resonance "indirect" saturation of bulk water (pool A) does not contribute to the CEST spectrum.

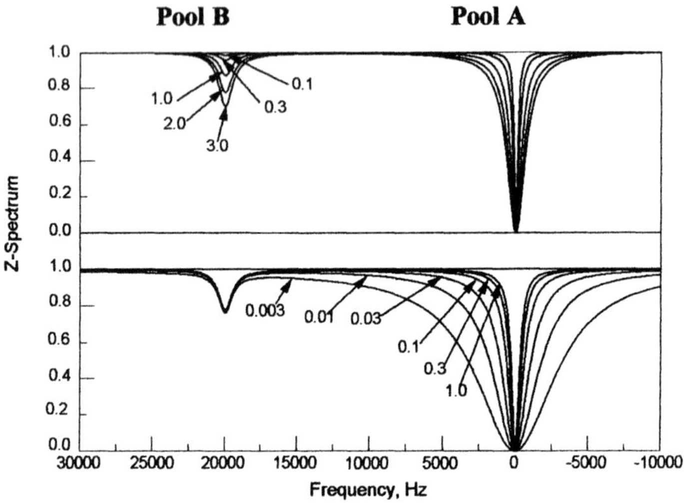

Figure 1. *Simulated Z-spectra of a two-pool exchange system seprated by 50 ppm illustrating the effect of relaxation times upon CEST efficiency. In the upper panel T_{2a} is fixed at 0.1 s and T_{1a} is varied; in the lower panel T_{1a} is fixed (2 s) and T_{2a} is varied. Other parameters are fixed (T_{1b} = 0.1 s, T_{2b} = 0.08 s, C_b = 3000 s^{-1}, M_0^b/M_0^a = 0.0007272, and B_1 = 128 Hz) (reproduced from reference 35).*

Conversely, other simulations show that the relaxation times of the hyperfine shifted water protons have less effect on the Z-spectra. This may be important in PARACEST agent design because the paramagnetic ions that induce the largest shifts in the bound water resonance are typically the same ions that relax those protons most efficiently (*see* Table 1, *i.e.*, Tb^{3+}, Dy^{3+}).

Some experimental CEST spectra and best fit simulated curves are illustrated in Figure 2.[36] These two CEST spectra differ dramatically in the chemical shift separation between the bound and bulk water resonances (50 ppm for the Eu^{3+} complex *versus* -800 ppm for the Dy^{3+} complex) and the simulations show that water exchange also differs substantially in these two complexes (τ_M ~95 μs for Eu^{3+} *versus* ~22 μs for Dy^{3+}). For the Dy^{3+} complex, an approximate 5% decrease in bulk water intensity can be achieved using 1 mM agent. This change translates into a comparable image contrast effect as that produced by 1 mM Gd^{3+} as a T_1 relaxation agent. These data illustrate that lanthanides other than Gd^{3+} may find novel applications as MRI contrast agents in the future.

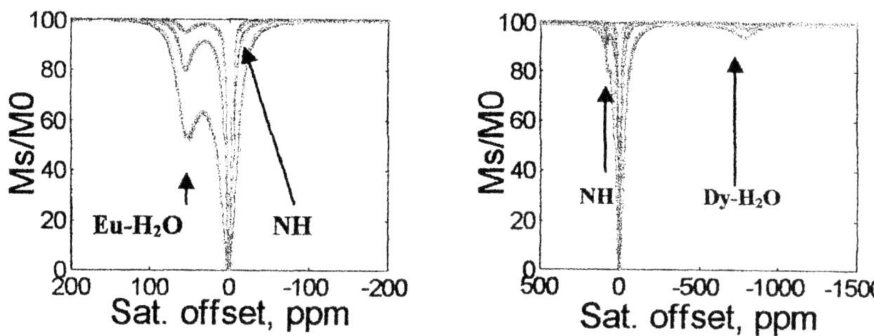

Figure 2. *Z-spectral data (individual points) and best fit simulations to Bloch theory (solid lines) for* ***10 mM*** *Eu(10)*$^{3+}$ *at* B_1 *values of 229, 505 and 1020 Hz (left, in descending order in the graph) and* ***1 mM*** *Dy(10)*$^{3+}$ *at* B_1 *values of 505, 1020 and 2000 Hz (right (see ligand structure in Figure 4).* [36]

(See page 2 of color insert.)

Comparison of the sensitivity limits of Gd^{3+} relaxation agents *versus* PARACEST agents with optimal water exchange

The modified Bloch equations for two site exchange predict a simple relationship between net magnetization of bulk water ($M_Z{}^a$) and concentration of PARACEST agent (equation 1).[35]

$$\frac{M_Z^a}{M_0^a} = \frac{\tau_a}{\tau + T_{1a}} = \left(1 + \frac{Cq}{55.5}\frac{T_{1a}}{\tau_M}\right)^{-1} \qquad (1)$$

where C is the concentration of agent and q is the number of bound inner-sphere water molecules in the complex. This relationship allows one to predict the lower detection concentration limit for any agent with a known bound water lifetime (τ_M) and T_{1a} (bulk water). Figure 3 shows a plot of net magnetization *versus* concentration of an agent with a τ_M of 3 μs, a bulk water T_1 = 1 s, and assuming that a frequency-selective presaturation sequence can be implemented on the clinical scanner that fully saturates the bound water resonance within acceptable SAR limits. This predicts that the bulk water signal intensity can be reduced by ~37% with ~100 μM agent and by ~5% with ~10 μM agent. The latter concentration is well below the detection limit of a low molecular weight Gd^{3+}-based contrast agent with a typical relaxivity (r_1) of 4-5 $mM^{-1}s^{-1}$.[37]

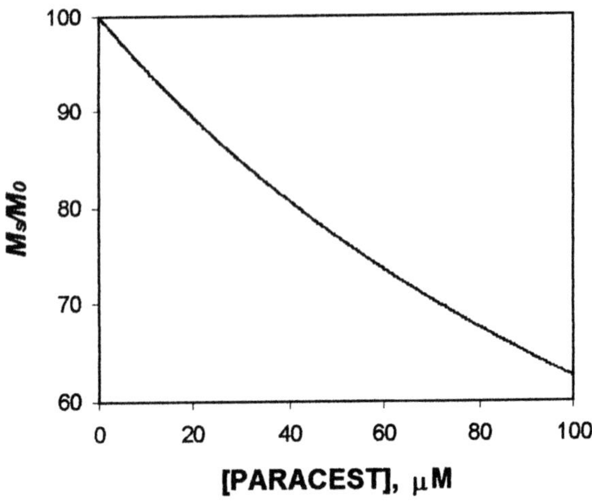

Figure 3. *Plot of the fractional decrease in bulk water signal intensity expected for a PARACEST agent having a bound water lifetime (τ_M) of 3 μs and a bound water (^1H) paramagnetic shift of 500 ppm (eqn. 1).*

This suggests that PARACEST agents have the potential of being as sensitive as or even more sensitive than Gd^{3+}-based T_1 agents assuming all practical aspects can be solved. Just as one can increase the relaxivity of Gd^{3+}-based agents substantially by conjugation to a polymer or formation of an aggregate, similar modifications of PARACEST agents could be done and this might even extend the lower detection limit into the sub-μM range. This was recently demonstrated for polymers containing large numbers of exchangeable -NH groups.[30, 38]

The chemical/physical features that affect τ_M

Now let's return to the topic of modulating the bound water exchange rate. Just as Gd^{3+} requires fast water exchange for optimal T_1 agents, PARACEST complexes have the requirement of slow-to-intermediate water exchange. As described above, NMR studies of the $LnDOTA^-$ complexes have shown that two, slowly inter-converting coordination isomers are present in solution, and the SAP isomer exhibits slower water exchange than the TSAP isomer. This suggests that complexes having the SAP geometry should make more suitable PARACEST agents. It has been shown that replacement of DOTA acetate groups with amide pendant arms has two combined effects on τ_M: 1) it alters the ratio of SAP/TSAP isomers in solution, usually in favor of the SAP isomer and 2) it increases the positive charge on the central lanthanide ion. Both effects contribute to slowing water exchange. This is nicely illustrated by a series in which the acetates of DOTA are replaced by amides ($CH_2CONHCH_2PO_3^{2-}$), the number of which (n) is increased from 0 – 4. Introducing just one amide into the complex significantly lengthens τ_M from 0.24 μs in DOTA to 1.3 μs (n = 1). Further increasing the number of amides lengthens τ_M yet more, 6.2 μs (n = 2) and 26 μs (n = 4), while the percentage of SAP present increases from 70 % (n = 1) to > 98 % (n = 4).

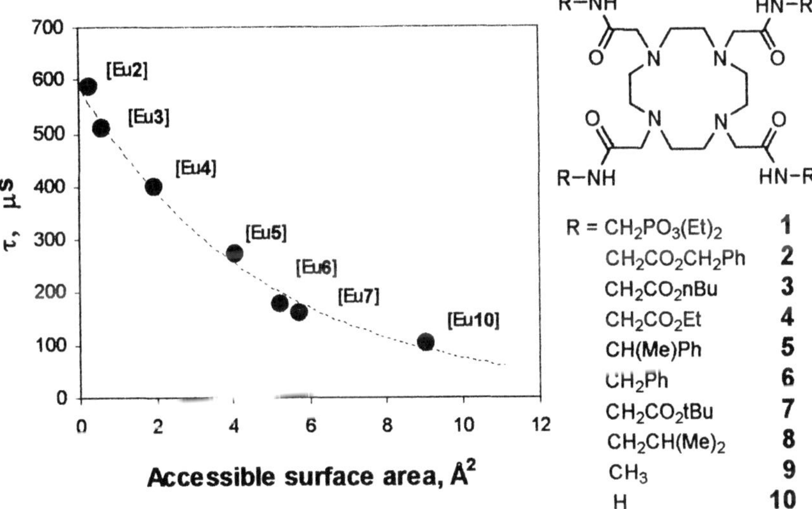

Figure 4. Relationship between coordinated water molecule lifetimes in Eu–tetraamide complexes versus the calculated solvent accessible surface around the bound water (adapted from ref.[16]). The solvent accessible surface areas have not been reported for compounds 1, 8 and 9.

The charge and polarity of the extended amide side-chain arms also seems to play a role in determining τ_M. Is has been proposed that τ_M correlates with a solvent accessability parameter, as estimated by molecular mechanics (Figure 4). [16] Given that water exchange in these tetra-amide complexes is thought to occur *via* a dissociative mechanism, then greater solvent accessability to the Eu^{3+}-bound water may aid in dissociation of the bound water by hydrogen bonding. In general, for EuDOTA-tetra(amide) complexes, the bound water lifetime correlates with polarity of the ligand side-chains, *i.e.*, phosphonate ester ≥ carboxylate ester >> alkyl groups ≥ simple amides. We have also observed that Eu^{3+}-bound water lifetimes for complexes that can be measured in more than one solvent are about 2-fold shorter in aqueous solution than in wet acetonitrile[12] and complexes with negatively charged side-chains have shorter τ_M values than their neutral counterparts. Both observations suggest that there is proton exchange contribution to τ_M in aqueous media.

The ionic radius of the Ln^{3+} also appears to be critically important. We have recently shown for the series $Ln(\mathbf{4})^{3+}$ that τ_M is longest for Eu^{3+} then falls off dramatically to either lighter or heavier Ln^{3+} series.[15] These data were similar to those reported earlier for a limited number of LnDTPA-BMA complexes[39] where the sharp changes in τ_M *versus* Ln^{3+} radii were ascribed to an altered water exchange mechanisms along the series. However, it is not known whether this same trend holds for all other LnDOTA-tetraamide systems, especially those that may exist as mixtures of SAP and TSAP coordination isomers in solution.

Kinetic inertness *versus* thermodynamic stability

Although the thermodynamic stabilities of LnDOTA-tetraamide complexes are known to be considerably lower than the corresponding LnDOTA complexes, their kinetic stabilities appear to be much more favorable. Replacing the all four acetates with the amide methylene phosphonate groups discussed above renders the resulting complexes surprisingly inert. Indeed simply replacing one of the acetates increases the kinetic stability of the complex with respect to acid catalysed dissociation. When the mixed acetate/amide series discussed previously was challenged with 2.5 M nitric acid the half-life of the complex increase in the order: 4.5 h (n = 0), 5.5 h (n = 1), 17 h (n = 2) and 9 days (n = 4). It has been shown for GdL contrast agents that the kinetic stability of complexes in acid correlates with *in vivo* toxicity. If this trend holds, then LnDOTA-tetraamide complexes should be even less toxic *in vivo* than current clinical contrast agents. Although we have not performed a complete toxicity evaluation of these systems, our experience with GdDOTA-$4AmP^{5-}$ (the tetra-phosphonate complex) and EuDOTA-4AmC (the analogous tetra-carboxylate complex, **4**) in rats indicates that large quantities of these complexes can be infused without overt toxicity. We also know that charge is important. Positively charged complexes such as those shown in Figure 4 are all quite toxic in animals while those with negatively charged appended side-chains, such as EuDOTA-4AmP^{5-} or EuDOTA-4AmC^{-}, are quite inert and safe for *in vivo* use.

Examples of "responsive" MRI agents based upon slow exchanging systems

Although the optimal fast exchange system was the major target in persuing the highest relaxivity Gd^{3+}-based agents, slow water exchange systems may prove to have advantages in the design of "responsive" agents.[40-46] It is the slow exchange that made it possible for the lifetime of bound water molecule or protons being modulated by certain external stimuli, such as pH, temperature, redox or metabolite levels. As prototypes, two slowly exchanged examples were shown here, one based on gadolinium and another on europium.

Imaging of tissue pH. Although GdDOTA-4AmP^{5-} has a single, slowly exchanging, inner-sphere water molecule (τ_M = 26 μs), protonation of the pendent phosphonate groups (four pK_a's over the physiological pH range) appears to initiate catalytic exchange of the bound water protons and bulk water protons. This imparts a pH dependency on the bulk water relaxivity of this complex, steadily increasing between pH 8.5 (r_1 = 4.5 $mM^{-1}s^{-1}$) and pH 6 (r_1 = 9.8 $mM^{-1}s^{-1}$).[40] Such a pH-dependent relaxivity is not a feature of current clinical Gd^{3+}-based CAs where water exchange is relatively fast, so the slow water exchange feature of GdDOTA-4AmP^{5-} makes this an attractive agent for mapping extracellular pH in tissues by MRI, Figure 5.[45]

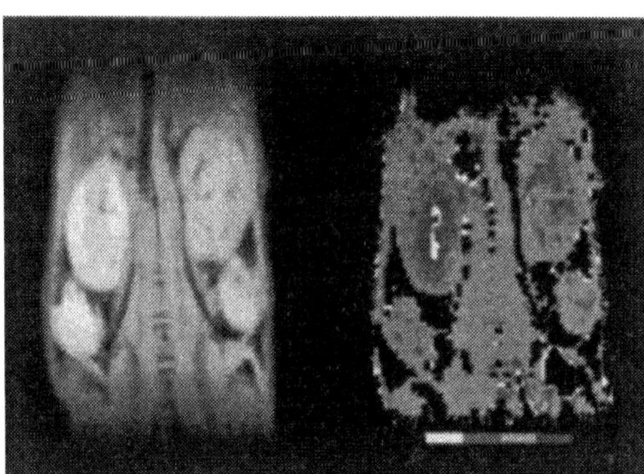

Figure 5. Standard T_1 weighted MRI of a mouse (left) and its pH map (right) derived by comparing clearance of GdDOTA-4AmP^{5-} versus GdDOTP^{5-}. The color code shows the measured pH gradient across the kidney.[45]

(See page 3 of color insert.)

Imaging of glucose. It is well-known that boronic acids can bind selectively and reversibly with structures containing *cis*-diols, a property that has been widely exploited in the design of new saccharide sensors.[47, 48] This property has previously been used to target Gd^{3+} chelates to glycated proteins to improve the relaxivity.[49, 50] With this background, we set out to build the prototype PARACEST agent for mapping tissue glucose (see structure in Figure 6).[44] A Eu^{3+}-bound water peak, usually observed between 50-55 ppm in complexes such as this, is not observed in the absence of glucose but is observed upon addition of excess glucose. This indicates that binding of glucose to this Eu-bis(boronic acid) does indeed slow water exchange in this system. A phantom consisting of four plastic tubes (ID 4 mm) each containing 10 mM Eu-bis(boronic acid) and different amounts of glucose at pH 7.1 was prepared. MR images of these phantoms collected at 4.7T show clear differences in image intensities between the four samples (Figure 6). This is an exciting result because it illustrates that one can detect and probably quantify glucose at physiologically relevant concentrations using a standard clinical imaging MR system and this unique CEST agent. It is important to emphasize that these images were collected not by direct detection of glucose (typically **5 mM** in blood) but rather *via* the bulk water signal (**55 M**) as normally done in a standard MRI diagnosis. This is an important advance for the field of molecular imaging because it demonstrates for the first time that one should be able to design CEST imaging agents for monitoring a wide variety of important metabolites in tissue. One can envision applications of such Eu-Bis(boronic acid) complexes for imaging the distribution of glucose in livers of diabetic patients *in vivo* (for overproduction of glucose), for monitoring glucose consumption in tumors (analogous to FDG in PET imaging), and for detecting metabolic activation of the brain during functional stimulation (fMRI).

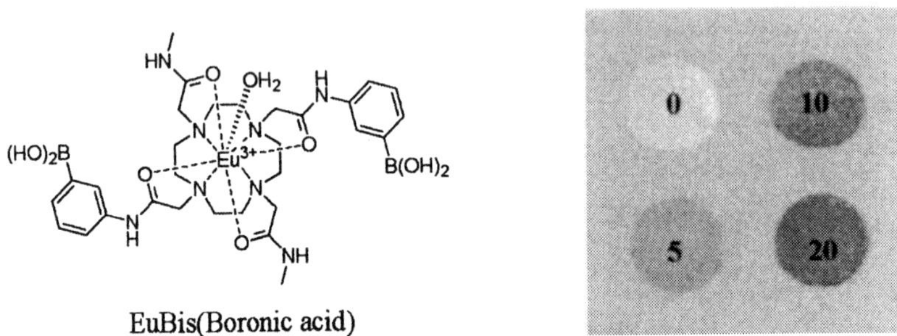

Figure 6. *CEST images of phantoms containing 10 mM EuBis(Boronic acid) plus either 0, 5, 10 or 20 mM glucose, which was obtained by subtraction the image by saturating at 50 ppm from that at 30 ppm.* [44]

Conclusions

In summary, we have demonstrated that the coordination geometry of a DOTA-type lanthanide complex may be controlled by careful consideration of the chirality when substituting both the macrocycle and the pendent arms. The bound water lifetime observed for TSAP geometry (15 ns) is close to the theoretical optimum value, which will be an attractive start point towards the maximum relaxivity for gadolinium-based agents. Conversely, it has also shown that lanthanide complexes that exchange water only slowly may provide advantages as well in design of certain "responsive" agents. A new method, based upon the chemical exchange saturation transfer mechanism, has been introduced which permits the use lanthanide complexes other than gadolinium as MRI contrast agents. PARACEST agents appear to be particularly versatile for creating imaging agents sensitive in physiology or metabolism and may provide an novel platform for novel molecular or cellular imaging agents.

Acknowledgments: This work was supported in part by grants from the Robert A. Welch Foundation (AT-584), the National Institutes of Health (CA-84697 and RR-02584) and the Texas Advanced Technology Program.

Reference:

[1] Merbach, A.E.; E. Toth, *The Chemistry of Contrast Agents in Medical MRI.* **2001**, Wiley, Chichester.

[2] Lauffer, R.B. *Chem. Rev.* **1987**, *87*, 901.

[3] Caravan, P.; Ellison, J.J.; McMurry, T.J.; Lauffer, R.B., *Chem. Rev.* **1999**, *99*, 2293.

[4] Aime, S.; Fasano, M.; Terreno, E.; *Chem. Soc. Rev.* **1998**, *27*, 19.

[5] Peters, J.A.; Huskens, J.;. Raber, D.J.; *Prog. Nucl. Magn. Reson. Spect.* **1996**, *28*, 283.

[6] Springer, C.S. Jr, *Ann. Rev. Biophys. Biophys. Chem.* **1987**, *16*, 375.

[7] Bansal, N.; Germann, M.J.; Seshan, V.; Shires, G.T. III; Malloy, C.R.; Sherry. A.D, *Biochem.* **1993** *32*, 5638.

[8] Aime, S.; Barge, A.; Botta, M.; de Sousa, A.S.; Parker, D., *Angew. Chem. Int'l. Ed.*, **1998**, *37*, 2673.

[9] Aime, S.; Barge, A.; Bruce, J.I.; Botta, M.; Howard, J.A.K.; Moloney, J.M.; Parker, D.; de Sousa, A.S.; Woods, M., *J. Am. Chem. Soc.* **1999**, *121*, 5762.

[10] Batsanov, A.S.; Beeby, A.; Bruce, J.I.; Howard, J.A.K.; Kenwright, A.M.; Parker. D., *Chem. Commun.* **1999**, 1011.

[11] Dunand, F.A.; Aime, S.; Merbach. A.E., *J. Am. Chem. Soc.* **2000**, *122*, 1506.

[12] Zhang, S.; Wu, K.; Biewer, M.C.; Sherry, A.D., *Inorg. Chem.* **2001**, *40*, 4284.

[13] Zhang, S.; Wu, K.; Sherry, A.D., *Invest. Radiol.* **2001**, *36*, 82.

[14] Dunand, F.A.; Dickins, R.S.; Parker, D.; Merbach, A.E., *Chem. Eur. J.* **2001**, *7*, 5160.

[15] Zhang, S.; Wu, K.; Sherry, A.*D.; J. Am. Chem. Soc.* **2002**, *124*, 4226.

[16] Aime, S.; Barge, A.; Batsanov, A.S.; Botta, M.; Castelli, D.D.; Fedeli, F.; Mortillaro, A.; Parker, D.; Puschmann, H.; *Chem. Commun.* **2002**, 1120.

[17] Zhang, S.; Winter, P.; Wu, K.; Sherry, A.D., *J. Am. Chem. Soc.* **2001**, *123*, 1517.

[18] Zhang, S; Sherry, A.D.*; J. Solid State Chem.* **2003**, *171*, 38.

[19] Meyer, M.; Dahaoui-Gindrey, V.; Lecomte, C.; Guilard, R.; *Coord. Chem. Rev.* **1998**, *178-180*, 1313.

[20] Hoeft, S; Roth, K.; *Chem. Ber.* **1993**, *126*, 869.

[21] Powell, D.H.; Ni Dhubhghaill, O.M.; Pubanz, D.; Helm, L.; Lebedev, Y.S.; Schlaepfer, W.; Merbach, A.E.*; J. Am. Chem. Soc.* **1996**, *118*, 9333.

[22] Xu, J.; Franklin, S.J.; Whisenhunt, D.W., Jr.; Raymond, K.N., *J. Am. Chem. Soc.* **1995**, *117*, 7245; Doble, D.M.J.; Botta, M.; Wang, J.; Aime, S.; Barge, A.; Raymond, K.N., *J. Am. Chem. Soc.* **2001**, *123*, 10758; Thompson, M.K.; Botta, M.; Nicolle, G.; Helm, L.; Aime, S.; Merbach, A.E.; Raymond, K.N., *J. Am. Chem. Soc.* **2003**, *125*, 14274.

[23] Woods, M.; Aime, S.; Botta, M.; Howard, J.A.K.; Moloney, J.M.; Navet, M.; Parker, D.; Port, M.; Rousseaux, O., *J. Am. Chem. Soc.* **2000**, *122*, 9781.

[24] Ranganathan R.S.; Raju, N.; Fan, H.; Zhang, X.; Tweedle M.F.; Desreux, J.F; Jacques, V., *Inorg. Chem.* **2002**, *41*, 6856.
Ranganathan R.S.; Radhakrishna, K.P.; Raju, N.; Fan, H.; Nguyen, H.; Tweedle, M.F.; Desreux J.F.; Jacques, V.; *Inorg. Chem.* **2002**, *41*, 6846.

[25] Woods, M.; Kovacs, Z.; Kiraly, R.; Brücher, E.; Zhang, S.; Sherry, A.D., Submitted to *Inorg. Chem.* **2004**

[26] Woods, M.; Kovacs, Z.; Zhang, S.; Sherry, A.D.; *Angew. Chem. Int'l. Ed* **2003**, *42*, 5889.

[27] Wolf, S.D.; Balaban, R.S.; *Magn. Reson. Med.* **1989**, *10*, 135

[28] Henkelman, R.M.; Stanisz, G.J.; Graham, S.J., *NMR in Biomedicine* **2001**, *14*, 57.

[29] Ward, K.M.; Aletras, A.H.; Balaban, R.S., *J. Magn. Reson.* **2000**, *143*, 79.

[30] Goffeney, N.; Bulte, J.W.M.; Duyn, J.; Bryant, L.H.; van Zijl, P.C.M., *J. Am. Chem. Soc.* **2001**, *123*, 8628.

[31] Zhou, J.; Payen, J.-F.; Wilson, D.A.; Traystman, R.J.; van Zijl,. P.C.M., *Nature Med.* **2003**, *9*, 1085.

[32] Sherry, A.D.; Zhang, S.; Wu, K., *PCT Int. Appl.* **2002**, WO 0243775. p. 76.
[33] Grad, J.; Bryant, R.G., *J. Magn. Reson.* **1990**, *90*, 1.
[34] Zhang, S.; Merritt, M.; Woessner, D.E.; Lenkinski, R.E.; Sherry, A.D., *Acc. Chem. Res.* **2003**, *36*, 783.
[35] Weossner, D.E.; Zhang, S.; Merritt, M.E.; Sherry, A.D., *Magn. Reson. Med.*, under review.
[36] Zhang, S.; Weossner, D.E.; Merritt, M.E.; Sherry, A.D., *ISMRM*, **2004**, Japan.
[37] Ahrens, E.T.; Rothbacher, U.; Jacobs, R.E.; Fraser, S.E., *Proc. Nat'l. Acad. Sci. USA* **1998**, *95*, 8443.
[38] Aime, S.; Castelli, D.D.; Terreno, E., *Angew. Chem. Int'l. Ed.* **2003**, *42*, 4527.
[39] Pubanz, D.; Gonzalez, G.; Powell, D.H.; Merbach, A.E., *Inorg. Chem.* **1995**, *34*, 4447.
[40] Zhang, S.; Wu, K.; Sherry, A.D.; *Angew. Chem. Int'l. Ed.*, **1999**, *38*, 3192.
[41] Aime, S.; Barge, A.; Castelli, D.D.; Fedeli, F.; Mortrillaro, A.; Nielsen, F.U.; Terreno, E., *Magn. Reson. Med.* **2002**, *47*, 639
[42] Aime, S.; Castelli, D.D.; Terreno, E., *Angew. Chem. Int'l. Ed.* **2002**, *41*, 4334.
[43] Aime, S.; Castelli, D.D.; Fedeli, F.; Terreno, E., *J. Am. Chem. Soc.* **2002**, *124*, 9364.
[44] Zhang, S.; Trokowski, R.; Sherry, A.D., *J. Am. Chem. Soc.* **2003**, *125*, 15288
[45] Raghunand, N.; Howison, C.; Sherry, A.D.; Zhang, S.; Gillies, R.J., *Magn. Reson. Med.* **2003**, *49*, 249.
[46] Zhang, S.; Michaudet, L.; Burgess, S.; Sherry, A.D., *Angew. Chem. Int'l. Ed.* **2002**, *41*, 1919.
[47] James T.D.; Shinkai, S., *Top. Curr. Chem.* **2002**, *218*, 159.
[48] Striegler, S., *Curr. Org. Chem.* **2003**, *7*, 81.
[49] Aime, S.; Botta, M.; Dastru, W.; Fasano, M.; Panero, M., *Inorg. Chem.* **1993**, *32*, 2068.
[50] Rohovec, J.; Maschmeyer, T.; Aime, S.; Peters, J.A., *Chem. Eur. J.* **2003**, *9*, 2193.

Chapter 11

Targeted Molecular Imaging with MRI

Peter Caravan

EPIX Medical, Inc., 71 Rogers Street, Cambridge, MA 02142–1118 (email: pcaravan@epixmed.com; fax: 617–250–6127)

What is molecular imaging?

To a chemist, this is an odd title – what does one image if not molecules? For instance, the optical absorbance of indocyanine green, the gamma ray emission of ^{99m}Tc in pertechnatate, and the nuclear magnetic resonance of ^{23}Na have all been exploited to generate images that serve as maps of the distribution of these compounds. The term molecular imaging comes from radiology. Radiologists see imaging in terms of anatomy (e.g. X-rays), and to some extent physiology (e.g. cardiac wall motion with MRI or ultrasound). Molecular imaging refers to the ability to obtain functional, metabolic, or biochemical information. The promise of another level of knowledge beyond anatomy and physiology has generated excitement in the clinical imaging field.

Weissleder and Mahmood *(1)* defined molecular imaging as "the in vivo characterization and measurement of biologic processes at the cellular and molecular level". Luker and Piwnica-Worms *(2)* go further and describe it as "the characterization and measurement of biologic processes in living animals, model systems, and humans at the cellular and molecular level by using remote imaging detectors". This narrows the description to in vivo imaging, and using some remote detector indicates a relatively non-invasive procedure.

Figure 1 shows an MR image of the brain taken without an exogenous contrast agent. This is not molecular imaging; although the image is powerful and provides detailed anatomy, it does not give any information as to what is occurring in the brain on a cellular or molecular level. This paper will attempt to describe molecular imaging with MRI starting with an overview of the current interest in molecular imaging in general and moving to why MRI may be a valuable technique for molecular imaging. The article focuses on targeted

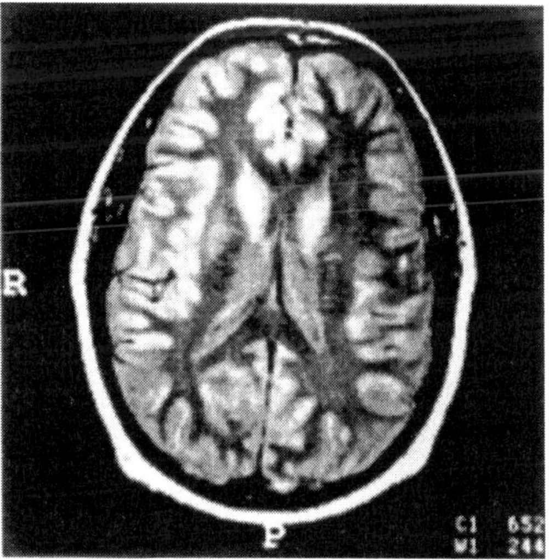

Figure 1. *Anatomical proton density magnetic resonance image (MRI) of human brain (image courtesy of EPIX Medical).*

molecular imaging using exogenous contrast agents. The challenges of designing targeted MRI contrast agents are set forth and some (not meant to be exhaustive) solutions and examples from the literature are provided.

Molecular imaging is part of a paradigm shift in radiology where the field is evolving to take advantage of advances in both science and technology. This is illustrated in Figure 2. The inner circle in Figure 2 represents current clinical practice and the outer circle summarizes directions where practice is headed. While there are some modalities utilizing techniques on the outer circle, this is becoming more widespread. For example, digital imaging is part of MRI but not X-ray; yet there are moves afoot to capture all images digitally. The inner circle in Figure 2 should be viewed as a foundation: there will always be a need for imaging in diagnosis, but imaging will also be used to help monitor and guide therapy. Likewise, the strength of molecular imaging to gain knowledge of specific metabolic function is enhanced by a detailed anatomical map of where the metabolism is occurring.

The title of this piece is "targeted molecular imaging with MRI". Targeted is meant to distinguish work done with exogenous contrast agents as opposed to using an endogenous contrast mechanism. Figure 3 shows an example of using differing iron levels in the brain to characterize dementia. Iron can be a very potent relaxor of water protons, especially shortening T_2. This results in negative image contrast where iron is present *(3,4)*. This is discussed in more detail in the chapter by Schenck on high field magnetic resonance imaging of brain iron. Other examples of using endogenous markers for

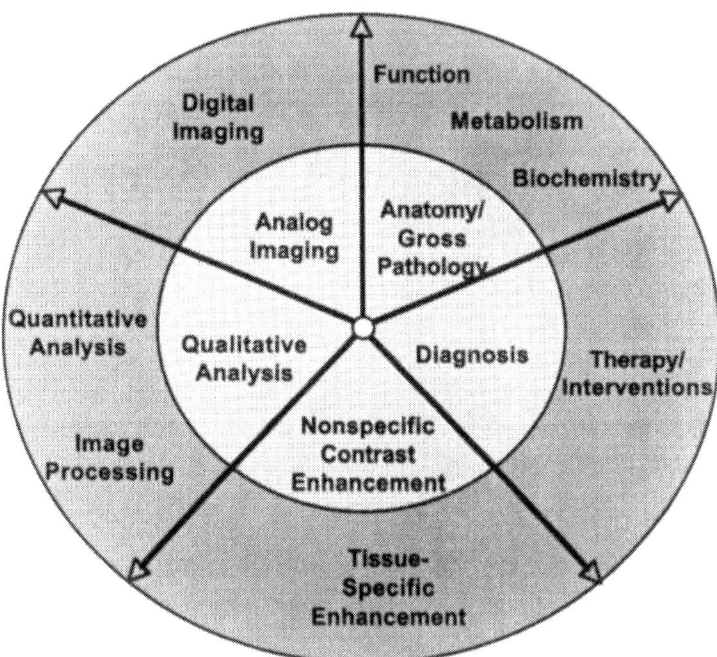

Figure 2. *Cartoon depicting the evolution of clinical practice in radiology. The inner circle represents existing practice and the outer circle shows how radiological practice is expanding technologically and scientifically.*

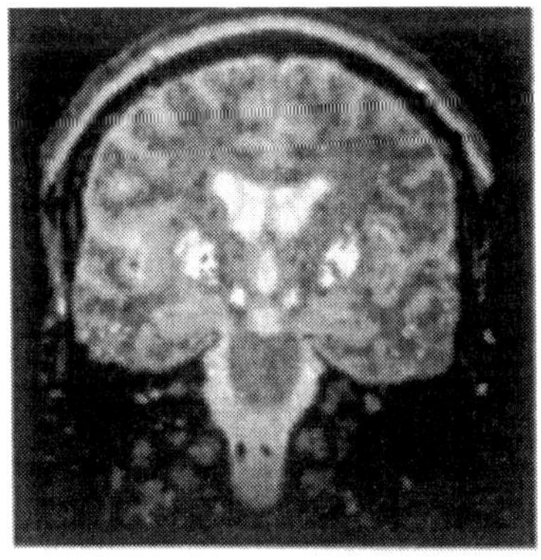

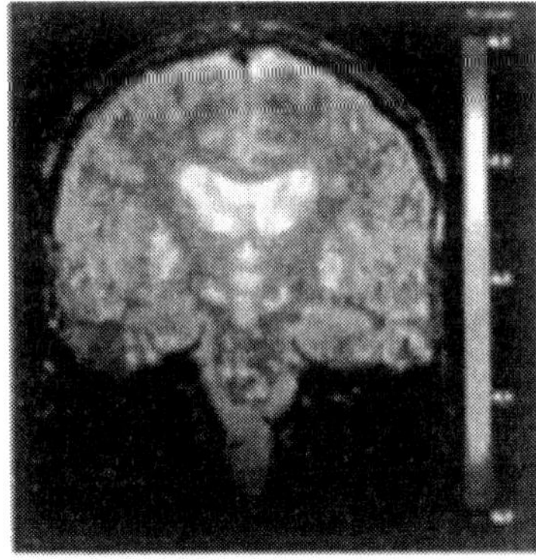

Figure 3. *Non-targeted molecular imaging with MRI: brain iron distribution as a potential biomarker for neurodegenerative disease. Thin-slice, T2-weighted coronal images demonstrate increased distribution of pixels with short effective T2 (Fe content) in brain of dementia patient as compared to normal volunteer. Work of Dr. D. Alsop, Beth Israel Deaconess Medical Center and Drs. J. Schenck and A. Alyassin, GE Global Research Center; used with permission.*

(See page 3 of color insert.)

molecular imaging include using ^{31}P spectroscopy and imaging to follow phosphorylated metabolites *(5)*, or determining choline levels by in vivo NMR and using these to distinguish benign from malignant breast lesions *(6)*.

There are many modalities suited to molecular imaging, Table 1. The nuclear medicine techniques positron emission tomography (PET) and single photon emission computed tomography (SPECT) are well suited to molecular imaging because of their excellent sensitivity *(7)*. Nanomolar receptors can be

Table 1. Imaging modalities employed in molecular imaging and means of detection.

Molecular imaging modalities	Means of detection
Computed tomography (CT)	X-rays
Single photon emission computed tomography (SPECT)	γ-ray
Positron emission tomography (PET)	Positrons
Magnetic resonance imaging (MRI)	Radiofrequency waves
Optical	Near IR light
Magnetic resonance spectroscopy (MRS)	Radiofrequency waves
Ultrasound	Ultrasound

routinely imaged with targeting vectors (antibodies, small molecules) tagged with γ-ray or positron emitting nuclei. The fusion of PET with X-ray computed tomography (CT) allows molecular imaging information (PET) to be superimposed on anatomical information (CT). Near infrared optical agents *(8,9)* and targeted ultrasound microbubbles *(10)* are also being explored, though at a lower level of research activity.

Benefits and challenges of MRI for molecular imaging.

MRI has many advantages as a diagnostic imaging modality. It is noninvasive, delivers no radiation burden, and has excellent (submillimeter) spatial resolution. Soft tissue contrast is superb and MRI readily yields anatomical information. Moreover, there are many techniques that can provide contrast in MRI resulting in markedly different images form the same anatomical region. For instance, pulse sequences can be weighted to highlight differences among tissues that have different proton density, T_1 or T_2 relaxation times, different rates of water diffusion, or different chemical shifts (water vs lipids) *(11)*.

The overriding challenge with MRI is its relatively low sensitivity. Consider clinical imaging: what is primarily observed are hydrogen atoms from water that are present in tissue at ~90M. In order to induce additional contrast, a substance is required that will affect some property of this 90M water to such an extent that an observable effect is achieved. Such substances are called MRI contrast agents. They are paramagnetic, superparamagnetic, or ferromagnetic compounds that catalytically shorten the relaxation times of bulk water protons.

All contrast agents shorten both T_1 and T_2. However it is useful to classify MRI contrast agents into two broad groups based on whether the substance increases the transverse relaxation rate ($1/T_2$) by roughly the same amount that it increases the longitudinal relaxation rate ($1/T_1$) or whether $1/T_2$ is altered to a much greater extent. The first category is referred to as T_1 agents because, on a percentage basis, these agents alter $1/T_1$ of tissue more than $1/T_2$ owing to the fast endogenous transverse relaxation in tissue. With most pulse

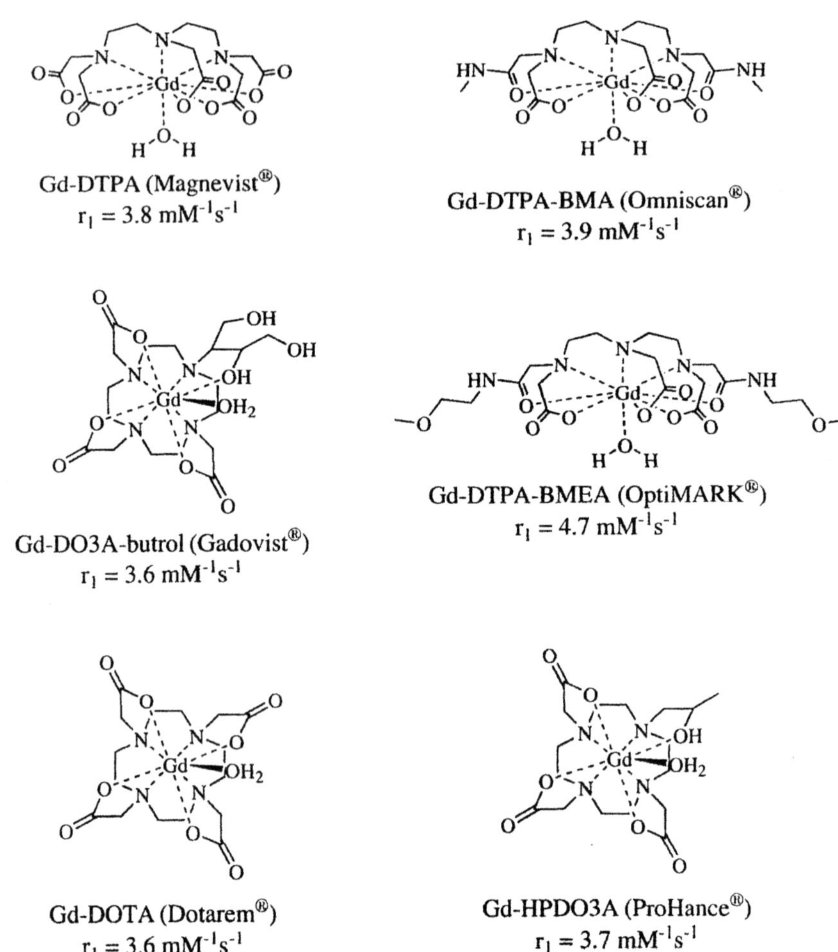

Figure 4. *Some commercial MRI contrast agents and their relaxivities (from ref. 12) at 0.47 tesla (20 MHz), 37 °C.*

sequences, this dominant T_1 lowering effect gives rise to increases in signal intensity; these are positive contrast agents. The T_2 agents largely increase the $1/T_2$ of tissue selectively and cause a reduction in signal intensity; these are negative contrast agents. Paramagnetic gadolinium based contrast agents *(12)* are examples of T_1 agents, while ferromagnetic large iron oxide particles are examples of T_2 agents *(13)*.

The ability of a contrast agent to change a relaxation rate is represented quantitatively as relaxivity, r_1 or r_2, where the subscript refers to either the longitudinal ($1/T_1$) or the transverse rate ($1/T_2$). Relaxivity is simply the change in relaxation rate after the introduction of the contrast agent ($\Delta 1/T_1$) normalized to the concentration of contrast agent or metal ion (M), eq. 1.

$$r_1 = \frac{\Delta 1/T_1}{[M]} \qquad (1)$$

With its seven unpaired electrons and slow electronic relaxation rate, gadolinium is a very potent relaxor of protons. It is the basis of several commercial MRI contrast agents (Figure 4), all of which have one coordinated water molecule that is in rapid exchange *(12)* with bulk water ($k_{ex} \sim 10^6\ s^{-1}$).

To return to the sensitivity issue, in order to induce observable contrast in a robust clinical exam, a relaxation rate change of at least about 1 s^{-1} is required. For commercial contrast agents with a relaxivity of ~4 $mM^{-1}s^{-1}$, this means a concentration of ~250 μM. For targeted imaging and assuming a 1:1 binding stoichiometry (Gd:target molecule), this would require a biological target present at 250 μM. For absolute sensitivity, a more rigorous analysis by

Wedeking et al. *(14a)* gave a limit of detection of 30 μM in mouse skeletal muscle for the contrast agent Gd-HPDO3A. This limits the number of potential biological targets for imaging using current technology *(14b)*.

Despite the sensitivity problem, there is a strong desire to perform molecular imaging with MRI because of the excellent spatial resolution and inherent anatomical information that is obtained. Figure 5 illustrates some possible technical approaches that could be used to image low concentration targets. The first is to design a better magnetic catalyst to give much higher relaxivity, and hence sensitivity. The second is to append multiple copies of the signal generating moiety to the targeting vector (high molecular relaxivity). Binding to a target would bring multiple paramagnetic ions.

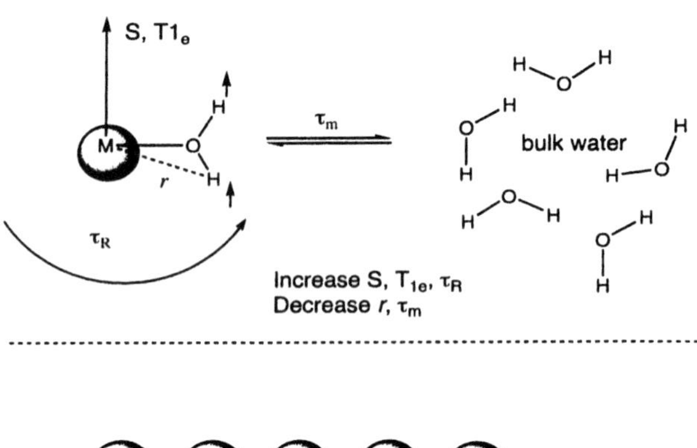

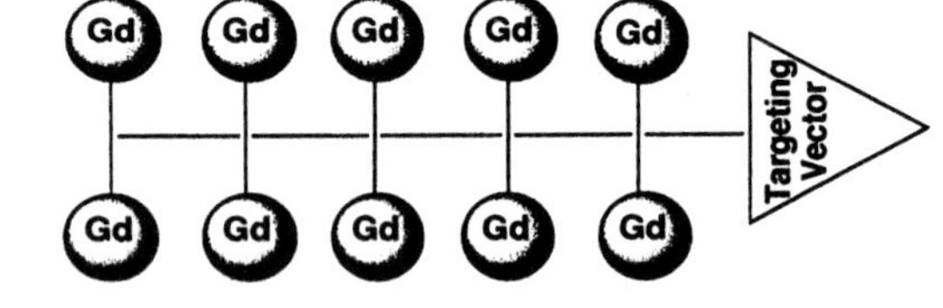

Figure 5. *Cartoon to illustrate approaches to improving the sensitivity of MRI contrast agents. Top: design a better magnetic catalyst by tuning the molecular parameters that influence relaxivity. Bottom: append multiple copies of the signal generating moiety to a targeting vector to increase the overall, molecular relaxivity.*

Increasing relaxivity of the ion.

Relaxivity depends on a host of factors *(12)*: the spin quantum number (S), the distance between the ion and the hydrogen (r), how fast the molecule tumbles in solution ($1/\tau_R$), how fast the water molecules exchange with the bulk ($1/\tau_m$), and how fast the electrons in the ion relax ($1/T_{1e}$). Early on, Lauffer et al. *(15)* showed that, in addition to having a coordinated exchangeable water molecule, the rate at which the complex rotated was the single most important factor in improving relaxivity. This was demonstrated by covalently linking either EDTA or DTPA ligands to the slow tumbling protein bovine serum albumin. The relaxivity of the metal-ligand-protein construct was much higher than the metal-ligand complex alone. This was directly analogous to earlier work using paramagnetic ions as probes of calcium and zinc binding sites on proteins where the solvent relaxation rates increase when the metal ion binds to the protein, the so-called proton relaxation enhancement or PRE effect *(16)*. Lauffer then posited that if small molecules were non-covalently targeted to proteins or other biopolymers, then the resultant adduct would have a much higher relaxivity than the unbound form *(17)*. This was dubbed the RIME (Receptor Induced Magnetization Enhancement) effect. Using the RIME effect, contrast at the target site is increased both by accumulation of contrast agent through targeting, and by increased relaxivity of the contrast agent when it is bound.

The RIME effect was used in designing MS-325, a rationally designed contrast agent for blood vessel imaging *(18)*. MS-325 comprises the thermodynamically stable GdDTPA core and is further derivatized with a moiety that provides binding to serum albumin and good pharmacokinetic properties, Figure 6.

By binding to albumin in plasma, the relaxivity is markedly increased compared to GdDTPA *(19)*. The RIME effect is also shown in Figure 6 where the relaxivity of MS-325 is amplified 9-fold on going from buffer to plasma. The high relaxivity enables brighter images at lower doses of the contrast agent compared with existing agents. Figure 7 shows an example of MS-325 enhanced blood pool imaging. This image was taken during the dynamic phase where the contrast agent is present in the abdominal aorta and renal arteries but has not yet transited the body to enhance the vena cava yet.

The contrast agent helps to delineate vascular abnormalities such as the stenosis in the renal artery shown in Figure 7. There is an ongoing effort in several laboratories to tweak the molecular parameters to create an optimum high relaxivity compound *(20)*. Figure 8 shows imaging data from a contrast agent targeted to DNA *(21)* that exhibits very good relaxivity at 1.5 tesla (the most common clinical imaging field). Figure 8 illustrates that high relaxivity is necessary if the goal is to image low concentration targets. Here 4 test tubes are imaged together. One tube contains only buffer and the other three contain 50 μM Gd but the Gd is present in different contrast agents. The test tubes are imaged using a T_1-weighted sequence. The high relaxivity agent is obviously much more detectable at 50 μM while the commercial agent ($r_1 = 4\ mM^{-1}s^{-1}$) is only slightly brighter than buffer at this concentration of gadolinium.

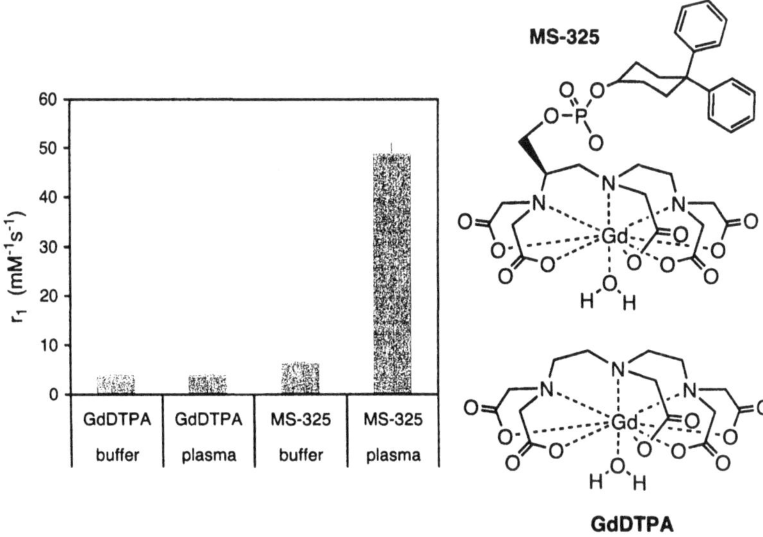

Figure 6. *Comparison of the relaxivities of MS-325 and GdDTPA in phosphate buffered saline or human plasma at 37 °C, 0.47 tesla. The high relaxivity of MS-325 in plasma demonstrates the RIME effect and the ability to increase relaxivity by altering the molecular properties (in this case rotation) of the compound.*

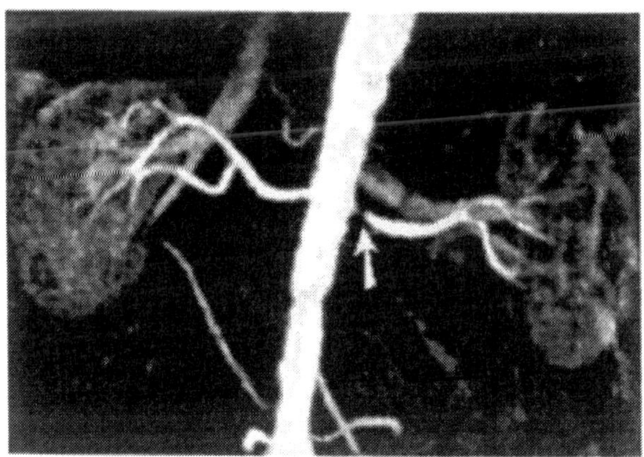

Figure 7. *MS-325 enhanced dynamic MR image of the renal arteries and kidneys of a patient exhibiting a stenosis (narrowing) of the proximal left renal artery (see arrow). Image courtesy of Dr. R. Guzman, St. Boniface General Hospital, Winnipeg, Canada and EPIX Medical.*

Increasing the local metal concentration.

The sensitivity problem in MRI for targeted contrast was recognized at an early stage in the development of MRI contrast agents *(14)*. There have been numerous reports over the last 20 years of using multimeric metal chelates to increase molecular relaxivity. Multiple gadolinium complexes have been coupled to bovine serum *(15, 22)*, poly-L-lysine *(23)*, dextrans *(24, 25)*, dendrimers *(26, 27)*, or the complexes have themselves been polymerized into co-block polymers *(28)*. In addition, gadolinium complexes derivatized with lipid groups self assemble in aqueous solution to form micelles *(29, 30)*, liposomes *(31 – 33)*, or other aggregates. Challenges with these types of compounds or assemblies include difficulty in manufacturing related to controlling and characterizing the polydispersity, as well as solubility issues with respect to the high concentration formulations required for MRI. In addition, many of these compounds are slowly excreted. One example of a gadolinium multimer that has entered clinical trials is Gadomer®, a dendritic compound containing 24 gadolinium chelates that was designed for blood pool imaging *(26, 34, 35)*. Gadomer® is described in more detail in the chapter by Platzek and Schmitt-Willich.

There have been numerous efforts to target these multimeric gadolinium ensembles. For example, Wiener and coworkers linked folate to a dendrimer functionalized with gadolinium chelates and used this compound to target the overexpressed folate receptor in certain tumors *(36 – 38)*. Bednarski and colleagues *(39)* described chemically cross-linked polymerized liposomes that could be derivatized with paramagnetic complexes. Targeting was introduced by incorporating biotin groups into head groups of the liposome in addition to GdDTPA moieties. Antibodies for a given target could be derivatized with avidin and this resulted in a liposome-biotin-avidin-antibody-target conjugate (schematically shown in Figure 9). This was demonstrated with antibody targeting to the overexpressed adhesion molecule ICAM-1 in a model of encephalitis *(40)* or to the $\alpha_v\beta_3$ integrin in a model of angiogenesis *(41)*. Work from a group at Washington University has centered on large (micron sized) lipid based emulsions with a perfluorcarbon core. Gadolinium complexes derivatized with lipid chains can be noncovalently incorporated into the emulsion *(42)*. The particles were targeted using the biotin-avidin-antibody approach described above to fibrin *(43)* in blood clots and to the $\alpha_v\beta_3$ integrin overexpressed during angiogenesis *(44)*. Angiogenesis targeting was also achieved by using a targeting vector (in this case a peptidomimetic that recognizes the $\alpha_v\beta_3$ integrin *(45, 46)*) directly linked to a lipid and incorporated into the emulsion.

The literature on targeted gadolinium polymers and particles has focused primarily on demonstrating efficacy. There has been little work on how these compounds are excreted and how fast they are excreted.

Iron oxide particles are another type of contrast media *(13)*. These are often referred to as small particles of iron oxide (SPIO) or ultrasmall particles of iron oxide (USPIO). Certain formulations have been approved for liver imaging. The SPIOs have large magnetic susceptibilities and function by

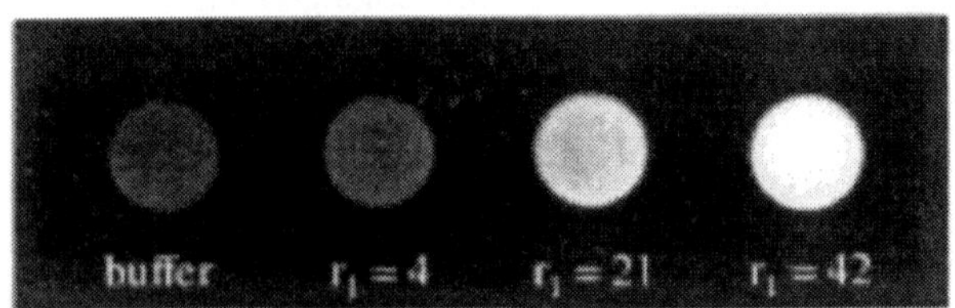

Figure 8. *Images of 4 test tubes containing either buffer only (solution a) or 48 μM of a contrast agent with a relaxivity of b) 4 $mM^{-1}s^{-1}$; c) 21 $mM^{-1}s^{-1}$; d) 42 $mM^{-1}s^{-1}$. Images obtained at 1.5 tesla with a T_1-weighted imaging sequence (3D SPGR TR = 40, TE = 4.0, α = 40°).*

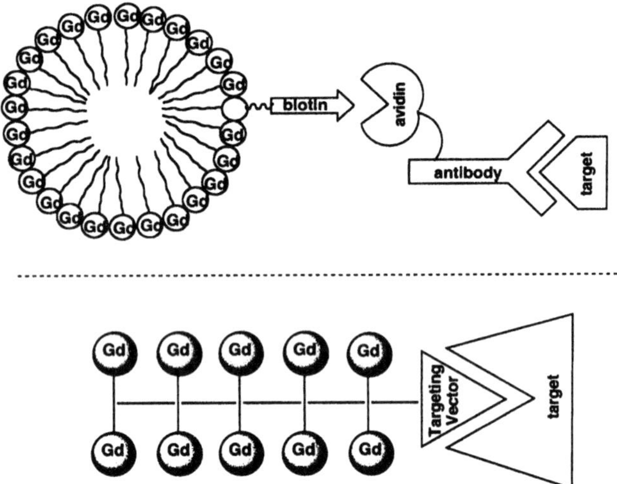

Figure 9. *Targeting approaches employed for using targeted metal assemblies for MRI. Left: Incorporation of biotin into the metal assembling and conjugation of avidin to a target seeking antibody; targeting can be done in a two step approach where the antibody first localizes at the site of interest and the metal-biotin conjugate then binds to the antibody via the biotin-avidin interaction, or the metal-biotin-avidin-antibody is administered directly. Right: Direct incorporation of a targeting vector(s) onto the metal assembly*

shortening T_2 and T_2*, thus providing negative contrast. The liver agents are recognized by the reticuloendothelial system and taken up in Kupfer cells in the liver. The iron is not excreted, but assimilated into the body's store of iron.

The iron particles are coated with organic molecules such as dextran or citrate. The coatings have been modified to direct biodistribution. An early example was using arabinogalactan as a coating to target the particles to the asialoglycoprotein receptor on hepatocytes *(47)*. The Weissleder group has developed chemistry to modify the dextran surface of SPIOs to prepare iron oxide particles with functional groups on the surface *(48)* which they call cross-linked iron oxides (CLIO). The CLIOs have been used to create targeted imaging agents that report on, for example, gene expression (transferrin receptor *(49, 50)*), inflammation (E-selectin *(51)*), or apoptosis (annexin V *(52)*). Both intra- and extracellular targets have been probed.

Small(er) molecule approaches.

Potential solutions to the sensitivity problem in MRI are often incomplete. To obtain positive (bright) contrast in an image, gadolinium complexes are usually required. However, for diagnosis it is critical that the gadolinium be completely eliminated from the patient and many polymeric gadolinum chelate derivatives and particles have long residency times in animal models. The iron particles can be metabolized and the iron taken up in the body's iron pool; however iron agents act by causing negative contrast which may be difficult to accurately detect for small targets.

Some recent work at EPIX Medical has focused on targeting fibrin with gadolinium based contrast agents *(53 – 55)*. Fibrin is present in many disease states: for example deep vein thrombosis, pulmonary embolism, coronary thrombis, and high risk atherosclerotic plaque. Fibrin is an insoluble polymer formed by the action of the enzyme thrombin on plasma circulating fibrinogen. The concentration of fibrinogen in plasma is about 7.5 μM (3 g/L). Thus the concentration of fibrin monomer in a blood clot is at least 7.5 μM and this makes fibrin a target that may be present in sufficient concentration to be detected by an MRI contrast agent. Fibrin is a useful target for detecting clot – it is present in all thrombi resulting in good potential sensitivity, but fibrin is not present in circulating blood resulting in good specificity for disease. Some low molecular weight peptides that bind specifically to fibrin were identified by phage display *(56)*.

If the peptide is labeled with a fluorescent tag (e.g. tetramethyl-rhodamine, TMR) then fibrin present in clot can be visualized by fluorescence microscopy. Figure 10 shows a confocal microscopy image of a platelet rich plasma clot that has been treated with a TMR-tagged fibrin binding peptide.

These peptides can also be functionalized with gadolinium chelates. In one example, EP-1873, the peptide has 4 GdDTPA chelates appended to it. EP-1873 binds to two equivalent sites on human fibrin with a dissociation constant

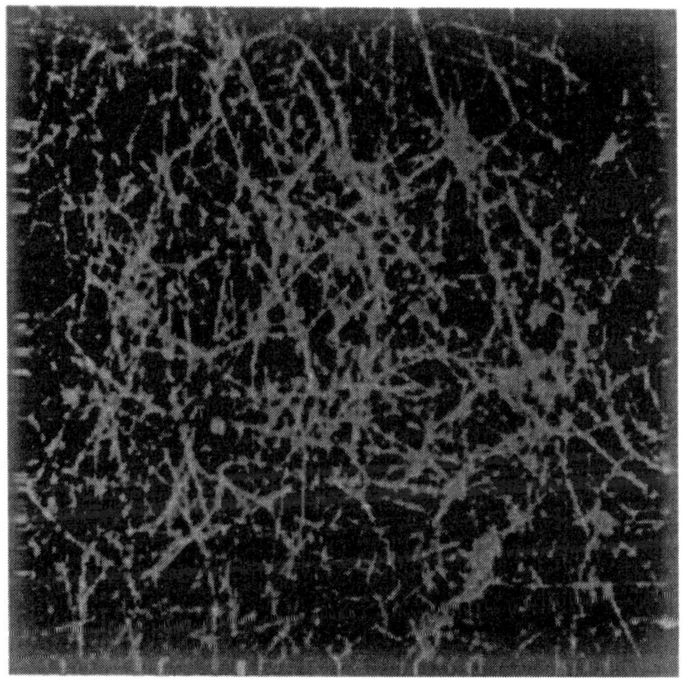

Figure 10. *Confocal fluorescence microscopy image of a human platelet rich plasma diluted 1:5 and clotted with 10 μg/mL of thrombin. After 2.5 minutes, the TMR-labeled petide is added and the clot is imaged. The fibrin network is apparent in the image. Image courtesy of Prof. P. Janmey, University of Pennsylvania.*

(See page 4 of color insert.)

of 2 μM. The complex specifically binds to fibrin and not fibrinogen (K_d > 200 μM). The relaxivity of EP-1873 is consistent with this result.

Figure 11a shows relaxivity per Gd ion for EP-1873 in different media at 20 MHz, 37 °C. The relaxivity in human plasma or human plasma enriched in fibrinogen is the same and not much different than the relaxivity in buffer only. This very weak RIME effect is consistent with a weak interaction with plasma proteins and/or fibrinogen. In contrast, the observed relaxivity in fibrin gel is enhanced 150% relative to buffer. This strong RIME effect is consistent with a large fraction of EP-1873 being bound to fibrin.

Figure 11b demonstrates the molecular relaxivity of EP-1873 in comparison with GdDTPA and MS-325. These data are shown at 1.5 tesla, the most common imaging field strength. Figure 11b shows that the relaxivity gain made with MS-325 relative to GdDTPA has been maintained to a large extent with EP-1873; since EP-1873 has 4 chelates appended, the molecular relaxivity is ~4X MS-325 and greater than 20X that of GdDTPA. This high molecular relaxivity provides sufficient sensitivity to image thrombi in vivo.

Botnar et al. *(57)* used EP-1873 to image thrombi in a rabbit model of atherosclerotic plaque rupture. These authors used a model known to produce atherosclerotic lesions in the aorta of the rabbit. Plaque rupture was then induced using Russell's viper venom and histamine. The ruptured plaque induces thrombus formation. One hour after plaque rupture, EP-1873 was administered intravenously (2 μmol/kg). MR imaging post EP-1873 readily

A

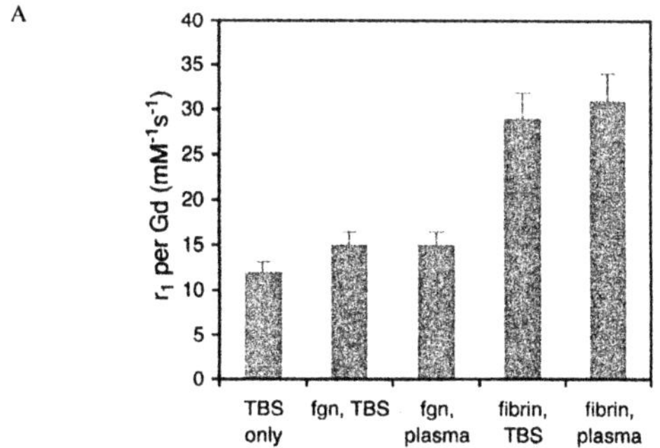

B

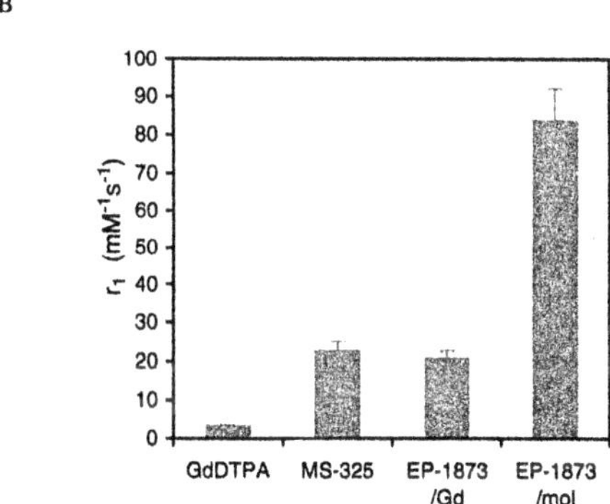

Figure 11. *A) Comparison of per Gd relaxivities of EP-1873 (50 μM) in different media at 0.47 tesla, 37 °C. This very weak RIME effect in human plasma or fibrinogen (fgn, 10 mg/mL) is consistent with a weak interaction r_1 per Gd ($mM^{-1}s^{-1}$) with plasma proteins and/or fibrinogen. In contrast, the observed relaxivity in fibrin gel (10 mg/mL) is enhanced 150% relative to buffer. B) Comparison of EP-1873 molecular and per Gd relaxivity in fibrin gel (10 mg/mL) compared with MS-325 or GdDTPA in human plasma at 1.5 tesla, 37 °C. Gains in relaxivity made by MS-325 are maintained at 1.5 T for EP-1873, but the sensitivity is greater because of the 4 gadolinium ions in EP-1873.*

identified all the aortic thrombi in all the animals (verification via correlation with histology).

In a separate rabbit study, a crush injury model was used to induce thrombus formation in the carotid artery. The injury produces a non-occlusive thrombus. Figure 12 shows T1-weighted MRIs of an injured carotid artery. The main images are maximum-intensity projections; insets are single axial slices through the site of injury. On the left is an image obtained 30 minutes after administration of the fibrin-binding contrast agent EP-2104R (a close analog of EP-1873). Arrows indicate the site of injury. On the right is the same animal, imaged using a blood pool contrast agent. The blood pool image demonstrates the patency of the vessel, as well as diffuse injury in the model, but does not specifically demonstrate the location or extent of thrombus. This demonstrates the specificity of molecular imaging – the fibrin seeking agent not only dentifies a lesion, it also characterizes the composition. Positive image contrast makes the lesion readily apparent in a projection image.

Conclusion

Molecular imaging is a rapidly evolving field that is expanding the information available to physicians through imaging. The high resolution and soft tissue contrast of MRI coupled with its lack of radiation burden make it an attractive modality for molecular imaging. Combining the excellent inherent anatomical contrast of MRI with a contrast agent reporting on molecular activity is a valuable tool. The highest hurdle at the moment is to improve the sensitivity

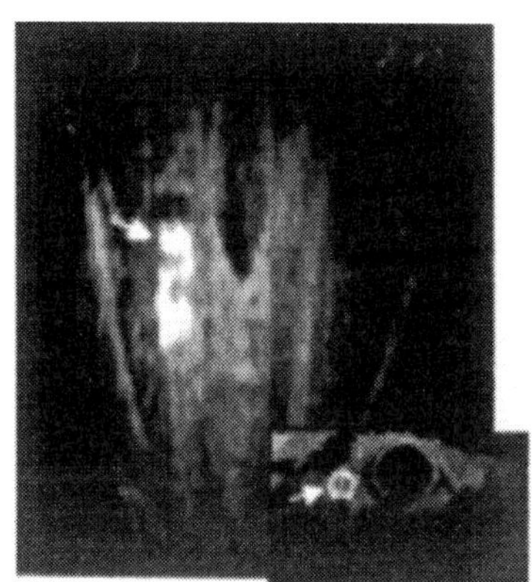

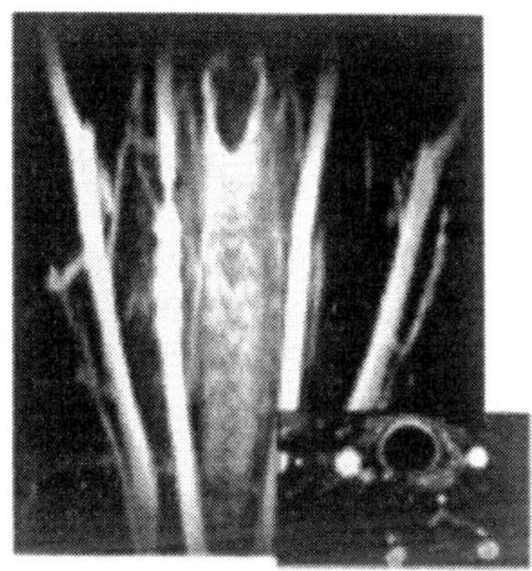

***Figure 12.** T_1-weighted MRIs of a non-occlusive thrombus in the carotid artery of a rabbit. The main images are maximum-intensity projections; insets are single axial slices through the site of injury. Left: image obtained 30 minutes post administration of the fibrin-binding contrast agent EP-2104R. Arrows indicate the site of injury – region of thrombosis clearly delineated. Right: same animal, imaged using a blood pool contrast agent. The blood pool image demonstrates patency of the vessel, as well as diffuse injury in the model, but does not specifically demonstrate the location or extent of thrombus.*

of the MRI reporter. There are several techniques that have been shown to be feasible to create new targeted MRI contrast agents. There are small molecule approaches that seek to optimize the relaxivity (signal) of the contrast agent, and nanoparticle constructs that bring together 10s to 1000s of paramagnetic reporters. A key challenge in molecular imaging with MRI is the identification of an appropriate target. For large assemblies of gadolinium or iron, the target must be accessible from the vasculature. For small molecules, the target must be present at a sufficiently high concentration. Fibrin is one such example: a peptide – multichelator construct optimized for relaxivity can provide positive image contrast in animal models of thrombosis.

Acknowledgement

The entire research and development group at EPIX Medical is thanked for providing data, figures, and useful scientific discussions. Dr. Mark Carvlin is acknowledged for Figure 2 and Dr. John Schenck and GE Global Research for permission to use Figure 3. Prof. Paul Janmey is thanked for the confocal microscopy image in Figure 11.

References:

(1) Weissleder, R.; Mahmood, U. *Radiology* **2001**, *219*, 316-333.
(2) Luker, G. D.; Piwnica-Worms, D. *Acad Radiol* **2001**, *8*, 4-14.
(3) Schenck, J. F. *J. Neurol. Sci.* **2003**, *207*, 99-102.

(4) Pomerantz, S.; Siegelman, E. S. *Magn. Reson. Imaging Clin. N. Am.* **2002**, *10*, 105-120.

(5) Beer, M.; Seyfarth, T.; Sandstede, J.; Landschutz, W.; Lipke, C.; Kostler, H.; von Kienlin, M.; Harre, K.; Hahn, D.; Neubauer, S. *J. Am. Coll. Cardiol.* **2002**, *40*, 1267-1274.

(6) Bolan, P. J.; Meisamy, S.; Baker, E. H.; Lin, J.; Emory, T.; Nelson, M.; Everson, L. I.; Yee, D.; Garwood, M. *Magn. Reson. Med.* **2003**, *50*, 1134-1143.

(7) Blasberg, R. G.; Gelovani, J. *Mol. Imaging* **2002**, *1*, 280-300.

(8) Rolfe, P. *Annu. Rev. Biomed. Eng.* **2000**, *2*, 715-754.

(9) Bremer, C.; Ntziachristos, V.; Weissleder, R. *Eur. Radiol.* **2003**, *13*, 231-243.

(10) Dayton, P. A.; Ferrara, K. W. *J. Magn. Reson. Imaging* **2002**, *16*, 362-377.

(11) Edelman, R. R.; Hesselink, J. R.; Zlatkin, M. B. *MRI: Clinical Magnetic Resonance Imaging*; 2nd ed.; Saunders: Philadelphia, 1996.

(12) Caravan, P.; Ellison, J. J.; McMurry, T. J.; Lauffer, R. B. *Chem. Rev.* **1999**, *99*, 2293-2352.

(13) Muller, R. N.; Roch, A.; Colet, J.-M.; Ouakssim, A.; Gillis, P. In *Chemistry of Contrast Agents in Medical Magnetic Resonance Imaging*; Toth, E., Merbach, A. E., Eds.; Wiley: New York, 2001, pp 417-435.

(14) (a) Wedeking, P.; Shukla, R.; Kouch, Y. T.; Nunn, A. D.; Tweedle, M. F. *Magn. Reson. Imag.* **1999**, 17, 569-575. (b) Nunn, A. D.; Linder, K. E.; Tweedle, M. F. *Q. J. Nucl. Med.* **1997**, *41*, 155-162.

(15) (a) Lauffer, R. B.; Brady, T. J. *Magn. Reson. Imaging* **1985**, *3*, 11-16. (b) Lauffer, R. B.; Brady, T. J.; Brown, R. D., III; Baglin, C.; Koenig, S. H. *Magn. Reson. Med.* **1986**, *3*, 541-548.

(16) see for example, Dwek, R. A. *Nuclear Magnetic Resonance in Biochemistry*, Clarendon Press: Oxford, 1973.

(17) Lauffer, R. B. *Magn. Reson. Med.* **1991**, *22*, 339.

(18) Lauffer, R. B.; Parmclee, D. J.; Dunham, S. U.; Ouellet, H. S.; Dolan, R. P.; Witte, S.; McMurry, T. J.; Walovitch, R. C. *Radiology* **1998**, *207*, 529-538.

(19) Caravan, P.; Cloutier, N. J.; Greenfield, M. T.; McDermid, S. A.; Dunham, S. U.; Bulte, J. W.; Amedio, J. C., Jr.; Looby, R. J.; Supkowski, R. M.; Horrocks, W. D., Jr.; McMurry, T. J.; Lauffer, R. B. *J. Am. Chem. Soc.* **2002**, *124*, 3152-3162.

(20) Aime, S.; Cabella, C.; Colombatto, S.; Geninatti Crich, S.; Gianolio, E.; Maggioni, F. *J Magn Reson Imaging* **2002**, *16*, 394-406.

(21) Caravan, P.; Greenwood, J. M.; Welch, J. T.; Franklin, S. J. *Chem. Commun.* **2003**, 2574-2575.

(22) Ogan, M. D.; Schmiedl, U.; Moseley, M. E.; Grodd, W.; Paajanen, H.; Brasch, R. C. *Invest. Radiol.* **1987**, *22*, 665-671.

(23) Schuhmann-Giampieri, G.; Schmitt-Willich, H.; Frenzel, T.; Press, W. R.; Weinmann, H. J. *Invest. Radiol.* **1991**, *26*, 969-974.

(24) Meyer, D.; Schaefer, M.; Bouillot, A.; Beaute, S.; Chambon, C. *Invest. Radiol.* **1991**, *26*, S50-S52.

(25) Gibby, W. A.; Bogdan, A.; Ovitt, T. W. *Invest. Radiol.* **1989**, *24*, 302-309.

(26) Raduchel, B.; Schmitt-Willich, H.; Platzek, J.; Ebert, W.; Frenzel, T.; Misselwitz, B.; Weinmann, H.-J. *Polym. Mater. Sci. Eng.* **1998**, *79*, 516-517.

(27) Wiener, E. C.; Brechbiel, M. W.; Brothers, H.; Magin, R. L.; Gansow, O. A.; Tomalia, D. A.; Lauterbur, P. C. *Magn. Reson. Med.* **1994**, *31*, 1-8.

(28) Kellar, K. E.; Henrichs, P. M.; Hollister, R.; Koenig, S. H.; Eck, J.; Wei, D. *Magn. Reson. Med.* **1997**, *38*, 712-716.

(29) Glogard, C.; Hovland, R.; Fossheim, S. L.; Aasen, A. J.; Klaveness, J. *J. Chem. Soc., Perkin Trans. 2* **2000**, 1047-1052.

(30) Andre, J. P.; Toth, E.; Fischer, H.; Seelig, A.; Macke, H. R.; Merbach, A. E. *Chem. Eur. J.* **1999**, *5*, 2977-2983.

(31) Torchilin, V.; Babich, J.; Weissig, V. *Journal of Liposome Research* **2000**, *10*, 483-499.

(32) Unger, E.; Shen, D.; Fritz, T.; Wu, G.-L.; Kulik, B.; New, T.; Matsunaga, T.; Ramaswami, R. *Invest. Radiol.* **1994**, *29*, S168-S169.

(33) Kabalka, G. W.; Davis, M. A.; Moss, T. H.; Buonocore, E.; Hubner, K.; Holmberg, E.; Maruyama, K.; Huang, L. *Magn. Reson. Med.* **1991**, *19*, 406-415.

(34) Misselwitz, B.; Schmitt-Willich, H.; Ebert, W.; Frenzel, T.; Weinmann, H.-J. *Magnetic Resonance Materials in Physics, Biology and Medicine* **2001**, *12*, 128-134.

(35) Nicolle, G. M.; Toth, E.; Schmitt-Willich, H.; Raduchel, B.; Merbach, A. E. *Chem. Eur. J.* **2002**, *8*, 1040-1048.

(36) Konda, S. D.; Aref, M.; Brechbiel, M.; Wiener, E. C. *Invest. Radiol.* **2000**, *35*, 50-57.

(37) Konda, S. D.; Wang, S.; Brechbiel, M.; Wiener, E. C. *Invest. Radiol.* **2002**, *37*, 199-204.

(38) Wiener, E. C.; Konda, S.; Shadron, A.; Brechbiel, M.; Gansow, O. *Invest. Radiol.* **1997**, *32*, 748-754.

(39) Storrs, R. W.; Tropper, F. D.; Li, H. Y.; Song, C. K.; Sipkins, D. A.; Kuniyoshi, J. K.; Bednarski, M. D.; Strauss, H. W.; Li, K. C. *J Magn Reson Imaging* **1995**, *5*, 719-724.

(40) Sipkins, D. A.; Gijbels, K.; Tropper, F. D.; Bednarski, M.; Li, K. C. P.; Steinman, L. *J. Neuroimmunol.* **2000**, *104*, 1-9.

(41) Sipkins, D. A.; Cheresh, D. A.; Kazemi, M. R.; Nevin, L. M.; Bednarski, M. D.; Li, K. C. *Nat Med* **1998**, *4*, 623-626.

(42) Winter, P. M.; Caruthers, S. D.; Yu, X.; Song, S. K.; Chen, J.; Miller, B.; Bulte, J. W.; Robertson, J. D.; Gaffney, P. J.; Wickline, S. A.; Lanza, G. M. *Magn Reson Med* **2003**, *50*, 411-416.

(43) Flacke, S.; Fischer, S.; Scott, M. J.; Fuhrhop, R. J.; Allen, J. S.; McLean, M.; Winter, P.; Sicard, G. A.; Gaffney, P. J.; Wickline, S. A.; Lanza, G. M. *Circulation* **2001**, *104*, 1280-1285.

(44) Anderson, S. A.; Rader, R. K.; Westlin, W. F.; Null, C.; Jackson, D.; Lanza, G. M.; Wickline, S. A.; Kotyk, J. J. *Magn Reson Med* **2000**, *44*, 433-439.

(45) Winter, P. M.; Morawski, A. M.; Caruthers, S. D.; Fuhrhop, R. W.; Zhang, H.; Williams, T. A.; Allen, J. S.; Lacy, E. K.; Robertson, J. D.; Lanza, G. M.; Wickline, S. A. *Circulation* **2003**, *108*, 2270-2274.

(46) Winter, P. M.; Caruthers, S. D.; Kassner, A.; Harris, T. D.; Chinen, L. K.; Allen, J. S.; Lacy, E. K.; Zhang, H.; Robertson, J. D.; Wickline, S. A.; Lanza, G. M. *Cancer Res* **2003**, *63*, 5838-5843.

(47) Weissleder, R.; Reimer, P.; Lee, A. S.; Wittenberg, J.; Brady, T. J. *AJR Am J Roentgenol* **1990**, *155*, 1161-1167.

(48) Wunderbaldinger, P.; Josephson, L.; Weissleder, R. *Acad Radiol* **2002**, *9*, S304-306.

(49) Weissleder, R.; Moore, A.; Mahmood, U.; Bhorade, R.; Benveniste, H.; Chiocca, E. A.; Basilion, J. P. *Nat Med* **2000**, *6*, 351-355.

(50) Moore, A.; Josephson, L.; Bhorade, R. M.; Basilion, J. P.; Weissleder, R. *Radiology* **2001**, *221*, 244-250.

(51) Kang, H. W.; Josephson, L.; Petrovsky, A.; Weissleder, R.; Bogdanov, A., Jr. *Bioconjug Chem* **2002**, *13*, 122-127.

(52) Schellenberger, E. A.; Hogemann, D.; Josephson, L.; Weissleder, R. *Acad Radiol* **2002**, *9*, S310-311.

(53) Barrett, J. A.; Kolodziej, A. F.; Caravan, P.; Nair, S.; Looby, R. J.; Witte, S.; Costello, C. R.; Melisi, M. A.; Drezwecki, L.; Chesna, C.; Pratt, C.; McMurry, T. J.; Lauffer, R. B.; Yucel, E. K.; Weisskoff, R. M.; Carpenter, A. P.; Graham, P. B. In *75th AHA Scientific Sessions*: Chicago, IL, 2002, p 603.

(54) Caravan, P.; Kolodziej, A. F.; Greenwood, J. M.; Witte, S.; Looby, R. J.; Zhang, Z.; Spiller, M.; McMurry, T. J.; Weisskoff, R. M.; Graham, P. B. In *Proc. ISMRM*: Honolulu, HI, 2002, p 217.

(55) Lauffer, R. B.; Graham, P. B.; Lahti, K. M.; Nair, S.; Caravan, P.; Kolodziej, A. F. In *73rd AHA Scientific Sessions*: New Orleans, LA, 2000, p 1831.

(56) Wescott, C. R.; Nair, S. A.; Kolodziej, A.; Beltzer, J. P. U. S. Patent Application US2003143158, 2003.

(57) Botnar, R. M.; Perez, A. S.; Witte, S.; Wiethoff, A. J.; Laredo, J.; Quist, W.; Parsons, E. C.; Vaidya, A.; Kolodziej, A. F.; Barrett, J. A.; Graham, P. B.; Weisskoff, R. M.; Manning, W. J.; Johnstone, M. T. *Circulation* **2004**, *109*, 2023-29.

Chapter 12

Synthesis and Development of Gadomer: A Dendritic MRI Contrast Agent

Johannes Platzek[1] and Heribert Schmitt-Willich[2,*]

[1]Process Research A and [2]Research Center Europe, Schering AG, 13342 Berlin, Germany

A novel dendritic contrast agent for magnetic resonance imaging (MRI) has been synthesized. Gadomer consists of 24 macrocyclic gadolinium chelates covalently linked to a dendritic L-lysine skeleton. Due to its high molecular weight (17.5 kDa) the compound has excellent blood pool properties and remains in the intravascular space for a prolonged period of time after i.v. administration without relevant extravasation. This is the first dendritic gadolinium-based contrast agent to have entered clinical trials. The drug proved to be safe and its pharmacological/toxicological profile as well as the imaging properties are excellent. The development of a reproducible and robust synthesis led to the preparation of several multi kg batches for tox studies and clinical trials.

Gadolinium-based contrast agents play a major role in Magnetic Resonance Imaging (MRI) (*1*). Due to its unique paramagnetic properties the Gd(III)-ion is by far the most efficient metal ion in terms of reduction of T1 relaxation times of the surrounding water. Conventional extracellular gadolinium chelates are of great value in many clinical examinations. However, the use of these low molecular weight chelates is limited in MR angiography (MRA). Since these agents quickly 'leak out' of the blood stream, the imaging time window is restricted to 'first-pass' measurements within the first seconds after the injection. Hence, for angiographic applications, e.g. coronary angiography, a longer residence time of the contrast agent is required in order to perform equilibrium or so-called 'steady-state' measurements.

Polymeric Blood Pool Agents

Since as early as the mid-80's, many research groups investigated the use of high molecular weight structures as contrast agents featuring prolonged circulation in the intravascular space. This was achieved by attachment of the chelates to proteins, e.g. albumin (*2-4*), oligosaccharides, e.g. dextran (*5, 6*) or polyamino acids like poly(L-lysine) (*7-11*) or pegylated poly(L-lysine) (*12*), respectively.

None of these experimental macromolecular agents have ever entered human clinical trials. Besides the prerequisites all potential contrast agents have to fulfill like e.g. low toxicity, high water-solubility and high metal content, there are two almost contradictory prerequisites in the case of potential blood pool agents: [1] sufficient retention in the intravascular compartment and [2] quantitative renal elimination of the gadolinium-containing polymer. As discussed by Schuhmann-Giampieri et al. (*7*), even the most widely used experimental blood pool agent among the above mentioned agents, i.e. Gd-DTPA-polylysine, does not meet both of these criteria due to unavoidable heterogeneity of the backbone.

More recently, Gd-DTPA and Gd-DO3A dendrimers have been synthesized (*13-19*). They were shown to have much higher homogeneity than conventional linear backbones. However, in the case of polyamidoamines (*13-15*) it is known that defect structures occur during the repetitive Michael addition and aminolysis reactions (*20*). In contrast, a so-called Gd-DTPA-cascade-polymer (*16-19*) was based on a different synthetic approach which led ultimately to the design of Gadomer (*21*).

Gadomer

This 17.5 kDa 'gadolinium dendrimer', initially called Gadomer-17, is characterized by the target structure "Gadomer 24" (Figure 1) which contains 24 gadolinium complexes, plus limited amounts of related dendrimers containing 16-26 macrocyclic gadolinium chelate moieties and varying numbers of lysine units. The target structure ‚Gadomer 24‘ has 24 macrocyclic Gd-chelates covalently linked to the 24 amino groups of 12 L-lysyl residues at the surface of the dendritic skeleton (Figure 1).

For the sake of simplicity, the more general term Gadomer is used throughout the text if it is not a specific matter of defect structures.

Gadomer attracted quickly some attention for its presumed first application of a dendritic molecule as a drug (*22-24*). The quantitative renal elimination together with high acute tolerance and very favorable pharmacokinetics (*25*) made Gadomer a promising blood pool agent (*26-30*). Water exchange kinetics, rotational dynamics and electronic relaxation have also been discussed in terms of their implications for proton relaxivity of this dendrimer (*31*).

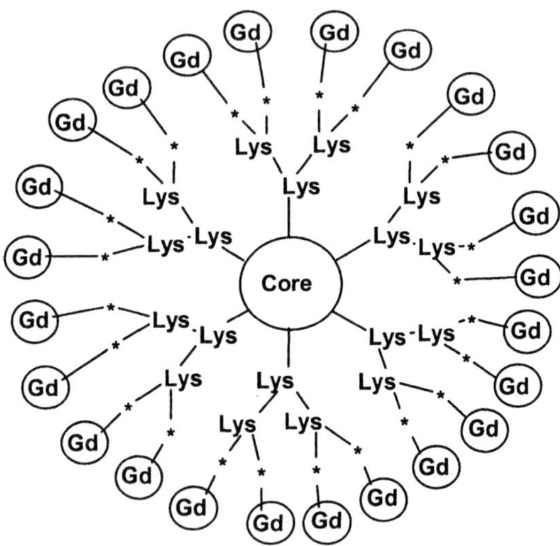

Figure 1. Simplified target structure of Gadomer (1). For details see Figure 2 and 7.

Synthesis of Gadomer

The structure of Gadomer consists of two main components, the dendritic skeleton with 24 amino groups at the surface and the macrocyclic cyclen-based gadolinium chelates. These two parts were synthesized separately and combined in the final steps.

The dendritic skeleton

The key intermediate building block containing the dendritic part of the molecule is the fully protected Z-24-amine 2 (see Figure 2). All 24 primary amino functions at the surface of 2, resulting from 12 L-lysyl residues, are protected by Z (benzyloxycarbonyl) groups. Three different synthetic routes have been explored during the upscaling process and the development of Gadomer.

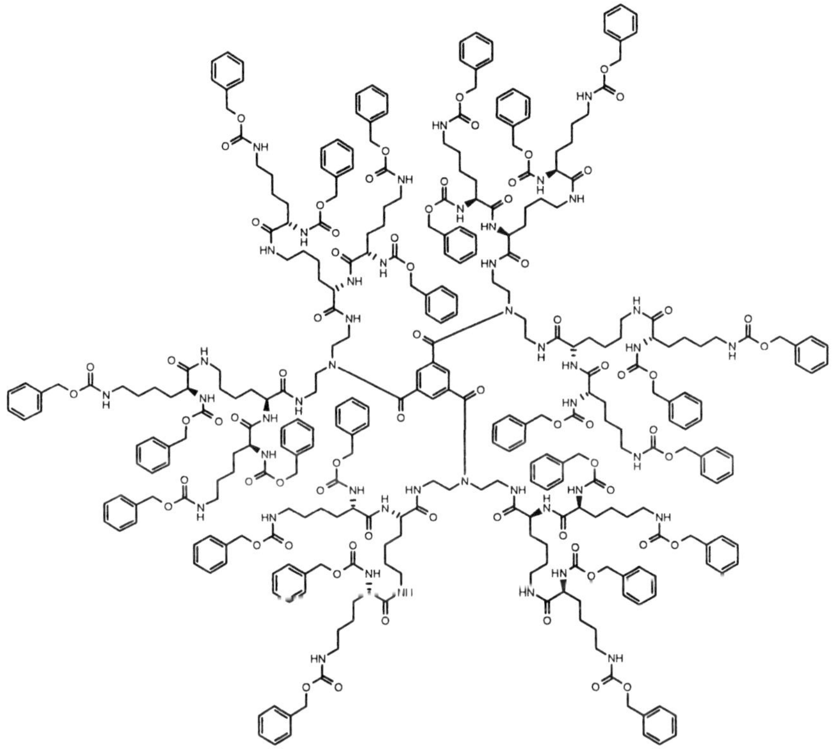

Figure 2. Structure of Z-24-amine 2

Figure 3 shows our first approach. Diethylene triamine was reacted with two equivalents of benzyloxycarbonylnitrile, a selective acylation agent that can be directed selectively towards the protection of primary amino groups. The resulting Z-2-amine 3 was isolated in crystalline form and high yield. Subsequently, three equivalents of 3 were reacted with trimesoyl trichloride (benzene-1,3,5-tricarboxylic acid trichloride) in DMF / triethylamine to give Z-6-amine 4 in high purity. Deprotection under standard conditions (HBr/glacial acetic acid) gave the hexabromide 5 which was finally reacted with 6 equivalents of N_α, N_ε-bis(dibenzyloxycarbonyl-L-lysyl)-L-lysine ("trilysine") building block 6.

Trilysine 6 was easily obtained by reaction of L-lysine hydrochloride in DMF with 2 equivalents of N_α, N_ε-dibenzyloxycarbonyl-L-lysine para-nitrophenyl ester (Z-Lys(Z)-ONp).

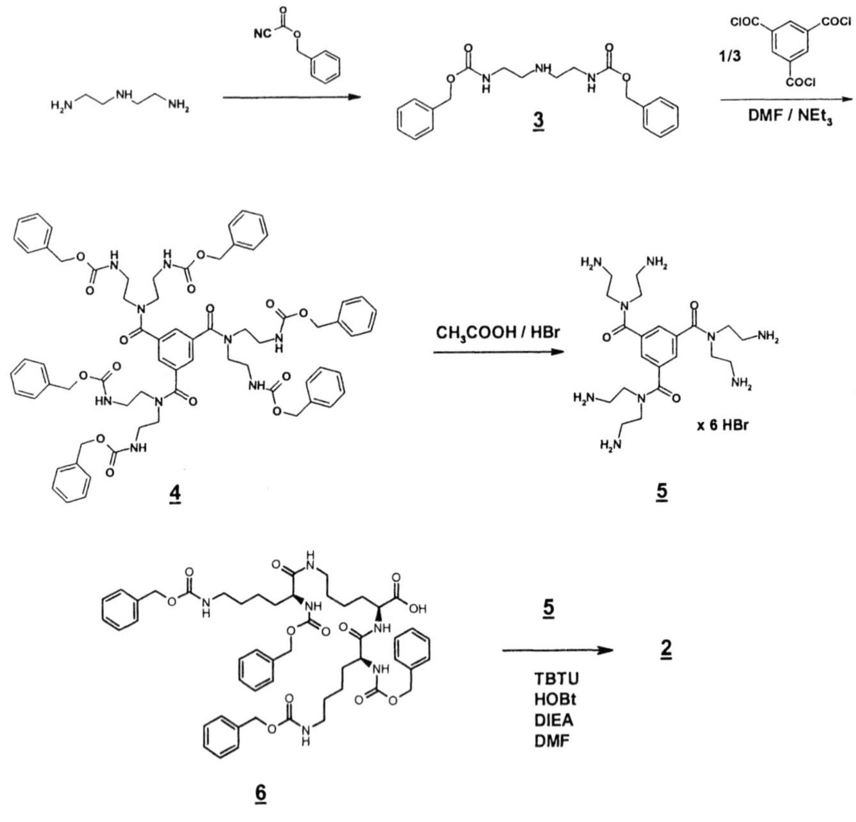

Figure 3. First synthesis of Z-24-amine 2

6 was activated by TBTU (O-[Benzotriazol-1-yl]-N,N,N',N'-tetramethyluronium tetrafluoroborate) in the presence of 1-Hydroxybenzotriazole (HOBt) and then reacted with 5 in DMF and excess of diisopropyl ethyl amine (DIEA) to yield 2 after several chromatographic steps (ethyl acetate/ethanol 2:1). As known from the activation of dipeptides (*32*), this coupling strategy via trilysine resulted in some epimerization of the activated L-lysine.

In the second approach (Figure 4) the two generations of lysine residues were introduced one after the other. Hexabromide 5 was reacted with Z-Lys(Z)-ONp in DMF/triethylamine to build up the first generation of six lysine units in Z-12-amine 7. After deprotection (HBr/glacial acetic acid) the second generation was introduced by coupling again Z-Lys(Z)-ONp under the same conditions as before. The target compound 2 was isolated after only one chromatographic step without any D-Lys detectable.

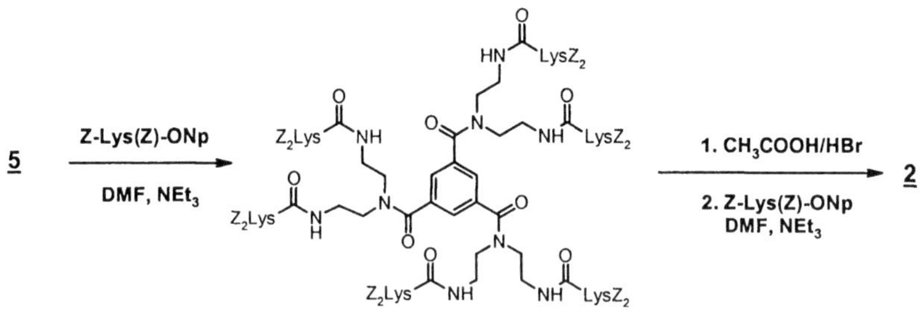

Figure 4. Synthesis of Z-24-amine 2 via Z-12-amine 7

Upscaling of the synthesis of 2, however, had still to deal with some critical issues concerning safety of the reagents and the number of synthetic steps. First, in order to replace the highly toxic und expensive benzyloxycarbonylnitrile, an alternative pathway was studied in more detail (see Figure 5).

The reaction of diethylene triamine with two equivalents of ethyl trifluoroacetate gave the crystalline bis(trifluoroacetyl)amide 8 in nearly quantitative yield. Trimerization with trimesoyl trichloride resulted in the desired TFA-6-amine 9. Under the alkaline saponification conditions, however, the free hexaamine underwent a partial rearrangement. This could be demonstrated by addition of Z-Cl (benzyloxycarbonylchloride) in one pot, which led to a mixture of four species 4, 4a, 4b and 4c with the target molecule 4 as the main compound. The separation of this mixture was very difficult on large scale and so no further activities have been invested in this route.

The third route leading to Z-24-amine 2 is depicted in Figure 6. Diethylene triamine was converted to 10 by acylation with 2 equivalents of Z-Lys(Z)-ONp. This so-called Z-4-amine 10 was obtained in crystalline form and high yield. The reaction was clean and easy to scale up. Subsequent trimerization with trimesoyl trichloride, again in DMF/triethylamine, gave the already known Z-12-amine 7

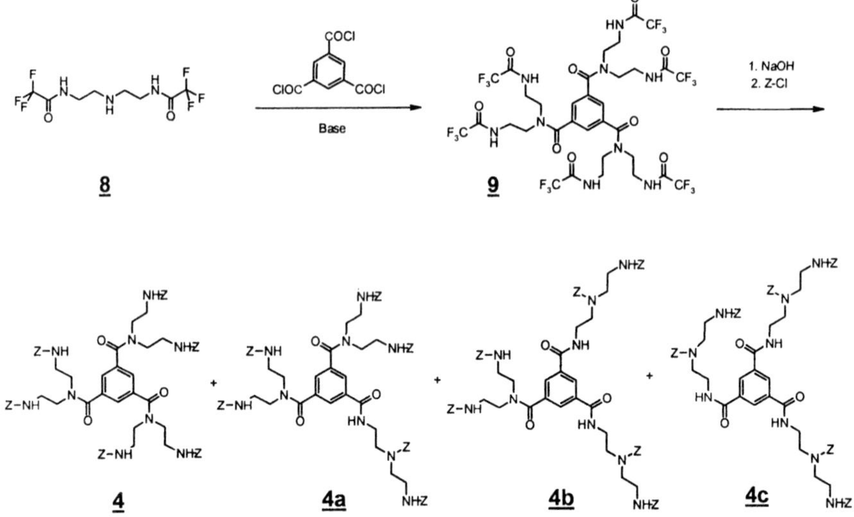

*Figure 5. Attempted Synthesis of Z-6-amine **4** via **9***

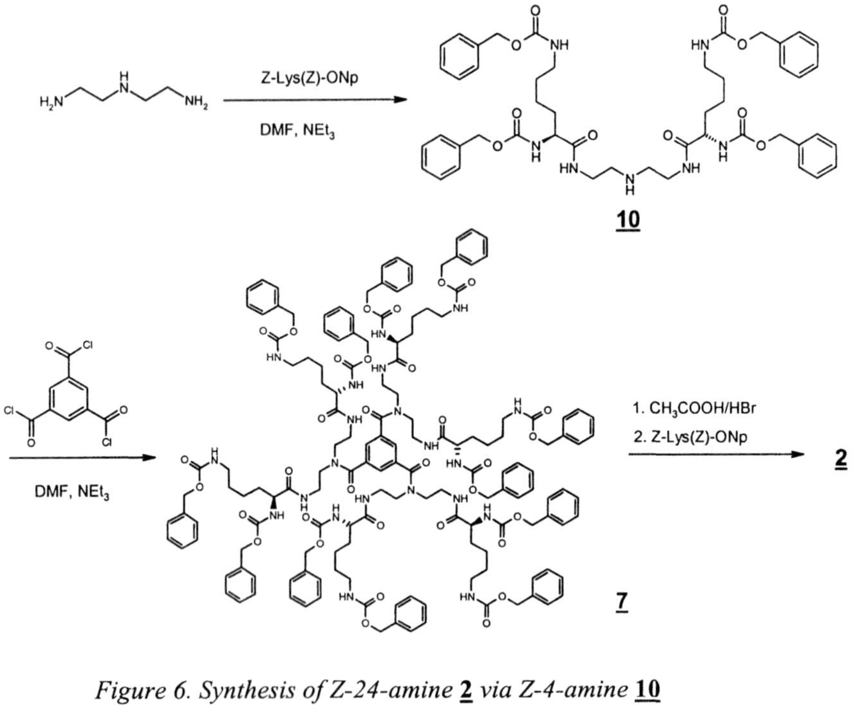

*Figure 6. Synthesis of Z-24-amine **2** via Z-4-amine **10***

in high yield and excellent purity. Deprotection of the Z-groups (HBr/glacial acetic acid) and introduction of the second lysine generation by reaction of Z-Lys(Z)-ONp was performed as described above. High quality Z-24-amine **2** was isolated in high overall yield after chromatographic purification with dichloromethane/methanol (18:1).
This third route accomplished the synthesis of **2**, which contains the full dendritic skeleton of Gadomer with a molecular weight of nearly 6000 Da, in only four synthetic steps and two chromatographic purification steps. The procedure was successfully scaled up and **2** has been prepared in kilogram quantities in the pilot plant.

The chelator

The chelator was prepared according to Figure 7. Glycine benzyl ester p-toluenesulfonate salt was acylated with 2-bromopropionic acid chloride in dichloromethane/triethylamine. The resulting amide **11** was reacted in chloroform with an excess of cyclen (1,4,7,10-tetraazacyclododecane) to give the mono-functionalized cyclen derivative **12**. The organic phase was washed with water in order to remove the excess of cyclen. **12** was then treated, without further purification, with an excess of *t*-butyl bromoacetate, in the presence of acetonitrile/sodium carbonate to yield the protected chelator **13**. Finally, the benzyl ester was removed by catalytic hydrogenation in isopropanol to give the target compound **14**, which was obtained in crystalline form and in high overall

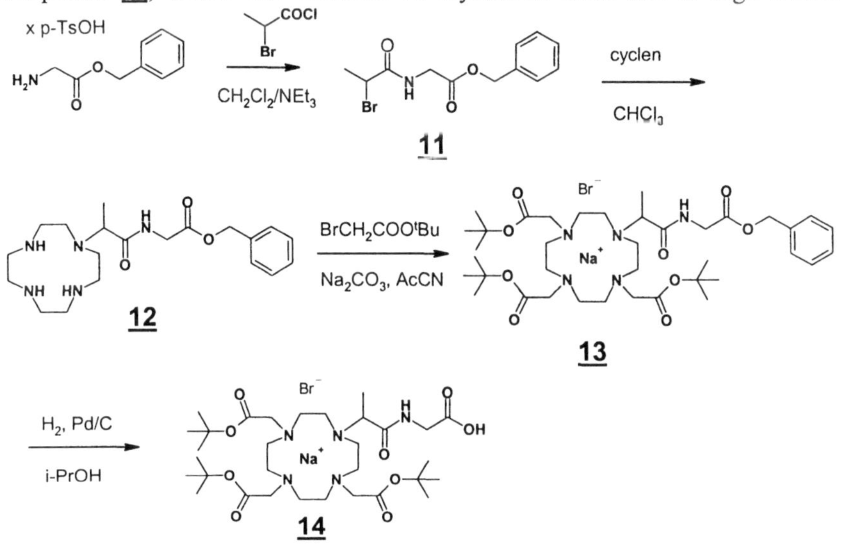

Figure 7. Synthesis of the protected chelator **14**

yield without the need for chromatography. This synthesis was scaled up to the multi kg range in the pilot plant.

Final steps

The first synthesis of Gadomer was accomplished by the following steps: [1] attachment of **14** onto the dendritic skeleton, [2] deprotection, and [3] complexation.

In more detail, Z-24-amine **2** was deprotected under standard acidic conditions (HBr/acetic acid). The resulting H-24-amine x 24 HBr (**15**) was subsequently reacted with the protected chelator **14** under peptide coupling conditions using TBTU/ HOBt in DMF. The fully *t*-butyl protected Gadomer precursor was not isolated but treated with trifluoroacetic acid for complete deprotection. The raw polymacrocyclic chelating acid was then complexed with gadolinium oxide in water (80°C). The solution was filtered from remaining metal oxide and then purified by an ultrafiltration (cut off 5,000 Da) in water in order to remove salts and small molecular weight impurities. Finally, the aqueous retentate was treated alternately with cation and anion exchange resins in order to remove charged by-products. Gadomer remains unaffected by this procedure as the target structure is a neutral molecule.

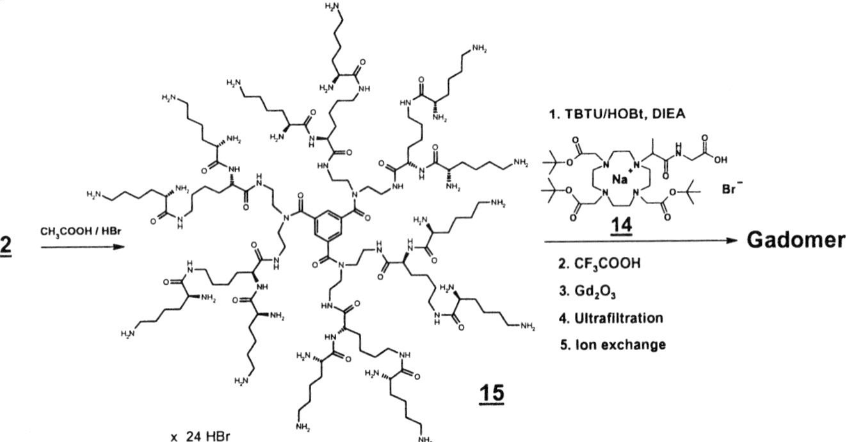

Figure 8. First synthesis of Gadomer

Although the Matrix Assisted Laser Desorption Ionization - Time of Flight (MALDI-TOF) mass spectrum of this material showed some dendritic impurities with related masses (see Figure 11, below), we considered this synthesis as a success.

It turned out that the other byproducts resulted from an uncomplete acylation of the amino groups of H-24-amine 15. Therefore, defect structures have been observed that contain mainly 23, 22 and 21 Gd-chelates, respectively. We assumed that the steric hindrance of the close *t*-butyl groups was responsible for the incomplete reaction. Therefore we tried to avoid conventional protecting groups and to couple the Gd chelate directly to 15. In this strategy, the metal atom itself would act as a 'protecting group' of the carboxylates. To the best of our knowledge no amide coupling of a Gd chelate was known in the literature when we tried this strategy. So the main question was under which conditions the hydrophilic Gd chelate could be coupled to the dendritic amine 15. A major obstacle for the chemistry of Gd chelates is the poor solubility of these compounds in organic solvents.

The Gd-chelate 16 (Gd-GlyMeDOTA) was synthesized from 14 by deprotection of the *t*-butyl groups (HBr/glacial acetic acid) and introduction of gadolinium (Gd_2O_3, 80°C) in water (Figure 9). Crystalline Gd-GlyMeDOTA was obtained in high purity from ethanol. The complexation has been done on a multi-kg scale in the pilot plant without any problems.

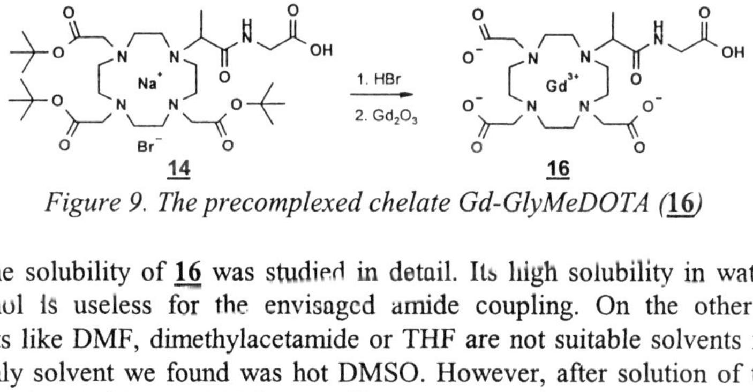

Figure 9. The precomplexed chelate Gd-GlyMeDOTA (16)

The solubility of 16 was studied in detail. Its high solubility in water and methanol is useless for the envisaged amide coupling. On the other hand, solvents like DMF, dimethylacetamide or THF are not suitable solvents for 16. The only solvent we found was hot DMSO. However, after solution of the Gd complex in DMSO at 100°C it crystallizes out on cooling to room temperature. Only the addition of alkali metal salts to the solution in DMSO stabilizes the solution and prevents crystallization. In our hands the use of LiCl, LiBr or NaBr gave the best results (*33*, *34*). The N-hydroxysuccinimide (NHS) ester of 16 is easily obtained under standard conditions (dicyclohexylcarbodiimide) in DMSO with LiCl as additive.

Figure 10 shows the final synthesis of Gadomer. H-24-amine x 24 HBr (15) was converted to its free amine by treatment with an anion exchange resin in water. The resin was filtered off and the free poly-amine 15a isolated by freeze drying. In the final reaction step, this amine was added to a solution of the freshly generated N-hydroxysuccinimide ester of 16 in DMSO. The reaction was complete after 6 hours and crude Gadomer was precipitated by the addition of ethyl acetate. The solid was collected by filtration and dissolved in water for the final purification steps by ultrafiltration (*35*).

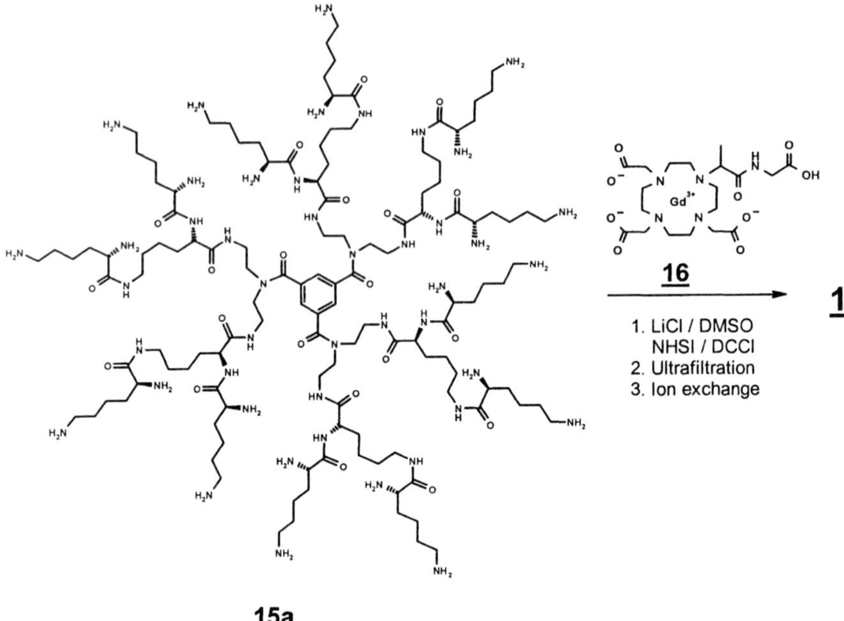

Figure 10. Final synthesis of Gadomer (1)

First, an ultrafiltration membrane with a cut off of 5 kDa was used to remove salts and small molecular weight impurities. In a second ultrafiltration step using a 30 kDa membrane the eluate was collected. In this step high molecular weight contaminants and endotoxins were removed. As in the procedure described above, the aqueous solution was treated alternately with cation and anion exchange resins in order to remove charged by-products. Finally, 1 was isolated by freeze drying and obtained as a white fluffy powder.

Analytical characterization

This dendrimer is difficult to characterize due to its size and due to the presence of the paramagnetic gadolinium. Therefore, conventional 1H-NMR und 13C-NMR cannot be used for its structure confirmation. Nevertheless, several other analytical methods proved to be suitable. Mass spectroscopic techniques were of utmost importance in the characterization of Gadomer.

MALDI-TOF

The Matrix Assisted Laser Desorption Ionization-Time of Flight (MALDI-TOF) technique is by far the best qualitative method for the characterization of Gadomer. Two MALDI spectra of Gadomer resulting from different synthetic pathways are presented in Figure 11 and 12, respectively. The first spectrum (Figure 11) shows the presence of defect structures resulting from the incomplete coupling during the final step of our first synthesis (see Figure 8 for synthetic details).

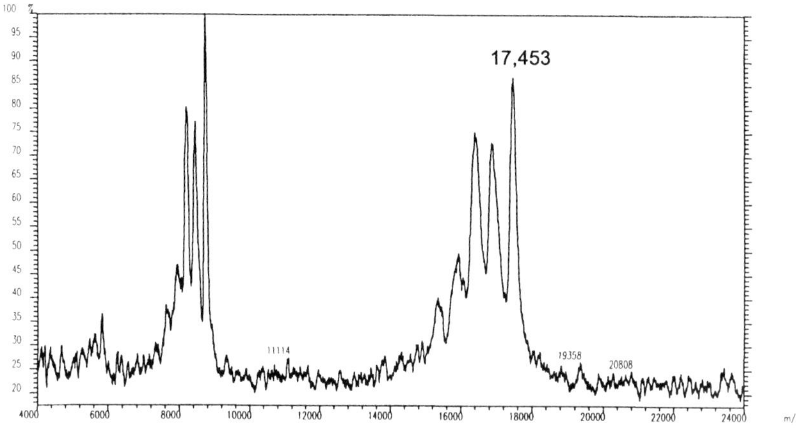

Figure 11. MALDI-TOF MS of Gadomer, first synthetic approach

The second spectrum (Figure 12) demonstrates the breakthrough with the new synthesis (see Figure 10 for synthetic details) where the target molecule Gadomer with its 24 chelates and a theoretical mass of 17,453 Da is the main product.

However, the half width of this MALDI peak is relatively broad. As natural gadolinium consists of several isotopes the 24mer Gadomer has a statistical distribution of molecular weights. Figure 13 shows a simulation of this effect. The theoretical half width is approximately 25 Da but experimentally, peaks with half widths of > 100 Da have been found. This has of course some consequences for the structure elucidation of byproducts which show only minor differences in mass. Simulation experiments show that dendrimers which differ in about 40 Da will even show an overlap of the peaks if the experimental half width is only 25 Da.

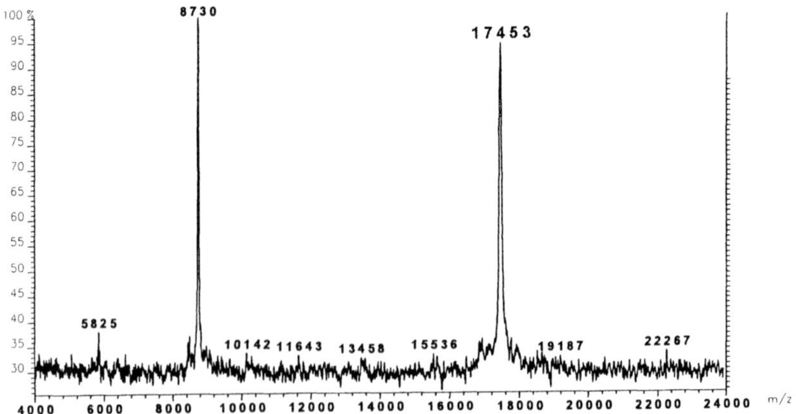

Figure 12. MALDI-TOF MS of Gadomer, final synthesis

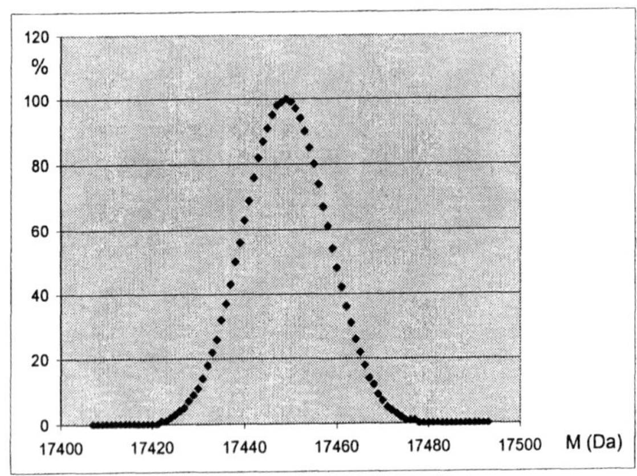

Figure 13. Simulation of MALDI-TOF MS of Gadomer

HPLC

The purity of Gadomer batches was routinely checked with HPLC: The standard method used a Polymer-amine Phase 5μm (Advanced Separation Technologies Inc.); mobile phase: H_2O/HCOOH/acetonitrile ($\lambda = 200$ nm).

Figure 14 shows a typical HPLC chromatogram for a Gadomer batch prepared by the final synthesis procedure (see Figure 10).

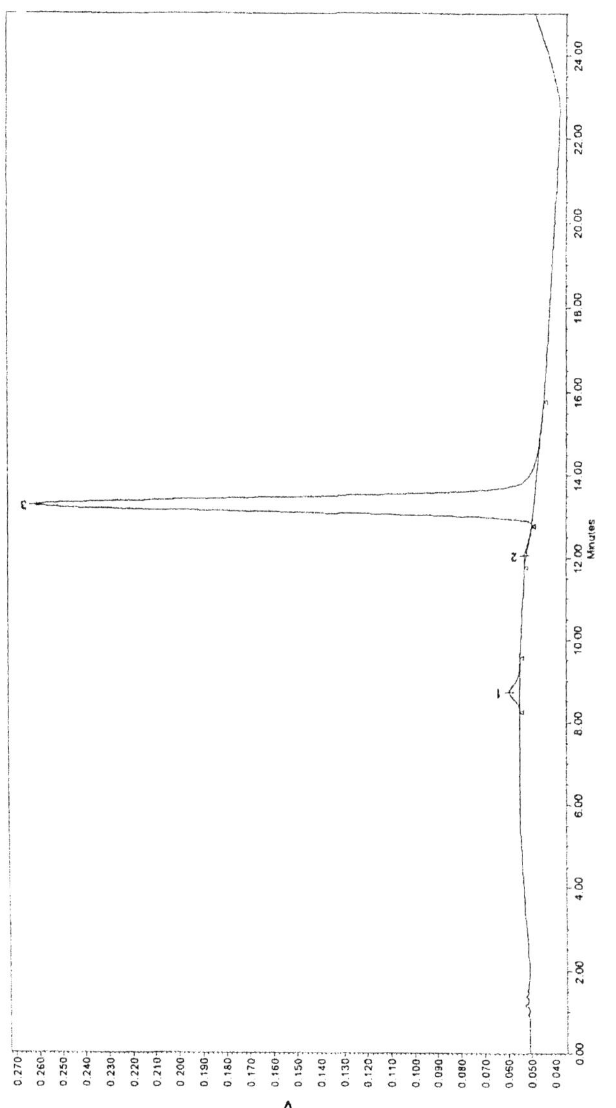

Figure 14. HPLC of Gadomer

The purity range according to HPLC was between 92 and 97%. The major byproduct in Gadomer is the so-called 'Gadomer 23' which lacks one Gd-GlyMeDOTA complex due to incomplete coupling in the final step.

Other Methods

Elemental analysis (C,H and N), determination of gadolinium by Atomic Absorption Spectroscopy (AAS) and Inductively Coupled Plasma (ICP) have been used routinely. Additionally, gel permeation chromatography, gel and capillary electrophoresis as well as isoelectric focussing are useful tools for the characterization of Gadomer.

Discussion

24 Amide bonds are formed during the final coupling step. A simple calculation shows that the yield per single amide bond must be very high in order to obtain a reasonable purity of the dendritic product. Figure 15 shows the correlation between the yield per amide bond and the corresponding purity of Gadomer.

Based on our HPLC studies which show a Gadomer purity of up to 97%, the calculated yield in such a case is > 99.85% of the theory per single bond. This is remarkably high for an organic reaction. In order to achieve such high yields large excesses of reagents have to be used like in solid phase peptide synthesis.

Amplification of impurities

In order to reach the high quality standards of a drug substance applied to humans it is mandatory to have a strong specification that defines the upper limits of impurities and byproducts of all intermediates throughout the whole synthesis. Impurities with chemically active functional groups might compete with the target reagent and this will result in complex mixtures in the following step. Due to the dendritic i.e. multifunctional nature of the molecule an amplification effect towards impure products is observed. This means that the percentage of a reactive impurity (imp) in the (non-dendritic) starting material leads to a much higher occurence in the (dendritic) product. Figure 16 shows this effect for the final coupling step.

The calculation reveals that a quality of the Gd complex acid **16** that contains 1% of an impurity which possesses a COOH group (Imp) will result in about 20% of the defect structure '$Gd_{23}Imp_1$-Gadomer' (Figure 17). Even if the amount of such an impurity Imp is only about 0.1%, the resulting defect structure will be present to about 2.3% of the total amount. Additionally, with an

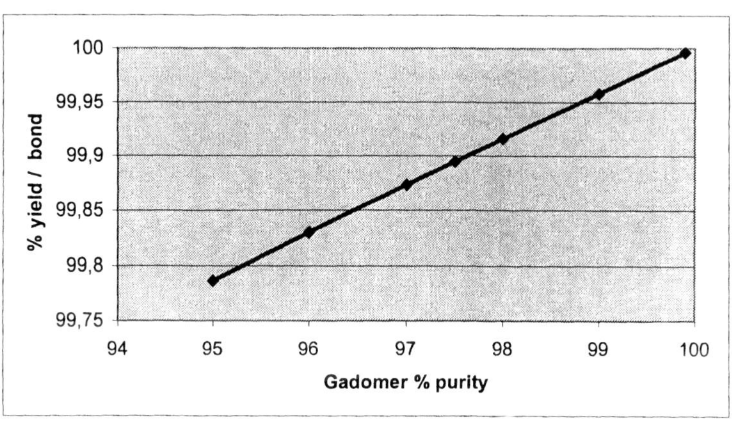

Figure 15. Purity as function of coupling yield/amide bond

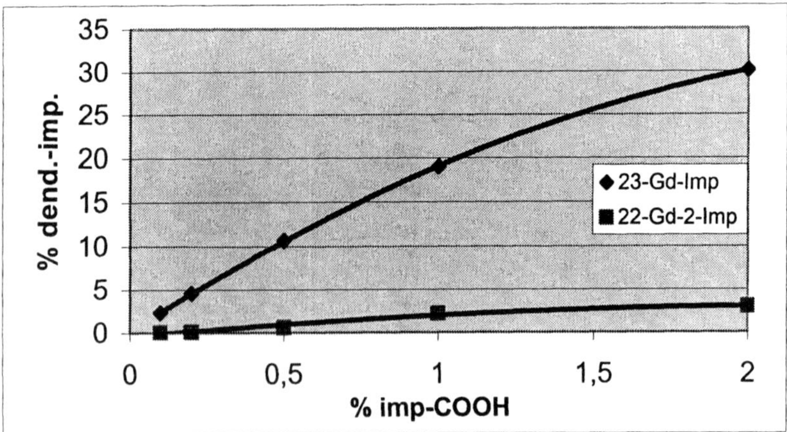

Figure 16. Percentage of defect structures as function of a reactive impurity

increasing amount of Imp further defect structures will be obtained which show two or more impurity units as part of the dendrimer.

A thorough HPLC quality control ensures that GdGlyMeDOTA (**16**) used in the final coupling step has a very high purity (>> 99%) and has no potentially active byproducts.

However, it should be noted that this effect of amplification of impurities is an inherent difficulty during numerous steps of this synthesis. Another example is the formation of 'Gadomer 25' (Figure 19).

Gadomer 25

This molecule contains an additional lysine. In order to understand the formation of this product one has to look at the starting material N_{α}-N_{ε}-dibenzyloxycarbonyl-L-lysine (Z-Lys(Z)-OH). This is prepared by the reaction of benzyloxycarbonyl-chloride (Z-Cl) with L-lysine. However, if the product Z-Lys(Z)-OH reacts with an additional Z-Cl, a mixed anhydride is formed which is able to react with a free amine group present in the reaction mixture. The regioisomeric dipeptides A and/or B might be formed as impurities (Figure 18). As Z-12-amine **7** after deprotection is treated with Z-Lys(Z)-OH to give Z-24-amine **2** (Figure 6), traces of A or B will lead to Z-25-amine, the precursor of Gadomer 25. It was empirically found that the content of A+B present in Z-Lys(Z)-OH has to be lower than 0.02% in order to limit the content of this byproduct Z-25-amine to below 1%.

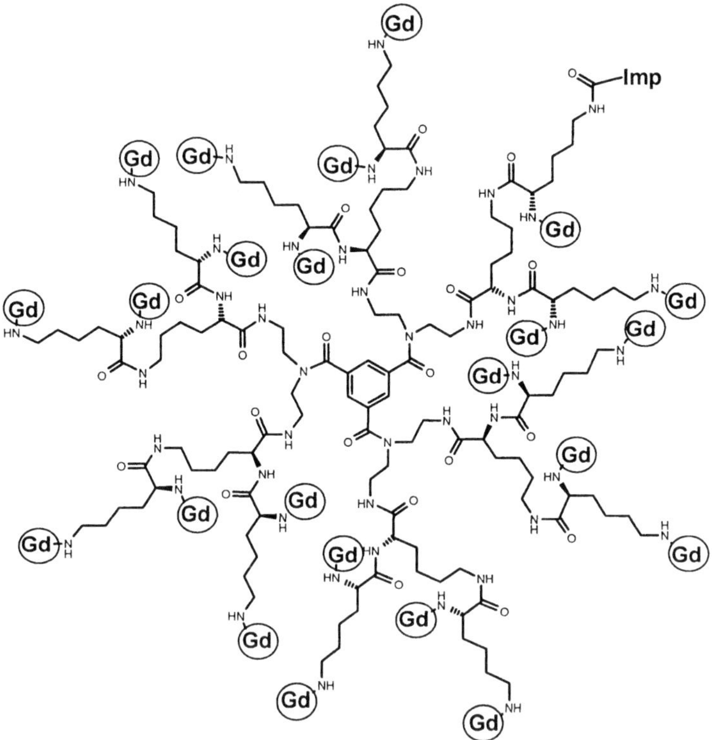

Figure 17. Gd_{23}-Imp_1-Defect structure (position arbitrarily chosen)

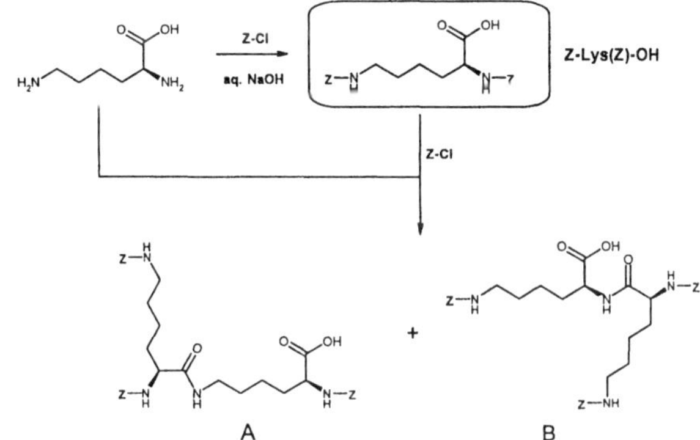

Figure 18. Side reaction in the synthesis of Z-Lys(Z)-OH

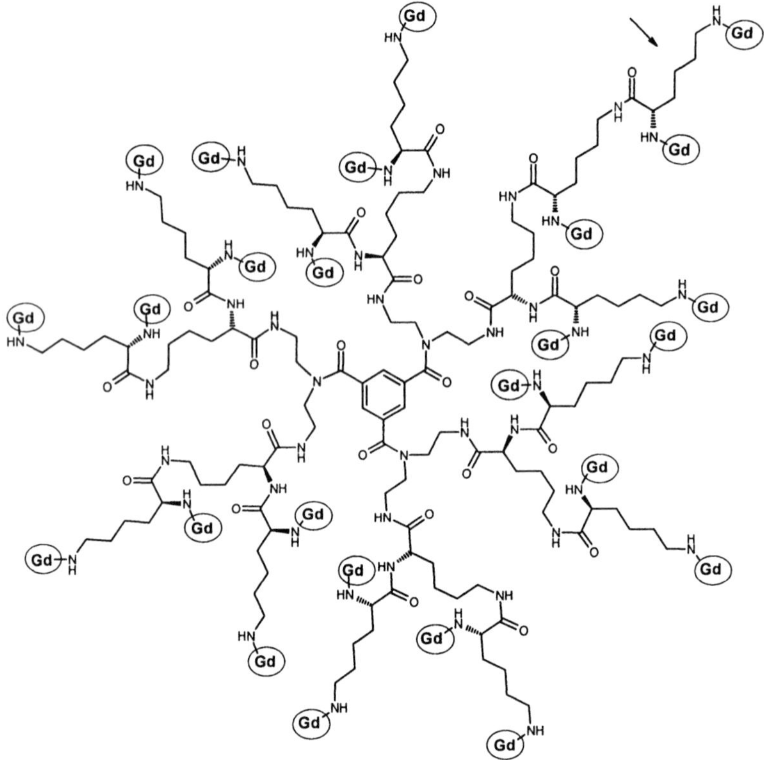

Figure 19. Gadomer 25 (position of 25th lysine arbitrarily chosen).

Active minor components

It can be concluded from the preclinical and clinical investigations that Gadomer is a very safe and well tolerated drug. The small amount of gadolinium containing byproducts present in the drug substance has nearly the same physicochemical properties in terms of relaxivity and pharmacokinetics. These related dendrimers which have the same efficacy and safety profile as the target molecule 'Gadomer 24' are defined as 'active minor components'. The predominant single species within this group is the so-called 'Gadomer 23' with one free amino group. This minor component results from an incomplete coupling of the Gd complex **16** in the final coupling step (Figure 10).

The kinetics of this process is shown in Figure 20. Two equivalents of **16** are used per amino group of H-24-amine **2**. The reaction is highly advanced after 15–30 minutes as only Gadomer 23 and the target molecule Gadomer 24 are present at this time point. In order to reach a nearly complete conversion the reaction time has to be six hours. However, even after this time there is still some Gadomer 23 present (2-4%). Attempts to use a higher excess of <u>**16**</u> in order to drive the reaction to completion have failed. We assume that the remaining amino group in some of the Gadomer 23 isomers is sterically hindered and is therefore prevented from complete acylation.

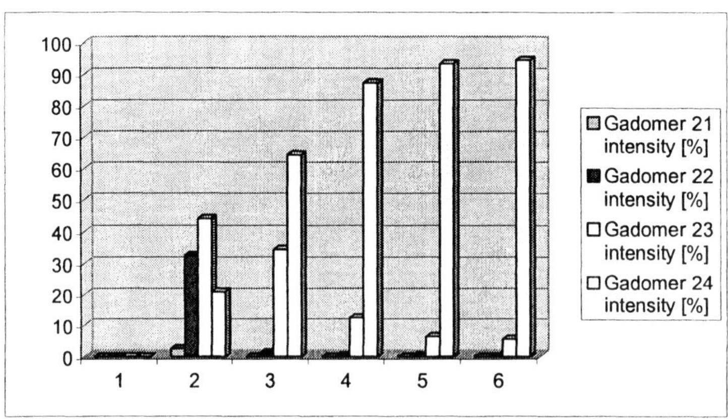

Figure 20. Percentage of Gadomer intermediates at different time points: (1) 0 min, (2) 5 min, (3) 15 min, (4) 30 min, (5) 60 min, (6) 120 min.

Drug product

Gadomer is developed as a sterile aqueous solution for i.v. administration. The concentration is 0.5 M based on gadolinium.

The formulation (drug product) also contains the macrocyclic calcium complex CaNaButrol (Figure 21) which is approved as additive in Gadovist® *(37)*.

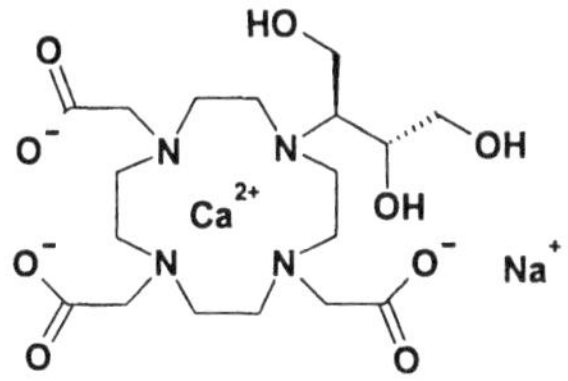

Figure 21. CaNaButrol

Stability of the drug product

Stability tests indicate that Gadomer is a very stable compound. The weakest functional groups are the amide bonds. Five different types out of a total of 69 amide bonds in the molecule can be differentiated: [1] between the macrocycle and glycine, [2] between glycine and the 2nd generation of lysines, [3] between the 1st and 2nd generation of lysines, [4] between the inner lysine and a diethylene triamine arm, and [5] the core trimesoyl amide.

Interestingly, the main degradation reaction that has been observed is the cleavage of one Gd-GlyMeDOTA resulting in Gadomer 23 as an active minor component. The high stability of the formulation and of the drug substance was shown by stress tests. Taking into account that only the outer amides may have access to the aqueous environment this is in good agreement with a paper published by R. M. Smith et al. (*36*) who demonstrate that the half life of an amide bond in water (pH 7.0 / 37°C) is about 200 years.

Conclusion

Gadomer is a new promising agent for MRI. Due to its intravascular retention within the first minutes after i.v. administration, a variety of clinical applications are anticipated, most interestingly its use in cardiac imaging. The compound is safe, very well tolerated and completely excreted from the body. It is the first gadolinium containing dendrimer in pharmaceutical development.

A highly effective and reproducible synthesis of this 17.5 kDa dendrimer has been developed which provides the drug substance in good quality and multi kg quantities.

The chemistry of the formation of several byproducts during the synthesis is well understood. The drug product is very stable and the aqueous solution can be stored for many years.

Acknowledgment

We thank Werner Behrendt and Uwe Rettig for the skillful assistance in the preparation of Gadomer on the laboratory scale, Günther Baude for the MALDI-TOF characterization and Norbert Schwarz for his outstanding assistance in all patent related issues.

The Gadomer Development Team is acknowledged for its engaged contribution: Dr. Wolfgang Beckmann, Dr. Peter Blaszkiewicz, Dr. Andreas Brandt, Dr. Karlo Büche, Dr. Klaus-Dieter Graske, Dr. Peter Hölscher, Dr. Jan Hübner,

Dr. Julius Leonhardt, Gregor Mann, Ronald Mertins, Dr. Harribert Neh, Dr. Orlin Petrov, Dr. Schulz, Dr. Christine Schwer and Uwe Thiel.

References

1. Caravan, P.; Ellison, J.J.; McMurry, T.J.; Lauffer, R.B. *Chem. Rev.* **1999**, *99*, 2293-2352.
2. Brasch, R.C. *Magn. Res. Med.* **1991**, *22*, 282-287.
3. Lauffer, R.B.; Brady, T.J. *Magn. Reson. Imaging,* **1985**, *3*, 11-16.
4. Schmiedl, U.; Ogan, M.D.; Moseley, M.E.; Brasch, R.C. *Am. J. Roentgenol.* **1986**, *147*, 1263-1270.
5. Armitage, F. E.; Richardson, D.E.; Li, K.C.P. *Bioconjugate Chem.* **1990**, *1*, 365-374.
6. Rebizak, R.; Schaefer, M.; Dellacherie, E. *Bioconjugate Chem.* **1997**, *8*, 605-610.
7. Schuhmann-Giampieri, G.; Schmitt-Willich, H.; Frenzel, T.; Press, W.R.; Weinmann, H.-J. *Invest. Radiol.* **1991**, *26*, 969-974.
8. Marchal, G.; Bosmans, H.; Van Hecke, P.; Speck, U.; Aerts, P.; Vanhoenacker, P.; Baert, A.L. *Am. J. Roentgenol.* **1990**, *155*, 407-411.
9. Berthezène, Y.; Vexler, V.; Price, D.C.; Wisner-Dupon, J.; Moseley, M.E.; Aicher. K.P.; Brasch, R.C. *Invest. Radiol.* **1992**, *27*, 346-351.
10. Böck, J.C.; Pison, U.; Kaufmann, F.; Felix, R. *JMRI* **1994**, *4*, 473-476.
11. Vexler, V.S.; Clément, O.; Schmitt-Willich, H.; Brasch, R.C. *JMRI* **1994**, *4*, 381-388.
12. Bogdanov Jr., A.A.; Weissleder, R.; Brady, T.J. *Adv. Drug Delivery Rev.* **1995**, *16*, 335-348.
13. Wiener, E.C.; Brechbiel, M.W.; Brothers, H.; Magin, R.L.; Gansow, O.A.; Tomalia, D.A.; Lauterbur, P.C. *Magn. Res. Med.* **1994**, *31*, 1-8.
14. Tóth, E.; Pubanz, D.; Vauthey, S.; Helm, L.; Merbach, A.E. *Chem. Eur. J.* **1996**, *2*, 1607-1615.
15. Bryant Jr., L.H.; Brechbiel, M.W.; Wu, C.; Bulte, J.W.M.; Herynek, V.; Frank, J.A. *JMRI* **1999**, *9*, 348-352.
16. Platzek, J.; Schmitt-Willich, H. et al., U.S. Patent 5,911,971, **1999**.
17. Adam, G.; Neuerburg, M.D.; Spüntrup, E.; Mühler, A.; Scherer, K.; Günther, R.W. *JMRI* **1994**, *4*, 462-466.
18. Schwickert, H.C.; Roberts, T.P.L.; Mühler, A.; Stiskal, M.; Demsar, F.; Brasch, R.C. *EJR* **1995**, *20*, 144-150.
19. Roberts, H.C.; Saeed, M.; Roberts, T.P.L.; Mühler, A.; Brasch, R.C. *JMRI* **1999**, *9*, 204-208.
20. Tomalia, D.A. *Adv. Mater.* **1994**, *6*, 529-539.
21. Schmitt-Willich, H.; Platzek, J. et al., European Patent 0836485, **2002**.

22. Perkins, A.C. *Journal of Drug Targeting* **1998**, *6*, 79-84.
23. Fischer, M.; Vögtle, F. *Angew. Chem. Int. Ed.* **1999**, *38*, 884-905.
24. Stiriba, SE.; Frey, H.; Haag, R. *Angew. Chem.* **2002**, *114*, 1385-1390.
25. Misselwitz, B.; Schmitt-Willich, H.; Ebert, W.; Frenzel, T.; Weinmann, H.-J. *MAGMA* **2001**, *12*, 128-134.
26. Dong, Q.; Hurst, D.R.; Weinmann, H.-J.; Chenevert, T.L.; Londy, F.J.; Prince, M.R. *Invest. Radiol.* **1998**, *33*, 699-708.
27. Henderson, E.; Sykes, J.; Drost, D.; Weinmann, H.-J.; Rutt, B.K.; Lee, T.-Y. *JMRI* **2000**, *12*, 991-1003.
28. Clarke, S.E.; Weinmann, H.-J.; Dai, E.; Lucas, A.R.; Rutt, B.K. *Radiology* **2000**, *214*, 787-794.
29. Torchia, M.G.; Misselwitz, B. *Am. J. Roentgenol.* **2002**, *179*, 1561-1565.
30. Misselwitz, B.; Schmitt-Willich, H.; Michaelis, M.; Oellinger, J.J. *Invest. Radiol.* **2002**, *37*, 146-151.
31. Nicolle, G.M.; Tóth, E.; Schmitt-Willich, H.; Radüchel, B.; Merbach, A.E. *Chem. Eur. J.* **2002**, *8*, 1040-1048.
32. Goodman, M.; Glaser, C. In *Peptides. Proceedings 1st Am. Peptide Symp. 1968;* Weinstein, B., Ed.; Marcel Dekker, Inc.: New York, NY, 1970; pp 267-335.
33. Platzek, J.; Schmitt-Willich, H. et al., WO/24774, **1998**.
34. Schmitt-Willich, H.; Platzek, J. et al., WO/24775, **1998**.
35. Platzek, J.; Schmitt-Willich, H. et al., U.S. Patent 6,576,222, **2003**.
36. Smith, R.M.; Hansen, D.E. *J. Am. Chem. Soc.* **1998**, *120*, 8910-8913.
37. Platzek, J.; Blaszkiewicz, P.; Gries, H.; Luger, P.; Michl, G.; Müller-Fahrnow, A.; Radüchel, B.; Sülzle, D. *Inorg. Chem.* **1997**, *36*, 6086-6093.

Chapter 13

Amyloid Precursor Protein and Ferritin Translation: Implications for Metals and Alzheimer's Disease Therapeutics

Jack T. Rogers

Genetics and Aging Research Unit, Department of Psychiatry, Massachusetts General Hospital (East), Harvard Medical School, Room 3850, Building 114, Charlestown, MA 02129

ABSTRACT:

Iron was shown to closely regulate the expression Alzheimer's Amyloid Precursor Protein (APP) gene at the level of message translation. Intracellular levels of APP holoprotein were shown to be modulated by a mechanism that is similar to the pathway by which iron controls the translation of the ferritin L- and H mRNAs by Iron-responsive Elements in their 5'untranslated regions (5'UTRs)[48]. More recently APP gene transcription was found to be responsive to copper deficit where the Menkes protein depleted fibroblasts of copper, an event that suppressed transcription of APP through metal regulatory and copper regulatory sequences upstream of the 5' cap site [1].

Genetic and biochemical evidence has also linked the biology of metals (Fe, Cu and Zn) to Alzheimer's disease. The genetic discovery that alleles in the hemochromatosis gene accelerate the onset of disease by five years [2] has certainly validated interest in the model wherein metals (iron) accelerate the course of AD. Biochemical measurements demonstrated elevated levels of copper zinc and iron in the brains of AD patients [3]. Current models have to account for the fact that the Aβ-amyloid precursor protein (APP) of Alzheimer's disease is a copper binding metalloprotein [4]. From *in vitro* experiments, copper zinc and iron accelerated the aggregation the Aβ peptide and enhanced metal catalyzed oxidative stress associated with amyloid plaque formation [5]. These amyloid associated events remain the central pathological hallmark of AD in the brain cortex region. The involvement of metals in the plaque of AD patients and the demonstration of metal dependent translation of APP mRNA have encouraged the development of chelators as a major new therapeutic strategy for the treatment of AD, running parallel to the development of a vaccine.

We screened 1,200 FDA pre-approved drugs for lead compounds that limited APP 5'UTR directed translation of a reporter gene, and identified several chelators as lead compounds. In a cell based secondary assay we found that the APP 5'UTR directed drugs desferrioxamine (Fe^{3+} chelator), tetrathiolmobdylate (Cu^{2+} chelator) and dimercaptopropanol (Pb^{2+}, Hg^{2+} chelator) each suppressed APP holoprotein expression (and lowered Aβ peptide secretion) [6, 7]. These agents will be tested as potentially relevant anti amyloid drugs for controlling the progression of Alzheimer's disease. Our findings have practical implications for Alzheimer's disease therapy but also support the hypothesis that APP is a metalloprotein with an integral role to play in Fe, Cu and Zn metabolism.

Abbreviations.

APP = Amyloid Precursor Protein, P97 = mellanotransferrin, PS-1 = Presenillin-1 and PS-2 = Presenillin–2, ADAM-10 = A Disintegrin and Metalloprotease Domain-10, and ADAM-17 = = A Disintegrin and Metalloprotease Domain 17, TACE-1 = Tumor Necrosis Factor alpha Converting Enzyme, ORF = Open Reading Frame.

1. ALZHEIMER'S DISEASE PATHOLOGY.

Introduction.

This article focuses on how the marked increase in the steady-state levels of metals (iron, copper and zinc) in the AD brain contributes to gene expression with deleterious consequences for neuronal survival [3]. Certainly APP mRNA translational control by iron (Rogers et al., 2002) and APP gene transcriptional control by copper [1] each provide new genetic support for the model that APP is a metalloprotein with an integral role in metal metabolism.

Both extracellular amyloid plaques and intracellular neurofibrillary tangles are the predominant pathological features characterizing the clinical onset of AD (both early onset and late onset AD). A proximal pathological feature of AD is the formation of neurofibrillary tangles by Tau even though mutations to the Tau gene cause hereditary fronto-temporal dementia [8, 9]. Iron (Fe (III)) binds with hyperphosphorylated Tau [10], and inhalation of aluminum dust was observed to cause a mild cognitive disorder which might prelude AD among foundry workers in northern Italy [11]. However the chromosome 21 gene encoding Amyloid Precursor Protein (APP) remains central to our understanding Alzheimer's disease progression and developing therapeutic agents for AD. We will discuss the relationships of metals to AD, referring to APP and amyloid formation [12].

APP is a Metalloprotein cleaved by the α, β and γ-secretases.

The abundant and ubiquitously expressed APP is a metalloprotein that spans membranes in the endoplasmic reticulum, trans-Golgi and on the cell surface of most cell-types. Originally APP was shown to be a redox active copper-zinc binding protein of unknown function during normal health [4]. We model that APP may also bind iron. During health a group of metalloproteases, the alpha-secretases (ADAM-10, ADAM-17 and TACE), cleave the Aβ precursor protein to generate the neuroprotective 90 kDa ectodomain of APP (APPs) that is released from cells into the cortex, cerebrospinal fluid and the bloodstream.

During the course of Alzheimer's disease APP is the single substrate which is proteolysed to produce the 40-42 amino acid Aβ peptide that aggregates to form the main component of amyloid plaques [13]. The pathogenic Aβ peptide encompasses amino acids 672-714 of APP770, and the amino terminus of Aβ includes 28 external amino acids and also comprises the additional first residues of the transmembrane region of nascent APP. A decade of research from multiple laboratories showed that the γ-secretase multimer (Pen2, presenillin, APH1 and nicastrin) [14] and BACE [15] cleave APP to generate the pathogenic 40 to 42 amino acid Aβ peptide[16].

It is worth noting that only 10% of cases of AD are familial. These cases are associated with autosomal dominant inheritance of APP [17, 18] and PS-1 and PS-2 [19] gene mutations [20]. Mutations within the APP gene on chromosome 21 cause altered cleavage of the Aβ-peptide from the APP holoprotein and have been genetically linked with early onset "Familial Alzheimer's Disease"(FAD) [17, 18, 21]. APP and Ps gene mutations independently enhance cleavage of the precursor into the Aβ peptide [Aβ(1-40) and Aβ(1-42)] that accumulates in the brain as amyloid plaques [23]. Trisomy of the APP gene is associated with Down's syndrome [22].

Late Onset Alzheimer's Disease (Iron as a Risk Factor).

The current level of understanding of late onset AD was increased by the discovery that individuals homozygous for the E4 allele of apolipoprotein E (ApoE) are at increased risk for developing Alzheimer's Disease [24]. Other genome scans have identified several other genetic loci where mutations to their encoded proteins increased the risk of developing AD. These include alpha 2 macroglobin for Aβ peptide clearance [25], and insulin degrading enzyme [26] and alpha-1 antichymotrypsin [27].

Iron imbalance appears to be a genetic factor that causes late onset sporadic AD. Sampietro et al., (2001) were the first authors to compare the involvement of the genotype of the HFE gene in hemochromatosis (iron overload) in 107 patients with sporadic late-onset AD and in 99 age-matched non-demented controls [2]. They observed that patients carrying the mutant HFE-H63D allele had a mean age of onset at 71.7 +/- 6.0 years versus 76.6 +/- 5.8 years of those who were homozygous for the wild-type allele ($p = 0.001$). Further recent studies confirmed the central role of heavy metals in the etiology of AD. Jeffries et al., (2003) showed that the known iron transporter, P97, is a potential early detection biomarker for AD [28, 29]. Trace copper in the water supply dramatically accelerated amyloid plaque formation in APP transgenic mice [30]. Thompson et al., (2003) provided evidence that null ferritin heterozygous mutants developed oxidative features in the cortex reminiscent of PD and AD [31].

2. IRON HOMEOSTASIS: FERRITIN AND APP EXPRESSION (FROM ANEMIA TO ALZHEIMER'S DISEASE).

Iron Metabolism and AD Pathology.

Iron imbalance, like Inflammation, has been recognized genetically in the etiology of neurological disorders. Consistent with these observations, the acute phase protein, alpha-1 antichymotrypsin [32] is associated with amyloid plaques and the iron storage protein ferritin is present in neuritic plaques [33, 34].

An elucidation of the mechanisms of brain iron homeostasis, as outlined in figure 1, will help our understanding of AD especially since iron binds to Aβ-peptide and enhances beta-amyloid toxicity [35-38]. Excess iron accumulation is a consistent observation in the AD brain. As discussed above, patients with hemochromatosis are at risk developing AD at an earlier age [2]. Brain autopsy samples from AD patients have elevated levels of ferritin iron, particularly in the neurons of the basal ganglia [39] and most amyloid plaques contain iron and ferritin-rich cells [40]. Clinically there is a reported decrease in the rate of decline in AD patients who were treated with the intramuscular iron chelator, desferrioxamine [41]. Iron enhances cleavage of the Aβ-peptide domain of APP by the metalloprotease alpha secretase [42, 43]. Part of the protective effect of the major cleavage product of APP, APP(s), may derive from its capacity to scavange metals to diminish metal-catalyzed oxidative stress to neuronal cells [44]. APP is, itself, a metalloprotein [4].

Like ferritin L- and H-mRNAs, the leader of the mRNA for APP is unique to encode a stable secondary structure that positively regulates 40S ribosome scanning and potentiates the onset of APP synthesis (Figures 3 and 4). Our data supports the hypothesis that IL-1 actively stimulates the translation of the APP transcript by a pathway that is similar to the translational regulation of the mRNAs encoding the L- and H-subunits of ferritin [45]. The mRNA for FMR1 (Fragile X) presents as a more typical eukaryotic mRNA in which RNA secondary structure suppresses protein synthesis [46]. However our model for APP mRNA translational regulation was extended when we discovered that the APP 5'UTR sequences also encodes functional sequences that are 75% similar to the iron-responsive element (IRE) in the mRNAs for the ferritin L- and H-subunits (Ferritin IREs) (Figure 3). We will review the common features of ferritin and APP translation by iron and Interleukin-1 [47, 48].

Lessons from the genetic control of translation of ferritin mRNAS and transferrin receptor mRNA stability can be applied to the control of APP expression by iron. This information will be relevant to Alzheimer's disease pathology after applying these models of post-transcriptional control to APP gene expression.

Iron Homeostasis and Ferritin and Transferrin Receptor Expression.

Ferritin is the intracellular iron storage protein composed of two subunits (L and H-subunits) that co-assemble to form the 240,000 dalton iron storage shell found in all tissues of the body [49](Figure 1). Ferritin is cytoprotective [50] because the H-subunit oxidizes Fe^{2+}, the bioavailable form of iron that causes tissue damage via hydroxyl radicals (Fenton Reaction). Ferric iron then becomes stored as an Fe^{3+}-$P0_4$ crystal inside the internal cavity of the ferritin shell [49]. Transferrin receptor (TfR) is present at the surface membrane of all cells and tissues. TfRs have a high affinity for binding iron-loaded transferrin (90,0000 dalton protein), thereby facilitating the uptake of iron from the bloodstream into tissues [49]. One of the major markers of any inflammation is lowered serum iron levels accompanied by a reduction in the steady-state levels of blood transferrin levels [51]. There is also an increase in serum ferritin levels seen during inflammation [52].

In the next section we will outline briefly how IRPs control intracellular iron homeostasis by modulating ferritin mRNA translation and transferrin receptor mRNA stability [53, 54].

Iron-responsive Elements and Transferrin receptor mRNA Stability.

Transferrin receptors (TfRs) control iron uptake, and TfR mRNA expression is controlled via five distinct Iron-responsive Element RNA stem loops (IREs) in the 3'UTR of TfR mRNA. In fact IREs modulate intracellular iron uptake by mediating both TfR mRNA and Divalent Metal Ion Transporter (DMT-1) mRNA stability.

Iron-responsive Elements and Ferritin Translation.

The iron regulatory proteins, IRP-1 and IRP-2, bind at high affinity (K_d 40-100 pM) to highly conserved RNA stemloops, the Iron-responsive Elements (IREs) at the 5' cap sites of the L- and H-ferritin mRNAs [55-57] (Figure 2). The IRE/IRP interaction impedes the access of the small 40S ribosome subunit to the 5' end of the ferritin mRNAs and thereby suppresses L- and H-ferritin mRNA translation [58-60]. The potency of the IRE to regulate translation is position dependent [61]. Iron influx relieves repression of ferritin translation by removing IRP-1 from the ferritin IREs; IRP-1 is simultaneously interconverted to a cytoplasmic cis-aconitase with an iron sulfur cluster, and IRP-2 is degraded by iron influx [53]. Removal of IRPs from the ferritin IREs restores recruitment of the 40S ribosome at the 5' cap sites. Ferritin translation can take place after eIF4E promotes eIF4G binding to eIF3 and hence 40S ribosome recruitment [62] (see Figure 2). The 40S ribosome then scans ferritin mRNAs before the eIF2-α dependent 60S ribosome subunit "joining step" occurs, after which protein synthesis begins at the optimal start codon [63]. The ratio of IRP-1 and IRP-2 is tissue specific, but IRP-2 is more abundant than IRP-1 in pituitary cells [64].

Hormonal signaling via IRPs certainly modulates intracellular ferritin levels, and appears to be cell line specific. Interleukin-1 (IL-1) does not change steady-state levels of IRP-1 and IRP-2 but appears to enhance binding to IREs [65]. This later finding would suggest that IL-1signalling may negate the predicted IRP functional inhibition of ferritin translation even though the cytokine increases IRP binding to IREs. This model is required since IL-1α and IL-1β also act through the downstream acute box domain in ferritin mRNA 5'untranslated regions (see below). In this regard both Epidermal Growth Factor (EGF) and TRH (Thyroid Releasing Hormone) increased the phosphorylation of IRP-1 and IRP-2 and binding of IRPs to IREs, an event that was shown not always to be correlated with the direction of ferritin translation [64].

Oxidative stress modulates the IRE/IRP regulatory system as an additional control point of ferritin translation. For example, H_2O_2 was shown to rapidly activate IRP-1 binding to ferritin IRE stemloops [67]. Coordinate transcriptional and translational regulation of ferritin was observed in response to oxidative stress [66]. These data confirmed that rapid responses in the gene expression of ferritin (and TfR) to oxidative stress was mediated by Iron-regulatory Proteins [68]. Toth et al., (1999) showed that hypoxia altered IRP-1 binding to IREs and modulated cellular iron homeostasis in human hepatoma and erythroleukemia cells [69]. Intracellular iron chelation with desferrioxamine and hypoxia blocked the degradation of both the wild-type and mutant IRP-2 proteins suggesting a model whereby 2-OG-dependent dioxygenase activity may be involved in the oxygen and iron regulation of IRP-2 protein stability [70]. Iwai et al., (1998) previously showed that iron-mediated degradation of IRP-2 requires the oxidation of key cysteines that reside within a 73-amino acid unique region [55]. However Hanson et al., (2003) showed that a mutant IRP-2 protein lacking this 73-amino acid region degraded at a rate similar to that of wild-type IRP-2 [70]. These results have key implications for Alzheimer's disease since oxidative stress is important during AD and the APP mRNA has an active Iron-responsive Element (Figure 3).

Translation of Ferritin RNA by Interleukin-1 responsive Acute Box Domains.

We generated a molecular model for iron sequestration (as observed during anemia of chronic disease) by employing HepG2 cells grown in tissue culture (see below). The anemia of the acute phase response was first observed during inflammation and disease progression by Cartwright in 1946 [74]. Here, an enlargement of tissue ferritin pools was seen to be preceded by a marked increase of hepatic and spleen ferritin mRNA translation and subsequent ferritin gene transcription [75]. Several groups then showed that enlarged ferritin pools

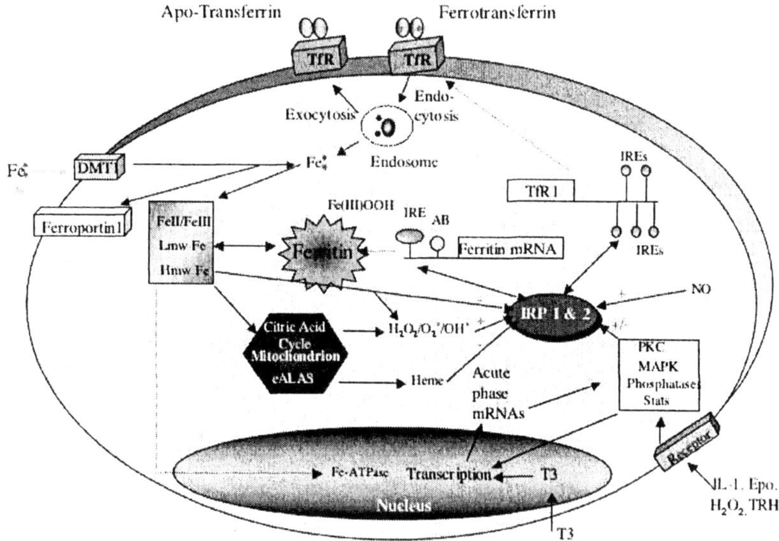

Figure 1. ***Intracellular Iron Homeostasis.*** *Iron transit across the cell surface membrane is mediated by (i) ferrotransferrin internalization by the transferrin receptor (TfR), (ii) DMT-1, (iii) ferroportin mediated iron efflux from the duodenum into the blood. Ferritin mRNA translation is regulated by the modulated interaction between the IRPs and the IREs in the 5'UTR of ferritin mRNA. MAP kinase signaling events influence ferritin translation and transferrin receptor activity and expression.*

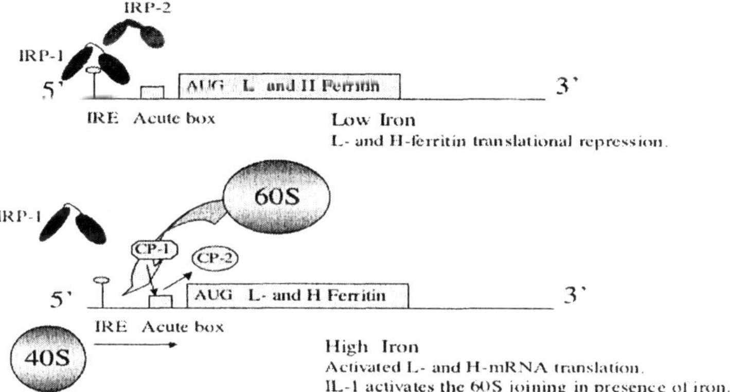

Figure 2. ***Model for Ferritin mRNA Translational Control.*** *Iron releases IRP-1/IRP-2 from suppressing ferritin mRNA translation at the Iron responsive Element stemloops (IREs) specific to the L- and H- mRNA 5' cap sites. In the diagram, IRP-1 and IRP-2 are depicted as two domains separated by a hinge region (line). Our preliminary data suggests that the RNA binding protein (Poly C-binding proteins, CP-1 and CP-2) interact with the ferritin mRNA acute box(AB) domain (box) downstream from the IRE (Thomson et al., In revision).*

could cause iron sequestration from the bloodstream onto the rat liver and spleen for tissue storage [52]. Like tissue ferritin, serum ferritin may provide cytoprotection both to oxidized low density lipoprotein in vitro and to heme-aggravated damage [50].

For our model system, hepatic ferritin gene expression was monitored in response to interleukin-1 (IL-1) [71]. IL-1 was found to significantly increase ferritin translation by signaling though a novel translation enhancer element, the acute box motif, downstream from the IREs at the 5' cap site [45, 72]. This acute box motif is located immediately downstream from Iron-responsive Element in the 5'untranslated regions of both the L- and H-ferritin transcripts [73], and was also found to be present in front of the start codon in the APP transcript [47]. This observation was consistent with the pattern of both APP and ferritin expression in response to inflammation during both Alzheimer's disease and the anemia associated with chronic disease.

Most recently we found that the acute box element in the 5' leaders of the ferritin transcripts selectively interacted with the PolyC-binding Proteins (Thomson et al., 2003, Submitted to JBC). During erythropoeisis both CP-1 and CP-2 are known to be the mediators of globin mRNA stability [80] and Lipoxygenase (1:5) mRNA translation [81]. After the 40S ribosome has scanned the 5'untranslated regions of ferritin mRNAs IL-1 increases ferritin translation at the 60S ribosome subunit "joining step" when protein synthesis begins at the optimal start codon [82]. IL-1 certainly appears to act through the ferritin H-mRNA acute box (AB) domain by altering the interaction between the AB and Poly C binding proteins (CP-1, CP-2).

Since poly(C)-binding protein affect ferritin translation via a 5'UTR sequence similar to that found in the APP transcript it is worth speculating that poly(C)-binding protein isoforms may have a role in determining the expression of APP and ferritin during mini-strokes associated with AD [83]. Zhu et al., (2002) used immunostaining to show that poly(C)-binding protein 1, but not poly(C)-binding protein 2, expression was increased in the ischemic boundary zone (penumbra) of the frontal cortex after 90 min of ischemia, and persisted for at least 72 h after reperfusion [84]. These results demonstrated that poly(C)-binding protein 1 and poly(C)-binding protein 2 in cortical neurons are differentially affected by hypoxic/ischemic insults, suggesting that there are functional differences between poly(C)-binding protein isoforms. We are currently investigating the role of poly(C)-binding proteins relevant to the regulation of APP mRNA translational control.

As a biological/clinical correlate serum ferritin appears at markedly increased levels in the bloodstream after the onset of an inflammation [76]. Serum ferritin is L- rich and is known to be immunosuppressive protein [77]. Clinically the average amounts of serum ferritin are between 2 and 450 mg/liter. Serum ferritin levels are clinically used after trauma as an index of tissue damage and as a predictor of disease (for Hepatitis B see [78]). During disease progression the concentration of iron rises sharply, although to a level less than that observed during conditions of iron overload such as during hemochromatosis (1000 μg/ml). Inflammation reduced hepatic transferrin gene expression [79], reflecting the levels of transferrin bound iron versus free iron present in the bloodstream during inflammation (the anemia of chronic disease).

An Iron-responsive Element in Transcripts of Proteins of Iron Metabolism.

It is crucial to note that Iron-responsive Elements are involved in the post-transcriptional regulation of several genes that control intracellular iron homeostasis. The Alzheimer's disease brain harbors dysregulated binding of IRPs to IREs [85], an event that would be predicted to have implications for the expression of any of these iron-related proteins. Certainly the generic Iron-responsive Element RNA stemloop may be an important site to cause mis-regulation of these key proteins during the course of AD. The absence of Iron Regulatory Protein-2 (IRP-2), which controls iron homeostasis (Figure 1), was associated with a mis-regulated iron metabolism and ferritin translation and TfR mRNA stability in both the gut mucosa and the central nervous system [86].

Like ferritin, the other transcripts that encode active IREs in their 5'UTR include:- (i) the mRNA for erythroid aminoluvenyl synthase (eALAS). Erythroid ALAS is the rate limiting enzyme that controls heme biosynthesis in the mitochondria of red blood cells [87, 88]. (ii) transferrin (Tf) transports iron throughout the bloodstream to tissues, and translation of Tf is activated by binding of IRPs to the 5'untranslated region of the transferrin transcript [89]. (iii) Thomson et al., (1999) reviewed the mechanism of action of 5'UTR specific IREs that control the translation of IREG-1 (ferriportin), which is responsible for iron efflux from the duodenum into the bloodstream and macrophage iron efflux [90, 91]. (iii) APP mRNA is the most recently characterized member of the family of genes that encodes a functional Iron-responsive Element (Rogers et al., 2002).

Like the transferrin receptor, the divalent metal ion translporter (DMT-1) is known to be regulated by an IRE in its 3'UTR as one of the alternatively spliced DMT-1 transcripts [57]. DMT-1 is responsible for uptake of iron and other divalent cations from the gut into the bloodstream into enterocytes [92].

An Iron-responsive Element in Alzheimer's APP Transcript.

IL-1-and iron-responsive sequences were demonstrated to be active RNA regulatory elements in the 5' untranslated region of the Alzheimer's APP transcript. A novel IL-1 responsive acute-box RNA enhancer was immediately in front of the start codons of both APP and ferritin mRNAs [47]. Since this finding our purpose has been to define the function of the newly discovered iron regulatory translational enhancer in APP mRNA in the context of IL-1-dependent APP translation (Figure 3). Like ferritin, APP is a metalloprotein [47, 48].

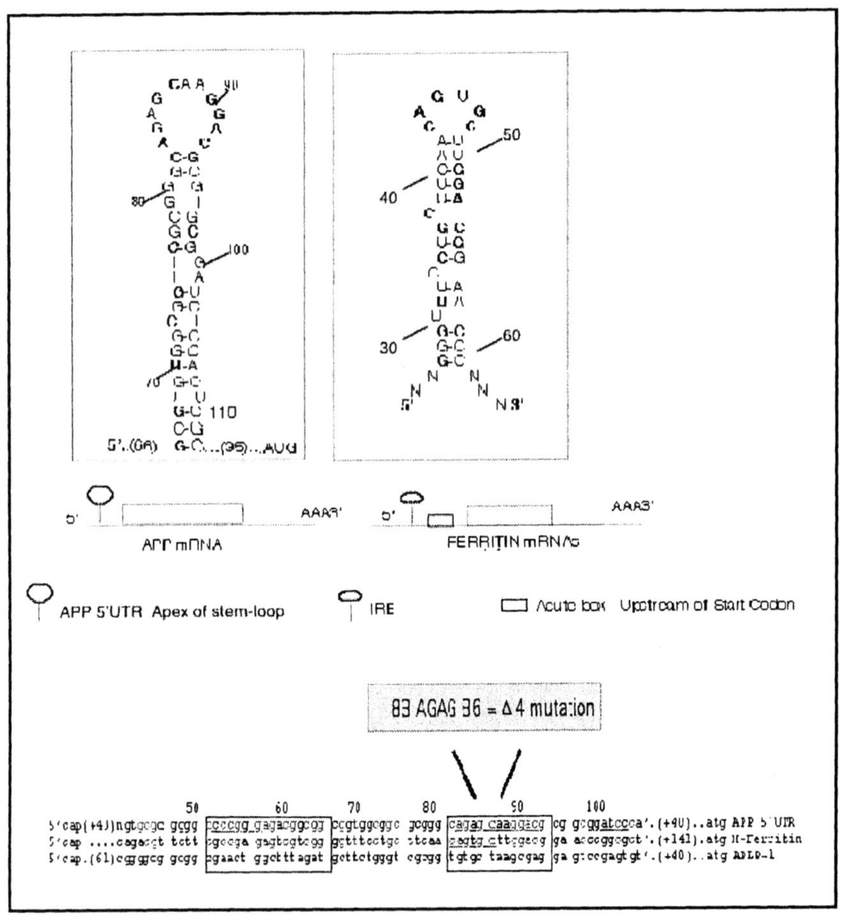

***Figure 3. An Iron responsive Element (Type II) in the 5'UTR of APP mRNA. Top Panel:** The APP 5'UTR was predicted to fold into the stable RNA stemloop similar to the 5'UTR-specific IRE in H-ferritin transcript ("APP 5'UTR" ΔG = -54 kCal/mol) (Zuker et al., 1989). **Bottom Panel**: The APP mRNA IRE (Type II) was identified after alignment with the ferritin Iron-responsive Element (shown in two clusters of > 70% sequence similarity)(Adapted from Rogers et al., 2002).*

(See page 5 of color insert.)

To characterize the location of iron responsive domain in the APP 5'UTR we performed computer-alignments of APP 5'UTR sequences with the known IRE in H-ferritin mRNA. Figure 3 shows the sequence alignment generated by the "Gap " sequence alignment program (GCGDefs, Univ Wisconsin) [48]. To ensure that these alignments were specific we showed that 5'UTR sequences in the APP and APLP-1 transcripts only exhibited 25% percent homology. There was found to be an overall 67% sequence identity between APP 5'UTR sequences (+51 to +94) and the 44 nt Iron-Responsive Element (IRE) in H-ferritin mRNA (+12 to +59) (red lettering is used to depict sequence identity). Two clusters within this APP 5'UTR IRE-Type II domain showed >70% identity with the ferritin IRE sequences. First an 18 base sequence in APP mRNA (+51 to +66) was found to be 72% similar to 5' half of the H-mRNA IRE (+12 to +29). Second APP sequences (+82 to +94) (a 13 base cluster) were found to be 76% identical to the loop domain of Ferritin IRE (+43 to +55). The IRE alignment shown in figure 3 demonstrated that the IRE -Type II sequence in the APP 5'UTR (+51 to +94) was sited immediately upstream of the IL-1 responsive acute box domain in the APP 5'UTR (+100 to -146).

This Iron-Responsive Element (Type II) in the 5' of APP mRNA was fully functional as assessed by multiple separate transfection assays [48]. RNA gel-shift experiments showed that the mutant version of the APP 5'UTR cRNA probe no longer binds to Iron-regulatory Proteins (IRP) (shaded box in Figure 3) [48]. Using RNA electrophoretic mobility shift assays (REMSA) we performed many controls to demonstrate that IRP-1 specifically binds to the stemloop that is predicted to fold from APP 5'untranslated region sequences [48]. Our preliminary data also confirmed that IRP-2 selectively interacted with the APP 5'UTR to the same extent as originally observed for IRP-1.

We concluded that IRPs interact selectively with the APP transcript. These data provided strong genetic support for an integral role for the involvement APP holoprotein in iron metabolism. The fact that IRP-1/IRP-2 are the cognate APP 5'untranslated region binding partners will prove useful for screening novel lead compounds that limit APP translation, particularly if IRP-1 and IRP-2 expression can be functionally linked to APP expression.

APP mRNA Stability and Translation.

Cytokines, which are physiologically relevant to Alzheimer's disease, change the efficiency of APP mRNA translation and APP mRNA stability by signaling through RNA sequences [47,93] . By this route APP gene expression is increased at the post transcriptional level to generate more template for the deposition of more amyloid Aβ [94]. Two cis-acting elements in the APP 3'UTR regulate the stability of the precursor transcript where TGFβ induces a 68 kDa protein to bind to an 81nt motif and thus stabilize the APP mRNA [93]. Addition of serum growth factors overrides the action of another 3'UTR element, a 29-base sequence (2285-2313), that normally destabilizes APP mRNA in endothelial cells (and peripheral blood lymphocytes) [95] (Figure 4).

Addition of serum growth factors overrides the action of another 3'UTR element, a 29-base sequence (2285-2313), that normally destabilizes APP mRNA in endothelial cells (and peripheral blood lymphocytes) [95] (Figure 4).

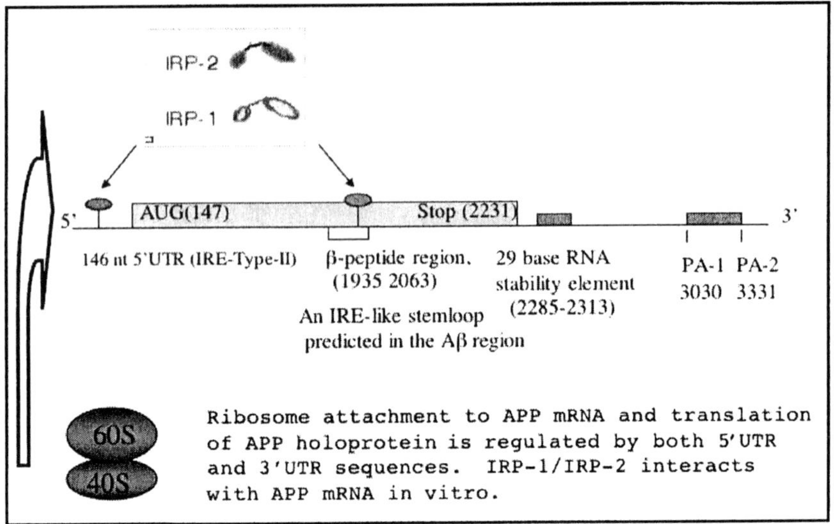

Figure 4. ***Post-Transcriptional regulatory domains mapped to the APP Transcript.*** *The 3 kb APP transcript is controlled at the level of message translation by the action of 5'UTR regulatory domains (IL-1 (47), iron (48)). A TGFβ transcriptional regulatory element was reported to be in DNA sequences encoded by the APP 5'UTR (Lahiri et al., 2003). The 3'untranslated region is alternatively polyadenylated, and the longer APP transcript is translated more efficiently than the shorter transcript (97). A 29 nt RNA destabilizing element was mapped to the 3'UTR of APP mRNA (95).*

In astrocytes we originally found that APP gene expression is up-regulated by IL-1 at the translational level [47]. Up-regulation occurs through 5'UTR sequences by a pattern of gene expression similar to that for the universal iron storage protein, ferritin [45]. This is in keeping with the findings of de Sauvage et al., (1991) who reported that polyadenylation site selection in the APP mRNA 3'UTR is critical for efficient translation in xenopus oocytes [96]. The longer 3.3 Kb APP mRNA was found to be translated 3-fold more efficiently than the shorter 3.042 Kb transcript. Mbella et al (2000) then observed that these 3'UTR sequences enhanced APP mRNA translation in mammalian (CHO) cells, and more closely mapped two guanosine residues that are crucial for this action [97]. These authors used RNA-electrophoretic mobility shift assays (REMSA) to determine that a translational repressor protein interacts with the shorter transcript

[97]. However, the presence of the 258 nucleotide poly(A) regulatory region (PAR) (nt +3042 to +3300) removes binding of this APP mRNA repressor and facilitates longer APP mRNA translation [97]. We conclude that APP 3' untranslated regions operate in conjunction with the APP 5'UTR to establish the post-transcriptional control of APP gene expression (and Aβ production) [97]. (Figure 4).

In future studies we will rigorously assess how APP mRNA 5'UTR sequences and 3'UTR interact to effectively regulate the translation of the Aβ-precursor protein. Certainly, when mRNAs compete for ribosome binding, neither the cap structure nor the poly(A) tail alone is enough to drive efficient translation, but together they synergize and direct ribosome entry to the 5' end [98-100]. The elegant studies of Chen et al., (1995) demonstrated that initiation of protein synthesis by the eukaryotic translational apparatus includes the formation of circular RNAs where the 3' end interacts with the 5' end of any given message [101]. In this regard, an APP mRNA circularization step may be a crucial for drug targeting particularly since chelators inhibit APP translation (see Section 4).

3. METALS AND ALZHEIMER'S DISEASE.

Overview of iron and copper chelators as therapeutic agents for AD treatment.

We will describe how iron and copper chelators appear to operate at two levels as therapeutic agents for the treatment of AD. In a clinical trial that was run in Toronto (Canada), desferrioxamine provided clear and effective therapeutic relief to Alzheimer's patients as registered by measurement of cognitive performance [41]. Notably desferrioxamine, which is a highly specific iron chelator (Kd Fe^{3+} $=10^{-31}$ M), suppressed intracellular levels of APP holoprotein by inhibiting translation of APP mRNA from it's 5' untranslated region 5'UTR) [102]. We will show that the broad specificity copper-zinc-iron chelator, clioquinol, patently dissolved extra cellular fibrillar Aβ, and thereby diminished plaque burden [103]. Clioquinol has been tested in a Phase II clinical trial in Melbourne Australia, wherein the metal binding drug improved cognition in severe AD patients [104].

Iron and AD (APP Expression is Regulated by Iron).

The integral role iron plays in the etiology of AD pathology was directly supported from an MRI imaging study of the brains of Alzheimer's patients that revealed elevated levels of iron, particularly in the neurons of the basal ganglia [39]. Confirming these results, an elegant spectroscopic study demonstrated that amyloid plaques harbor increased burden of iron copper and zinc [3]. Iron and zinc levels were measured to be present at concentrations as high as the 1 mM level in the vicinity of amyloid plaques. There has been controversy as to whether the iron-Aβ peptide interaction was neurotrophic or neuroprotective. Aβ-peptide, itself, was found to be neurotrophic under some circumstances [105], and perhaps A β toxicity is only evident in the presence of metals. Certainly iron enhanced beta-amyloid action to accelerate Aβ-induced neuroblastoma cell death, providing one direct link between excessive iron and the known loss of neuronal function seen in AD patients [38]. Bishop and Robinson (2003) reported that Aβ peptide reduced iron toxicity in rat cerebral cortex [106] although iron remains a candidate pathological mediator of AD [107].

Iron was further demonstrated to be relevant to neurodegenerative pathogenesis since the brains of AD patients exhibit disrupted iron distribution [108]. Ferritin is the universal iron storage protein, and the transferrin receptor (TfR) is the universal receptor responsible for transferrin-mediated iron transport. Ferritin and TfR are respectively regulated by modulated binding of Iron-regulatory Proteins (IRPs) to the 5' cap site-and 3'UTR specific Iron-responsive Element (IREs) RNA stemloops in each transcript [53]. It has been demonstrated that 30% of AD brain samples displayed a stronger interaction between the IRP-1 and IRP-2 and the IRE stemloops in the 5' untranslated region of the transcripts coding for ferritin and the 3' untranslated region of the transferrin receptor mRNA (TfR-mRNA) [85]. Since IRP binding to the IREs controls intracellular iron homeostasis, a change in this RNA binding affinity would be predicted to enhance iron transport into neuronal cells during AD, but decrease ferritin levels, and thus diminish the cellular iron storage capacity. In these circumstances, neuronal cells in the AD brain would harbor an enlarged cellular pool of dangerous unstored iron [85]. This model of reduced ferritin synthesis may be accurate at earlier stages in AD progression, but the hypothesis has to account for the fact that most amyloid plaques contain ferritin-rich cells. To account for this discrepancy, ferritin appears to be deposited into plaques at later stages in disease progression when iron homeostasis in neurons has been reset by the enhanced preceding increase in transferrin receptor activity. Certainly ferritin synthesis of microglial, rather than neuronal origin, may also account for most plaque associated ferritin in the AD brain [34].

We described a direct link between iron metabolism and AD pathogenesis when we reported that the 5' untranslated region of APP mRNA is related to the 5'UTRs of L- and H-ferritin mRNAs [47]. IRP-1/IRP-2 appear to be trans-activators of the APP 5'UTR, thereby to stimulate translation of APP holoprotein [102]. Figure 4 depicts the post-transcriptional regulatory domains in the APP transcript including the APP 5' untranslated region which encodes both the IL-1 and iron dependent enhancers of APP translation. The presence of a functional IRE in the APP 5'UTR is a further indication for the essential requirements of metal metabolism for the regular hitherto unknown function of APP. This view is supported by the finding that the APP cytoplasmic tail can also be interchanged with that of the transferrin receptor and maintain 50%

endocytosis of transferrin bound iron [109]. A downstream IRE-like sequence was also hypothesized to exist in the coding region of APP mRNA near the Aβ domain [110].

In keeping with a metal regulatory element (IRE-Type II) in the APP 5'untranslated region, we identified both iron and copper chelators to be a major class of APP 5'UTR-directed leads from our screen of a library of 1,200 FDA pre-approved drugs [111] (final section of this monograph). Drug screens targeted to the APP 5' untranslated region identified dimercaptopropanol (Hg and Pb chelators), and tetrathiolmobdylate (Cu chelator) as FDA pre-approved leads that limited APP holoprotein expression and (Aβ-peptide output) [7].

Iron chelators

Intramuscular Injection of Desferrioxamine.

Daily intramuscular injection of the intracellular Fe^{3+} chelator, desferrioxamine, was also shown to decrease the cognitive decline in a large cohort of AD patients [41]. Desferrioxamine (DFO) was at first thought to chelate aluminium, which was considered a risk factor for AD [112]. However desferrioxamine is commonly used as an iron chelator in the treatment of Sickle Cell disease [113], and for treatment of patients suffering from poisoning during acute iron overload [114-116]. Thus the mechanism of action of desferrioxamine in the treatment of AD is likely associated with iron chelation rather than aluminum chelation. Desferrioxamine appears to specifically suppresses APP mRNA translation (Figure 4) [102].

Desferrioxamine (Df) has a dissociation constant for binding to Fe^{3+} at 10^{-31}M which provides the very high specificity for the chelation of iron required for the treatment of patients with transfusion iron overload [114-116]. As such Df has been very closely monitored as clinical agent for 20 years. The main clinical drawback of desferrioxamine is that the chelator can only be administered by intramuscular injection (usually to Sickle Cell Disease patients in crisis). There has been an active program to introduce new orally active iron chelators [114-116].

Iron Chelators that supercede Desferioxamine (Deferiprone and Ferralex).

Deferiprone is the orally active iron chelator that has been most often used for the treatment of iron overload disorders [117]. It will be of interest to test whether deferiprone provides efficacy for AD based on the experimental capacity of chelators to limit APP translation, as has been observed for both desferrioxamine (Fe^{3+} chelator) and dimercaptopropanol (Hg2+, Cu2+ chelator)(see Table I). There is controversy concerning the use of deferiprone as an agent to replace desferrioxamine since the chelator has caused hepatic fibrosis in cases of transfusion iron overload [116].

Iron Chelation and Tau Phosphorylation.

Aside from the amyloid based model for AD, iron was also demonstrated to enhance phosphorylation of Tau and the formation of the neurofibrillary tangles associated with the AD brain [10]. In this report a novel trivalent cationic chelator, Feralex, effectively dissociated binding of aluminum and iron to hyperphosphorylated Tau (neurofibrillary tangle) pertinent to its use as a therapeutic agent for AD.

Future Directions for Iron Chelators.

New oral iron chelators are rapidly being developed to combat disorders of iron overload such as sickle cell anemia, transfusional iron overload, including (a) the hexadentatephenolicaminocarboxylate HBED [n,N'bis(2-hydroxybenzyl)ethylenediamine-N,N-diacetic acid], (b) the tridentate desferrithiocin derivative 4'-OH-dadmDFT[4'dihydroxy-(S)-deszaddesmethyl-desferrithiocin, (S)-4,5-dihydro-2-(2,4-dihydroxylphenyl)-4-thiazolecarboxylic acid], (c) the tridentate triazole ICL670A[CGP72 670A; 4-[3,5-bis-(hydroxylphenyl)-1,2,4triazol-1-yl]-benzoic acid], and (d), the bidentate hydroxypyridin-4-one deferoprone [L1,CP20: 1,2-dimethyl-3-hydroxypyridin-4-one] [114]. The successful use of new oral iron chelators for blood-associated diseases will predict their therapeutic capacity for the treatment of AD. Their efficacy will be measured using the same both histochemical and nuclear magnetic resonance imaging techniques that originally identified that the brain cortex of Alzheimer's disease patients displayed an abnormal iron distribution of [39].

Metal Dependent Amyloid Formation (Oxidative Stress and Neurotoxicity).

Metals (Fe, Cu, Zn) clearly provide one of the ultrastructural requirements needed for polymerization of Aβ peptide [118] in addition to the reported presence of pathological chaperones such as alpha-1 antichymotrypsin (ACT) or ApoE [27]. Copper (Cu) was demonstrated to be a potent natural binding partner of Aβ with a dissociation constant at the attomolar affinity level [119, 120]. The metals, Fe and Cu, were demonstrated to be very important in mediating the neurotoxic action of Aβ protofibrils [5]. At the molecular level, the histidine mediated zinc binding site was mapped to a contiguous sequence between positions 6 and 28 of the Aβ sequence [35]. Amyloid Aβ is normally secreted from cells grown in culture as a monomer, but also as a covalently linked dimer in the ratio 55: 30: 15 monomers: dimers: trimers [121]. Human amyloid derived from oligomeric Aβ results from a tyrosine cross-linked oligomerization that is induced by metal catalyzed oxidation systems.

For long time oxidative stress has been strongly associated with the neurotoxicity in Alzheimer's disease [122-124]. It has been recently shown that over-expression of superoxide dismutase -1 protected neurons against beta-amyloid peptide toxicity [125]. The antioxidant, Co-enzyme Q, has already been tested for efficacy in animal models receiving increased oxidative stress, and is currently being tested for efficacy to AD [126].

One important mechanism to link Aβ peptide induced neurotoxicity and oxidative damage came from the finding that Aβ itself, is a generator of metal-catalyzed oxidative stress (Aβ(1-42) > Aβ(1-40) [36, 127]. Electron spin trap measurements have shown that Aβ expressed in C-elegans does generate superoxides in solution [128]. However, Fe, Cu and Zn bind strongly with Aβ peptide [119]. Huang et al., (1999) then proposed that Cu and Fe are reduced by Aβ peptides, and that this catalytic reaction transfers electrons to molecular oxygen thereby generating neurotoxic H_2O_2 and (superoxide ions ($O_{2)}$. [5]. Using in-vitro assays, the presence of iron or copper (100 nM) with Aβ (10 μM) was found to catalyze H_2O_2 production [36]. Redox interactions between Aß and Cu(II) and Fe(III) engender production of reduced metal ions, Cu(I) and Fe(II), and consequential generation of ROS- , H_2O_2 and OH•. Like Cu(II), Zn(II) precipitates Aβ in-vitro (Aβ1-42 > Aβ 1-40 > rat Aβ (1-40).

Certainly a new model for amyloid induced neuronal loss in AD [5] can be explained by the binding of Aβ-peptides to copper and iron to catalyze the generation of toxic H_2O_2. Equation 1 shows the standard Fenton reaction by which iron or copper react with hydrogen peroxide and superoxides to generate the toxic and deadly hydroxyl radicals that mediate neuronal loss during AD [129]. The presence of reduced copper (Cu^{1+}) and iron (Fe^{2+}) also implied that even more damaging hydroxyl radical would be formed by the Fenton and Haber-Weiss chemical reactions [118, 130] (Equation 1, next page).

Equation 1:

$$Fe^{+2}\ (Cu^{2+}) + H_2O_2 \longrightarrow Fe^{+3}\ (Cu^{1+}) + OH^{-} + OH^{\cdot}$$

Equation 1: *Metal-catalyzed neuronal oxidative damage via hydroxyl radicals.*

An appealing model suggests that the pathologically damaging production of hydrogen peroxide generated by Aβ is the result of a corruption of a beneficial superoxide dismutase activity associated with Aβ. In fact, Aβ dimer may form a copper - zinc protein (8 kDa) protein that has intrinsic superoxide dismutase activity that is capable of converting O_2^{-} into H_2O_2. [131]. This hydrogen peroxide can then be converted into two water molecules by catalase to complete an antioxidant function. One possible model for a pathogenic corruption of Aβ function may occur when oxygen is consumed as a substrate instead of O_2^- radicals, thereby changing a potentially cytoprotective peptide into a pro-oxidant which has deleterious activity.

At a therapeutic level, it is worth remembering that Zn^{2+} is redox-inert and appears to play an inhibitory role in H_2O_2-mediated Aβ toxicity by copper. After co-incubation of zinc with Cu(II), Zn(II) was found to rescue primary cortical and human embryonic kidney 293 cells that were exposed to Aβ1-42, correlating with the effect of Zn(II) in suppressing Cu(II)-dependent H_2O_2 formation from Abeta1-42 [132]. Other antioxidants will also prove beneficial to counteract the toxicity of the redox interaction between Fe and Cu and Aβ peptide. For example coenzyme coenzyme Q10 is a well known nutraceutical that may serve this important antioxidant purpose for AD therapeutics [133].

Chelators (Cu Zn) to dissolve amyloid plaques and reduce oxidative burden.

It is clear that the strategy of using chelators to eliminate both Aβ fibrilization and Aβ-dependent metal oxidative stress neurotoxicity could well provide a major new therapeutic impact on AD progression. Certainly the copper chelator, clioquinol, has been successfully used as an inhibitor of copper-dependent aggregation of Aβ peptide. Trace amounts of copper in water induce Abeta-amyloid plaques and learning deficits in a rabbit model of Alzheimer's disease [30]. Also clioquinol suppressed metal catalyzed H_2O_2 production by copper interacing with Aβ peptide [127]. Thus clioquinol both chelated metals and dissolved amyloid plaques, and continuous diet with the chelator to transgenic mice over-expressing APP have improved amyloid burden, and slowed the rate of cognitive decline [103]. Clioquinol underwent Phase II clinical trails for its therapeutic impact to Alzheimer's Disease patients [104].

There remains considerable controversy as to whether copper actually accelerates the onset of AD. Two studies demonstrated that copper appeared to lower the amyloid burden in the AD brain of APP transgenic mice. In the first study Phinney et al., 2003 demonstrated an in vivo reduction of amyloid-Aβ by a mutant copper transporter [134]. They used a well-known spontaneous mutation of a special strain of mutant toxic-milk mice that accumulate too much copper. Eventually those animals get a liver disease which is a facsimile of a human disorder called Wilson's disease. In direct contrast to the results of Sparks and Scheur (2003) [30], this study demonstrated that an excess of copper in the brain actually reduced of amyloid burden.

Superoxide dismutase-1 activity was found to be stabilized by excess dietary copper and reduced amyloid Aβ production in APP23 transgenic mice [135]. These data suggested that copper may not be is a causative risk factor in AD since their mice do show that an excess of dietary copper can be beneficial. On the other hand copper deficiency associated with the Menkes disease protein (copper pump) reduced APP mRNA levels and APP proximal promoter activity in fibroblasts [1]. This latter finding was consistent with the model that copper may enhance APP production, be amyloidogenic and act as a promoter of Alzheimer's Disease. To accommodate these controversies at the therapeutic level it may be possible to screen for drugs that alter copper transport and distribution in patients. These experimental lines will help us understand the interface between copper biology and APP biology. Interestingly copper and iron metabolism are closely linked [90].

It is intriguing to speculate that iron-specific chelators (porphyrins) may be considered a viable therapeutic option in aging and AD [136]. These findings are in keeping with an a earlier therapeutic strategy for AD that was designed to inhibit Aβ fibrilization, and thus block amyloid plaque formation in the pathogenesis of AD [137]. To monitor drug efficacy in reducing the extent of Aβ fibrilization, scientists at Smith Kline and Beecham developed a monoclonal antibody-based assay (DELFIA) that detected fibril formation over the presence of monomer (dimer Aβ). Using this assay, hemin and related porphyrins were found to effectively inhibit Aβ-amyloid aggregation [136]. The rank order of fibril inhibition in-vitro was hemin > hematin > zinc > protoporphyrin IX in the absence of cytotoxic action of the compounds [136].

Porphyrins and phthalocyannines also inhibited protease resistant prion protein formation relevant to Bovine Spongiform Encephalopathy (BSE) and the appearance of Creutzfeldt-Jakob disease in humans [138, 139]. Scrapie is transmitted by intra-peritoneal injection of protease sensitive prion (PrP -sens), and porphyrin-specific suppression of PrP conversion to a pathogenic protease resistant PRP (PrP-res) occurs in the periphery before crossing the blood brain barrier to cause encephalopathy[138]. By contrast, during the progression of AD amyloid plaques form in the brain cortex whereas heme and proteoporphyrin do not cross the blood-brain barrier, and thus it is uncertain that the use of porphyrines will have therapeutic impact in this case. Also shorter protofibrils of prefibrillar, diffusible assemblies of Aβ peptide are deleterious agents during the progression of Alzheimer's disease. These facts might argue against identifying drugs, like heme, targeted to inhibit the formation of larger fibrils as a treatment for AD [140].

In summary, brain levels of Zn, Cu, and Fe, and their binding proteins are dysregulated in AD [34, 40]. Levels of Zn and Fe are increased to concentrations as high as 1 mM in cerebral amyloid plaques (Cu is at 400 mM) [3]. Normally Zn, Cu, and Fe are concentrated in the temporal cortex at relatively lower levels [eg: 7.94 mM Fe / mg protein, in the temporal cortex and 20 mM Fe / mg in the motor cortex [40]. The distributions of both the brain iron storage proteins, transferrin and ferritin, are also altered in the white matter versus the gray matter in the brains of AD patients compared to age matched controls [33, 142]. Iron also causes neural damage, behavioral changes and microgliosis in mouse behavioral models.

4. SMALL MOLECULES TARGETED TO THE ALZHEIMER'S APP 5'UNTRANSLATED REGION.

Screens of FDA pre-approved drugs to limit APP 5'UTR conferred translation.

Our goal has been to screen a drug library for APP 5'UTR directed compounds that limit APP translation and ultimately amyloid-Aβ-peptide output from neuronal cell culture systems. Figure 3 shows that the APP 5'UTR folds into a stable RNA secondary structure (Gibbs free energy: $\Delta G = -54.9$ kcal/ mol [146]. The APP 5'UTR was found to be different to typical eukaryotic mRNAs such as FMR mRNA (Fragile X syndrome). In this regard FMR 5'UTR sequences are structured [46] and conform to the "Kozak model" by suppressing 40S ribosome translation scanning [147]. Instead, the APP 5'UTR increases baseline translation of reporter mRNAs. Of significance, the APP 5'UTR is an important riboregulator of the amount of APP translated in response to IL-1 [148] and iron [102]. Iron levels and interleukin-1 regulate translation of APP through the 5'untranslated region (APP 5'UTR) of the precursor transcript [47]. Aβ peptide is cleaved from the amyloid precursor protein (APP) to fibrilize and form the plaque, which is an underpinning pathological hallmark of Alzheimer's disease. We, therefore, reasoned that drugs that interact with the APP 5'UTR could well be useful drug candidates for AD therapeutics.

Since the APP 5'UTR is responsive to both metals and IL-1 we predicted that drug screens to the APP 5'untranslated region (see Figure 5) would lead to identification of novel metal chelators and NSAID drugs as therapeutic agents for AD [111]. As shown in Figure 5 (below) our laboratory used a transient transfection based assay to screen a library of 1,200 FDA pre-approved drugs, and identified 17 drug "hits" that >95% suppressed translation of the hybrid APP 5'UTR-Luciferase transcript in neuroblastoma (SY5Y) cells (n=5). Our drug hits were commonly used FDA drugs which distributed over 5 distinct classes of drugs [111]. These classes were (i) blockers of receptor ligand interactions, (ii) bacterial antibiotics, (iii) statins, (1V) metal chelators, (v) mutagens, and (vi) detergents.

During a secondary screen we validated 6 of the 17 hits for their capacity to selectively reduce luciferase expression translationally driven from the APP 5'UTR but maintain co-expression of GFP with an internally controlled viral Internal Ribosome Entry Site (Dicistronic construct) [7]. For this purpose SY5Y neuroblastoma cells were stably transfected with the dicistronic construct (pJR1). The FDA drugs, dimercaptopropanol, paroxetine, azithromycin, quinoline-gluconate, tamsulosin and atorvastatin each suppressed luciferase reporter mRNA translation via the 146 nt APP 5'UTR sequence [111]. The iron chelator, desferrioxamine [102], and the anticholinesterase, phenserine (AchEi) [149], also reduced APP holoprotein expression by suppressing APP 5'UTR directed translation of the Aβ-amyloid precursor. The rank order of effectiveness of FDA drugs to suppress translation conferred by the APP 5'UTR was:- Desferrioxamine (Fe^{3+} chelator > Dimercaptopropanol (Hg^{2+}, Pb^{2+} chelator) >> Paroxetine (SSRI) > >phenserine.

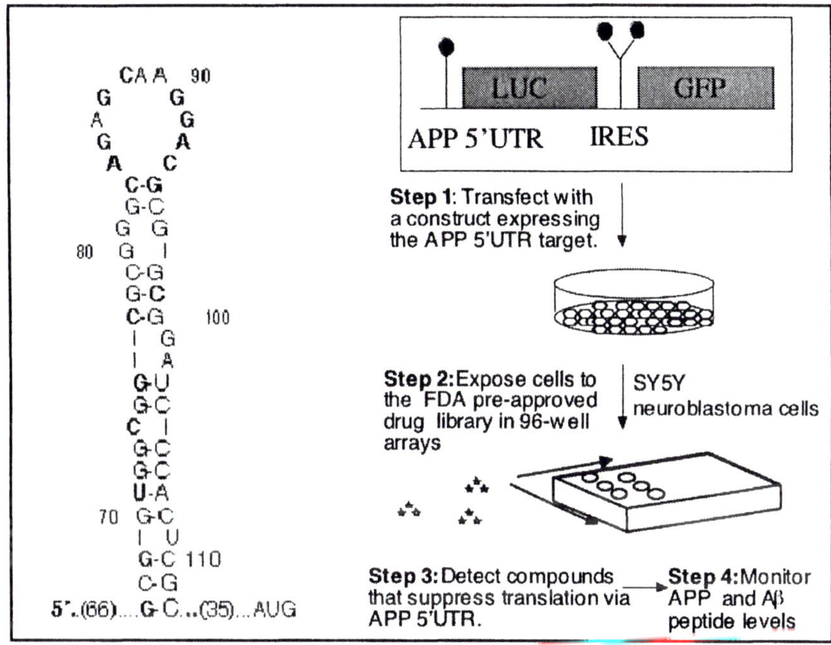

Figure 5. Primary screen for drugs targeted to the APP 5'untranslated region. *In a screen of a library of 1,200 FDA-pre-approved compounds phenserine(ACHEi) and the potent intracellular iron chelator, desferrioxamine, were the validated positive control drugs that suppressed APP mRNA translation through APP 5'UTR sequences. In this transfection based screen several lead APP-5'UTR-directed drugs were identified to limit luciferase gene expression driven by the APP 5'UTR. As an internal selectivity control for this screen, downstream dicistronic GFP gene expression (at the translational level by a viral internal ribosome entry site (IRES)) was unresponsive to drug action. Several leads (i.e., dimercaptopropanol) were identified to be chelators as described.*

As a control for specificity during screening we checked that none of the drug leads suppressed dicistronic GFP gene expression (GFP was translationally driven from another RNA structure, a viral Internal Ribosome Entry Site (IRES) upstream from the GFP start codon)(see (Payton et al., 2003). The internal ribosome entry side of poliovirus is an excellent screening control since the basis of the assay is to identify small molecules that only recognize the APP 5'UTR but exhibit no inhibitory action on the downstream ORF translationally controlled by the IRES RNA structure. Therefore the dicistronic construct in Figure 5 presents for an internally controlled screening step with great RNA targeting selectivity for the APP 5'untranslated region.

We further characterized the pharmacological action of the most promising drugs that were identified from our APP 5'UTR drug screen of 1,200 FDA pre-approved drugs. Using western blot analysis we showed that dimercaptopopanol (DMP) and paroxetine were the leads that optimally limited APP holoprotein expression [7].

Several APP 5'UTR-directed FDA Drugs were found to be metal chelators.

Our identification of several chelators as compounds to limit APP holoprotein synthesis was not surprising since an Iron-responsive Element was shown to be active in the APP 5'UTR [102]. Western blot analysis provided a secondary assay of five APP 5'UTR-directed FDA drugs (with chelator activity) that were particularly active to limit the steady-state levels of intracellular APP holoprotein in neuroblastoma (SY5Y) cells (Table I). The following list describes how dimercaptopropanol (Hg^{2+}, Pb^{2+} chelator), desferrioxamine (Fe^{3+} chelator), tetrathiolmobdylate (Cu^{2+} chelator), phenserine (AchEI) and paroxetine (SSRI) suppressed APP holoprotein translation and limited Aβ-peptide secretion:-

1. Dimercaptopropanol (DMP) is a Hg^{2+} and Pb^{2+} chelator that suppressed neuroblastoma APP levels by >50% compared to untreated neuroblastoma counterparts. The effective therapeutic range for DMP is very low (10 nM to 100 nM). APP levels were diminished by 36% (n=3) at the lower dose of dimercaptopropanol (0.01 μM) without toxicity to cells. Higher concentrations (0.1 μM) of dimercaptopropanol were found to be toxic to cells.
2. Desferrioxamine (DFO) harbors a dissociation constant Kd = 10^{-21} M as an Fe^{3+} chelator. DFO effectively suppressed APP holoprotein expression after >48 hour treatment of neuroblastoma (SY5Y) cells with chelator (1-5 μM dose range) [102]. In SY5Y cells the effect of DFO was highly selective since β-actin and APLP-1 levels were unchanged. Intramuscular administration of the iron chelator, desferrioxamine (DFO), clinically slows the rate of decline of daily living skills of patients afflicted with AD [41]. This finding was originally attributed to the role of desferrioxamine as a chelator of aluminum from the brains of AD patients, even though DFO is

Table I

Several lead drugs that had been screened from a library of 1,200 FDA pre-approved drugs for their capacity to limit APP 5'UTR directed translation of a luciferase reporter gene were found to be metal chelators. These compounds were secondary screened for their capacity to decrease the steady state levels of APP holoprotein in neuroblastoma (SY5Y) cells, and to limit the secretion of Aβ from B33 lens epithelial cells.

DRUG	APP INHIBITION %x+/-SEM	Aβ INHIBITION
Dimercaptopropanol (Hg)	>50% +/- 10% (n =3)	Payton et al., (2003)
Desferrioxamine (Fe)	80% +/- 15% (n=7)	50%
Tetramobdylate (Cu)	50% +/- 15% (n=4)	50%
Paroxetine (Fe, secondary)	35% +/- 15% (n=3)	Payton et al., (2003)
Phenserine (Fe, Secondary)	50% +/- 10% (n=4)	Shaw et al., (2001)

Conditioned medium was collected from drug- and untreated human B3 cells. Aβ(1-40) and Aβ(1-42) levels were measured using a standard sandwich ELISA assay where 2G3 and 21F12 antibodies captured Aβ1-40 and Aβ1-42, respectively. 266B antibody was used for detection.

primarily used as an iron chelator in several disease conditions associated with iron overload, and is used to counteract iron poisoning [113]. Oral derivatives of desferrioxamine may prove more clinically useful to effectively target amyloid buildup by limiting the expression of the APP holoprotein. Desferrioxmine is the drug lead that most effectively provides proof-of-concept that the APP 5'UTR remains an attractive target for the selection of novel medicinal drugs that can selectively block APP mRNA translation and the production of Aβ-peptide. This is a required step before amyloid plaque formation.

3. Tetrathiolmobdylate is a highly selective copper chelator [150] which we found suppressed APP expression in neuroblastoma cells, although at higher relative dose than the iron chelator desferrioxamine.
4. Phenserine is an anti-cholinesterase that is undergoing phase III clinical trials as a cognitive enhancer useful for AD therapeutics (Axonyx NY, NY). Phenserine reduced APP levels significantly in drug-exposed neuroblastoma cells and reduced Aβ in the conditioned medium of SY5Y cells [149]. Transfection studies demonstrated that phenserine operated more effectively to suppress APP 5'UTR conferred translation of a luciferase reporter gene under conditions of partial intracellular iron chelation with desferrioxamine.
5. Paroxetine (paxil) is a serotonin reuptake inhibitor (SSRI). Paroxetine is an FDA pre-approved drug that influences the intracellular iron Balance. SY5Y cells treated with paxil (at concentrations of 1-20 μM) exhibited reduced steady-state levels of APP holoprotein (by 20%, n=6), whereas equivalent levels of APLP-1 were present in both treated and untreated SY5Y counterparts [7]. Paroxetine significantly reduced Aβ secretion from lens epithelial B3 cells [7]. Like desferrioxamine, we found that paroxetine administration (1-100 μM) increased TfR-mRNA levels in neuroblastoma cells (Morse et al., 2004, In Press). Steady-state levels of TfR mRNA is known to be an internal marker for changes in iron homeostasis. The presence of TfR mRNA concentrations inside the cell is a very sensitive indicator of changes in intracellular iron levels because the desferrioxamine responsive IRP/IRE interaction regulates steady-state levels of TfR mRNA [54]. We proposed that paroxetine reduced APP translation and thus Aβ peptide levels by the indirect mechanism of changing intracellular iron pools. The pathway by which paroxetine acts to alter intracellular iron pools may be a consequence of the drug acting as, itself, a chelator or alternatively by activating iron efflux or iron redistribution in the cell.

APP 5'UTR-Directed Drugs Limited Aβ peptide Production.

We tested the capacity of dimercaptopropanol, desferrioxamine and paroxetine (paxil) (APP 5'UTR screened drugs that reduced translation of APP holoprotein) to lower secreted Aβ peptide levels. For this purpose, lens epithelial cells (B3) were an excellent endogenous cell line since B3 cells generate greater levels of basal Aβ relative to neuroblastoma cells. ELISA data in Table I shows that paroxetine lowered Aβ peptide secretion into the conditioned medium of lens B3 cells by up to 50% after a 4 day treatment (15 μM dose paroxetine). Treatement with dimercaptopropanol also reduced Aβ(1-40) and Aβ(1-42) levels by 30% to 50% relative to control treated B3 cells (72 hour treatment of the cells). In the future we will test whether drugs in the same class as paroxetine (i.e., prozac, another serotonin re-uptake blocker) also reduce Aβ secretion. This experiment will determine whether paroxetine operates to perturb iron metabolism by its SSRI activity or intrinsic pharmacology related to its chemical structure. In sum, our current results showed that drugs that lower APP holoprotein translation (i.e., paroxetine and dimercaptopropanol) also reduced the secretion of Aβ(1-40) and Aβ(1-42).

Selectivity and Targeting the APP 5'UTR as a Therapeutic Strategy for AD.

A sufficiently specific therapeutic compound would be expected to leave expression of the related APLP-1 and APLP-2 genes unchanged before any given drug can be tested in-vivo as a therapeutic agent for Alzheimer's Disease. Limiting APP expression would be predicted to have low negative physiological impact since APLP-1 and APLP-2 should provide for functional redundancy for APP as illustrated in the good health of APP knockout mice (152).

The identification of two metal chelators (dimercaptopropanol and desferrioxamine) as APP 5'UTR-directed compounds that lower APP levels but leave APLP-1 levels unchanged (therapeutic action) is consistent with the presence of an Iron-responsive Element (Type II) in the 5'UTR of the APP transcript but not in the 5'UTR of the APLP-1 mRNA. Methods for transfected cell based drug screens are described by Rogers et al (2002b) and Payton et al., (2003). In future screens we will include a selectivity control to ensure specificity wherein chosen drugs will have to be counter-screened against the 5'UTRs of APLP-1 and APLP-2 transcripts [151, 152].

We propose that the APP 5'UTR target will prove very useful to identify novel chelators and NSAIDS from a large drug library (i.e., 65,000 compounds in the library of the Laboratory of Drug Discovery in Neurodegeneration (LDDN library) (Dr. Ross Stein, Brigham and Women's Hospital, Boston, MA). The APP 5'UTR is a valid target that is unique but also harbors IL-1, TGFβ, and iron responsive domains. Therefore drugs targeted to this APP 5'UTR should include both new chelators and NSAIDs, in addition to other drug classes reported in our pilot FDA library screen [7]. Larger novel (non-FDA) Libraries of combinatorial low molecular weight compounds can now also be screened for their capacity to block 5'UTR dependent translation of APP (and hence Aβ peptide production).

The challenge of our screening approach has been to identify new drugs (i.e., chelators and NSAIDs) that suppress expression of APP holoprotein translation, and thereby limit Aβ peptide output using tissue culture models. The use of transgenic mouse models of AD is the subsequent step for testing our APP directed lead drugs. We have set up a transgenic mouse model to test the efficacy of lead APP 5'UTR-directed drugs (chelators) to limit amyloid burden. The CRND8 trangenic mouse model encodes the human APP transgene (717 London mutation), which is expressed under the control of the natural APP 5'untranslated region (SmaI-NruI fragmentof the APP gene). Therefore the use of the CRND8 transgenic mice could well be an a good match as an animal model to test APP 5'UTR-directed drugs, unlike other mouse models of AD [153].

Conclusions.

Chelators are currently proven to be of therapeutic significance for AD because the Cu/Zn chelator clioquinol was effective in a Phase II clinical trial [104]. The results of the clioquinol trial were consistent with a pervious human subjects study that showed desferrioxamine to have efficacy for the treatment of AD [41]. These two studies have stimulated an international effort to discover novel chelators for the treatment of AD. We reported that the chelators desferroxamine and dimercaptopropanol to be the most potent APP 5'UTR directed compounds from our transfection based drug screen targeted to the APP 5'untranslated region. These chelators validated our discovery that there is an authentic Iron-responsive Element in the 5'untranslated region of the Alzheimer's APP transcript [48]. Certainly, the iron-responsive APP 5'untranslated region appears to be an excellent drug target for the purpose of screening for novel (non FDA) chelators from large combinatorial drug libraries for small molecules. Drug design and development of hits would be predicted to tailor interesting lead chelators of therapeutic value for the treatment of Alzheimer's disease.

ACKNOWLEDGMENTS:

The author is grateful to the support of Dr. Rudolph Tanzi and Dr. Ashley Bush. Dr. Andrew Thomson was helpful with the model for iron metabolism. JTR was supported from an NIA grant AG21081, ISOA and Alzheimer's Association (IIRG-02-3524).

REFERENCES

1. Bellingham, S.A., et al., *Copper depletion down-regulates expression of Alzheimer's disease Amyloid-beta precursor protein gene.* J Biol Chem, 2004.
2. Sampietro, M., et al., *The hemochromatosis gene affects the age of onset of sporadic Alzheimer's disease.* Neurobiol Aging, 2001. **22**(4): p. 563-8.
3. Lovell, M.A., et al., *Copper, iron and zinc in Alzheimer's disease senile plaques.* J Neurol Sci, 1998. **158**(1): p. 47-52.
4. Multhaup, G., Schlicksupp, A., Hesse, L., Beher, D., Ruppert, T., masters, C., Beyreuther, K., *The Amyloid Precursor protein of Alzheimer's disease in the reduction of copper(II) to copper(I).* Science, 1996. **271**: p. 1406-1409.
5. Huang, X., et al., *Cu(II) potentiation of alzheimer abeta neurotoxicity. Correlation with cell-free hydrogen peroxide production and metal reduction.* J Biol Chem, 1999. **274**(52): p. 37111-6.
6. Rogers, J.T., et al., *Alzheimer's disease drug discovery targeted to the APP mRNA 5'untranslated region.* J Mol Neurosci, 2002. **19**(1-2): p. 77-82.
7. Payton, S., et al., *Alzheimer's Disease Drug Discovery Targeted to the APP mRNA 5'Untranslated region; paroxetine and dimercaptopropanol are drug hits.* Journal of molecular Neuroscience, 2003. **20**: p. 267-275.
8. Kosik, K., et al., *Discovery of compounds that will prevent tau pathology.* J Mol Neurosci, 2002. **19**(3): p. 261-6.
9. Lu, M. and K. Kosik, *Competition for microtubule-binding with dual expression of Tau missense and splice isoforms.* Mol Biol Cell, 2001. **12**(1): p. 171-84.
10. Shin, R., et al., *A novel trivalent cation chelator Feralex dissociates binding of aluminum and iron associated with hyperphosphorylated tau of Alzheimer's disease.* Brain Res, 2003. **961**(1): p. 139-46.
11. Polizzi, S., et al., *Neurotoxic effect of aluminum among foundry workers and Alzheimer's disease.* Neurotoxicology, 2002. **23**(6): p. 761-774.
12. Selkoe, D.J., *Amyloid beta-protein and the genetics of Alzheimer's disease.* J Biol Chem, 1996. **271**(31): p. 18295-8.
13. Haass, C., et al., *Amyloid beta-peptide is produced by cultured cells during normal metabolism.* Nature, 1992. **359**(6393): p. 322-5.
14. LaVoie, M., et al., *Assembly of the gamma-secretase complex involves early formation of an intermediate of APh-1 and nicastrin.* J. Biol. Chem, 2003. **278**(39): p. 37213-22.
15. Vassar, R., et al., *Beta-secretase cleavage of Alzheimer's amyloid precursor protein by the transmembrane aspartic protease BACE [see comments].* Science, 1999. **286**(5440): p. 735-41.

16. De Strooper, B., *Aph-1, Pen, and Nicastrin with presenilin generate an active gamma secretase complex.* Neuron, 2003. **38**(10): p. 9-12.
17. Goate, A., et al., *Segregation of a missense mutation in the amyloid precursor protein gene with familial Alzheimer's disease [see comments].* Nature, 1991. **349**(6311): p. 704-6.
18. Mullan, M., Crawford, F., Axelman, K., Houlden, H., Lilius, L., Winblad, B., Lannfelt, L., *A pathogenic mutation for probable Alzheimer's disease in the APP gene at the N-terminus of β-amyloid.* Nature Genetics, 1992. **1**: p. 345-347.
19. Levy-Lahad, E., et al., *Candidate gene for the chromosome 1 familial Alzheimer's disease locus.* Science, 1995. **269**(5226): p. 973-7.
20. Wong, P.C., et al., *Presenilin 1 is required for Notch1 and Dll1 expression in the paraxial mesoderm.* Nature, 1997. **387**(6630): p. 288-92.
21. Van Broeckhoven, C., et al., *Amyloid beta protein precursor gene and hereditary cerebral hemorrhage with amyloidosis (Dutch).* Science, 1990. **248**(4959): p. 1120-2.
22. Wisniewski, K.E., et al., *Alzheimer's disease in Down's syndrome: clinicopathologic studies.* Neurology, 1985. **35**(7): p. 957-61.
23. Wolfe, M.S., et al., *Two transmembrane aspartates in presenilin-1 required for presenilin endoproteolysis and gamma-secretase activity.* Nature, 1999. **398**(6727): p. 513-7.
24. Corder, E., Saunders, AM., Strittmatter, MD., Schmechel, DE., Gaskell, PC., Rimmler., MS., Locke, BA., Conneally, PM., Schmader, KE., Tanzi, RE., Gusella, J., Small, GW., Roses, AD., Pericak-Vance, MA., Haines, JL., *Apolipoprotein E survival in Alzheimer's disease patients, and the competing risks of death and Alzheimer's disease.* Neurology, 1995. **45**: p. 1323-1328.
25. Qiu, Z., et al., *Alpha2-macroglobulin enhances the clearance of endogenous soluble beta- amyloid peptide via low-density lipoprotein receptor-related protein in cortical neurons.* J Neurochem, 1999. **73**(4): p. 1393-8.
26. Farris, W., et al., *Insulin-degrading enzyme regulates the levels of insulin, amyloid beta protein, and the beta amyloid precursor protein intracellular domain in vivo.* PNAS, 2003. **100**(7): p. 4162-7.
27. Nilsson LN, B.K., DiCarlo G, Gordon MN, Morgan D, Paul SM, Potter H., *Alpha-1-antichymotrypsin promotes beta-sheet amyloid plaque deposition in a transgenic mouse model of Alzheimer's disease.* J Neurosci, 2001. **21**(5): p. 1444-51.
28. Ujiie, M., D. Dickstein, and W. Jeffires, *p97 as a biomarker for Alzheimer's disease.* Front Biosci, 2002. **7**(e): p. 42-7.
29. Jeffries, W., D. Dickstein, and M. Ujiie, *Assessing p97 as an Alzheimer's disease serum biomarker.* J Alzheimers Dis, 2001. **3**(3): p. 339-344.

30. Sparks, D. and B. Schreurs, *Trace amounts of copper in water induce beta-amyloid plaques and learning deficits in a rabbit model of Alzheimer's disease.* PNAS, 2003. **100**(19): p. 11065-9.
31. Thompson, K., et al., *Mouse brains deficient in H-ferritin have normal iron nconcentration but protein profile of iron deficiency and increased oxidative stress.* J Neuroscie Res, 2003. **71**(1): p. 46-63.
32. Abraham, C.R., et al., *Alpha 1-antichymotrypsin is present together with the beta-protein in monkey brain amyloid deposits.* Neuroscience, 1989. **32**(3): p. 715-20.
33. Robinson, S., Noone, DF., Kril, J., Haliday, GM., *Most amyloid plaques contain ferritin rich cells.* Alzheimer's Research, 1995. **1**(4): p. 191-193.
34. Grundke-Iqbal, I., et al., *Ferritin is a component of the neuritic (senile) plaque in Alzheimer dementia.* Acta Neuropathol, 1990. **81**(2): p. 105-10.
35. Bush, A., Pettingell, WH., Multhaup, G., Paradis, .D., Vonsattel, Gusella, JF., Beyreuther, K., Masters, CL., Tanzi, RE., *Rapid induction of Azheimer A-beta amyloid formation by zinc.* Science, 1994. **265**: p. 1464-1467.
36. Huang, X., et al., *The A beta peptide of Alzheimer's disease directly produces hydrogen peroxide through metal ion reduction.* Biochemistry, 1999. **38**(24): p. 7609-16.
37. Mantyh, P.W., et al., *Aluminum, iron, and zinc ions promote aggregation of physiological concentrations of beta-amyloid peptide.* J Neurochem, 1993. **61**(3): p. 1171-4.
38. Schubert, D. and M. Chevion, *The role of iron in beta amyloid toxicity.* Biochem Biophys Res Commun, 1995. **216**(2): p. 702-7.
39. Bartzokis, G., et al., *In vivo evaluation of brain iron in Alzheimer disease using magnetic resonance imaging.* Arch Gen Psychiatry, 2000. **57**(1): p. 47-53.
40. Connor, J.R., et al., *A quantitative analysis of isoferritins in select regions of aged, parkinsonian, and Alzheimer's diseased brains.* J Neurochem, 1995. **65**(2): p. 717-24.
41. Crapper McLachlan, D., Dalton, AJ., Kruck, TPA., Bell, MY., Smith, WL., Kalow, W., Andrews, DF., *Intramuscular desferrioxamine in patients with Alzheimer's disease.* Lancet, 1991. **337**: p. 1304-1308.
42. Buxbaum, J.D., et al., *Evidence that tumor necrosis factor alpha converting enzyme is involved in regulated alpha-secretase cleavage of the Alzheimer amyloid protein precursor.* J Biol Chem, 1998. **273**(43): p. 27765-7.
43. Bodovitz, S., Falduto, MT., Frail, DE., Klein, WL, *Iron levels modulate alpha-secretase cleavage of amyloid precursor protein.* J. Neurochem, 1995. **64**: p. 307-315.
44. Bush , A.I., Atwood, C., Goldstein, L., Huang, X., Rogers, JT, *Could Aß and ßPP be antioxidants?* J. Alzheimer's Disease, 2000. **2**: p. 83-84.

45. Rogers, J.T., *Ferritin translation by interleukin-1: the role of sequences upstream of the start codons of the heavy and light subunit genes.* Blood, 1996. **87**(6): p. 2525-37.
46. Feng, Y., et al., *Translational suppression by trinucleotide repeat expansion at FMR1.* Science, 1995. **268**(5211): p. 731-4.
47. Rogers, J.T., et al., *Translation of the alzheimer amyloid precursor protein mRNA is up- regulated by interleukin-1 through 5'-untranslated region sequences.* J Biol Chem, 1999. **274**(10): p. 6421-31.
48. Rogers, J.T., et al., *An iron-responsive element type II in the 5'-untranslated region of the Alzheimer's amyloid precursor protein transcript.* J Biol Chem, 2002. **277**(47): p. 45518-28.
49. Rogers, J., *Genetic regulation of the iron transport and storage genes: Links with the acute phase response. In: Iron and Human diseases.* CRC Press, Boca Raton., 1992. Lauffer R, ed.(77-104).
50. Balla G., J.H., Balla, J., Rosenberg, M., Nath, K., Apple, F., Eaton,JW., Vercellotti, GM, *Ferritin: A cytoprotective antioxident strategem of endothelium.* J. Biol. Chem., 1992. **267**: p. 18148-18153.
51. Gordeuk, V., Prithviraj, P., Dolinar, T., Brittenham, GM., *Interleukin-1 administration in mice produces hypoferremia despite neutropenia.* J. Clin. Invest, 1988. **82**: p. 1934-1938.
52. Tran, T.N., et al., *Secretion of ferritin by rat hepatoma cells and its regulation by inflammatory cytokines and iron.* Blood, 1997. **90**(12): p. 4979-86.
53. Thomson, A.M., J.T. Rogers, and P.J. Leedman, *Iron-regulatory proteins, iron-responsive elements and ferritin mRNA translation.* Int J Biochem Cell Biol, 1999. **31**(10): p. 1139-52.
54. Klausner, R., Rouault, TA., Harford, JB., *Regulating the fate of mRNA: The control of cellular iron metabolism.* Cell, 1993. **72**: p. 19-28.
55. Iwai, K., et al., *Iron-dependent oxidation, ubiquitination, and degradation of iron regulatory protein 2: implications for degradation of oxidized proteins.* Proc Natl Acad Sci U S A, 1998. **95**(9): p. 4924-8.
56. Kaldy, P., et al., *Identification of RNA-binding surfaces in iron regulatory protein-1.* Embo J, 1999. **18**(21): p. 6073-83.
57. Gunshin, H., et al., *Iron-dependent Regulation of the Divalent Metal Ion Transporter.* FEBS Letters, 2001. **509**: p. 309-316.
58. Eisenstein, R. and E. Theil, *Combinatorial mRNA Regulation; Iron regulatory Protein and Iso Iron-responsive elements.* J. Biol. Chem., 2000. **275**(52): p. 40659-40662.
59. Leibold, E., Munro, HN., *Cytoplasmic protein binds in-vitro to a highly conserved sequence in the 5' untranslated region of ferritin heavy- and light-subunit mRNAs.* PNAS, 1988. **85**: p. 2171-2175.
60. Leedman, P., Stein, AR., Chin, WW., Rogers, JT., *Thyroid hormone modulates the interaction between the iron regulatory protein and ferritin mRNA Iron-responsive elements.* J. Biol. Chem., 1996. **271**: p. 12017-12023.

61. Goossen, B. and M.W. Hentze, *Position is the critical determinant for function of iron-responsive elements as translational regulators.* Mol Cell Biol, 1992. **12**(5): p. 1959-66.
62. Muckenthaler, M., N.K. Gray, and M.W. Hentze, *IRP-1 binding to ferritin mRNA prevents the recruitment of the small ribosomal subunit by the cap-binding complex eIF4F.* Mol Cell, 1998. **2**(3): p. 383-8.
63. Wu, S. and R.J. Kaufman, *Double-stranded (ds) RNA binding and not dimerization correlates with the activation of the dsRNA-dependent protein kinase (PKR).* J Biol Chem, 1996. **271**(3): p. 1756-63.
64. Thomson, A.M., J.T. Rogers, and P.J. Leedman, *Thyrotropin-releasing hormone and epidermal growth factor regulate iron- regulatory protein binding in pituitary cells via protein kinase C- dependent and - independent signaling pathways [In Process Citation].* J Biol Chem, 2000. **275**(41): p. 31609-15.
65. Pinero, D.J., et al., *Interleukin-1beta increases binding of the iron regulatory protein and the synthesis of ferritin by increasing the labile iron pool [In Process Citation].* Biochim Biophys Acta, 2000. **1497**(3): p. 279-88.
66. Tsuji, Y., et al., *Coordinate transcriptional and translational regulation of ferritin in response to oxidative stress.* Mol Cell Biol, 2000. **20**(16): p. 5818-27.
67. Caltagirone, A., G. Weiss, and K. Pantopoulos, *Modulation of cellular iron metabolism by hydrogen peroxide. Effects of H2O2 on the expression and function of iron-responsive element- containing mRNAs in B6 fibroblasts.* J Biol Chem, 2001. **276**(23): p. 19738-45.
68. Pantopolulos, K., Hentze, MW., *Rapid responses to oxidative sress mediated by iron regulatory protein.* EMBO, 1995. **14**: p. 2917-2924.
69. Toth, I., et al., *Hypoxia alters iron-regulatory protein-1 binding capacity and modulates cellular iron homeostasis in human hepatoma and erythroleukemia cells.* J Biol Chem, 1999. **274**(7). p. 4467-73.
70. Hanson, E.S., M.L. Rawlins, and E.A. Leibold, *Oxygen and iron regulation of iron regulatory protein 2.* J Biol Chem, 2003. **278**(41): p. 40337-42.
71. Rogers, J.T., et al., *Translational control during the acute phase response. Ferritin synthesis in response to interleukin-1.* J Biol Chem, 1990. **265**(24): p. 14572-8.
72. Rogers JT, A.J., Lacroix L, Kasschau, KD, Durmowicz GP, Bridges KR., *Translational regulation of H-ferritin mRNA by Il-1β acts through 5' leader sequences distinct from the Iron Responsive Element.* Nucl. Acids. Research., 1994. **22**: p. 2678-2686.
73. Rogers, J.T., et al., *Translational enhancement of H-ferritin mRNA by interleukin-1 beta acts through 5' leader sequences distinct from the iron responsive element.* Nucleic Acids Res, 1994. **22**(13): p. 2678-86.

74. Cartwright, G., Lauritsen, MA., Humphreys, S., Jones, PJ., Merrill, IM., Wintrobe, MM., *Anemia associated with chronic infection.* Science, 1946. **103**: p. 72.

75. Konijn, A.M., et al., *Ferritin synthesis in inflammation. II. Mechanism of increased ferritin synthesis.* Br J Haematol, 1981. **49**(3): p. 361-70.

76. Jacobs, A., Woodworth, M, *Ferritin in Serum.* N. Engl .J. Med, 1975. **292**: p. 951-956.

77. Broxmeyer, H., Cooper, S. Arosio, P, *Mutated recombinant human heavy-chain ferrittins and myelosuppression imdvitnovivo: A link between ferritin ferroxidase activity and biological function.* PNAS, 1991. **88**: p. 770-774.

78. Lustlader, E., Hann, E-WL., Blumberg, BS, *Serum Ferritin as a predictor of host response to hepatitis Bvirus infection.* Science, 1983. **220**: p. 423-425.

79. Ramadori, G., Sipe, JD., Dinarello, CA., Mizel, B. and Colten, HR., *Pretranslational modulation of acute phase hepatic protein synthesis by murine recombinant interleukin I (IL-1) and purified human IL-1.* J. Exp. Med., 1985. **162**: p. 930-942.

80. Chkheidze, A.N., et al., *Assembly of the α-Globin mRNA Stability Complex Reflects Binary Interaqxtion between thre Pyrimidine-Rich 3' Untranslated Region Determinant and Poly(C) Binding Protein αCP.* Mol Cell Biol., 1999. **19**(7): p. 4572-4681.

81. Ostareck, D.H., et al., *mRNA silencing in erythroid differentiation: hnRNP K and hnRNP E1 regulate 15-lipoxygenase translation from the 3' end.* Cell, 1997. **89**(4): p. 597-606.

82. Hentze, M., Kuhn, LC., *Molecular control of vertebrate iron mertabolism: mRNA-based regulatory circuits operated by iron, nitric oxide, and oxidative stress.* PNAS, 1996. **83**: p. 8175-8182.

83. Fujioka, M., et al., *Magnetic resonance imaging shows delayed ischemic striatal neurodegeneration.* Ann Neurol, 2003. **54**(6): p. 732-47.

84. Zhu, Y., et al., *Expression of poly(C)-binding proteins is differentially regulated by hypoxia and ischemia in cortical neurons.* Neuroscience, 2002. **110**(2): p. 191-8.

85. Pinero, D.J., J. Hu, and J.R. Connor, *Alterations in the interaction between iron regulatory proteins and their iron responsive element in normal and Alzheimer's diseased brains [In Process Citation].* Cell Mol Biol (Noisy-Le-Grand), 2000. **46**(4): p. 761-76.

86. LaVaute, T., et al., *Targeted deletion of the gene encoding iron regulatory protein-2 causes misregulation of iron metabolism and neurodegenerative disease in mice.* Nat Genet, 2001. **27**(2): p. 209-214.

87. Melofors, O., Goossen, B., Johansson, B., Stripecke, R., Gray, NK. Hentze, MW., *Translational control of 5-aminolevulinate synthase mRNA by Iron-responsive Elements in erythroid cells.* J. Biol. Chem., 1993. **268**: p. 5974-5978.

88. Cox, T.C., et al., *Human erythroid 5-aminolevulinate synthase: promoter analysis and identification of an iron-responsive element in the mRNA.* Embo J, 1991. **10**(7): p. 1891-902.
89. Cox, L.A. and G.S. Adrian, *Posttranscriptional regulation of chimeric human transferrin genes by iron.* Biochemistry, 1993. **32**(18): p. 4738-45.
90. Chung, J., D.J. Haile, and M. Wessling-Resnick, *Copper-induced ferroportin-1 expression in J774 macrophages is associated with increased iron efflux.* Proc Natl Acad Sci U S A, 2004. **101**(9): p. 2700-5.
91. Knutson, M.D., et al., *Iron loading and erythrophagocytosis increase ferroportin 1 (FPN1) expression in J774 macrophages.* Blood, 2003. **102**(12): p. 4191-7.
92. Gunshin, H., et al., *Cloning and characterization of a mammalian proton-coupled metal-ion transporter.* Nature, 1997. **388**(6641): p. 482-8.
93. Amara, F.M., et al., *TGF-beta(1), regulation of alzheimer amyloid precursor protein mRNA expression in a normal human astrocyte cell line: mRNA stabilization.* Brain Res Mol Brain Res, 1999. **71**(1): p. 42-9.
94. Rajagopalan, L.E. and J.S. Malter, *Growth factor-mediated stabilization of amyloid precursor protein mRNA is mediated by a conserved 29-nucleotide sequence in the 3'- untranslated region.* J Neurochem, 2000. **74**(1): p. 52-9.
95. Zaidi, S.H. and J.S. Malter, *Amyloid precursor protein mRNA stability is controlled by a 29-base element in the 3'-untranslated region.* J Biol Chem, 1994. **269**(39): p. 24007-13.
96. de Sauvage, F., Kruys, V., Marinx, O., Huez, O., Octave, JN., *Alternative polyadenylation of the amyloid protein precursor mRNA.* EMBO, 1992. **11**: p. 3099-3103.
97. Mbella, E.G., et al., *A GC nucleotide sequence of the 3' untranslated region of amyloid precursor protein mRNA plays a key role in the regulation of translation and the binding of proteins.* Mol Cell Biol, 2000. **20**(13): p. 4572-9.
98. Sachs, A.B., P. Sarnow, and M.W. Hentze, *Starting at the beginning, middle, and end: translation initiation in eukaryotes.* Cell, 1997. **89**(6): p. 831-8.
99. Preiss, T., M. Muckenthaler, and M.W. Hentze, *Poly(A)-tail-promoted translation in yeast: implications for translational control.* Rna, 1998. **4**(11): p. 1321-31.
100. Preiss, T. and M.W. Hentze, *Dual function of the messenger RNA cap structure in poly(A)-tail- promoted translation in yeast.* Nature, 1998. **392**(6675): p. 516-20.

101. Chen, C.Y. and P. Sarnow, *Initiation of protein synthesis by the eukaryotic translational apparatus on circular RNAs*. Science, 1995. **268**(5209): p. 415-7.

102. Rogers JT, R.J., Cahill CM, Eder PS, Huang X, Gunshin H, Leiter L, McPhee J, Sarang SS, Utsuki T, Greig N, Lahiri DK, Tanzi RE, Bush AI, Giordano T, Gullans S, *An iron-responsive element type II in the 5' untranslated region of the Alzheimer's amyloid precursor protein transcript*. J Biol Chem, 2002. **277,** p. 45518-45528.

103. Cherny, R., Atwood, CS., Xilinas, ME., Gray, DN., Jones, WD., McLean, CA., Barnham, KJ., Volitakis, I., Fraser, FW., Kim, Y., Huang, X., Goldstein, LE., Moir, RD., Lim, JT.., Beyreuther, K., Zheng, H., Tanzi, RE., Masters, CL., Bush AI., *Treatment with copper-zinc chelator markedly and rapidly inhibits beta-amyloid accumulation in Alzheimer's disease transgenic mice*. Neuron, 2001. **30**(3): p. 641-642.

104. Ritchie, C.W., et al., *Metal-protein attenuation with iodochlorhydroxyquin (clioquinol) targeting Aβ amyloid deposition and toxicity in Alzheimer's disease: a pilot phase 2 clinical trial*. Archives of Neurology, 2003. **60**: p. 1685-1691.

105. Yankner, B.A., L.K. Duffy, and D.A. Kirschner, *Neurotrophic and neurotoxic effects of amyloid beta protein: reversal by tachykinin neuropeptides*. Science, 1990. **250**(4978): p. 279-82.

106. Bishop, G. and S. Robinson, *Human Abeta1-42 reduces iron-induced toxicity in rat cerbral cortex*. J Neuroscie Res., 2003. **73**(3): p. 316-323.

107. Bishop, G., et al., *Iron: a pathological mediator of Alzheimer's disease?* Dev. Neurosci, 2002. **24**(2-3): p. 184-7.

108. Connor, J.R., et al., *Regional distribution of iron and iron-regulatory proteins in the brain in aging and Alzheimer's disease*. J Neurosci Res, 1992. **31**(2): p. 327-35.

109. Lai, A., Sisodia, SS., Trowbridge, IS., *Characterization of the sorting signals in the beta-amyloid precursor protein cytoplasmic domain*. J. Biol. Chem., 1995. **270**(8): p. 3565-3573.

110. Tanzi, R.E. and B.T. Hyman, *Alzheimer's mutation*. Nature, 1991. **350**(6319): p. 564.

111. Rogers JT, R.J., Eder PS, Huang X, Bush AI, Tanzi RE, Venti A, Payton SM, Giordano T, Nagano S, Cahill CM, Moir R, Lahiri DK, Greig N, Sarang SS, Gullans SR., *Alzheimer's disease drug discovery targeted to the APP mRNA 5'untranslated region*. J Mol Neurosci, 2002. **19**: p. 77-82.

112. Campbell, A. and S.C. Bondy, *Aluminum induced oxidative events and its relation to inflammation: a role for the metal in Alzheimer's disease [In Process Citation]*. Cell Mol Biol (Noisy-Le-Grand), 2000. **46**(4): p. 721-30.

113. Halliwell, B., Hedlund, B, *Therapeutic Strategies to inhibit iron-catalyzed tissue damage*. Iron and Human Disease, 1992: p. 477-500.

114. Brittenham, G., *Iron chelators and iron toxicity.* Alcohol, 2003. **30**(2): p. 151-8.
115. Brittenham, G., et al., *Deferiprone versus desferrioxamine in thalassaemia, and T2 validation and utility.* The Lancet, 2003. **361**(9352).
116. Brittenham, G., et al., *Deferiprone and hepatic fibrosis.* Blood, 2003. **101**(12): p. 5089-5091.
117. Konoghiorghes, G., K. Neocleous, and A. Kolnagou, *Benefits and risks of deferiprone in iron overload in Thalassaemia and other conditions: comparison of epidemiological and therapeutic aspects with deferoxamine.* Drug Safety, 2003. **26**(8): p. 553-84.
118. Bush, A.I., *The Metallobiology of Alzheimer's Disease.* Trends in Neurosciences, 2003. **26**: p. 207-214.
119. Atwood, C.S., et al., *Dramatic aggregation of Alzheimer abeta by Cu(II) is induced by conditions representing physiological acidosis.* J Biol Chem, 1998. **273**(21): p. 12817-26.
120. Atwood, C., Huang, X., Khatri, A., Scarpa, R., Kim, Y-S., Moir, RD., Tanzi, RE., Roher, AE., Bush, AI., *Copper catalyzed oxidation of Alzheimer Aβ.* Cellular and Molecular Biology, 2000. **46**(4): p. 777-783.
121. Masters, C.L., et al., *Amyloid plaque core protein in Alzheimer disease and Down syndrome.* Proc Natl Acad Sci U S A, 1985. **82**(12): p. 4245-9.
122. Smith, M.A., et al., *Widespread peroxynitrite-mediated damage in Alzheimer's disease.* J Neurosci, 1997. **17**(8): p. 2653-7.
123. Smith, M.A., et al., *Heme oxygenase-1 is associated with the neurofibrillary pathology of Alzheimer's disease.* Am J Pathol, 1994. **145**(1): p. 42-7.
124. Smith, C.D., et al., *Excess brain protein oxidation and enzyme dysfunction in normal aging and in Alzheimer disease.* Proc Natl Acad Sci U S A, 1991. **88**(23): p. 10540-3.
125. Celsi, F., et al., *Overexpression of superoxide dismutase 1 protects against beta=amyloid peptide toxicity: effect of estrogen and copper chelators.* Neurochem Int, 2004. **44**(1): p. 25-33.
126. Beal, M.F., *Mitochondria, oxidative damage, and inflammation in Parkinson's disease.* Ann N Y Acad Sci, 2003. **991**: p. 120-31.
127. Opazo, C., et al., *Metalloenzyme-like activity of Alzheimer's disease beta-amyloid. Cu-dependent catalytic conversion of dopamine, cholesterol, and biological reducing agents to neurotoxic H(2)O(2).* J Biol Chem, 2002. **277**(43): p. 40302-8.
128. Yatin, S.M., et al., *Temporal relations among amyloid beta-peptide-induced free-radical oxidative stress, neuronal toxicity, and neuronal defensive responses.* J Mol Neurosci, 1998. **11**(3): p. 183-97.
129. Halliwell, B and Gutteridge., JMC., *Oxygen toxicity, oxygen radicals, transition metals and disease.* Biochem. J., 1984. **219**: p. 1-14.

130. Gutteridge, J.M., *Hydroxyl radicals, iron, oxidative stress, and neurodegeneration.* Ann N Y Acad Sci, 1994. **738**: p. 201-13.
131. Curtain, C.C., et al., *Alzheimer's disease amyloid-beta binds copper and zinc to generate an allosterically ordered membrane-penetrating structure containing superoxide dismutase-like subunits.* J Biol Chem, 2001. **276**(23): p. 20466-73.
132. Cuajungco, M.P., et al., *Evidence that the beta-amyloid plaques of Alzheimer's disease represent the redox-silencing and entombment of abeta by zinc.* J Biol Chem, 2000. **275**(26): p. 19439-42.
133. Beal, M.F. and C.W. Shults, *Effects of Coenzyme Q_{10} in Huntington's disease and early Parkinson's disease.* Biofactors, 2003. **18**(1-4): p. 153-161.
134. Phinney, A., et al., *In vivo reduction of amyloid-β by a mutant copper transporter.* PNAS, 2003. **100**(24): p. 14193-14198.
135. Bayer, T., et al., *Dietry Cu stabilizes brain superoxide dismutase 1 activity and reduces amyloid Aneta production in APP23 transgenic mice.* PNAS, 2003. **100**(24): p. 14187-14192.
136. Howlett, D., et al., *Hemin and related porphyrins inhibit beta-amyloid aggregation.* FEBS Lett, 1997. **417**(2): p. 249-51.
137. Harper, J.D. and P.T. Lansbury, Jr., *Models of amyloid seeding in Alzheimer's disease and scrapie: mechanistic truths and physiological consequences of the time-dependent solubility of amyloid proteins.* Annu Rev Biochem, 1997. **66**: p. 385-407.
138. Priola, S.A., A. Raines, and W.S. Caughey, *Porphyrin and phthalocyanine antiscrapie compounds [see comments].* Science, 2000. **287**(5457): p. 1503-6.
139. Caughey, W.S., et al., *Inhibition of protease-resistant prion protein formation by porphyrins and phthalocyanines.* Proc Natl Acad Sci U S A, 1998. **95**(21): p. 12117-22.
140. Walsh, D.M., et al., *The oligomerization of amyloid beta-protein begins intracellularly in cells derived from human brain.* Biochemistry, 2000. **39**(35): p. 10831-9.
141. Shohum, S., Youdim, MBH, *Iron involvement in neural damage and microgiosis in models of neurodegenerative diseases.* Cellular and Molecular Biology, 2000. **46**: p. 743-760.
142. Connor, J.R., et al., *A histochemical study of iron, transferrin, and ferritin in Alzheimer's diseased brains.* J Neurosci Res, 1992. **31**(1): p. 75-83.
143. Buxbaum, J.D., et al., *Cholinergic agonists and interleukin 1 regulate processing and secretion of the Alzheimer beta/A4 amyloid protein precursor.* Proc Natl Acad Sci U S A, 1992. **89**(21): p. 10075-8.
144. Nitsch, R.M., et al., *Release of Alzheimer amyloid precursor derivatives stimulated by activation of muscarinic acetylcholine receptors.* Science, 1992. **258**(5080): p. 304-7.

145. Hooper, N. and A. Turner, *The search for alpha-secretase and its potential as a therapeutic approach to Alzheimer's disease.* Curr Med Chem, 2002. **9**(11): p. 1107-19.
146. Zuker, M., *On finding all suboptimal foldings of an RNA molecule.* Science, 1989. **244**(4900): p. 48-52.
147. Svitkin, Y.V., et al., *The requirement for eukaryotic initiation factor 4A (elF4A) in translation is in direct proportion to the degree of mRNA 5' secondary structure.* RNA, 2001. **7**(3): p. 382-94.
148. Nilsson, L., Rogers, J., Potter, H, *The Essential Role of in Iflammation and Induced Gene Expression in the Pathogenic Pathway of Alzheimer's Disease.* Frontiers in Bioscience, 1998. **3**: p. d436-446.
149. Shaw, K.T., et al., *Phenserine regulates translation of beta -amyloid precursor protein mRNA by a putative interleukin-1 responsive element, a target for drug development.* Proc Natl Acad Sci U S A, 2001. **98**(13): p. 7605-10.
150. Mandinov, L., et al., *Copper chelation represses the vascular response to injury.* Proceedings of the National Academy of Sciences of the United States of America, 2003. **100**(11): p. 6700-5.
151. Wasco, W., et al., *Isolation and characterization of APLP2 encoding a homologue of the Alzheimer's associated amyloid beta protein precursor.* Nat Genet, 1993. **5**(1): p. 95-100.
152. von Koch, C.S., et al., *Generation of APLP2 KO mice and early postnatal lethality in APLP2/APP double KO mice.* Neurobiol Aging, 1997. **18**(6): p. 661-9.
153. Chishti, M.A., et al., *Early-onset amyloid deposition and cognitive deficits in transgenic mice expressing a double mutant form of amyloid precursor protein 695.* J Biol Chem, 2001. **276**(24): p. 21562-70.

Chapter 14

Anticopper Therapy with Tetrathiomolybdate for Wilson's Disease, Cancer, and Diseases of Inflammation and Fibrosis

George J. Brewer

Department of Human Genetics and Department of Internal Medicine, University of Michigan Medical School, Ann Arbor, MI 48109–0534

Tetrathiomolybdate (TM) a copper-lowering drug, has efficacy in Wilson's disease, cancer, disease of neovascularization, and potential efficacy in diseases of inflammation and/or fibrosis. In Wilson's disease TM has been developed for initial therapy, the first two to four months of treatment, to quickly and safely gain control of copper toxicity. TM is efficacious in cancer and other diseases of angiogenesis because it has antiangiogenic effects on a number of angiogenic cytokines and promoters. TM also inhibits fibrotic and inflammatory cytokines, and is effective in a number of animal models of inflammation and fibrosis.

Background on Wilson's Disease and its Therapy.

The average human diet, at least in the U.S., contains about 1.0mg of copper per day *(1)*. Adults require about 0.75mg per day *(1)*. The excess of 0.25mg per day is normally excreted by the liver, the organ responsible for copper balance, into the bile for loss in the stool *(2)*.

In Wilson's disease, mutations *(3-5)* in both copies of one of the genes (ATP7B) of the biliary excretory pathway causes this pathway to be inoperative, and these patients accumulate a little copper every day of their lives. The liver is capable of storing large quantities of copper, but sometime in childhood, this storage capacity is exceeded, and the liver begins to be damaged. The patient may present clinically with liver disease in the second or third decades of life, or sometimes even later *(6-8)*. Alternatively, the liver disease may not present clinically and excess copper accumulates elsewhere, with the brain generally the next most sensitive organ. Patients who present with neurologic disease present with a movement disorder, with symptoms such as tremor, dysarthria, incoordination, and dystonia *(9)*. Many patients who will present neurologically have behavioral problems that may precede their neurologic symptoms by a year or two *(10, 11)*.

Anticopper treatment of Wilson's disease has evolved considerably in recent years. For several decades, anticopper agents were limited to penicillamine, introduced in 1956 *(12)*, and trientine *(13)*, introduced in 1982. These drugs are copper chelators that, act primarily by increasing the urinary excretion of copper. Penicillamine causes neurologic worsening about half of the time if given as initial treatment to patients who present neurologically *(14)*. It appears to do so by mobilizing hepatic copper stores, temporarily elevating blood copper and probably the copper in the brain. In addition penicillamine has a long list of toxicities, listed in reference 1, and provided in detail in the Physicians Desk Reference. In brief, they include a 25% incidence of an initial hypersensitivity reaction, many acute and subacute toxicities such as bone marrow depression, proteinuria, and increased susceptibility to infection, and many chronic toxicities such as autoimmune disorders, collagen abnormalities, and negative cosmetic effects on the skin. Trientine is less toxic, but shares some of penicillamine's toxicities, and prior to our work, had not been tried in patients who presented neurologically. In 1997, zinc which acts by inducing intestinal cell metallothionein and blocking copper absorption, was approved as maintenance therapy by the U.S. FDA, based primarily upon our data *(1, 7, 8, 15-17)*. Zinc is essentially non-toxic and is rapidly becoming the maintenance therapy of choice. However, we judge zinc to be too slow acting for the initial therapy of acutely ill neurologically presenting patients. To fill this therapeutic need, we have introduced tetrathiomolybdate (TM) therapy.

Background on Tetrathiomolybdate (TM).

The discovery and development of TM unfolded over a period of about 60 years. First it was noted that ruminants, but not non-ruminants, feeding on certain pastures in Australia and New Zealand, developed a serious disease *(18-20)*. After a time it was realized the disease was due to copper deficiency, and then it was realized that the soil of these pastures was high in molybdenum content. Molybdenum fed to ruminants reproduced the copper deficiency syndrome, but molybdenum had no effect on non-ruminants *(21-23)*. Finally it was realized that the rumen is rich in sulfur metabolism, and was generating thiomolybdates *(24)*. Thiomolybdates given to non-ruminants produced copper deficiency, with the tetra-substituted compound, tetrathiomolybdate (TM), being the most potent *(25-27)*.

Experimental studies with TM have shown it to be a very potent anticopper compound *(25-28)*. It forms a stable tripartite complex with copper and protein *(26-28)*. Given with food, it prevents the absorption of copper. Given away from food, it is absorbed into the blood and complexes available copper and albumin. The copper in this complex is unavailable for cellular uptake. This complex accumulates in the blood, and is slowly cleared by the liver, where it is excreted in the bile *(29, 30)*.

Tetrathiomolybdate Therapy in Wilson's Disease.

We introduced TM into the treatment of the neurologic presentation of Wilson's disease because of the therapeutic need discussed earlier. Our first study was an open label study which eventually involved 55 patients, each given an eight week course of TM *(31-34)*. The dose with meals was generally 20mg three times daily, but the away from food dose was varied from 60 to 200mg. We found that higher doses seemed to offer no benefit and settled on 60mg as the away from food dose. We used quantitative neurologic and speech tests to evaluate preservation of neurologic function. Only 2 of the 55 patients (3.6%) reached our criteria for neurologic worsening *(34)*, compared to the estimated 50% worsening rate for penicillamine. Patients went on zinc maintenance therapy and most showed substantial neurologic recovery over the next two years.

We next compared TM to trientine for treating neurologically presenting patients in a double blind study. The rationale for testing trientine was that it had never been tried for initial treatment of these patients, and is a somewhat gentler chelator than penicillamine. Trientine was given in a standard dose of 1.0g daily for eight weeks, and TM was given 20mg 3 times daily with meals and 60mg at bedtime away from food, for eight weeks. Zinc therapy at 50mg 2X daily was given in both arms. This study is not yet completed, but in preliminary data trientine has about a 20% incidence of worsening while TM continues to have a low incidence of 3-4% *(35)*. Thus, based on preliminary evidence, it seems that TM may be the treatment of choice for initial therapy of neurologically presenting patients.

Side effects of TM in Wilson's disease include about a 10-15% incidence of mild anemia and/or leukopenia, and about a 10-15% incidence of a mild increase in serum transaminase enzymes. The anemia/leukopenia is due to overtreatment bone marrow depletion of copper, since copper is required for cellular synthesis and replication. The cause of the elevation of transaminase enzymes is unknown but likely has something to do with TM movement of copper in the liver, since we haven't seen transaminase elevations in TM treatment of other diseases, such as cancer.

Both side effects tend to occur four to five weeks into treatment, and both respond to halving the TM dose to 10mg 3X daily with meals and 30mg at bedtime. For this reason we are carrying out another double blind study of TM in the neurologic presentation. In this study we are comparing the original TM regimen (120mg daily for eight weeks) with a new regimen consisting of 120mg for two weeks, then 60mg for 14 weeks. In this manner we hope to retain the excellent efficacy of TM in the original regimen, but eliminate, or greatly reduce, the side effects.

Finally, we are now initiating a double blind three arm study of TM, trientine, and penicillamine in the hepatic failure presentation of Wilson's disease. There are reasons to think TM will work well in this setting – it is life saving to copper poisoned sheep, who usually die of acute liver failure. In one patient we have studied, restoration of liver function tests to normal was about twice as fast with TM as we typically see with the best current regimen, a combination of trientine and zinc *(36)*. This study is just beginning.

Background on Angiogenesis and Cancer, and Copper and Angiogenesis.

Folkman pioneered the concept that tumor growth is dependent on angiogenesis, while angiogenesis is relatively unessential for adults *(37-39)*. This suggested that agents targeting angiogenic factors might be effective anticancer drugs. In the last 15 years this field has seen a great deal of activity, with many pharmaceutical companies searching for and developing antiangiogenic agents *(40, 41)*. The results have been somewhat disappointing *(42-44)*.

The reason for the disappointment probably relates primarily to the large number of angiogenic promoters that exist, and are generally available to be recruited by the tumor *(40, 41)*. Pharmaceutical agents generally target one angiogenic promoter. If the tumor is at least partially dependent on that factor, cessation of growth may occur. But the cessation is usually quite temporary, because the tumor, particularly large and advanced tumors, with billions of cells, many constantly mutating, find ways around the block by recruitment of additional angiogenic promoters, and the disease progresses.

In the early 1980s work was initiated showing that copper was required for angiogenesis. One model that was used was the cornea of rabbits made modestly copper deficient with a combination of penicillamine and low copper diet *(45)*. Angiogenic substances, such as prostaglandin E_1, placed in the cornea produced angiogenesis in control animals, but angiogenesis was markedly less in copper deficient animals.

Brem et al *(46)* took this a step further and showed that growth of tumors explanted to the brains of rats or rabbits was markedly less in animals made modestly copper deficient, again with a combination of penicillamine and low copper diet. Brem et al *(47)* also showed that invasion of normal brain tissue by blood vessels was absent in the copper deficient animals, but was occurring aggressively in control animals.

Tetrathiomolybdate in Cancer Therapy.

Being aware of the connections discussed in the preceding section, and coming along with a new, potent, nontoxic anticopper drug in TM, we decided to try TM as a cancer therapy. After a pilot successful initial study of MCA (methylcholanthrene-induced) sarcoma in the mouse *(Brewer, unpublished)*, I joined forces with an oncologist, Dr. Sofia Merajver. She designed an elegant study in Her 2/neu transgenic mice, all of which develop breast cancer during the first year of life. This study was strongly positive in that none of TM treated animals developed visible cancer while most of the controls did during the 221 days of followup *(48)*. Histologic examination of the breasts of TM treated animals revealed small, avascular, clusters of cancer cells. Release of TM treated animals from treatment resulted in growth of typical mammary cancers resembling those of untreated controls.

We have gone on to study several other rodent tumor models, including another breast *(48)*, and prostate *(49)*, head and neck *(50, 51)*, and lung *(52)*, all with positive results. TM in combination with either radiation therapy *(52)* or chemotherapeutic agents *(49, 53)* leads to at least additive effects. A canine study (pet dogs presenting with a variety of advanced and metastatic cancers), has been positive in that almost half of the TM treated animals had unexpectedly long periods of freedom from progression *(54)*.

We have carried out a phase I/II* clinical trial involving 42 patients with a

*Phase I studies primarily look at toxicity of a drug, while phase II studies look at indications of efficacy. A phase I/II study combines the two.

variety of types of advanced and metastatic cancer *(55 and unpublished)*. In 18 evaluable patients, freedom from progression averaged 11 months, quality of life stabilized, there were occasional exceptional results, with one patient with metastatic chondrosarcoma now out five years without progression.

In the meantime, at the University of Michigan and collaborating institutions, we have initiated nine phase II studies of TM treatment in specific cancers. For the most part these involve therapies of advanced disease, some using TM in combination with other agents, some using TM as sole therapy. One paper has been published, on renal cell cancer, with only modest potentially positive results *(56)*. It is too early to report on the others, but results in some seem to be encouraging.

My own predictions are that TM will be best in micrometastatic disease (such as modeled by the Her 2/neu study), or at least metastatic disease where the tumor has been greatly reduced by prior or concomitant therapy. An example would be breast cancer with positive nodes, treated by lumpectomy and chemotherapy, and then treated with TM. TM could also play a role in chemoprevention in patients at high risk for cancer.

It appears that TM may be a more global inhibitor of angiogenesis than agents developed so far by pharmaceutical companies *(48, 57)*. The reason appears to be that many angiogenic promoters appear to require copper at normal levels. The mechanism of the copper dependence may vary. In some cases the molecule contains copper, in other cases the molecule requires copper for functioning, and in other cases transcription of the molecule appears to be inhibited by lowered copper levels. It is particularly interesting that activity of NFκB, a master regulator of transcription of many cytokines, is copper dependent *(48)*.

Cancer is not the only disease dependent on neovascularization. For example, retinopathy is another angiogenic disease. In one animal model of retinopathy, the retinopathy of prematurity, we have been able to strongly inhibit neovascularization with TM *(Elner, S, Elner, V, and Brewer, unpublished results)*.

Background on Cytokines and Inflammatory and Fibrotic Diseases.

Organ injury of almost any type can lead to excessive and dysregulated inflammation followed by an excessive and dysregulated fibrotic response. Common examples are various liver injuries (virus, chemical, drug, autoimmune) causing hepatitis and leading to cirrhosis, lung injuries (infection, smoke, near drowning, chemical, autoimmune) leading to pulmonary fibrosis, and kidney injuries of a similar varieties of types, including diabetes, leading to interstitial renal fibrosis *(58)*.

The excessive inflammation is mediated by inflammatory cytokines, especially tumor necrosis factor alpha (TNFα) and interleukin-1-beta (IL-$1_β$). The excessive fibrotic response is mediated by transforming growth factor beta ($TGF_β$), which activates connective tissue growth factor (CTGF), which activates a whole cadre of collagen genes and other profibrotic genes. In recent years, considerable effort by pharmaceutical companies has been aimed at finding ways to inhibit these cytokines for therapeutic purposes. Considerable success has been obtained with antibodies directed at TNFα or its receptor. Two commercial TNFα antibody products have been proven very successful at treating certain diseases associated with excessive TNFα levels, such as rheumatoid arthritis *(59-61)*, Crohn's disease *(62, 63)*, and psoriasis *(64-66)*.

Tetrathiomolybdate in Treatment of Inflammatory and Fibrotic Diseases.

This work started because it seemed to us that the $TGF_β$ and CTGF cytokines involved in fibrotic diseases had a high likelihood of sharing the copper dependence of angiogenic cytokines. As a first test of this hypothesis, we studied TM treatment of the bleomycin mouse model *(67)*. In this model, bleomycin instilled into the trachea results in an explosive inflammatory reaction, primarily mediated by TNFα, which peaks at seven days. This is followed by a progressive fibrosis which is severe by 21 days, is dependent on $TGF_β$ function and is associated with weight loss. The fibrosis can be visualized histologically, or quantitated biochemically by measuring hydroxyproline, a major amino acid constituent of collagen.

TM treatment completely prevented the fibrosis as measured by hydroxyproline and as seen visually, and prevented the weight loss *(68)*. The bleomycin animals had marked increases in $TGF_β$ in the lungs, and this increase was largely and significantly prevented by TM therapy *(69)*.

We also observed that TNFα expression (in RNA levels) increased markedly in the lungs of bleomycin animals at seven days, and this increase was largely and significantly prevented by TM therapy *(69)*. We were able to show that the $TGF_β$ inhibition was independent of prior inhibition of TNFα and inflammation by starting TM therapy after the seven day peak of TNFα and inflammation, and still show inhibition of $TGF_β$ and fibrosis.

We have extended these studies to the liver, where we are able to show inhibition of the hepatitis from injection of concanavalin A by TM *(70, 71)*, and inhibition of hepatitis and cirrhosis from carbon tetrachloride injection for 12 weeks *(71)*.

These observation open up the possibility of TM therapy for human diseases of excessive inflammation and/or excessive fibrosis, both of which are very numerous, and some of which are quite common. A clinical trial in idiopathic pulmonary fibrosis has been initiated and others are being planned for psoriasis, primary biliary cirrhosis, scleroderma, and spondyloarthropathies.

Acknowledgements and Financial Disclosure

This work has been supported by grants FD-R-000179 and FD-U-000505 from the US Food and Drug Administration's Orphan Products Office, and by the General Clinical Research Center of the University of Michigan Hospitals, supported by the National Institutes of Health (grant number MO1-RR00042).

Dr. Brewer is supported in part by a grant from Attenuon LLC, San Diego, CA. The University of Michigan has recently licensed the antiangiogenic uses of TM to Attenuon LLC, and Dr. Brewer has equity in and is a paid consultant to Attenuon LLC.

References

1. Brewer G.J., Yuzbasiyan-Gurkan V. *Medicine* **1992**, *71*, 139-164.
2. Ravestyn A.H. *Acta. Med. Scand.* **1944**, *118*, 63-196.
3. Bull P.C., Thomas G.R., Rommens J.M., Forbes J.R., Cox D.W. *Nature Genet.* **1993**, *5(4)*,327-337.
4. Tanzi R.E., Petrukhin K., Chernov I., Pellequer J.L., Wasco W., Ross B., et al. *Nature Genet.* **1993**, *5(4)*, 44-50.
5. Yamaguchi Y., Heiny M.E., Gitlin J.D. *Biochem. Biophy. Res. Commun.* **1993**, *197*, 271-277.
6. Scheinberg IH., Sternlieb I. Wilson's disease. In: *Major Problems in Internal Medicine*. Editor L.S.;. W.B. Saunders Company Philadelphia, PA:, 1984 Vol. 23.
7. Brewer, G.J. *Wilson's Disease: A Clinician's Guide to Recognition, Diagnosis, and Management*; Kluwer Academic Publishers: Boston, 2001.
8. Brewer G.J. *Exp. Biol. Med.* **2000**, *223(1)*, 39-49.
9. Starosta-Rubinstein S, Young AB, Kluin K, Hill GM, Aisen AM, Gabrielsen T, Brewer GJ. Arch. *Neurol.* **1987**, *44*, 365-370.
10. Portala K. Psychopathology in Wilson's Disease. Acta Universitatis Upsaliensis, Uppsala, Thesis, 2001.

11. Akil, M, Brewer, GJ. Advances in Neurology, Vol. 65 -Behavioral Neurology of Movement Disorders. Weiner WJ, Lang AE (Eds). Raven Press, New York, **1995**, *171-178*.
12. Walshe J.M. *Am. J. Med.* **1956,** *21,* 487-95.
13. Walshe J.M. *Lancet* **1982,** *1,* 643-47.
14. Brewer G.J., Terry C.A., Aisen A.M., Hill G.M. *Arch. Neurol.* **1987,** *44,* 490-494.
15. Brewer G.J., Hill G.M., Prasad A.S., Cossack Z.T., Rabbani P. *Ann. Intern. Med.* **1983,** *99,* 314-320.
16. Brewer G.J., Dick R.D., Johnson V.D., Brunberg J.A., Kluin K.J., Fink J.K. *J. Lab. Clin. Med*.**1998,** *132,* 264-278.
17. Brewer G.J. *J. Trace Elements in Exp. Med.* **2000,** *13,* 51-61.
18. Ferguson W.S., Lewis A.L., Waterson S.J. *J. Agri. Sci.* **1943,** *33,* 44.
19. Dick A.T, Bull L.B. *Aust. Vet. J.* **1945,** *21,* 70.
20. Miller R.F., Engel R.W. *Fed. Proc.* **1960,** *19,* 666.
21. Macilese Ammerman C.B., Valsecchi R.M., Dunavant B.G., Davis G.K. *J. Nutr.* **1969,** *99,* 177.
22. Mills C.F., Monty K.J., Ichihara A., Pearson P.B. *J. Nutr.* **1958,** *65,* 129.
23. Cox D.H., Davis G.K., Shirley R.L., Jack F.H. *J. Nutr.* **1960,** *70,* 63.
24. Dick A.T., Dewey D.W., Gawthorne J.M. J. *Agr. Sci.* **1975,** *85,* 567.
25. Mills C.F., El-Gallad T.T, Bremner I. *J. Inorg. Biochem.* **1981,** *14,* 189.
26. Bremner I., Mills C.F., Young B.W. *J. Inorg. Biochem.* **1982,** *16,* 109.
27. Mills C.F., El-Gallad TT., Bremner I., Wenham G. *J. Inorg. Biochem.* **1981,** *14,* 163.
28. Gooneratne S.R., Howell J.M, Gawthorne J.M. *Br. J. Nutr.* **1981,** *46,* 469.
29. Mason J., Lamand M., Hynes M.J. *J. Inorg. Biochem.* **1983,** *19,* 153.
30. Hynes M., Lamand M., Montel G., Mason J. *Br. J. Nutr.* **1984,** *52,* 149.
31. Brewer G.J., Dick R.D., Yuzbasiyan-Gurkan V., Tanakow R., Young A.B., Kluin K.J. *Arch. Neurol.* **1991,** *48,* 42-47.
32. Brewer G.J., Dick R.D., Johnson V., Wang Y., Yuzbasiyan-Gurkan V., Kluin K.J., Fink J.K., Aisen A. *Arch. Neurol.* **1994,** *51,* 545-554.
33. Brewer G.J., Johnson V., Dick R.D., Wang Y., Kluin K.J., Fink J.K., Brunberg J.A. *Arch. Neurol.* **1996,** *53,* 1017-1025.
34. Brewer G.J., Hedera P., Kluin K.J., Carlson M, Askari F., Dick R.B., Sitterly J., Fink J.K. *Arch. Neurol.* **2003,** *60(9),* 1303-6.
35. Brewer G.J., Schilsky M., Hedera P., Carlson M.D., Fink J.K., Askari F.K., Kluin K., Lorincz M., Shneider B., Velickovic M., Blitman S., Mojica J. and Vila N. *J. Investig. Med.* **2003,** *51*(S 2).

36. Askari F.K., Greenson J., Dick R.D., Johnson V.D. and Brewer GJ. *J. Lab. Clin. Med.* **2003**, *142(6)*, 385-390.
37. Folkman J. *N. Eng. J. Med.* **1971**, *285*, 1182-1186.
38. Folkman J., Klagsburn M. *Science* **1987**, *235*, 442-447.
39. Folkman J. *Nature Med.* **1995**, *1*, 27-31.
40. Brem S. *Canc. Control* **1999**, *6*, 436-458.
41. Beckner M.E. *Canc. Invest.* **1999**, *17*, 594-623.
42. Susman E. Consensus panel: Antiangiogenesis drugs unlikely as single agents. Oncology Times June, **2001**.
43. Susman E. San Antonio Breast cancer symposium. Prediction: Shutting down cancer blood vessels still years away. Oncology Times April, **2002**.
44. Dotts T. Is angiogenesis inhibition the future of cancer treatment? Hem/Onc. Today, March **2002**.
45. Ziche M., Jones J., Gullino P.M. *J. Natl. Cancer Inst.* **1982**, *69*, 475-48.
46. Brem S.S., Zagzag D., Tsanaclis A.M.C, Gatley S., Elkouby M.P., Brein S.E. *Am. J. Path.* **1990**, *137*, 1121-1147.
47. Brem S., Tsanaclis A.M., Zagzag D. *Neurosurgery* **1990**, *26*, 391-396.
48. Pan Q., Kleer C., van Golen K., Irani J., Bottema K., Bias C., De Carvalho M., Mesri E., Robins D., Dick R.B., Brewer G.J., and Merajver S.D. *Cancer Res.* **2002**, *62*, 4854-4859.
49. van Golen K., Bao L., Brewer G.J., Pienta K., Karadt J., Livant D., Merajver S.D. *Neoplasia* **2002**, *4(5)*, 373-379.
50. Cox C.D., Teknos T.N., Barrios M., Brewer G.J., Dick R.D., Merajver S.D. *Laryngoscope* **2001**, *111*, 696-701.
51. Cox C., Merajver S.D., Yoo S., Dick R.D, Brewer G.J., Lee J.S., Teknos T.N. *Arch. Otolaryngol. Head Neck Surg.* **2003**, *129(7)*, 781-5.
52. Khan M.K., Miller M.W., Taylor J., Navkiranjit K.G., Dick R.D., van Golen K., Brewer G.J., Merajver S.D. *Neoplasia* **2002**, *4(2)*, 1-7.
53. Pan Q., Bao L.W., Kleer C.G., Brewer G.J., and Merajver S.D. *Mol. Cancer Ther.* **2003**, *2*, 617-622.
54. Kent M.S., Madewell B.R., Dank G., Dick RD., Merajver S.D., Brewer G.J. *J. Trace Elem. in Exp. Med.* (in press).
55. Brewer G.J., Dick R.D., Grover D.K., LeClaire V., Tseng M., Wicha M., Pienta K., Redman B.G., Thierry J., Sondak V.K., Strawderman M., LeCarpentier G., Merajver S.D. *Clin. Cancer Res.* **2000**, *6*, 1-10.
56. Redman B.G., Esper P., Pan Q., Dunn R.L., Hussain H.K., Chenevert T., Brewer G.J., Merajver S.D. *Clin. Cancer Res.* **2003**, *9(5)*, 1666-72.
57. Brewer G.J. *Exp. Biol. Med.* **2001**, *226*, 665-673.
58. Border W.A., Noble N.A. *N. Engl. J. Med.* **1994**, *331*, 1286-92.

59. Elliott M.J., Maini R.N., Feldmann M., Long-Fox A., Charles P., Brennan F.M., et al. *Arthritis Rheum.* **1993**, *36,* 1681–1690.
60. Charles P.J., Elliott M.J., Davis D., Potter A., Kalden J.R., Antoni C, et al. *J. Immunol.* **1999,** *163,* 1521–1528.
61. Shanahan C.S., St. Clair E.W. *Clin. Immunol.* **2002,** *103(3),* 231-242.
62. Present D.H., Rutgeerts P., Targan S., Hanauer S.B, Mayer L., van Hogezand R.A., et al. *N. Engl. J. Med.* **1999,** *340,* 1398-1405.
63. D'haens G., van Deventer S., van Hogezand R., Chalmers D., Kothe C, Baert F., et al. *Gastroenterology* **1999,** *116,* 1029-1034.
64. Ogilvie A.L., Atoni C., Dechant C., Manger B., Kalden J.R., Schuler G., et al. *Br. J. Dermatol.* **2001,** *144,* 587-589.
65. Chaudhari U., Roman P., Mulcahy L.D., Dooley L.T., Baker D.G., Gottlieb A.B. *Lancet* **2001,** *357,* 1842-1847.
66. Mease P.J., Goffe B.S., Metz J., VanderStoep A., Finck B., Burge D.J. *Lancet* **2000,** *35,* 385-390.
67. Phan S.H., Kunkel S.L. *Exp. Lung Res.* **1992,** *18,* 29-43.
68. Brewer G.J., Ullenbruch M.R., Dick R.D., Olivarez L., Phan S.H. *J. Lab. Clin. Med.* **2003,** *141(3),* 210-216.
69. Brewer G.J., Dick R.D., Ullenbruch M.R., Jin H. and Phan S.H. *J. Lab. Clin. Med.* (Submitted for publication).
70. Askari F.K., Greenson J., Dick R.D., Johnson VD. and Brewer G.J. *J. Lab. Clin. Med.* **2003**; *142(6),* 385-90.
71. Askari F.K., Dick R.D., Mao M., Brewer G.J. *Exp. Biol. Med.* (Submitted for publication).

Chapter 15

Heme Detoxification in Malaria: A Target Rich Environment

Clare Kenny Carney[1], Lisa Pasierb[2], and David Wright[1]

[1]Department of Chemistry, Vanderbilt University, Nashville, TN 37235
[2]Department of Chemistry and Biochemistry, Duquesne University, Pittsburgh, PA 15282

A novel target for the development of new antimalarial treatments is the detoxification pathway of free heme released during the catabolism of host Hb in the digestive vacuole of the malaria parasite *Plasmodium falciparum*. We have synthesized and examined two peptide dendrimers (BNT I and II) based on the tandem repeat motif of the histidine rich protein II (HRP II) from *P. falciparium* for their abilities to both bind heme substrates and to form the critical detoxification aggregate hemozoin. Each template was capable of binding significant amounts of the natural substrate Fe (III)PPIX along with alternate substrates such as Zn (II)PPIX and the metal free protoporphyrin IX. Further, it has been demonstrated that the dendrimeric biomineralization templates were capable of supporting the aggregation of hemozoin. New applications of this template for drug screening as well as the elucidation of antimalarial drug mechanisms are discussed.

Hemoglobin Catabolism and Heme Detoxification

Once hailed as the "king of diseases" (*1*), malaria is one of the oldest cross-cultural infections of our time (*2*). The most common and deadly species of the human malaria parasite, *Plasmodium falciparum*, is responsible for 1.5-2.7 million deaths and 300-500 million acute illnesses each year. These statistics combined with the development of drug resistant parasitic strains and vectors place a two billion dollar annual economic burden on precisely those populations least capable of shouldering it (*3-6*).

Human infection begins with the bite of the female Anopheles mosquito, who passes the parasite sporozoites in her saliva to the host. The sporozoites travel to the liver where they infect hepatic cells and undergo multiple asexual fission, resulting in merozoites. After 9-16 days in the liver, merozoites are released into the vasculature and invade erythrocytes. Inside the red blood cells they multiply and degrade host hemoglobin to obtain requisite amino acids for development (*7*). After 48 hours, the infected erythrocytes are lysed, releasing merozoites into the bloodstream to infect more red blood cells. To complete the cycle, another feeding mosquito bites the infected human and ingests the gametocytes which undergo sexual reproduction resulting in sporozoites that travel to its salivary gland where they are passed to another human host (*1*).

The recent unraveling of the *P. falciparum* genome (*8*) revealed the critical role of metalloproteins in the parasite's lifecycle (Figure 1). These metalloproteins offer a unique set of drug targets (*9*). For example, in the parasitic cytosol, nucleotide biosynthesis can be disrupted by the inhibition of

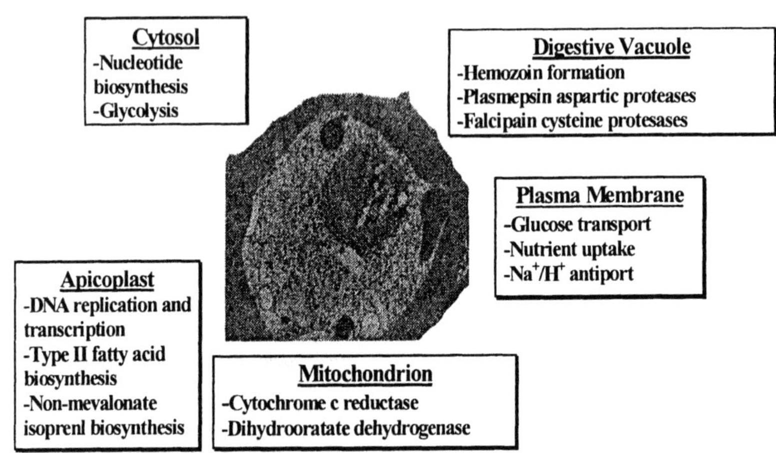

Figure 1: Sites of drug action. Electron micrograph showing the parasite inside a red blood cell reproduced with permission from D. E. Goldberg

ribonucleotide reductase, the enzyme involved in the first step in DNA biosynthesis. Here, chelating agents, such as polyhydroxyphenyl and hydroxamic acid derivatives, disrupt the di-iron core rendering the enzyme inactive (*9,10*). Atovaquone inhibits electron transport in the mitochondrial bc_1 complex of the parasite by acting on cytochrome c reductase, an enzyme essential for mitochondrial respiration (*1*). Chloroquine inhibits the formation of hemozoin (Figure 2), a unique dimer of heme units formed as a detoxification byproduct of hemoglobin catabolism (*11*).

The catabolism of hemoglobin is an ordered process occurring in the parasite's acidic secondary lysosome called the digestive food vacuole (*12*). The degradation process involves an orchestrated collection of aspartic proteases, a metalloprotease, a cysteine protease, a dipeptidyl peptidase and aminopeptidases (50). The first player is plasmepsin I, an aspartic protease, which clips the hinge region of hemoglobin releasing iron protoporypyrin IX (Fe(III)PPIX). Plasmepsin II works to further degrade the protein, preferring to cleave the acid-denatured globin. Together, the two unravel the tertiary structure leaving an apoprotein that is cleaved by falcipain, a cysteine protease. These protein fragments are subsequently broken down into peptidyl fragments by the metalloprotease, falcilysin (*14*). The release of the amino acids from the small peptide fragments is accomplished by dipeptidyl-peptidase I and several aminopeptidases. The proteolysis of hemoglobin results in the release of Fe(II) heme which is rapidly oxidized to the Fe(III) state. Lacking heme oxygenase, the first enzyme in the mammalian heme degradation pathway, levels of free heme could accumulate up to 400 mM (*13*). Associated with high vascular levels of free heme is the inhibition of digestive vacuole proteases, catalysis of lipid peroxidation, generation of oxidative free radicals and eventual lysis of the parasite (*15,16*).

Since the parasite ingests more than its own weight in host blood every few days, the potential damaging effects of free heme are significant (*17*). To circumvent this challenge, a detoxification pathway has evolved whereby the free heme is aggregated into hemozoin, an insoluble, nontoxic biomineral. Pagola *et al.* (*18*) have shown that the structure of hemozoin is a dimer of five coordinate ferric protoporphyrin IX (Fe(III) PPIX) linked by reciprocating monodentate carboxylate-linkages from one of the protoporphyrin IX's propionate moieties. The biomineral is composed of an extended network of these dimeric units hydrogen bonded together via the second propionic acid group of protoporphyrin IX. Such an aggregate form occludes the vast majority of potentially reactive iron within the crystallite (*17,19-21*). Inhibition of hemozoin formation precludes the parasite's ability to detoxify free heme and, therefore, its ability to survive, making hemozoin a unique target of drug action.

While there remains considerable controversy over the role of biological molecules in the formation of hemozoin, (*9,22,23*) an intriguing target is the ribonucleotide reductase, the enzyme involved in the first step in DNA biosynthesis. Here, chelating agents, such as polyhydroxyphenyl and hydroxamic acid derivatives, disrupt the di-iron core rendering the enzyme inactive (*9,10*). Atovaquone inhibits electron transport in the mitochondrial bc_1 complex of the parasite by acting on cytochrome c reductase, an enzyme essential for mitochondrial respiration (*1*). Chloroquine inhibits the formation of hemozoin (Figure 2), a unique dimer of heme units formed as a detoxification byproduct of hemoglobin catabolism (*11*).

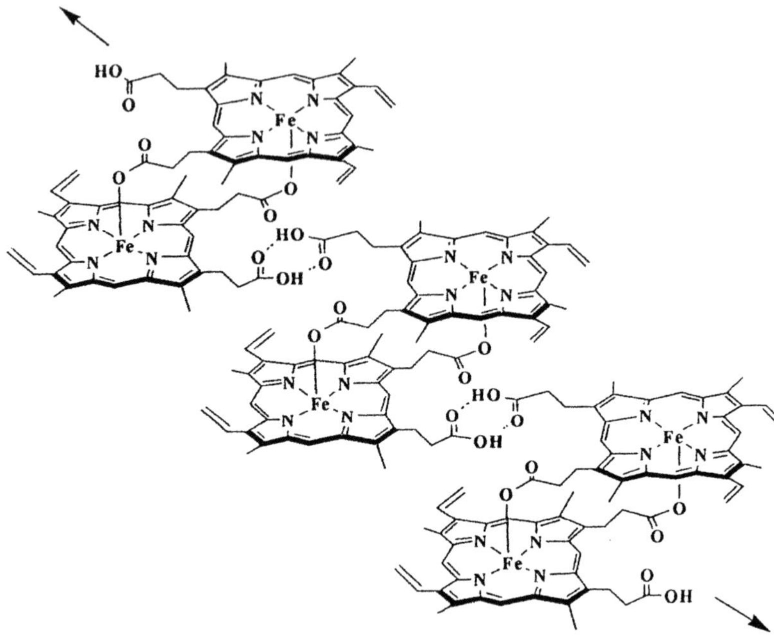

Figure 2: Structure of Hemozoin

The catabolism of hemoglobin is an ordered process occurring in the parasite's acidic secondary lysosome called the digestive food vacuole (*12*). The degradation process involves an orchestrated collection of aspartic proteases, a metalloprotease, a cysteine protease, a dipeptidyl peptidase and aminopeptidases (50). The first player is plasmepsin I, an aspartic protease, which clips the hinge region of hemoglobin releasing iron protoporypyrin IX (Fe(III)PPIX). Plasmepsin II works to further degrade the protein, preferring to cleave the acid-denatured globin. Together, the two unravel the tertiary structure leaving an apoprotein that is cleaved by falcipain, a cysteine protease. These protein fragments are subsequently broken down into peptidyl fragments by the metalloprotease, falcilysin (*14*). The release of the amino acids from the small peptide fragments is accomplished by dipeptidyl-peptidase I and several aminopeptidases. The proteolysis of hemoglobin results in the release of Fe(II) heme which is rapidly oxidized to the Fe(III) state. Lacking heme oxygenase, the histidine rich protein II (HRP II) of *P. falciparum*. Goldberg and coworkers, using monoclonal antibodies to probe the proteins of the digestive vacuole (pH 4.8-5.0) of *P. falciparum*, first identified a class of histidine-rich proteins (HRP II and HRP III) that mediate the formation of hemozoin (*24*). HRP II (M_r 35 kD) contains 51 repeats of the sequence Ala-His-His (76% of the mature protein is histidine and alanine (*25*). Biophysical characterization of HRP II demonstrated that the protein could bind 17 equivalents of heme and promote the formation of hemozoin at levels 20 times that of the background reaction (*26*). Additionally, hemozoin aggregation activity of HRP II was inhibited by chloroquine, destroyed by boiling, and had an activity profile with an optimal pH near 4.5 which slowed near pH 6.0 (*24*).

Biomimetic Templates

Studying HRP II as a potential drug target presents several challenges. There is the underlying fact that it comes from an infectious organism whose cultures must be kept synchronized. Further, the molecular biology is difficult due to high (80%) AT content of the parasite's DNA. Coupled to the highly repetitive nature of the HRP sequences, expression is difficult in many vectors. Finally, how is a meaningful site directed mutagenesis study to be designed for a protein composed of 76% two amino acids (*24*)? The presence of the Ala-His-His repeat motif is, however, reminiscent of a variety of nucleating scaffold proteins used in biomineralization. In these systems, the three-dimensional structure of the protein yields a preorganized, functionalized surface that serves as a template for nucleation. To investigate the mechanism of hemozoin formation and its subsequent inhibition, two peptide dendrimers templates (Figure 3) containing a minimal nucleating domain were designed and synthesized (*27*).

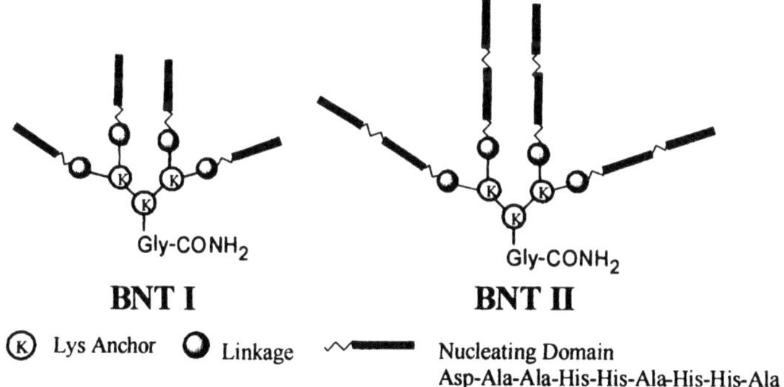

Figure 3: Generalized diagram of the dendrimeric bionucleating templates

An examination of the Ala-His-His repeats within HRP II reveals that a common organizational element consists of the 9-mer repeat Ala-His-His-Ala-His-His-Ala-Ala-Asp. Synthesis of the nucleating-domain peptide was accomplished using automated FMOC solid-phase techniques. The first bionucleating template (BNT I) incorporates four individual binding domains attached to the tetralysine dendrimer core for 8 tri-Ala-His-His repeats, while the second template (BNT II) couples two nucleating domains to each of the branches of the tetralysine core to generate a total of 16 tri-Ala-His-His repeats. The templates tested positive against monoclonal antibodies for HRP II using a commercial test (***Para***sight F) for *P. falciparum* malaria. Each template is capable of binding significant amounts of the natural substrate, Fe(III) protoporphyrin IX (Fe(III) PPIX). The templates are also able to support hemozoin formation in the presence of free heme under acidic conditions in proportion to the concentration of template. Furthermore, chloroquine inhibited the template-mediated synthesis of hemozoin consistent with previous reports (24) of HRP II inhibition by this drug (Figure 4).

It is reasonable to suggest that under these assay conditions the product may simply be the non-specific aggregation of Fe(III) PPIX and not hemozoin at all. To confirm that the identity of the end product is indeed hemozoin, the material is characterized by FTIR, SEM, and XRD and compared to native hemozoin (Figure 4). The IR spectrums contain fingerprint absorptions at 1664 cm^{-1} and 1211 cm^{-1} indicative of native hemozoin (*28*). The long, needle-like projections in the SEM demonstrate morphological similarity with authentic samples of native hemozoin (*19*). Finally, the XRD pattern matches the published fingerprint of native hemozoin (*29*) and verifies the absence of crystalline hemin. On a chemical and morphological level, the hemozoin product of the BNT II assay is identical to that produced by the parasite.

With a functional biomimetic template in hand, it is possible to begin to examine the potential mechanisms of action for antimalarials that disrupt hemozoin aggregation. Beginning with the hypothesis that HRP II indeed serves as a biomineralization template, there are several limiting cases in which inhibitors may disrupt the aggregation of hemozoin (Scheme 1). An inhibitor may bind the heme substrate in such a manner that the heme-inhibitor complex can not be recognized by the template. Alternately, a drug could interact with the template, blocking the heme binding site. Finally, a hemozoin aggregation inhibitor might trap the heme bound to the template, preventing formation of the dimeric unit or nucleation of the extended crystallite (*30,31*). Considering the potential routes for the disruption of hemozoin aggregation, it is often difficult to discern the precise mechanism of many antimalarial drugs.

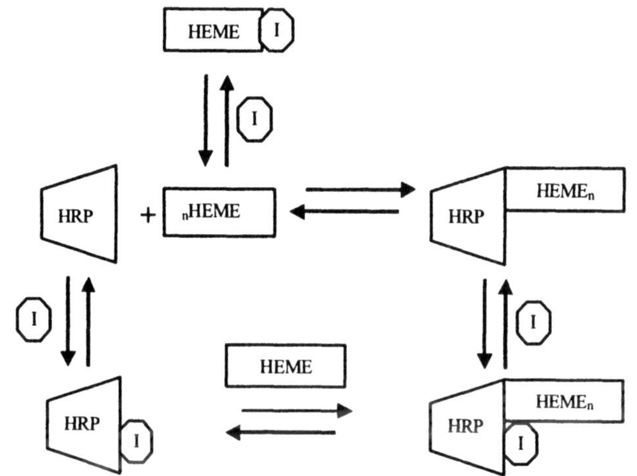

Scheme 1: Potential inhibition pathways to hemozoin aggregation

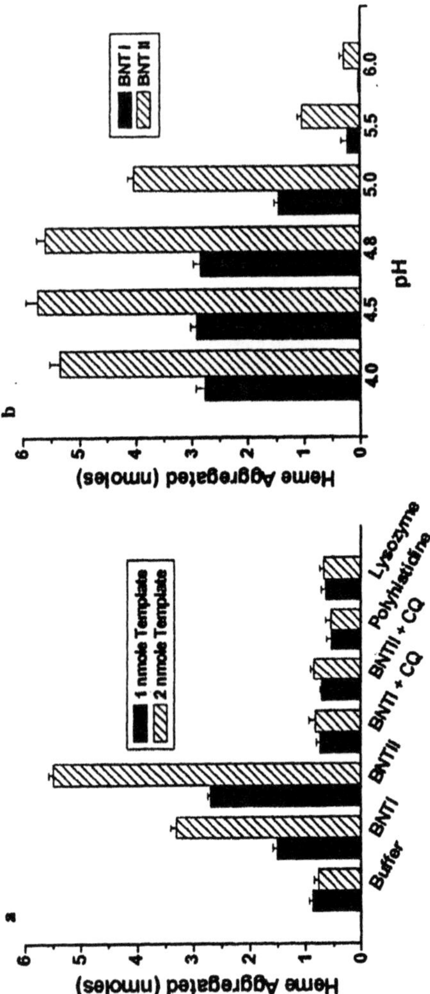
a
Heme Aggregated (nmoles)
0
1
2
3
4
5
6
1 nmole Template
2 nmole Template
Buffer
BNTI
BNTII
BNTI + CQ
BNTII + CQ
Polyhistidine
Lysozyme
b
Heme Aggregated (nmoles)
0
1
2
3
4
5
6
BNT I
BNT II
4.0
4.5
4.8
5.0
5.5
6.0
pH

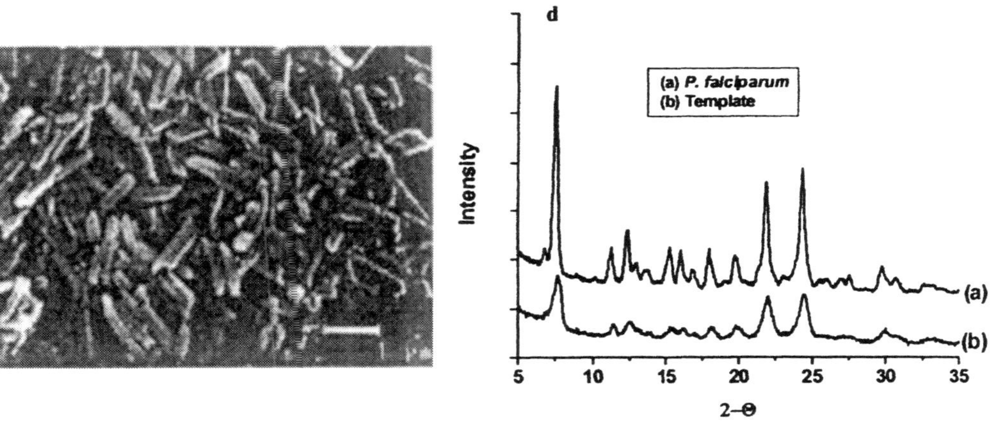

Figure 4: a) Hemozoin aggregation mediated by bionucleating templates b) pH dependence of the heme aggregation c) SEM and d) XRD characterization of aggregated material

Bionucleating Template Mediated Assay Systems

To screen inhibitor efficacy, investigators employ a battery of assays. Included among them are culture *in vivo* assays, trophozoite extract based assays, assays of hemozoin aggregation mediated by HRP II and the abiological spontaneous formation of β-hematin assay. We employed BNT II as an HRP II surrogate to screen a variety of known and experimental antimalarial compounds to explore the potential modes of interaction between the template, the heme substrate and the inhibitors. These include the "classic" antimalarial compound chloroquine, the hydroxythanones, currently under development by Riscoe, a novel class of Schiff-base inorganic complexes being developed by Sharma, and metalloporphyrins (not specifically as drug candidates, but as probes for possible modes of inhibitor interaction). Additionally, these experiments provide validation of the BNT II template in comparison to other assay methods.

Hydroxyxanthones

Among antimalarials (Figure 5), the hydroxyxanthones represent a class of compounds targeting heme aggregation whose efficacy is based on the formation of a strong heme-inhibitor complex through interactions between (1) the heme iron and the carbonyl oxygen of the tricyclic xanthone, (2) the two planar aromatic rings, and (3) the propionate groups of heme and the 4- and 5- position hydroxyls of the xanthone (*32*). As the formation of a heme-inhibitor complex is the principal mode of hemozoin inhibition, it is predicted that inhibitor efficacy should be similar for both *in vivo* assays and the spontaneous formation of β-hematin assay first described by Egan (43). When a series of hydroxyxanthones was assayed using BNT II rather than infected erythrocytes, the demonstrated effectiveness of the drugs were comparable. The mangostin negative control was inactive for all assays. The IC_{50} of dihydroxyxanthone was found to be 16, 14, and 25 uM for the *in vivo*, spontaneous formation, and the BNT II hemozoin aggregation assay, respectfully. The IC_{50} values for 2,3,4,5,6,7-hexahydroxyxanthone were similarly consistent at 0.1, 1.4 and 0.9 uM for the respective assays. The results verify the predicted similarity in IC_{50} values, based on the inhibitor's mode of action.

N_4O_2 Coordination Complexes

An unusual class of antimalarial agents targeted against heme aggregation is the N_4O_2 Schiff-base coordination complexes developed by Sharma. Stemming from their application against multidrug resistance in certain cancer cell lines (*33*), it was demonstrated that the pseudo-octohedral complexes with N_4O_2 donor

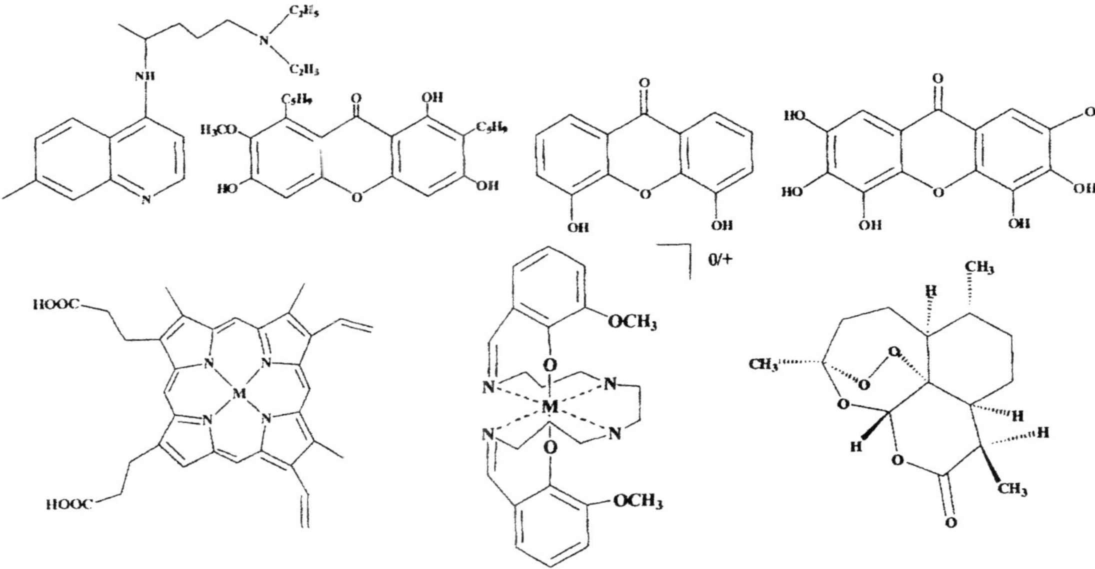

Figure 5: Common antimalarials. From top left, chloroquine, mangostin, dihydroyxyxanthone, 2,3,4,5,6,7-hexahydroxyxanthone, M-Protoporphyrin IX, N_4O_2 Schiff base complexes, and artemisinin. M=Fe, Ga, In, Al, or Mg.

sets disrupted heme aggregation in *P. falciparum*. A detailed analysis of crystal structures of the Fe(III) and Ga(III) complexes revealed a trans-configuration for the phenolic oxygens across the central metal atom (*34,35*). In contrast, molecular modeling predicted that the larger ionic radius of In(III) would force the ligand to adopt a cis-configuration. As the Fe(III), Ga(III) and Al(III) complexes are all more effectively transported than the In(III) complex, the obvious implication was that the *Plasmodial* analogue of MDR1 P-glycoprotein preferentially recognizes the trans-configuration of these drugs.

While certain structure activity relationships explain the observed efficacy of the N_4O_2 complexes in terms of biotransport and localization, there has been little discussion concerning the mode of action of these drugs. Screening the cationic $[Fe(III)ENBPI]^+$ and $[Ga(III)ENBPI]^+$ complexes and the neutral Mg(II)ENBPI complex in the BNT II assay system revealed that under conditions similar to those in an analogous HRP II mediated hemozoin aggregation assay, the cationic complexes demonstrated similar efficacies, while the neutral complex was entirely inactive. The IC_{50} was 4.5 uM for the iron complex, 50 uM for the gallium complex, and nonexistent for the neutral magnesium complex in an HRP II mediated assay. The IC_{50} for the BNT II mediated assay were found to be 9.0 and 58 uM for the iron and gallium complex respectively and again no activity was found for the magnesium complex (*36*). Considering the minimal hemozoin aggregation assay system employed, a readily envisioned mechanism for these complexes is that they interact with either the propionate moieties of the porphyrin ligand or the aspartic acid side chains of the BNT II template. Fluorescence quenching experiments revealed that under assay conditions, the drugs did not appreciable interact with the template (*36*). In contrast, Fe(III)PPIX quenched the weak fluorescence of the Fe(III) and Ga(III) ENBPI analogues, suggesting interaction between heme substrate and the inhibitor. The formation of such a complex between the cationic drug and the heme propionate group would limit the propionate's ability to form the requisite axial linkage of the repeating dimeric unit in the hemozoin aggregate.

If the mode of action for these antimalarial coordination complexes indeed originated from specific interactions with the heme's propionate moiety, then their efficacy would be sensitive to a competing receptor, such as acetate. When the acetate buffer concentration increased from 25 mM to 500 mM, the IC_{50} of the complexes increased until they no longer inhibited hemozoin aggregation. A control series of experiments performed at 25 mM acetate buffer concentration with increasing concentrations of NaCl revealed no effect on the IC_{50} values of the complexes (*36*). This suggested that the interactions between the drug complexes and heme propionate are not simple salt interactions, but specific to the carboxylate moiety. This observed sensitivity to acetate buffer concentration was in stark contrast to other antimalarials such as the quinolines, hydroxyxanthones and porphyrins, which are believed to act via a different mechanism (*36*). The elucidation of these novel interactions between drug and substrate highlights one of the advantages to using the reductionist BNT assay.

Porphyrin Probes

The BNT II assay may also be used to probe how drugs interact with the template. Following the maxim that a good starting point for the rational design of an inhibitor is a molecule that looks like the natural substrate, a number of groups have used metalloporphyrins as probes for the inhibition of hemozoin aggregation (*11*). By comparing the efficacy of a wide variety of metalloporphyrins using both the spontaneous aggregation assay and our BNT II based assay, the first major discrepancy between assay systems was noted. The results of these two assays are shown in Table I. Basilico and others have suggested that metalloporphyrin inhibitors interact with heme via π-π stacking interactions. While π-stacking can potentially cap the heme substrates, it is important to recognize that at inhibitor concentrations less than 2:1 (inhibitor to heme) an open face of heme remains. Thus, the effective interaction must be more than simply blocking of the faces of the heme substrate. Sullivan *et al.* have proposed that the π-stacking interactions drive the association of inhibitor with preformed dimers or smaller heme aggregates, blocking rapidly growing faces of small hemozoin crystallites (*37*). Alternately, in addition to π-stacking, hydrogen bonding between heme propionate and an appropriate donor from the inhibitor could inhibit β-hematin formation by, not only blockading faces but also by securing potential axial ligands from aggregation (*38,39*). Thus, while π-stacking may be crucial in establishing the geometry of inhibition, the precise mechanism likely involves the interplay of several different types of interactions in the spontaneous formation of β-hematin assay

In the presence of a bionucleating template, the affinity between the inhibitor and the template becomes an important consideration in how the drug may act (*40*). This is borne out by the significant perturbation of metalloporphyrin efficacies in assays using BNT II (Table I). Binding data between the template and inhibitors reveals that in almost every case, the inhibitor porphyrin is nearly as tightly associated with the template as the Fe(III)PPIX substrate. The metalloporphyrin efficacies presented here are similar to those reported by Slater and Cerami in inhibitory studies using trophozoite lysates with Zn(II)PPIX and Sn(IV)PPIX inhibitors (*41*). Recently, Chauhan et al. have demonstrated that a preformed heme/artemisin complex (hemart) not only binds to HRP II, but can displace heme already associated with the nucleating protein (*42*). Similarly, Choi *et al.* have identified new antimalarial lead compounds developed from several combinatorial libraries that show a direct correlation between parasite killing and interference with heme

Table I: Inhibition data of metalloporphyrin compounds for hemozoin aggregation

Porphyrin substrate	β-HI assay IC_{100} eq.	BNT II assay IC_{50} uM	Stoichiometry binding BNT II (eq.)
Fe(III)PPIX			13.0±1.0
PPIX	3.0	NS[a]	10.5±1.0
Co(III)PPIX	3.0	15.8±0.20	11.3±0.6
Cr(III)PPIX	3.0	19.9±0.14	10.8±0.4
Cu(II)PPIX	0.75	20.6±0.17	12.6±0.2
Mg(II)PPIX	0.5	53.2±0.07	10.4±0.3
Mn(III)PPIX	1.0	22.4±0.20	10.3±0.3
Sn(IV)PPIX	0.5	35.9±0.29	10.6±0.8
Zn(II)PPIX	0.5	82.7±0.04	15.2±0.6

[a] not soluble

binding to HRP II (*31*). For the metalloporphrins, the interplay of affinities for the nucleating template establishes a complex equilibrium between the hetero-association of the heme/inhibitor complex as well as the inhibitor/template complex resulting in a new ordering of inhibitor efficacies.

Conclusion

Effective assay systems designed to explore the inhibition of hemozoin formation must consider a variety of potential modalities for drug candidates. Hemin can completely aggregate in acetate buffer to yield β-hematin (*43*). There is a broad correlation between inhibition of this chemical reaction and intraerythrocytic antimalarial activity for malarial drugs, thus making it a useful reaction for model studies. Consequently, this abiological *in vitro* system has been the basis of several related assays for the discovery of drug candidates which disrupt β-hematin formation. Recently, Parapini *et al.* have advanced a standardized set of conditions for the β-hematin inhibition assay designed to screen for molecules capable of inhibiting β-hematin aggregation via π-π interactions with heme (*44*). The adaptation of the β-hematin inhibition assay to high-throughput screening methodologies represents an important development in the application of modern methods of drug discovery to finding new antimalarials (*44-47*). This screen is, however, predicated on the assumption that the potential drug's interaction with heme, and heme alone, is the basis for the disruption of heme aggregation. In the complex milieu of the digestive vacuole, this is unlikely.

Alternative assays for the inhibition of hemozoin aggregation are broadly based on *in vitro* trophozoite lysate methods (*48*). A significant disadvantage of lysate methods is the requirement to maintain synchronized cultures of *Plasmodium*. Another disadvantage is the heterogeneous nature of the lysate rendering preferred optical methods of analysis ineffective due to interferences and/or scattering. Consequently, several investigators have developed assays using biological agents believed to mediate the biomineralization of hemozoin within the digestive vacuole of the parasite, namely lipids (*49*) and the histidine-rich protein, HRP II (*24*). Despite the availability of commercial sources of several of the lipid classes capable of supporting β-hematin aggregation, the biphasic nature of the resulting assay mixture complicates their use in a high-throughput screen. Further, the isolation of HRP II and the need to adequately characterize the resulting protein contributes to the cost of its use in any drug screen.

We have obviated the need for isolating nucleating agents from the parasite by utilizing an entirely synthetic dendrimeric peptide bionucleating template (BNT II) based on the putative heme binding domain of HRP II that functions as a histidine-rich protein analog in such assays (*27*). The bionucleating templates mediate the formation of hemozoin under physiologically relevant conditions. The resultant hemozoin is chemically, morphologically, and crystallographically identical to native hemozoin. Additionally, the minimal assay system employed allows for the investigation of discrete interactions between antimalarial drugs, the nucleating template and the heme substrate to better understand the possible mechanisms of drug action involved in the disruption of hemozoin aggregation.

Acknowledgements

The many workers of the Wright group who have contributed to the development of the dendrimeric templates; particularly James Ziegler, Kelly Cole, and Rachel Linck. The *Para*sight F test was generously provided by Beckton-Dickinson Pharmaceutical. We thank the Rockefeller Brothers Fund's Charles E. Culpeper Biomedical Pilot Initiative (grant 01-272) for financial support.

References

1. Sherman,I.W.In *Malaria: Parasite Biology, Pathogenesis, and Protection*; Sherman, I.W. Ed; ASM Press: Washington D.C., 1998, pp 3-10.
2. Miller, R.L.; Ikram, S.; Armelagos, G.J.; Walker, R.; Harer, W.B.; Shiff, C.J.; Baggett, D.; Carrigan, M.; Maret, S.M. *Trans. R. Soc. Trop. Med. Hyg.* **1994**, *88*, 31-32.
3. Peters, W. *Chemotherapy and Drug Resistance in Malaria*, 2nd ed.; Academic Press: London **1987**.
4. Anderson, J.; MacLean, M.; Davies, M. *Malaria Research: An audit of International Activity*; Prism Report 7, The Welcome Trust: London **1996**.
5. *Tropical Disease Research News*, (News from the WHO Division of Tropical Diseases), **1999**, Vol. XXX.
6. *Tropical Disease Research News,* (News from the WHO Division of Tropical Diseases), **2002**, Vol. 94.
7. Goldberg, D.E. *Semin. Cell. Biol.* **1993**, *4*, 355-358.
8. Gardner, M.J.; Hall, N.; Fung, E.; White, O. et al *Nature* **2002**, *419*, 498-511.
9. Ridley, R. *Nature* **2002**, *415*, 686-693.
10. Holland, K.P.;Elford, H.L.;Bracchi, V.;Annis, C.G.; Schuster, S.M.; Chakrabarti, D. *Antimicrob. Agents Chemother.* **1998**, *42*, 2456-2458.

11. Ziegler, J.; Linck, R.; Wright, D.W. *Curr. Med. Chem.* **2001**, *8*, 171-189.
12. Olliaro, P.L.; Goldberg, D. E. *Parasitol. Today* **1995**, *11*, 294-297.
13. Goldberg, D.E.; Slater, A. F. G.; Cerami, A.; Henderson, G.B. *Proc. Natl. Acad. Sci. USA* **1990**, *87*, 2931-2935.
14. Goldberg, D. E. *216th ACS National Meeting, Boston* **1998**.
15. Tappel, A. L. *Arch. Biochem. Biophys.* **1953**, *44*, 378-395.
16. Green, M.D.; Ziao, L.; Lal, A.A. *Mol. Biochem. Parisitol.* **1996**, *83*, 183-188.
17. Oliveira, M. F.; D'Avila, J.C.P.; Torres, C.R.; Oliveira, P.L.; *et al. Mol. & Biochem. Parasitol.* **2000**, *111*, 217-221.
18. Pagola, S; Stephens, P.W.; Bohle, D.S.; Kosar, A.D.; Madsen, S.K. *Nature* **2000**, *404*, 307-310.
19. Chen, M.M.; Shi, L.; Sullivan, D.J. *Mol. & Biochem. Parasitol.* **2001**, *113*, 1-8.
20. Oliveira, M.F.; Silva, J.R.; Dansa-Petretski, M.; De Souza, W.; Braga, C.M.S.; Masuda, H.; Oliveira, P.L. *FEBS Letters* **2000**, *477*, 95-98.
21. Slater, A.F.G; Cerami, A. *Nature* **1992**, *355*, 167-169.
22. Dorn, A.; Stoffel, R.; Matile, H.; Bubendorf, A.; Ridley, R.G. *Nature* **1995**, *374*, 269-271.
23. Dorn, A.; Vennerstrom, J.L.; Stoffel, R.; Matile, H.; Bubendorf,.; Ridley, R.G. *Biochem. Pharmacol.* **1998**, *55*, 737-747.
24. Sullivan D.J. Jr.; Gluzman, I.; Goldberg, D.E. *Science* **1996**, *271*, 219-222.
25. Wellems, T.E.; Howard, R.J. *Proc. Natl. Acad. Sci. USA* **1986**, *83*, 6065-6069.
26. Bligh, E. G.; Dryer, W.J. *Can. J. Biochem. Physiol.* **1959**, *37*, 911-917.
27. Ziegler, J.; Cole, K.A..; Wright, D.W. *J. Am. Chem. Soc.* **1999**, *121*, 2395-2400.
28. Slater, A.F.G.; Swiggard, W.J.; Orton, B.R.; Flitter, W.D.; Goldberg, D.E.; Cerami, A.; Henderson, G.B. *Proc. Natl. Acad. Sci. USA* **1991**, *88*, 325-329.
29. Bohle, D.S.; Dinnebier, R. E.; Madsen, S.K.; Stephens, P.W. *J. of Biol. Chem.* **1997**, *272*, 713-716.
30. Pandey, A.V.; Bisht, H.; Babbarwal, V.K.; Srivastava, J.; Pandey, K.C.; Chauhan, V.S. *Biochem. J.* **2001**, 355, 333-338.
31. Choi, C.Y.H.; Schneider, E.L.; Kimm, J.M.; Gluzman, I.Y.; Goldberg, D.E.; Ellman, J.A.; Marletta, M.A. *Chem. Biol.* **2002**, *9*, 881-889.
32. Ignatushchenko, M.V.; Winter, R.W.; Baechinger, H.P.; Hinrichs, D.J.; Riscoe, M.K. *FEBS Lett.* **1997**, *409*, 67-73.
33. Sharma, V.; Crankshaw, C. L.; Picwnica-Worms, D. *J. Med. Chem.* **1996**, *39*, 3483-3491.
34. Ito, T.; Sugimoto, M.; Ito, H.; Torimumi, K.; Nakayama, H.; Mori, W.; Sekizaki, M. *Chem. Lett.* **1983**, 121-123.

35. Tsang, B.W.; Mathias, C. J.; Green, M.A. *J. Med. Chem.* **1994**, *37*, 4400-4406.
36. Ziegler, J.; Schuerle, T.; Pasierb, L.; Kelly, C; Elamin, A.; Cole, K.A.; Wright, D.W. *Inorg. Chem.* **2000**, *39*, 3731-3733.
37. Sullivan, D.J. Jr.; Matile, H.; Ridley, R.G.; Goldberg, D.E. *J. Biol. Chem.* **1998**, *273*, 31103-31107.
38. Cole, K.A.; Ziegler, J.; Evans, C.A.; Wright, D.W. *J. Inorg. Biochem.* **2000**, *78*, 109-115.
39. Ziegler, J.; Cole, K.; Wright, D.W. *J. Inorg. Biochem.* **2000**, *78*, 109-115.
40. Ziegler, J.; Pasierb, L.; Cole, K.; Wright, D.W. *J. Inorg. Biochem.* **2003**, *96*, 478-486.
41. Martiney, J. A.; Ceramimi, A.; Slater, A.F.G. *Mol. Med.* **1996**, *2*, 236-246.
42. Kannan, R.; Sahal, D.; Chauhan, V.S. *Chem. & Biol.* **2002**, *9*, 321-332.
43. Egan, T.J. ; Ross, D.C.; Adams, P.A. *FEBS Lett.* **1994**, *352*, 54-57.
44. Parapini, S.; Basilico, N.; Pasini, E.; Egan, T.J.; Olliaro, P.; Taramelli, D.; Monti, D. *Exper. Parasitol.* **2000**, *96*, 249-256.
45. Basilico, N.; Pagani, E.; Monti, D.; Olliaro, P.; Taramelli., D. *J. Antimicrob. Chemother.* **1998**, *42*, 55-60.
46. Kurosawa, Y.; Dorn, A.; Kitsuji-Shirane, M.; Shimada, H.; Matile, H.; Hofheinz, W.; Kansy, R.; Ridley, R.G. *Antimicrob. Agents Chemother.* **2000**, *44*, 2638-2644.
47. Baelmans, R.; Deharo, E.; Munoz, V.; Sauvain, M.; Ginsburg, H. *Exper. Parasitol.* **2000**, *96*, 243-248.
48. Pandey, A.V.; Singh, N.; Tekwani, B.L.; Puri, S.K.; Chauhan, V.S. *J. Pharm. Biomed. Anal.* **1999**, *20*, 203-207.
49. Dorn, A.; Vippagunt, S.R.; Matile, H.; Bubendorf, A.; Vennerstrom, J.L.; Ridley, R.G. *Parisitol.* **1998**, *55*, 737-747.
50. Francis, S.E.; Sullivan, D.J.Jr.; Goldberg, D.E. *Annu. Rev. Microbiol.* **1997**, 51, 97-123.

Chapter 16

Heme as Trigger and Target of the Antimalarial Peroxide Artemisinin

Anne Robert[1], Françoise Benoit-Vical[1,2], and Bernard Meunier[1]

[1]**Laboratóire de Chimie de Coordination du CNRS, 205 route de Narbonne, 31077 Toulouse Cedex 4, France**
[2]**CHU Rangueil-Parasitologie, 1 Avenue Jean Poulhès, TSA 50032, 31059 Toulouse Cedex 9, France**

The present review is focused on the mechanism of action of artemisinin, a peroxide-containing antimalarial drug. Upon reductive activation by iron(II)-heme, artemisinin is transformed to a reactive C4-centered radical species able to efficiently alkylate the heme-macrocycle itself. On the basis of this reactivity, which is probably pharmacologically relevant, a family of new antimalarial drugs named Trioxaquines® have been synthesized. Trioxaquines combine, within a single molecule, two pharmacologically active moieties acting on the same target, heme, by two different mechanisms: a 4-aminoquinoline, present in the conventional antimalarial chloroquine, and an endoperoxide responsible for the activity of artemisinin. Several trioxaquines are active in vitro on chloroquine-resistant malaria parasite at nanomolar concentration, and are efficient to cure by oral administration malaria-infected mice at 20-50 mg/kg.

Metals in medicine: role of metal ions in drug-design

At the interface of inorganic chemistry and biology, the discovery and the rational design of metal-based drugs is a growing field among academic researchers with some spectacular large-scale medical applications (*cis*-platinum and contrast agents for magnetic resonance imaging). For these complexes which are currently used in medical applications, the metal center plays a key role and is responsible for the observed pharmacological and biological effects (*1, 2*). Besides this main category, some organic drugs should be activated in vivo by a metal center in order to generate the pharmacological reactive entities. This is true for organic drugs that have to be activated by cytochrome P450 enzymes, but also for few drugs that interact with a metal complex acting as trigger and target. This latter category of drugs is clearly exemplified by artemisinin and its mechanism of action related to the hemoglobin digestion in red blood cells infected by *Plasmodium*, the parasite responsible for malaria.

Heme in Malaria Infected Red Blood Cells

After a mosquito bite, the malaria parasites first invade the liver and replicate within red blood cells. Many antimalarial drugs are specifically active against the blood stage of the parasite life within humans (*3*).

Digestion of Hemoglobin

Within infected erythrocytes, *Plasmodium falciparum* (strain responsible for most fatal cases) digest the host hemoglobin with the aid of several specific parasite proteases (*4, 5*). Hemoglobin comprises 95% of the cytosolic protein of red blood cells and up to 80% of the 5 mM hemoglobin is degraded by the parasite (*6*). This intensive proteolysis releases free heme and amino-acids. These amino-acids are incorporated into parasitic proteins and this supply is essential for the survival of *Plasmodium*, a "true parasite", which has a limited capacity for the *de novo* amino-acid synthesis. Only a very limited amount of the free heme is metabolized by the parasite as iron source.

Heme Aggregation

In hemoglobin, iron is essentially in the ferrous state. Upon degradation of the globin and the release of heme, this iron(II)-protoporphyrin(IX) is able to transfer one electron to molecular oxygen to produce superoxide radicals, hydrogen peroxide and, finally, hydroxyl radicals, generating a lethal oxidative stress for the parasite which lacks heme oxygenase that vertebrates use for heme catabolism. As a detoxification process, heme is oxidized and aggregated into a redox inactive iron(III) crystalline material called hemozoin (*7*). Hemozoin is primarily formed by dimerization of heme, a carboxylate function of a heme molecule A being axial ligand of iron(III) of a second heme molecule B, and a carboxylate of B being axial ligand of the iron(III) of A. These dimers aggregate through hydrogen bonds of the remaining free carboxylates, leading to heme aggregation as a crystalline, insoluble material. Hemozoin is not a "covalent polymer", but an aggregate of dimers, and the parasitic proteins involved in its formation are not polymerases.

Any perturbation of this heme-detoxification process that is unique to *Plasmodium* is expected to have drastic consequences for the parasite survival. The inhibition of heme aggregation is the targeted action of conventional quinoline based antimalarial drugs such as quinine, chloroquine and mefloquine.

Reactivity of artemisinin in the presence of heme

Artemisinin, a sequiterpene with a trioxane entity (see Figure 1 for structure), exhibits a chemical structure very different compared to that of the other known antimalarial drugs and is thus likely to have a different mechanism of action. Among antimalarial drugs, artemisinin (*8*) (and its hemisynthetic derivatives artemether, arteether, and artesunate, obtained by reduction of the lactone to the lactol, dihydroartemisinin, followed by functionalization) retains efficacy against multidrug resistant parasite strains, and no clinically relevant drug-resistance has yet been reported, despite an increased use over the past two decades (*9*). Artemisinin derivatives have been proven to be safe, even for young children and pregnant women. However, when used as monotherapy, they are associated with a high recrudescence rate (that should not be mistaken for inherent parasite resistance) due to their short half-lives (*10*). Artemisinin derivatives are therefore being used in combination with longer half-life drugs to increase the efficacy of the treatment (*11, 12*).

The pharmacological activity of artemisinin derivatives lies in their peroxide function, the deoxyartemisinin being completely devoid of biological activity (*13*). When malaria parasites are incubated in the presence of [^{14}C]artemisinin, the radioactivity is associated with covalent drug adducts with hemozoin (*14*) and with a few number of specific parasite proteins (*15, 16*). One of these alkylated proteins is a malarial translationally controlled tumor protein (TCTP) homolog of yet unknown function (*17*). The selective inhibition PfATP6, a sarco/endoplasmic reticulum Ca^{2+}-ATPase, by artemisinin has been recently reported (*18*). Interestingly, the reaction of these proteins with artemisinin depends on the presence of iron(II)-heme in the case of TCTP, or on an unidentified iron(II) species in the case of PfATP6.

Heme-artemisinin covalent adducts

The importance of alkylating species generated by iron-mediated homolytic cleavage of the endoperoxide function of artemisinin, in particular the alkyl radical centered at position C4 of artemisinin or related trioxanes, was early proposed (*19*). After preliminary experiments with synthetic metalloporphyrins (*20*), we have reported that, when iron(III)-protoporphyrin(IX) was incubated with artemisinin in the presence of glutathione, heme was readily converted in high yield to heme-artemisinin covalent adducts, resulting from alkylation of the four *meso* positions of the heme ligand by the C4 alkyl radical derived from artemisinin (Figure 1) (*21-23*). After demetallation of the heme moiety, complete NMR characterization of these heme-artemisinin covalent adducts was obtained (*24*).

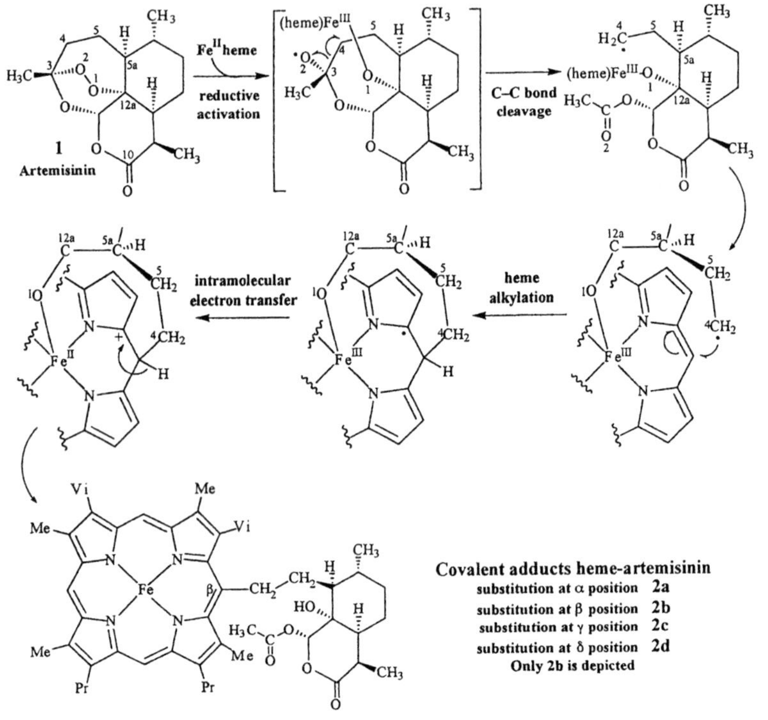

Figure 1: Alkylation of heme by artemisinin.

Similar alkylation of the heme moiety has been observed in the presence of artemether (*21*). In addition, by studying the reactivity of a large series of trioxanes (*20, 25, 26*), it has been possible to correlate their alkylating ability toward a heme model, manganese(II) tetraphenylporphyrin, and their pharmacological activity: Efficient antimalarial drugs behave as good alkylating agents and most of the inactive drugs were unable to alkylate the porphyrin ring. Furthermore, it appeared that trioxanes bearing a bulky methyl substituent at C4 on the α face of the endoperoxide (i.e. on the same side than the peroxide with respect to the mean drug plane), were at the same time inactive on infected red blood cells and unable to alkylate the porphyrin cycle (Figure 2) (*25*).

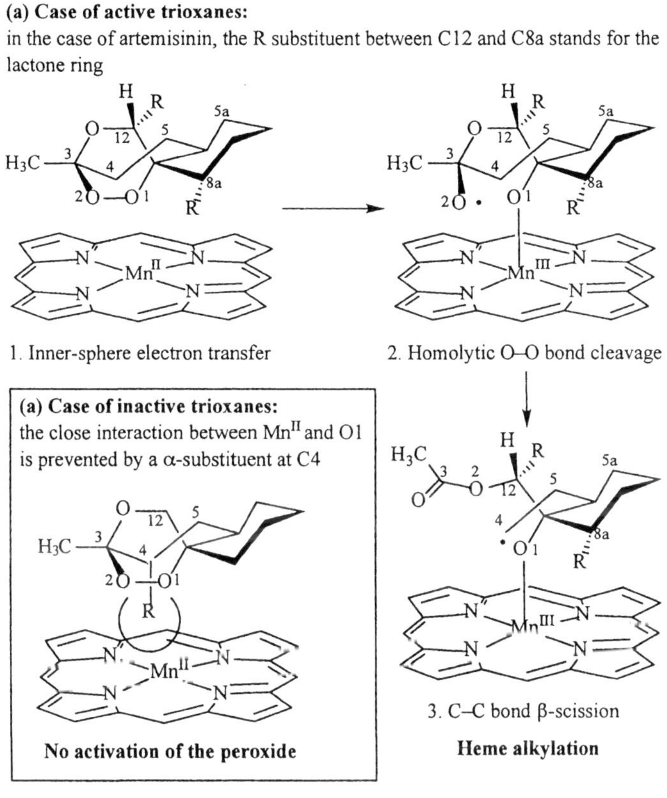

(a) Case of active trioxanes:
in the case of artemisinin, the R substituent between C12 and C8a stands for the lactone ring

1. Inner-sphere electron transfer

2. Homolytic O–O bond cleavage

3. C–C bond β-scission

Heme alkylation

(a) Case of inactive trioxanes:
the close interaction between Mn^{II} and O1 is prevented by a α-substituent at C4

No activation of the peroxide

Figure 2: Activation of trioxanes via an inner-sphere electron tranfer. Possible correlation between pharmacological activity and alkylating ability. Reprinted with permission from reference 23. Copyright (2002) American Chemical Society.

An analog trioxane with a methyl-C4 on the β-face, was active on malaria parasites and, at the same time, able to alkylate $Mn^{II}TPP$. Trioxanes without substituent at C4, whatever the substituents on other positions (for example methoxy at C12) were α or β, were pharmacologically active and able to alkylate $Mn^{II}TPP$. This indicates that the difference of reactivity of epimers at

C4 is probably due to the difference of interaction with the metal center, more than to other possible factors, say transport, competitive detoxification…

This data confirmed that (i) a close interaction between the reducing metal center and the peroxide is required, suggesting that this activation occurred through an inner-sphere electron transfer, (ii) only active endoperoxides alkylate heme or metalloporphyrin while inactive ones do not, suggesting that this alkylation reaction is probably important in parasite killing.

In addition, these artemisinin-heme adducts appear to be generated in artemisinin-treated parasites (*14*).

Possible pharmacological role of the heme-artemisinin adducts

How can the alkylation of heme by artemisinin kill the parasite? *Plasmodium falciparum*'s histidine-rich protein (PfHRP-II) promotes the formation of hemozoin. This protein contains repeats of the sequence His-His-Ala, together histidine and alanine making up 76% of the mature protein *(27)*. HRP is able to bind approximately 50 molecules of heme at pH 7 (*28*), and 17 at pH 4.8, thus acting as a scaffold for heme, important in the initiation of hemozoin chains (*29*). Recent studies suggest that the heme-artemisinin adducts (Figure 1) are able to bind to PfHRP-II with a higher affinity than heme itself, displace heme from PfHRP-II, and that either low pH or chloroquine dissociates heme but not heme-artemisinin adducts from PfHRP-II (*30*).

The binding of heme-artemisinin adducts to HRP-II with high affinity constants may result in the inhibition of hemozoin formation leading to the building-up of the concentration of free heme within the parasite.

At micromolar concentrations, artemisinin inhibits hemozoin formation (*30, 31*), but this has only been demonstrated in cell-free conditions, and remains controversial: artemisinin treatment of living parasites caused no measurable change in hemozoin content (*32*). However, the concentration of the heme pool (hemoglobin + free heme) that can be accumulated within the digestive vacuole during hemoglobin degradation can be as high as 400 *milli*molar (*33*). One has to keep in mind that free heme at *micro*molar concentration can damage cellular metabolism (*6*). Consequently, a very small portion of heme that escapes to the aggregation process (for example, 1 heme molecule over 10^4 or 10^5) should be sufficient to kill the parasite without having detectable effect on the hemozoin accumulation.

Heme-quinoline interactions

It has been underlined that the antimalarial activity of quinoline-based drugs (quinine, chloroquine, mefloquine) also depend on their interactions with heme, thus preventing aggregation of toxic heme released during proteolysis of hemoglobin by the parasite. In vitro, efficient quinolines block the aggregation of micromolar heme into hemozoin under approximated physiological conditions mediated by crude trophozoite lysates, seed crystals of hemozoin, or *Plasmodium falciparum*-derived HRPs. However, contrary to the covalent bond between heme and artemisinin, chloroquine and its congeners bind heme non covalently. Chloroquine has a high affinity for ferric heme (K_d = 3.5 nM), and ring-ring π–stacking of the quinoline nucleus with the porphyrin macrocycle has been observed by NMR (*34, 35*). In vitro, quinolines do not bind directly to the HRPs nor do they interact with isolated hemozoin. These results suggest that quinoline-heme complexes are incorporated into the growing aggregate to terminate the chain extension of hemozoin, blocking further sequestration of toxic heme and poisoning the parasite with its own waste (*6, 33*).

The interaction of chloroquine with iron(III)-heme requires an extensive degradation of the globin or, at least, release of the heme moiety from the protein. On the contrary, artemisinin may be activated by iron(II)-heme whenever one side of the heme nucleus is accessible for endoperoxide approach and activation. We recently found that alkylation of heme occurred in 25% yield by incubation of artemisinin with human ferrous hemoglobin A_0 in the absence of protease activity, under non denaturing conditions, or even when artemisinin was incubated with hemolyzed human blood (*36*).This is consistent with the antimalarial activity of artemisinin on early intra-erythrocytic stages of the parasite, whereas quinoline inhibition is specific for parasite stages that are actively degrading hemoglobin.

Other targets, other trigger(s) for artemisinin ?

In pioneer work, Meshnick and collaborators reported that, when *Plasmodium falciparum*-infected red blood cells was incubated with radiolabeled artemisinin derivatives, dihydroartemisinin, or arteether, drug derived radioactivity was found to be concentrated in the isolated parasites. In spite the difficulties to prepare protein-free hemozoin, radioactivity was associated, on one side, with heme and, on the other side, with specific parasitic membrane proteins, probably via alkylation. The reaction appears to be specific, since the alkylated proteins are not the most abundant in isolated parasites. It is likely that it is pharmacologically relevant, since all the active peroxides alkylate the same proteins and none of them were alkylated by the biologically inactive deoxyarteether (*15*). At this stage, the identities or functions of the alkylated proteins was unknown.

One of the most heavily labeled proteins was later isolated from parasite grown in the presence of [^{3}H]dihydroartemisinin (300 mM) and identified as a 25-kDa translationally controlled tumor protein (TCTP) homolog, that is able to bind heme with modest affinity (*17*). In vitro, the reaction of dihydroartemisinin with recombinant TCTP is clearly dependent on the presence of heme; the single cysteine of the protein also appears to be necessary for the reaction, probably serving as a source of electrons for the heme-mediated activation of the drug.

Although it is difficult to understand how alkylation of TCTP could kill the parasite because little is known about the physiological roles of this protein, the fact that this reaction occurs both in vitro and in vivo suggests that it is biologically relevant.

It was also reported that a high concentration of artemisinin (200 μM) inhibits the cysteine protease activity of vacuolar extracts of *Plasmodium*, involved in digestion of hemoglobin, but does not interfere with the activity of aspartic proteases that are responsible for the initial cleavage of the globin chain into large peptide fragments (*31*).

Recently, Ca^{2+}-dependent ATPase activity of PfATP6, the only SERCA-type Ca^{2+}-ATPase of *Plasmodium*, was shown to be abolished by active artemisinin derivatives, but not by deoxyartemisinin, quinine or chloroquine. On the contrary, artemisinin has no influence on the activity of a mammalian SERCA Ca^{2+}-ATPase, and several plasmodium proteins including a non-SERCA-type ATPase and a glucose transporter (*18*). PfATP6 was expressed in *Xenopus laevis* oocytes, a heme free medium. However, the iron chelator desferrioxamine significantly abrogated the inhibition of PfATP6 by artemisinin, suggesting that an unidentified iron species was this time responsible for the activation of artemisinin, leading to drug-derived radical species.

In all the reports on the interaction of artemisinin with biomolecules, there is strong evidence that drug-derived alkylating species are involved. There is both biological and chemical evidence for the role of heme in reductive activation of artemisinin, and the characterization of heme-artemisinin adducts shows that heme can be both the trigger and the target of artemisinin. But there is also evidence that heme is not the single target nor, maybe, the single trigger of antimalarial endoperoxides: other biological reducing iron species may play a similar role, leading to the alkylation of the protein(s) where they are located.

Other antimalarial drugs with the same mechanism ?

Synthetic trioxanes, simplified analogs of artemisinin, supposed to act in the same way, have also been developed (*37*). Future studies will provide information on the pharmacokinetics parameters of these artemisinin mimicks and tell us if these molecules have longer half-life times in plasma.

Recently, chloroquine has been successfully modified with a ferrocenyl entity incorporated within the side chain of the drug. This drug named ferroquine is active against chloroquine-resistant strains of the malaria parasite (*38*). This molecule is an example of the development of bioorganometallic chemistry in the design of drugs.

New molecules for an old target: Trioxaquines®

The combination of artemisinin derivatives with a second drug having a different mode of action is increasingly seen as the best way to prevent the emergence and spread of drug resistance, and to interrupt the transmission of *P. falciparum* (*11, 39*). Keeping in mind that chloroquine and artemisinin both interact with heme, but by two different mechanisms, we prepared new chimeric molecules, named trioxaquines®, by covalent attachment of a trioxane moiety to a 4-aminoquinoline (Figure 3) (*40*). Trioxaquines combine, within a single molecule, a aminoquinoline, known to easily accumulate within infected erythrocytes, and a peroxide acting as a potential alkylating species after reductive activation.

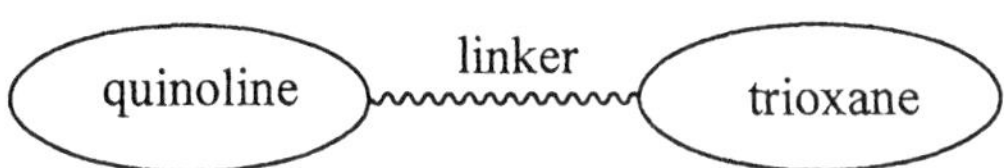

Figure 3. Trioxaquines are dual antimalarial molecules.

In order to get easily accessible molecules, a convergent synthesis based on classical reactions was used, starting from cheap materials. The synthesis of a second-generation trioxaquine (DU1301) is depicted in Figure 4.

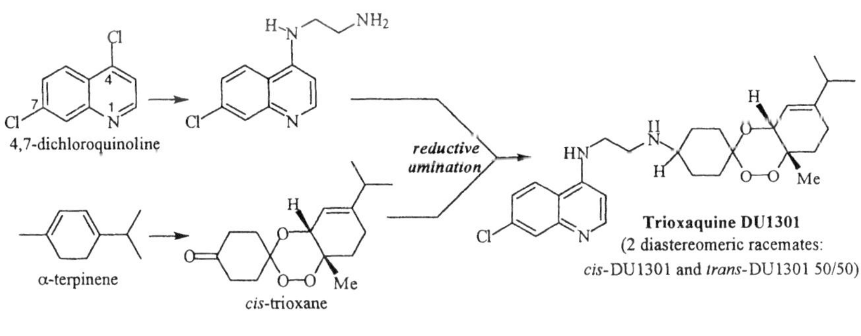

Figure 4. Convergent synthesis of Trioxaquine DU1301.

In trioxaquine DU1301, the amine and the peroxide substituents can be *trans* or *cis* with respect to the cyclohexane ring. The reductive amination reaction therefore provided two diastereomeric racemates *trans*-DU1301 and *cis*-DU1301 (50/50) (Figure 5).

For structure elucidation and biological evaluation, the two diastereomers of

DU1301 have been separated and their structures determined by X-ray diffraction. The structure of *trans*-DU1301 is depicted in Figure 6 (*41*).

Many modulations of the trioxaquine structure are possible (quinoline, diene, linker), and a number of them have been made, leading to a large family of new antimalarial compounds that are now under biological investigation.

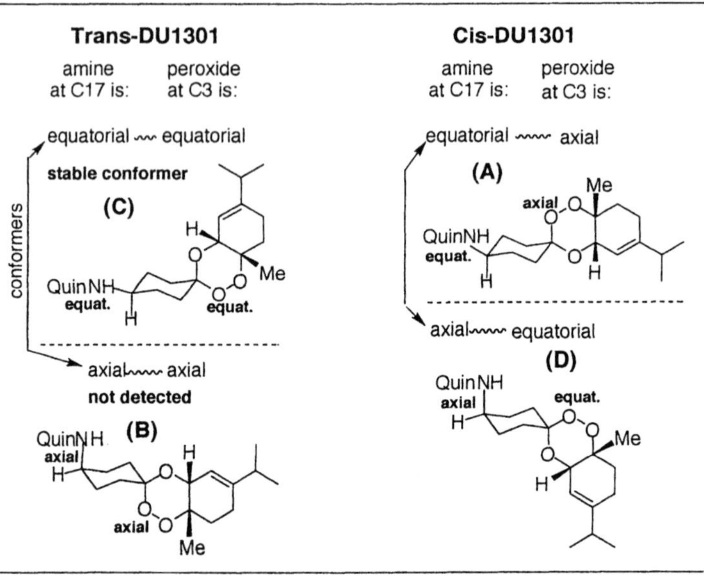

Figure 5. The two diastereomeric racemates of DU1301, the amine and the peroxide function being either axial or equatorial with respect to the cyclohexane ring. Quin-NH refers to the aminoquinoline residue.

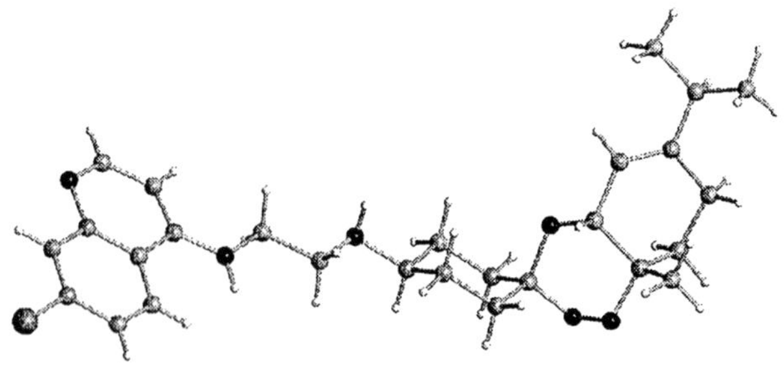

Figure 6. X-ray structure of trans-DU1301 (conformer C).

Antimalarial activity of trioxaquines

Trioxaquines have modular structures linking two different moieties. A study of the relation between the structure of the trioxaquines and their antiplasmodial activities in vitro and in vivo was done. All the trioxaquines tested in vitro present interesting antimalarial properties on strains of *Plasmodium falciparum* chloroquine-sensitive and chloroquine-resistant with IC_{50} inferior to 100 nM and around 10 nM for the best trioxaquines (*41*).

The trioxaquines DU1302 (citrate salt of the base form DU1301) and DU1102 (see ref. 40 for the structure of this first-generation trioxaquine) showed the most promising antimalarial activities. Their IC_{50} values ranged from 6 to 17 nM and from 22 to 27 nM respectively, whatever the chloroquine sensitivity of the strains used. These values are very similar to artemisinin activity (IC_{50} on the same strains: 5-8 nM) and largely better than chloroquine (62-174 nM).

The trioxaquine DU1102 is highly active against Cameroonian isolates as well (*42*). The mean IC_{50} obtained on 32 different *P. falciparum* isolates is 43 nM (ranging from 11 to 71 nM). There is no significant difference between the mean of DU1102 for chloroquino-sensitive (48 nM) and chloroquino-resistant (40 nM) isolates. There is no correlation between the response to DU1102 and chloroquine and a low correlation with pyrimethamine, suggesting independent modes of action of the trioxaquine against the parasite.

Because trioxaquine DU1302 showed the best antimalarial activity, more chemical and biological investigations have been performed (*41*). As explained above and in Figure 5 for trioxaquine DU1301, its citrate salt DU1302 exists as two diastereomeric racemates that have been independently tested in vitro on *P. falciparum*. The *trans*-DU1302, the *cis*-DU1302, and the 50/50 mixture of both diastereomers exhibit in vitro similar activities with IC_{50} values being 7-19 nM, 5-11 nM, and 6-17 nM, respectively. In addition, it should be noted that trioxaquines have better activities on highly chloroquine-resistant strains of the parasite.

DU1302 has been selected for the evaluation of its antimalarial activity in vivo on mice infected by *P. vinckei* in a 4-day suppressive test (*43*). DU1302 presents a potent antimalarial activity with ED_{50} values of 5 mg/kg/d and 18 mg/kg/d, by intraperitoneal and oral administration, respectively. These values are in the range of that reported for artemisinin itself. Moreover, complete cures of parasitemia without recrudescence have been observed at 20 (ip route) and 50 mg/kg/d (oral route) (i.e. no detected parasite after 60 days).

An absence of toxicity by oral route of DU1302 during 60 days has been observed both on non-infected mice treated with 100 mg/kg/d for three consecutive days and on infected mice treated with 120 mg/kg/d for four days.

Finally, the genotoxicity of DU1302 has been evaluated. Drugs that damage DNA induce systems of DNA repair such as the SOS-response, and the ability to induce this phenomenon in *Escherichia coli* has been shown to be correlated to the genotoxicity in human (*44*). The possible induction of the SOS-response to DU1302 in *E coli* (using the anti-cancer drug mitomycin C as control) has been studied. Whereas mitomycin C is found genotoxic at 3 μM, the trioxaquine DU1302 was unable to induce the SOS-response in *E. coli* up to 20 μM, a concentration 1000 times higher than its antimalarial IC_{50} values, indicating the absence of in vitro genotoxicity for this trioxaquine.

Furthermore, DU1302 is a stable compound, either at room temperature in air for more than six months, in an acidic solution for three days, or at 60 °C in air for 24 h.

In summary, the high efficacy in vitro and in vivo of DU1302 (in particular on chloroquine-resistant strains), its easy synthesis, its chemical stability, and its absence of toxicity and genotoxicity make this trioxaquine a promising drug-candidate for antimalarial therapy.

Conclusion

This report on the mechanism of action of antimalarial artemisinin derivatives and the design of trioxaquines is an example of the key role of a biological metal complex (heme in the present case) acting both as trigger and target for the drug.

Acknowledgements

All co-authors of the articles signed by AR, FBV and BM are warmly acknowledged for their key contributions on the mechanism of action of artemisinin derivatives and for the preparation and biological evaluation of trioxaquines. This work has been supported by the CNRS, the Région Midi-Pyrénées and PALUMED. This start-up company and SANOFI-SYNTHELABO are currently working on the development of trioxaquines. The authors are grateful to Jean-Paul Séguéla, Jean-François Magnaval and Antoine Berry (CHU-Rangueil, Departement of Parasitology) for many discussions on malaria therapy. Heinz Gornitzka (Université Paul Sabatier - CNRS Toulouse) is gratefully acknowledged for the X-ray determination of several trioxaquine and trioxane derivatives.

References

1. Guo, Z.; Sadler, P. J. *Angew. Chem. Int. Ed.* **1999,** *38,* 1512-1531.
2. Halpern, J.; Raymond, K. N. *Proc. Natl. Acad. Sci. USA* **2003,** *100,* 3562-3562.

3. Bannister, L.; Mitchell, G. *Trends Parasitol.* **2003,** *19,* 209-213.
4. Coombs, G. H.; Goldberg, D. E.; Klemba, M.; Berry, C.; Kay, J.; Mottram, J. C. *Trends Parasitol.* **2001,** *17,* 532-537.
5. Rosenthal, P. J.; Sijwali, P. S.; Singh, A.; Shenai, B. R. *Curr. Pharm. Des.* **2002,** *8,* 1659-1672.
6. Francis, S. E.; Sullivan, D. J.; Goldberg, D. E. *Annu. Rev. Microbiol.* **1997,** *51,* 97-123.
7. Pagola, S.; Stephens, P. W.; Bohle, D. S.; Kosar, A. D.; Madsen, S. K. *Nature* **2000,** *404,* 307-310.
8. Meshnick, S. R.; Taylor, T. E.; Kamchonwongpaisan, S. *Microbiol. Rev.* **1996,** *60,* 301-315.
9. Wongsrichanalai, C.; Pickard, A. L.; Wernsdorfer, W. H.; Meshnick, S. R. *Lancet Infect. Dis.* **2002,** *2,* 209-218.
10. Ittarat, W.; Pickard, A. L.; Rattanasinganchan, P.; Wilairatana, P.; Looareesuwan, S.; Emery, K.; Low, J.; Udomsangpetch, R.; Meshnick, S. R. *Am. J. Trop. Med. Hyg.* **2003,** *68,* 147-152.
11. Nosten, F.; Brasseur, P. *Drugs* **2002,** *62,* 1315-1329.
12. Ridley, R. G. *Nature* **2002,** *415,* 686-693.
13. Klayman, D. L. *Science* **1985**, *228,* 1049-1055.
14. Hong, Y.-L.; Yang, Y.-Z.; Meshnick, S. R. *Mol. Biochem. Parasitol.* **1994,** *63,* 121-128.
15. Asawamahasakda, W.; Ittarat, I.; Pu, Y.-M.; Ziffer, H.; Meshnick, S. R. *Antimicrob. Agents Chemother.* **1994,** *38,* 1854-1858.
16. Meshnick, S. R. *Int. J. Parasitol.* **2002,** *32,* 1655-1560.
17. Bhisutthibhan, J.; Pan, X.-Q.; Hossler, P. A.; Walker, D. J.; Yowell, C. A.; Carlton, J.; Dame, J. B.; Meshnick, S. R. *J. Biol. Chem.* **1998,** *273,* 16192-16198.
18. Eckstein-Ludwig, U.; Webb, R. J.; van Goethem, I. D. A.; East, J. M.; Lee, A. G.; Kimura, M.; O'Neill, P. M.; Bray, P. G.; Ward, S. A.; Krishna, S. *Nature* **2003,** *424,* 957-961.
19. Posner, G. H.; Oh, C. H.; Wang, D.; Gerena, L.; Milhous, W. K.; Meshnick, S. R.; Asawamahasakda, W. *J. Med. Chem.* **1994,** *37,* 1256-1258.
20. Robert, A.; Meunier, B. *J. Am. Chem. Soc.* **1997,** *119,* 5968-1569.
21. Robert, A.; Cazelles, J.; Meunier, B. *Angew. Chem. Int. Ed.* **2001**, *40*, 1954-1957.
22. Robert, A.; Coppel, Y.; Meunier, B. *Chem Commun.* **2002,** 414-416.
23. Robert, A.; Dechy-Cabaret, O.; Cazelles, J.; Meunier, B. *Acc. Chem. Res.* **2002,** *35,* 167-174.
24. Robert, A.; Coppel, Y.; Meunier, B. *Inorg. Chim. Acta* **2002,** *339,* 488-496.
25. Cazelles, J.; Robert, A.; Meunier, B. *J. Org. Chem.* **2002**, *67*, 609-616.
26. Robert, A.; Meunier, B. *Chem. Eur. J.* **1998,** *4,* 1287-1296.
27. Wellems, T. E.; Howard, R. J. *Proc. Natl. Acad. Sci. USA* **1986,** *83,* 6065-6069.
28. Choi, C. Y. H.; Cerda, J. F.; Chu, H.-A.; Babcock, G. T.; Marletta, M. A. *Biochemistry* **1999,** *38,* 16916-16924.

29. Sullivan, D. J.; Gluzman, I. Y.; Goldberg, D. E. *Science* **1996,** *271,* 219-222.
30. Kannan, R.; Sahal, D.; Chauhan, V. S. *Chem. Biol.* **2002,** *9,* 321-332.
31. Pandey, A. V.; Tekwani, B. L.; Singh, R. L.; Chauhan, V. S. *J. Biol. Chem.* **1999,** *274,* 19383-19388.
32. Asawamahasakda, W.; Ittarat, I.; Chang, C.-C.; McElroy, P.; Meshnick, S. R. *Mol. Biochem. Parasitol.* **1994,** *67,* 183-191.
33. Sullivan, D. J.; Gluzman, I. Y.; Russell, D. G.; Goldberg, D. E. *Proc. Natl. Acad. Sci. USA* **1996,** *93,* 11865-11870.
34. Moreau, S.; Perly, B.; Chachaty, C.; Deleuze, C. *Biochim. Biophys. Acta* **1985,** *840,* 107-116.
35. Leed, A.; DuBay, K.; Ursos, L. M. B.; Sears, D.; de Dios, A. C.; Roepe, P. D. *Biochemistry* **2002,** *41,* 10245-10255.
36. Selmeczi, K.; Robert, A.; Claparols, C.; Meunier, B. *FEBS Lett.* **2004,** *556,* 245-248.
37. Posner, G. H.; Paik, I.-H.; Sur, S.; McRiner, A. J.; Borstnik, K.; Xie, S.; Shapiro, T. A. *J. Med. Chem.* **2003,** *46,* 1060-1065.
38. Delhaes, L.; Biot, C.; Berry, L.; Delcourt, P.; Maciejewski, L. A.; Camus, D.; Brocard, J. S.; Dive, D. *ChemBioChem* **2002,** *3,* 418-423.
39. White, N. J.; Nosten, F.; Looareesuwan, S.; Watkins, W. M.; Marsh, K.; Snow, R. W.; Kokwaro, G.; Ouma, J.; Hien, T. T.; Molyneux, M. E.; Taylor, T. E.; Newbold, C. I.; Ruebush, T. K.; Danis, M.; Greenwood, B. M.; Anderson, R. M.; Olliaro, P. *Lancet* **1999,** *353,* 1965-1967.
40. Dechy-Cabaret, O.; Benoit-Vical, F.; Robert, A.; Meunier, B. *ChemBioChem* **2000,** *1,* 281-283.
41. Dechy-Cabaret, O.; Benoit-Vical, F.; Loup, C.; Robert, A.; Gornitzka, H.; Bonhoure, A.; Vial, H.; Magnaval, J.-F.; Séguéla, J.-P.; Meunier, B. *Chem. Eur. J.* **2004,** *10,* 1625-1636.
42. Basco, L. K.; Dechy-Cabaret, O.; Ndounga, M.; Meche, F. S.; Robert, A.; Meunier, B. *Antimicrob. Agents Chemother.* **2001,** *45,* 1886-1888.
43. Peters, W.; Portus, J. H.; Robinson, B. L. *Ann. Trop. Med. Parasitol.* **1975,** *69,* 155-171.
44. Hofnung, M.; Quillardet, P. *Ann. N. Y. Acad. Sci.* **1988,** *534,* 817-825.

Chapter 17

Administration of Mn Porphyrin and Mn Texaphyrin at Symptom Onset Extends Survival of ALS Mice

John P. Crow

Department of Pharmacology and Toxicology, University of Arkansas for Medical Sciences, Little Rock, AR 72205

Transgenic mice which overexpress the G93A human Cu, Zn superoxide dismutase (SOD1) mutant enzyme develop motor paralysis similar to amyotrophic lateral sclerosis (ALS) in humans. At 90-100 days of age, these mice develop hindlimb weakness which rapidly progresses to total body paralysis within 2-3 weeks. Daily doses of either a manganese porphyrin or a manganese texaphyrin, begun at the onset of disease in each mouse, enhanced survival up to 2.8-fold (relative to untreated mice) and preserved near-normal motor function until end-stage disease. These results are among the best ever seen for late (onset) administration of any pharmacological agent and suggest that administration of such agents at symptom onset, roughly equivalent to the earliest time when human treatment could begin, can fundamentally alter the course of the disease.

Introduction to ALS

Amyotrophic lateral sclerosis (ALS or Lou Gerhig's Disease) is a fatal neurodegenerative disease which typically strikes in the prime of life and effectively traps a healthy mind in a paralyzed body. ALS can begin as a weakness in a limb or as slurred speech and rapidly progresses to complete paralysis of all voluntary muscles. Loss of muscle control (and associated muscle wasting) is a direct result of the death of motor neurons within the spinal cord. The average age of onset follows a bell-shaped probability curve with a peak at approximately 45-50 years old and the time from the beginning of symptoms to death ranges from 1-6 years (1,2). With one notable exception (discussed later), there are no "predictors" of ALS and no way to know who is at risk prior to the onset of symptoms. Moreover, differential diagnosis often doesn't occur until weeks or months after the first symptoms. Thus, any potential treatments for ALS must be aimed at slowing disease progression and preserving the remaining spinal motor neurons.

Familial versus Sporadic ALS

Roughly 80% of ALS cases are sporadic (SALS), meaning that no specific etiology is known. Twenty percent of ALS cases are known to be inherited (familial ALS or FALS), although the exact genetic defects which underlie the disease are known in only a small fraction; 10% of the familial cases (or 2% of all ALS cases) result from mutations to Cu, Zn superoxide dismutase (SOD1) (2,3). It should be emphasized that these ALS classifications are based solely on whether or not a pattern of inheritance is evident in each case; there are no clinical or histopathological characteristics which permit any distinction or differential classification. Because SOD1 mutations represent the only known cause, it is intriguing to speculate that all forms of ALS could result from alterations to SOD1 and we have put forth such a molecular mechanism whereby this could occur (reference 4; briefly discussed later in chapter). Irrespective of the precise initiating event, it may well be that motor neuron loss occurs via common cell death pathways which can either be inhibited directly, or indirectly via scavenging of damaging oxidants and/or stimulation of cellular defenses.

Mouse Model of ALS

In 1993, Rosen et al. (3) reported the link between mutations to the Cu, Zn form of superoxide dismutase (SOD1 and human ALS. Shortly thereafter, Gurney et al. (5) showed that the presence of the mutated human SOD1 gene (where Alanine was substituted for Glycine at position 93, or G93A) was sufficient to produce motor neuron disease in transgenic mice. The term "motor neuron disease" (MND) is used to distinguish the paralytic syndrome in mice from true human ALS, which has some clinical differences. Nevertheless, MND in G93A mice is remarkably similar to human ALS and the G93A mouse is generally regarded as one of the best animal models of human neurodegenerative disease. As such, the G93A mouse is used extensively to identify compounds which may be of benefit in human ALS. Several human clinical trials have been initiated based solely on results obtained in this mouse model.

Classes of Agents Which Have Been Tested

Literally hundreds of agents, in many different chemical and therapeutic classes, have been tested in mice and/or humans for their ability to slow disease progression (ALS Therapy Development Foundation: http://www.als.net/default.asp). These include: antioxidants, glutamate antagonists, metal chelators, neural and non-neural growth factors, calcium regulators, anti-inflammatory agents, inhibitors of cell signaling and cell death pathways, energetic precursors (e.g., creatine), and a host of common dietary supplements, as well as immunoregulatory, cholinergic, dopaminergic, and anti-viral agents (reference 2 and ALS TDF website). Several compounds have shown at least some beneficial effects in mice, however, in many cases, human trials with these same agents did not prolong survival. While such negative results are disappointing, it may be incorrect—or at least premature—to conclude that the mouse model is not predictive.

Interpretation of Mouse Studies and Extrapolation to Humans

The vast majority of the mouse studies to date have involved administration of test agents as early as 30 days of age in mice that develop disease at ~ 90 days. Given the difference in average lifespan between mice and humans (where one mouse day roughly equals one human month), presymptomatic administration 60 days prior to symptom onset in mice equates to beginning treatment five years prior to onset of symptoms in humans. As discussed earlier, pre-symptomatic administration is currently not possible in human ALS patients, thus any delay in onset resulting from pre-symptomatic administration in mice would not be expected to extrapolate to humans. Thus far, only a few agents have been evaluated in an "age of onset" treatment paradigm (Crow et al., manuscript in preparation) and it remains to be seen whether this approach will yield better results in human trials. However, the utility of this treatment paradigm is compelling in terms of disqualifying potential therapies, i.e., those agents which fail to extend mouse survival when administered at onset would not be likely to benefit ALS patients.

Glutamate Excitotoxicity and Riluzole

Many different types of neurons—including motor neurons—are susceptible to glutamate-mediated excitotoxicity (6,7). Glutamate antagonism was the basis for testing of riluzole—a compound which proved to have a small, but reproducible survival benefit in both transgenic mice and human ALS patients (7-9). Riluzole extends survival of human ALS patients for only 2 months (9) and has no beneficial effect on muscle strength. Nonetheless, it remains as the standard of care (and the only FDA-approved drug) for ALS simply because no other agents have been found which are more effective.

Classic Antioxidants

Some of the first agents tested in ALS models were antioxidants. Correlative increases in oxidative markers in and around affected tissues strongly implicated oxidative injury as a component of overall neuronal loss in MND and ALS (4, 10, 11). Logic dictated that, even if oxidative injury is secondary to the primary insult, antioxidants could be beneficial, and a number of such agents were tried, including ascorbate (vitamin C), tocopherol (vitamin E), cysteine, and N-acetyl-cysteine (8,12-14). The fact that these compounds are relatively safe dietary supplements made them natural choices and they were among the first compounds tested in both mice and humans.

Results with these classical antioxidants were disappointing and led many to question the relevance of oxidative injury to neuronal death. However, the inability of these agents to alter the course of MND and ALS may have been more a function of their chemistry and pharmacology than a flawed rationale. That is, none of these endogenous antioxidants (or precursors like NAC) effectively scavenge the oxidants thought to be involved in neuronal death and injury. Also, these compounds already exist in significant concentrations in neurons, and normal physiologically processes keep them at prescribed levels; thus, even megadoses may not significantly alter their concentrations in the affected subset of spinal cord neurons. Only rarely (in these early studies) were attempts made to determine whether tissues levels of these antioxidants actually increased in response to administration and measurements in total spinal cord homogenates may not reflect intracellular levels in the affected neurons.

Novel Antioxidant Properties of Mn Porphyrins and Texaphyrins

New generation Mn porphyrins and texaphyrins (Figure 1) may circumvent the limitations of endogenous antioxidants in three important ways: 1) they react rapidly and directly with damaging biological oxidants, 2) they scavenge oxidants catalytically rather than sacrificially, and 3) they distribute to all major organs, including the central nervous system (CNS), following peripheral administration (Crow et al., unpublished pharmacokinetic studies).

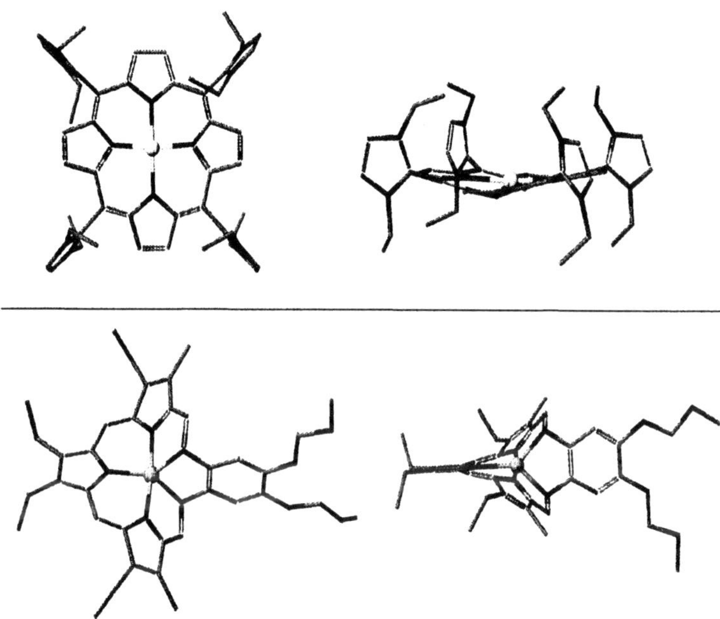

Figure 1. Top: Two Views of the Mn Porphyrin (AEOL 10150), and Bottom: Two Views of Mn Texaphyrin. *Geometries of compounds were optimized using a combination of semi-empirical and ab initio calculation methods in Quantum CAChe (Fujitsu). However, these structures are intended for illustration purposes only using the same geometric optimization methods; they are not offered as representations of true crystal or solutional structures. The Top and side views of both compounds reveal the relative planarity of the Mn porphyrin as compared with the "twisted", five-coordinate Mn nucleus of the texphyrin..*

This last property (point #3) means that tissue and cellular concentrations are proportional to dose and not subject to physiological regulation like ascorbate and tocopherol. Catalytic scavenging (point #2) of oxidants means that such compounds will be effective at much lower doses and that effective concentrations are not diminished by reaction with oxidants in the way that sacrificial antioxidants are. Further discussion of the importance of direct reactivity (point #1) with damaging oxidants requires a characterization of the oxidants in question.

Why Peroxynitrite?

A detailed review of the identity and potential sources of biological oxidants is beyond the scope of this chapter. However, when considering several factors such as the likelihood of formation, the lack of known endogenous enzymatic scavengers, and overall reactivitiy toward a host of important biomolecules, peroxynitrite has to be viewed as a major, endogenously produced cytotoxic agent (15,16). The rate constant for the reaction of nitric oxide (NO) with superoxide is close to the diffusion limit (10^{10} $M^{-1}sec^{-1}$, 17), meaning that virtually every collision between these two radicals will result in production of peroxynitrite anion. Peroxynitrite can carry out a variety of one and two electron oxidations of many different biomolecules as well as nitrating aromatic compounds like tyrosine, tryptophan, and guanine (18,19).

Peroxynitrite as a Cytotoxin

The cytotoxic potential of peroxynitrite has been demonstrated in many different cell and organ-based models (15,16). The precise mechanism(s) by which it kills cells has not been determined and may, indeed, differ depending on the cell type and the relative concentrations of target biomolecules. In general, it appears that vital components of mitochondrial energy production are prime targets for peroxynitrite and toxicity in many models can be explained on this basis (15,20). The mitochondrial form of superoxide dismutase (SOD2) is particularly susceptible to peroxynitrite-mediated inactivation (21,22) and a defect in this enzyme alone is sufficient to cause lethality (23). Thus, a generic mechanism whereby low-level peroxynitrite could kill mitochondria and the cells which depend on them, could occur as follows: some fraction of total SOD2 could be inactivated by peroxynitrite resulting in a rise in superoxide levels which drives additional peroxynitrite formation, and leads ultimately to inactivation of energy-producing components in mitochondria. Many different forms of oxidative cell and tissue injury can be explained by invoking this simple, positive feedback loop (20).

Mechanism of Peroxynitrite Scavenging and Quenching by Porphyrins and Texaphyrins

Irrespective of the precise mechanism of cytotoxicity, it is clear that scavenging of peroxynitrite can all but eliminate oxidative injury and death under many different conditions (24-26). Both iron and manganese porphyrins as well as manganese texaphyrins have been shown to catalytically decompose peroxynitrite *in vitro*, (27-30) using ascorbate as the intermediate reductant to regenerate the reduced state of the central metal. Charge attraction between the central, cationic metal ion and peroxynitrite anion likely contribute to high second rate constants for reaction between the oxidant and the scavengers.

Quenching of Peroxynitrite

Because of the fast reaction rates, both the porphyrins and texaphyrin can "channel" peroxynitrite into reaction pathways that are quenched more effectively by endogenous cellular antioxidants. That is, by intercepting peroxynitrite before it can react with biomolecules via the more irreversible (and potentially more damaging) two-electron oxidation pathways, the porphyrins and texaphyrin route its reactivity into less injurious one-electron pathways (31). The overall scavenging reaction can be viewed as the compounds effectively "splitting" peroxynitrite into two one-electron oxidants—hydroxyl radical and nitrogen dioxide radical—and directly reducing hydroxyl radical to hydroxide anion. The remaining "free" nitrogen dioxide is then rapidly reduced by recyclable cellular reductants such as ascorbate, tocopherol, and glutathione (31).

Manganese Oxidation State

The scavenging and quenching mechanism described above would apply to the redox cycle of Mn(III)/Mn(IV) porphyrin or Mn(II)/Mn(III) texaphyrin. The environment of the Mn in the texaphyrin stabilizes the +2 state and does not allow the Mn to become oxidized to +4 (32). Under *in vitro* conditions, the Mn in the porphyrin exists in the +3 state and does cycle to +4 upon reaction with peroxynitrite (27-31). However, in the presence of reductants (and at much lower oxygen tensions), as would exist intracellularly, the porphyrin Mn would likely exist in the +2 state; a redox cycle from +2 to +4 would allow it to completely reduce peroxynitrite and thereby totally quench its oxidative reactivity (31,30).

Peroxynitrite and Carbon Dioxide

Peroxynitrite reacts with carbon dioxide to form a nitrosoperoxocarbonate species with reactivities similar to carbonate radical and nitrogen dioxide radical—two very potent one-electron oxidants (30). The Mn porphyrin described here has been shown to react rapidly with carbonate radical (30). Thus, even the fraction of peroxynitrite which may escape initial capture by the porphyrin could still be intercepted as the carbonate radical and its biological reactivity quenched.

Rationale for Use of Novel Antioxidants in ALS Mice: SOD1 Mutants as Initiators of Oxidative Injury

Prior to discussing the effects of the Mn porphyrins and texaphyrin in ALS mice, it is essential to provide some mechanistic background on the model itself as well as how and why SOD1 mutants may be causing neuronal death. As the first proven cause of ALS, SOD1 mutants provided the first specific protein target, in a disease that previously had no known targets, receptors, or pathogonomic biochemical anomalies. Unlike many inborn errors of metabolism wherein a genetic mutation results in the loss of activity of a critical enzyme, most ALS-associated mutations to SOD1 (over 100 distinct missense mutations are currently known) have little, if any, effect on enzyme activity. Moreover, mice which possessed higher copy numbers of mutated human genes for SOD1 (and much higher total SOD1 activity) developed MND sooner than mice with fewer copies of the gene. Thus, it became apparent that SOD1 mutant-mediated was due to some gained toxic function rather than a loss of normal function. This realization paved the way for drug design approaches aimed at inhibiting the gained toxic function—whatever that turned out to be.

Lowered Zinc Binding Affinity of SOD1 Mutants

The slow pace of progress toward understanding how and why SOD1 mutants are toxic to motor neurons is due to the fact that significant differences between mutants and wild-type enzyme have been very difficult to uncover. All of the ALS-associated SOD1 mutants which don't have active site mutations—and have their full complement of both copper and zinc—have physical, chemical, and enzymatic properties which are virtually identical to wild-type enzyme[1]. However, changes in redox properties of bound copper and aggregatory properties of the overall protein become very apparent when zinc is absent, and the same is true for both mutants and wild-type enzyme (33-35). That is, *the major differences which have been described to date are more a function of whether or not zinc is bound than whether the enzyme is mutant or wild-type.* We have put forth a hypothesis based on formation of a zinc-deficient form as the common toxic species (4) and demonstrated that even wild-type SOD1, when zinc-deficient, is toxic to cultured motor neurons (33). Thus, the zinc-deficiency hypothesis could provide a framework for explaining how all forms of ALS may be related to malfunctions of SOD1.

Possible Toxic Mechanisms for SOD1 Mutants

Human SOD1 is a dimer of identical subunits, each of which has an essential copper atom in its active site and a zinc atom which is bridged to the copper via histidine-63 (Figure 2). The copper atom cycles between Cu(II) and Cu(I) during catalysis while the zinc is redox inactive and thought to play mainly a structural role.

We measured zinc and copper binding constants for five ALS-associated SOD1 mutant and wild-type and found that all mutants had lower zinc binding affinity than wild-type and that the mutant A4V, which causes the most rapid disease progression in humans, had the lowest binding affinity of the group (36)[2]. Other groups reported qualitative changes in mutant and wild-type zinc binding which were consistent with our quantitative determinations (37).

[1] Even mutants which have been shown to have lowered affinity for zinc will still bind it quite well in the absence of competition. Only when available zinc became limiting intracellularly would the differences in zinc binding affinity between mutants and wild-type enzyme make the mutants more likely to exist in a zinc-deficient form.

[2] It has been suggested that the differences in zinc binding affinity we reported were due to the use of mildly denaturing conditions (2 M urea) employed for chelator competition assays. However, it should be emphasized that, under these conditions, copper binding affinities for mutants and wild-type enzyme were virtually identical, whereas zinc binding affinities differed by as much as 30-fold. It is highly improbable that non-specific protein denaturation would selectively alter apparent zinc binding affinity while having no effect on copper binding.

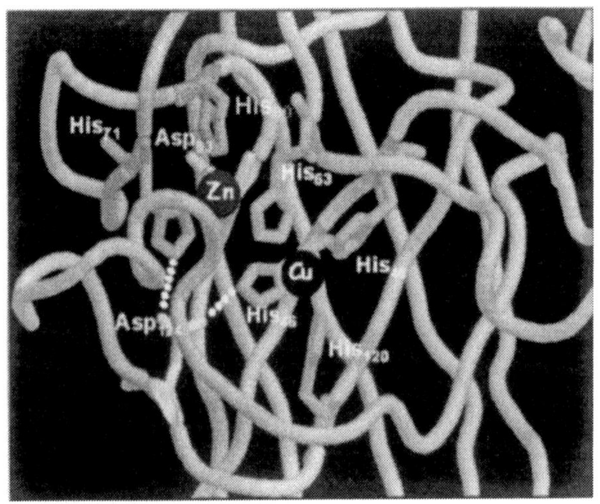

Figure 2. Active Site of Human Cu, Zn Superoxide Dismutase (SOD1). *Copper is bound by four histidines residues, one of which (His63) it shares with zinc. Aspartate124 is hydrogen-bonded to His46 and His71; mutation of Asp124 to a residue which does not H-bond results in lowering of zinc binding affinity by several orders of magnitude. This provides the structural basis to explain how mutations distal to the active site could alter zinc binding affinity. That is, any conformational perturbation which displaced Asp124 even a few Angstroms away from His46 and His71 would weaken or break the H-bonds, distort the geometry of the zinc binding ligands, and cause zinc binding affinity to decrease markedly .*

Renewed attention to the structure and function of SOD1 mutants and wild-type, and to the role of zinc, strongly suggests that the zinc ion helps maintain substrate specificity by tying together three otherwise independent polypeptide loops and "tightening up" the active site channel. This structural role of zinc, and the fact that it shares a binding ligand with copper, strongly suggest that its absence could influence both redox properties of bound copper as well as subunit folding and interactions with other subunits.

Abnormal Redox Activity of Enzyme-Bound Copper: "Peroxynitrite Synthase" Activity

In the absence of zinc, the active site copper ion in SOD1 becomes accessible to larger molecular weight cellular reductants such as ascorbate, NADH, urate, glutathione, etc.—substrates that would normally be excluded

(33). Abnormal reduction of the copper by cellular antioxidants, followed by re-oxidation, sets up a futile cycle of superoxide production (see Figure 3). However, this reverse reaction of SOD1 is quite slow (relative to the forward reaction) and significant pathology would be almost totally dependent on nitric oxide (NO); indeed, the rate of peroxynitrite production is directly proportional to NO concentration (33 and J. P. Crow, unpublished observations, 2001).

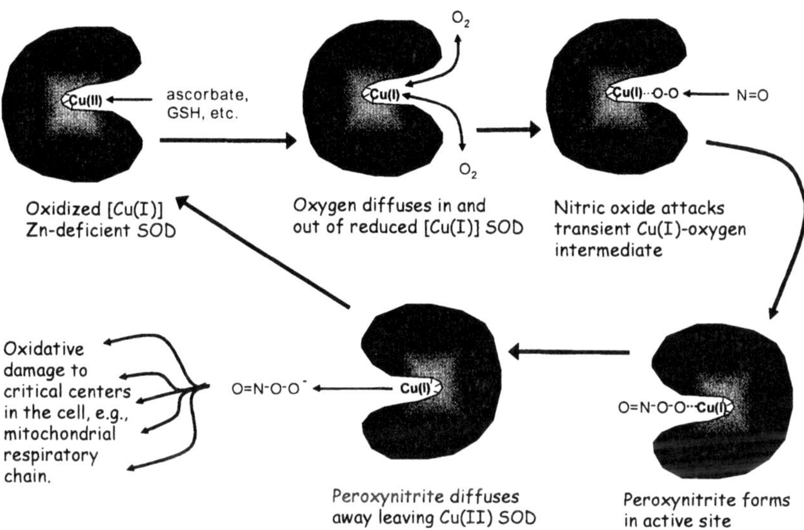

Figure 3. Proposed "Peroxynitrite Synthase" Activity of Zinc-Deficient SOD1. Zinc-deficient SOD1, whether derived from ALS-associated mutant or wild-type, will generate peroxynitrite in vitro in the present of micromolar concentrations of ascorbate and nanomolar concentrations of nitric oxide (33). When zinc is properly bound in the active site, ascorbate can no longer access bound copper and this activity ceases.

Superoxide reacts with NO five times faster than it is scavenged by SOD1 and it is this property (reaction rate) that allows NO to shift the equilibrium in favor of the reverse reaction and "coerce" the enzyme into making peroxynitrite. This abnormal redox activity of SOD1 has been criticized as being too slow to account for neuronal death. However, it must be recalled that manifestation of disease typically requires 50 years or so and then, is confined to motor neurons. It is reasonable to suggest that a more robust oxidant production from SOD1 mutants would manifest itself in earlier, more ubiquitous pathology. Also, NO-dependent peroxynitrite production from zinc-deficient SOD1 is catalytic and would, therefore, produce a sustained flux of peroxynitrite over time. Even so, given such a delayed onset of disease in only a subset of neurons, it is likely that additional factors would be needed to tip the balance in favor of pathology. For example, changes in cell physiology with age could result in a decrease in cellular defense and repair mechanisms and/or formation of more zinc-deficient SOD1 which could overwhelm such defenses. The current state of understanding of SOD1 mutant-mediated MND simply does not allow any specific mechanism to be unequivocally ruled out.

Aggregation of SOD1 Mutants

It is reasonable to postulate that accumulation of protein aggregates within neurons could lead to cell death by merely interfering with normal intracellular trafficking of other proteins, organelles, metabolites, etc. Such a "gridlock" hypothesis is particularly relevant to motor neurons, which must transport vital proteinaceous growth factors from the muscle they innervate back to the cell body via a densely packed axon up to one meter in length.

Toxicity from protein aggregation is consistent with the axonal inclusion bodies found in both transgenic mice and humans. Many investigators point to aggregation of Huntingtin Protein in Huntington's Disease, beta amyloid protein in Alzheimer's Disease, and misfolding of prion protein in bovine spongiform encephalopathy (Mad Cow Disease and other prion-mediated human diseases) as precedence for SOD1 misfolding/aggregation in ALS (38). However, in ALS it is not at all clear whether protein aggregates are the initial cause of cell injury or whether they form adventitiously during the process of cellular decline and death.

There are reports of neuronal death in the absence of detectable aggregates and conversely, the presence of aggregates with no associated cytotoxicity (39). One possible explanation for this lack of correlation is that discreet aggregates may not be needed. Continuous "misfolding" of SOD1 mutants could be toxic by merely tying up the chaperone proteins which help refold many other damaged or altered proteins and enzymes. Such folding chaperones or heat shock proteins (HSPs) have been shown to bind to SOD1 mutants (40). While current evidence favors the importance of misfolding and/or aggregation of SOD1 mutants in the pathogenesis of MND and ALS, the temporal sequence of such events, and whether they are a fundamental cause of neuronal pathology or secondary to some other initiating event, remains elusive.

Unifying Concepts: Zinc-Deficient SOD1 as The Common Toxic Species

If the known mutations to SOD1 serve mainly to destabilize zinc binding ligands, lower zinc binding affinity, and make them more likely to exist in a zinc-deficient, copper-containing form, then many of pieces of the ALS puzzle fall into place (see Figure 2 legend for additional rationale). That is, it would be possible to explain how 100 or so distinct, single amino acid mutations, spread out over large regions of the polypeptide, could all result in a common toxic species and produce toxicity via a common biochemical mechanism. Ultimate neuronal death from zinc-deficient SOD1 could result from the sustained, catalytic production of peroxynitrite and/or from the accompanying structural instability of the zinc-deficient form, and its tendency to misfold and aggregate (34,35).

The question of why only motor neurons are affected could be explained on the basis of zinc availability and the presence of a high density of neurofilament proteins which have been shown to bind up to 20 zinc ions per subunit (36). It is conceivable that conditions could exist (e.g., up-regulation of other zinc binding proteins like neurofilaments, metallothioneins, or glial fibrillary acid protein in glial cells) wherein incorporation of zinc into wild-type SOD1 could be inhibited as well. (Zinc-deficient wild-type SOD1 is as toxic to motor neurons as zinc-deficient mutant [33].) This would provide an etiological basis for all forms of ALS being caused by SOD1 and explain why no clinical or histopathological differences exist between familial and sporadic ALS. Much work remains to be done, but it is intriguing to speculate on such grand unification concepts.

Rational Drug Design: Targeting SOD1 Mutants

During our early studies of the properties of zinc-deficient SOD1, we discovered that Cu(I)-specific, heterocyclic chelators like bathocuproine and neocuproine bound to the Cu(I) in zinc-deficient SOD1 but not to the zinc-containing enzyme. That is, these chelators bound to SOD1 only when zinc was absent and the active site Cu(II) was reduced to Cu(I); adding zinc back prior to chelator addition or failure to pre-reduce Cu prevented chelator binding. These findings offered the hope of selective targeting of zinc-deficient SOD1 and we were awarded a patent for this process as a potential therapy in ALS (U.S. Patent 6,022,879). However, these compounds have properties which limited their utility as drugs and synthesis of more suitable analogs proved to be problematic. Moreover, parallel *in vitro* and *in vivo* findings of marked neuroprotection with iron and manganese porphyrins and a manganese texaphyrin suggested that scavenging of the toxic product—peroxynitrite—could be as effective as targeting of zinc-deficient SOD1 *per se*.

Scavenging of Secondary Oxidants and Up-Regulation of Cellular Defenses

The newest generation of manganese porphyrins is the result of many years of effort designed to produce low molecular weight SOD1 "mimetics" or "mimics" which would not have the therapeutic limitations of proteins. In the last 10 years, it became evident that the same redox properties of manganese porphyrins that endowed them with SOD-like activity also allowed them to catalytically decompose other oxidants such as peroxynitrite (41). Although the origins and rationale for development of the five-coordinate manganese texaphyrins was quite different (45), they too possess catalytic antioxidant properties which could prove therapeutically useful. *In vitro* studies of a close structural analog of the Mn texaphryin described herein suggest that, while it may not react as rapidly with peroxynitrite as some porphyrins, it may be more effective at utilizing cellular reductants to quench the secondary oxidants produced (32). The ability of the scavenger to quench secondary oxidants ranks very high among the criteria needed for successful *in vivo* scavenging of peroxynitrite (31).

Precedent-Setting Prototype Iron Porphyrin

Screening of several early generation Mn and Fe porphyrins in primary cultures of rat motor neurons revealed that one in particular, iron tetracarboxyphenylporphyrin (FeTCPP), kept the cells alive for an extended period even at sub-micromolar concentrations (46). Primary motor neurons are post-mitotic cells which can be kept alive via the addition of various growth factors to the media, most notably brain derived neurotrophic factor (BDNF). In the absence of BDNF, these cells induce nitric oxide synthase isoforms, leading to the accumulation of oxidative markers, and death via an apoptotic cell death pathway within 24-72 h. When FeTCPP was added to cell growth media at a concentration of 0.1 μM, the neurons lived for up to four weeks, almost twice as long as with BDNF treatment (46). These results prompted us to test FeTCPP in the G93A mice which, at that time of the cell culture findings, were already beginning to show signs of muscle weakness.

Despite the late, onset administration, FeTCPP (3 mg/kg/day via intraperitoneal injection) produced a significant survival effect and we were awarded a patent for the use of this class of compounds in the treatment of neurodegenerative disease (U.S. Patent 6,372,727). The onset administration paradigm was optimized and these early results were repeated at a dose of 1 mg/kg/day and published together with a study which involved presymptomatic administration of FeTCPP (42). The mean survival interval (SI), which is the average time from onset to death in each group, was 25.8 ±2.2 days for the onset administration FeTCPP-treated group and 16.7 ±1.3 days for controls (42). Comparing the mean survival interval (SI) of the drug-treated group to the mean SI of the untreated control group of mice from the same generation provides a -fold extension of lifespan that is very useful for comparing results from one study to the next or from one lab to the next. Such normalization is essential, given the inherent variability in onset and survival of different groups of mice.

Mn Porphyrin at Symptom Onset Extends Survival in G93A Mice

Results with the iron porphyrin FeTCPP helped establish the precedent that administration at symptom onset could, indeed, prolong survival. A newer Mn porphyrin analog, AEOL10150, which reacted with peroxynitrite at least 10-fold faster than FeTCPP, was administered to G93A mice at the onset of muscle weakness in each mouse. (Mice were alternately assigned to control or drug-treated groups as they developed hindlimb weakness, so that the mean age of onset in each group was identical.) At a dose of 5 mg/kg/day (intraperitoneal injection), AEOL10150 extended survival 2.6-fold relative to untreated controls run concurrently (Figure 4, right panel); this represents a marked improvement over the 1.6-fold seen with FeTCPP (42) and these results have been found to be highly reproducible.

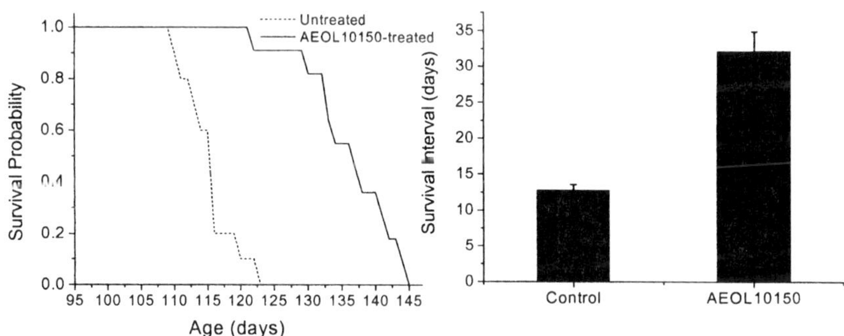

Figure 4. G93A Mice Treated with Manganese Porphyrin (AEOL10150) at Symptom onset. *Mn porphyrin (2.5 mg/kg/day following a 5.0 mg/kg loading dose on day one) was administered via intraperitoneal (IP) injection beginning on the first day of symptom onset in each mouse.* ***Left panel*** *shows classic Kaplan-Meier survival curves and* ***right panel*** *shows the mean survival interval (time from onset to death) in each group ± SE.*

The extension of lifespan was accompanied by a marked slowing of motor function loss. In almost all cases, drug-treated mice did not develop hindlimb paralysis—even during the period of extended survival; this contrasts sharply with untreated mice which typically did develop hindlimb paralysis within 12-14 days after onset. Thus, administration of porphyrin appeared to fundamentally alter the course of the disease

Cohort Studies Examining Mechanism of Action

Mouse studies which involve survival as an end-point are not suitable for comparing histologic, immunohistochemical, or biochemical indices; drug-treated mice may live up to three times longer but may eventually develop the same tissue pathology at end-stage disease as untreated mice. Thus, separate cohort studies must be carried out wherein drug-treated mice and age-matched controls are sacrificed at the same time points (after symptom onset).

Because untreated mice develop hindlimb paralysis as early as 12 days after onset, a time point of 10 days was chosen for tissue comparisons. At this time, control mice have lost significant motor function whereas drug-treated mice appear virtually the same as on the first day of onset. Histological examinations of spinal cord sections from cohort mice revealed better preservation of normal architecture and more neurons in Mn porphyrin-treated mice; this qualitative conclusion was verified quantitatively via counts of spinal cord neurons. Porphyrin-treated mice had 3489 ± 150 neruons/20 serial sections relative to 2131 ± 151 for untreated controls. Previous studies in both mice and humans have suggested that approximately 50% of spinal motor neurons must be lost before noticeable muscle weakness is seen. Once that threshold has been crossed, subsequent loss of even a few percent of the remaining motor neurons translates into significant new motor deficits.

Immunohistochemistry of Spinal Cord Tissue

Immunostaining was carried out on spinal cord sections of cohort mice using antibodies specific for nitrotyrosine—a marker of peroxynitrite, for glial fibrillary acidic protein (GFAP)—a marker of glial cell proliferation, and for heme oxygenase-1 (HO-1)—an enzyme associated with protection against oxidative injury. Immunostaining revealed that formation of protein-bound nitrotyrosine was largely inhibited by the porphyrin, particularly in the ventral horn region containing motor neuron cell bodies (Figure 5).

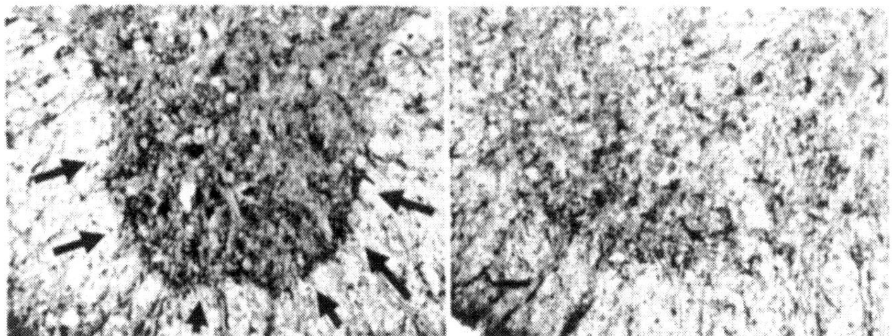

***Figure 5. Spinal Cord Sections Immunostained for Nitrotyrosine.** Mn Porphyrin (AEOL 10150)-treated and control cohort mice were sacrificed 10 days after symptom onset and spinal cord sections compared. **Left panel** shows nitrotyrosine immunoreactivity in spinal cord of G93A mouse and **right panel** is age-matched G93A mouse treated with Mn porphyrin. Note that staining in treated mouse (n = 7) is decreased relative to untreated (n = 6), particularly in region of ventral horn (see arrows), which contains motor neuron cell bodies.*

GFAP immunostaining was also lower in porphyrin-treated mice, suggesting that glial cells were sensing less injury to neurons[3]. Immunostaining for heme oxygenase-1 (HO-1) was more intense in porphyrin-treated mice suggesting that drug treatment somehow induced the expression of this protective enzyme. HO-1 (and other isoforms) carries out the initial ring-opening step in the mammalian metabolism of heme (iron protoporphyrin IX). Free heme is a signal to induce expression of HO-1 and it is possible that the Mn porphyrin is mimicking naturally-occurring heme in this regard.

Onset Administration of Mn Texaphyrin Extends Survival of G93A Mice

A Mn texaphyrin (Figure 1), which also possesses peroxynitrite scavenging activity, was tested at symptom onset in the G93A mice. The Mn texaphyrin has two polyethylene glycol functionalities which provide a good balance of water and lipid solubility and may aid in penetration into the central nervous system. Onset administration of 2.5 mg/kg/day (via IP injection) of Mn texaphyrin resulted in a 2.8-fold extension of lifespan of the G93A mice (Figure 6, right panel).

---

[3] Glial cells surrounding the motor neurons proliferate and induce the expression of many proteins, including GFAP, in response to neuronal injury. This process is often referred to as "reactive gliosis".

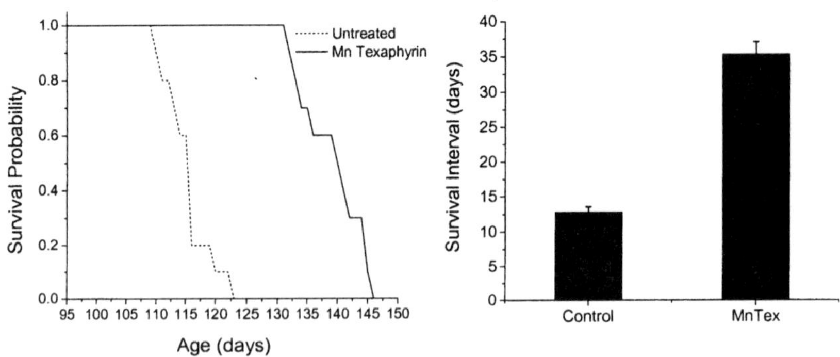

Figure 6. G93A Mice Treated with Manganese Texaphyrin at Symptom onset. *Mn Texaphyrin (2.5 mg/kg/day following 5.0 mg/kg loading dose on day one) was administered via intraperitoneal injection beginning on the first day of symptom onset in each mouse.* ***Left panel*** *shows classic Kaplan-Meier survival curves and* ***right panel*** *shows the survival interval (time from onset to death) in each group.*

Decreasing the frequency of Mn texaphyrin dosing to less than once per day significantly reduced the survival benefit. Aside from this, very little is known about the dose-response or dosing frequency required for efficacy with the Mn texaphyrin. However, it is clear that this compound produced a survival effect that ranks among the best ever reported—particularly for an onset administration paradigm—and that further studies are warranted.

Antioxidant Activity and Up-regulation of Cellular Defenses

Cohort studies with Mn texaphyrin have not yet been completed. However, preliminary HPLC analysis of end-stage Mn texaphyrin-treated mice indicate lower levels of nitrotyrosine in the spinal cord relative to untreated controls (Dr. Darren Magda, unpublished data 2003). Lowered nitrotyrosine levels, whether assessed quantitatively by HPLC or qualitatively by immunhistochemistry, are totally consistent with peroxynitrite scavenging by Mn porphyrin and Mn texaphyrin. However, indirect mechanisms involving activation of cellular defense and repair mechanisms may also be playing an important role.

Although the precise mechanism isn't known, induction of heme oxygenase-1 (HO-1) has been shown to correlate with the pre-conditioning response, wherein a mild oxidative stimulus pre-conditions the cell or tissue to survive a subsequent oxidative stress which would normally be lethal. Induction of HO-1 (also known as heat shock protein 32 or HSP32) by the Mn porphyrin (and likely the texaphyrin as well) suggests that the compound alone may be stimulating the neurons to protect themselves. HSP's are a family of chaperone proteins that help refold cellular proteins which become misfolded in response to many different types of stresses. Porphyrin analogs have been shown to induce expression of other heat shock proteins (HSP's) in other model systems (Dr. Balaraman Kalyanaraman, unpublished observations). Induction of other HSP's in neural tissue has not yet been examined, however, if it is occurring, this could help explain how they protect against SOD1 mutant-mediated toxicity, particularly with regard to the aggregation mechanism.

Summary: New Compounds With Novel, Catalytic Antioxidant Properties

Both the semi-planar, four coordinate Mn porphyrin and the non-planar five coordinate Mn texaphyrin are stable Mn compounds which are capable of using cellular reductants to catalytically decompose peroxynitrite. There is no indication that the *in vivo* effects of either compound are related to release of free Mn and the lack of therapeutic effect from administration of manganese chloride to G93A mice has been reported previously (44). In addition, we have seen no therapeutic effect of zinc or cobalt analogs of FeTCPP, suggesting that a redox active metal is required; such results are consistent with catalytic antioxidant properties.

The Mn porphyrin is also capable of decomposing hydrogen peroxide and possesses the highest SOD-like activity of any low molecular weight compound to date (30) (other potential antioxidant properties of the Mn texaphyrin have not yet been examined). While the catalase-like activity of the porphyrin could be important in the G93A model, it is highly unlikely that SOD-like activity is relevant, given the enormously high total SOD1 activity that already exists in this eight-fold SOD1 overexpressing mouse.

Summary of Mouse Results

Published results with the iron porphyrin FeTCPP (42) helped establish the precedent for extended survival with late, onset administration. Using the same experimental design, both the porphyrin and texaphyrin extend survival in the ALS mice considerably longer than FeTCPP. Indeed, the results reported herein compare favorably to the best results ever reported for a classical pharmacological agent—even those given presymptomatically. In addition, the course of the disease appears to be fundamentally altered: drug-treated mice rarely progress to total hindlimb paralysis, as compared to untreated mice which almost always do. Lowered nitrotyrosine levels in drug-treated mice are consistent with peroxynitrite scavenging by the porphyrin and the texaphyrin. In addition, indirect mechanisms involving activation of cellular defense (e.g., induction of HO-1) and repair mechanisms may also be playing an important role

Potential for Treating ALS in Humans

To the extent that the mouse results extrapolate to humans, this marked slowing of progression and preservation of motor function has profound implications for ALS patients. Mice live up to three times longer (after onset) with drug treatment and retain motor function at a level similar to the first day of onset until end-stage disease. Life expectancy for ALS patients ranges from 1-6 years after diagnosis, but the quality of life quickly diminishes due to rapid accumulation of multiple motor deficits. If ALS patients responded in a similar manner as the mice, it's possible that 3-18 years could be added to their life expectancy, during which time disease progression could be slowed to more manageable rates. The ability of these compounds to preserve motor function equivalent to that seen on the first day of onset suggests that early diagnosis and initiation of treatment will be essential to obtaining the best outcome.

Future Directions

Results seen thus far with the porphyrins and texaphyrin in the ALS mouse model suggest that these compounds may have some broad spectrum neuroprotective effects, with potential implications for other neurodegenerative diseases. Thus, it will be important to examine their effects in other disease models, particularly those where acute and/or chronic oxidative stress is known to play a role. Also, the essential pharmacophore needs to be identified by testing structural analogs possessing different central metal ions. Once the pharmacophore is established, functional groups can be varied to optimize pharmacological properties. These compounds are extremely well-suited for this type of structural optimization in that distal functional groups can be manipulated to alter biophysical properties without altering the central therapeutic properties.

References

1. M. Strong and J. Rosenfeld. Amyotrophic lateral sclerosis: a review of current concepts. Amyotroph.Lateral.Scler.Other Motor Neuron Disord. 4 (3):136-143, 2003.
2. K. E. Morrison. Therapies in amyotrophic lateral sclerosis-beyond riluzole. Curr.Opin.Pharmacol. 2 (3):302-309, 2002.
3. D. R. Rosen, T. Siddique, D. Patterson, D. A. Figlewicz, P. Sapp, A. Hentati, D. Donaldson, J. Goto, J. P. O'Regan, H. X. Deng, and . Mutations in Cu/Zn superoxide dismutase gene are associated with familial amyotrophic lateral sclerosis. Nature 362 (6415):59-62, 1993.
4. J. S. Beckman, A. G. Estevez, J. P. Crow, and L. Barbeito. Superoxide dismutase and the death of motoneurons in ALS. Trends Neurosci. 24 (11 Suppl):S15-S20, 2001.
5. M. E. Gurney, H. Pu, A. Y. Chiu, M. C. Dal Canto, C. Y. Polchow, D. D. Alexander, J. Caliendo, A. Hentati, Y. W. Kwon, H. X. Deng, and . Motor neuron degeneration in mice that express a human Cu,Zn superoxide dismutase mutation. Science 264 (5166):1772-1775, 1994.
6. T. L. Munsat and D. Hollander. Excitotoxins and amyotrophic lateral sclerosis. Therapie 45 (3):277-279, 1990.
7. J. D. Rothstein. Excitotoxic mechanisms in the pathogenesis of amyotrophic lateral sclerosis. Adv.Neurol. 68:7-20, 1995.
8. M. E. Gurney, F. B. Cutting, P. Zhai, A. Doble, C. P. Taylor, P. K. Andrus, and E. D. Hall. Benefit of vitamin E, riluzole, and gabapentin in a transgenic model of familial amyotrophic lateral sclerosis. Ann.Neurol. 39 (2):147-157, 1996.
9. R. G. Miller, J. D. Mitchell, M. Lyon, and D. H. Moore. Riluzole for amyotrophic lateral sclerosis (ALS)/motor neuron disease (MND). Amyotroph.Lateral.Scler.Other Motor Neuron Disord. 4 (3):191-206, 2003.
10. M. F. Beal. Mitochondria and oxidative damage in amyotrophic lateral sclerosis. Funct.Neurol. 16 (4 Suppl):161-169, 2001.
11. E. P. Simpson, A. A. Yen, and S. H. Appel. Oxidative Stress: a common denominator in the pathogenesis of amyotrophic lateral sclerosis. Curr.Opin.Rheumatol. 15 (6):730-736, 2003.
12. G. Kuther and A. Struppler. Therapeutic trial with N-acetylcysteine in amyotrophic lateral sclerosis. Adv.Exp.Med.Biol. 209:281-284, 1987.
13. O. A. Andreassen, A. Dedeoglu, P. Klivenyi, M. F. Beal, and A. I. Bush. N-acetyl-L-cysteine improves survival and preserves motor performance in an animal model of familial amyotrophic lateral sclerosis. Neuroreport 11 (11):2491-2493, 2000.

14. G. P. Paraskevas, E. Kapaki, G. Libitaki, C. Zournas, I. Segditsa, and C. Papageorgiou. Ascorbate in healthy subjects, amyotrophic lateral sclerosis and Alzheimer's disease. Acta Neurol.Scand. 96 (2):88-90, 1997.
15. V. C. Stewart and S. J. Heales. Nitric oxide-induced mitochondrial dysfunction: implications for neurodegeneration. Free Radic.Biol.Med. 34 (3):287-303, 2003.
16. C. Szabo. Multiple pathways of peroxynitrite cytotoxicity. Toxicol.Lett. 140-141:105-112, 2003.
17. R. Kissner, T. Nauser, P. Bugnon, P. G. Lye, and W. H. Koppenol. Formation and properties of peroxynitrite as studied by laser flash photolysis, high-pressure stopped-flow technique, and pulse radiolysis volume 10, number 11, november 1997, pp 1285-1292. Chem.Res.Toxicol. 11 (5):557, 1998.
18. J. S. Beckman and J. P. Crow. Pathological implications of nitric oxide, superoxide and peroxynitrite formation. Biochem.Soc.Trans. 21 (2):330-334, 1993.
19. C. Ducrocq, B. Blanchard, B. Pignatelli, and H. Ohshima. Peroxynitrite: an endogenous oxidizing and nitrating agent. Cell Mol.Life Sci. 55 (8-9):1068-1077, 1999.
20. L. A. MacMillan-Crow and D. L. Cruthirds. Invited review: manganese superoxide dismutase in disease. Free Radic.Res. 34 (4):325-336, 2001.
21. L. A. MacMillan-Crow, J. P. Crow, J. D. Kerby, J. S. Beckman, and J. A. Thompson. Nitration and inactivation of manganese superoxide dismutase in chronic rejection of human renal allografts. Proc.Natl.Acad.Sci.U.S.A 93 (21):11853-11858, 1996.
22. L. A. MacMillan-Crow, J. P. Crow, and J. A. Thompson. Peroxynitrite-mediated inactivation of manganese superoxide dismutase involves nitration and oxidation of critical tyrosine residues. Biochemistry 37 (6):1613-1622, 1998.
23. Li, T. T. Huang, E. J. Carlson, S. Melov, P. C. Ursell, J. L. Olson, L. J. Noble, M. P. Yoshimura, C. Berger, P. H. Chan, and . Dilated cardiomyopathy and neonatal lethality in mutant mice lacking manganese superoxide dismutase. Nat.Genet. 11 (4):376-381, 1995.
24. T. P. Misko, M. K. Highkin, A. W. Veenhuizen, P. T. Manning, M. K. Stern, M. G. Currie, and D. Salvemini. Characterization of the cytoprotective action of peroxynitrite decomposition catalysts. J.Biol.Chem. 273 (25):15646-15653, 1998.
25. S. Cuzzocrea, B. Zingarelli, G. Costantino, and A. P. Caputi. Beneficial effects of Mn(III)tetrakis (4-benzoic acid) porphyrin (MnTBAP), a superoxide dismutase mimetic, in carrageenan-induced pleurisy. Free Radic.Biol.Med. 26 (1-2):25-33, 1999.

26. S. Z. Imam, J. P. Crow, G. D. Newport, F. Islam, W. Slikker, Jr., and S. F. Ali. Methamphetamine generates peroxynitrite and produces dopaminergic neurotoxicity in mice: protective effects of peroxynitrite decomposition catalyst. Brain Res. 837 (1-2):15-21, 1999.
27. Stern, M. K., Jensen, M. J and Kramer, K. (1996) Peroxynitrite decomposition catalysts; J. Am. Chem. Soc. 118; 8735-8736.
28. J. B. Lee, J. A. Hunt, and J. T. Groves. Rapid Decomposition Of Peroxynitrite By Manganese Porphyrin- Antioxidant Redox Couples. Bioorganic &.Medicinal.Chemistry.Letters. 7 (22):2913-2918, 1997.
29. J. P. Crow. Manganese and iron porphyrins catalyze peroxynitrite decomposition and simultaneously increase nitration and oxidant yield: implications for their use as peroxynitrite scavengers in vivo. Arch.Biochem.Biophys. 371 (1):41-52, 1999.
30. G. Ferrer-Sueta, D. Vitturi, I. Batinic-Haberle, I. Fridovich, S. Goldstein, G. Czapski, and R. Radi. Reactions of manganese porphyrins with peroxynitrite and carbonate radical anion. J.Biol.Chem. 278 (30):27432-27438, 2003.
31. J. P. Crow. Peroxynitrite scavenging by metalloporphyrins and thiolates. Free Radic.Biol.Med. 28 (10):1487-1494, 2000.
32. R. Shimanovich, S. Hannah, V. Lynch, N. Gerasimchuk, T. D. Mody, D. Magda, J. Sessler, and J. T. Groves. Mn(II)-texaphyrin as a catalyst for the decomposition of peroxynitrite. J.Am.Chem.Soc. 123 (15):3613-3614, 2001.
33. A. G. Estevez, J. P. Crow, J. B. Sampson, C. Reiter, Y. Zhuang, G. J. Richardson, M. M. Tarpey, L. Barbeito, and J. S. Beckman. Induction of nitric oxide-dependent apoptosis in motor neurons by zinc-deficient superoxide dismutase. Science 286 (5449):2498-2500, 1999.
34. R. Rakhit, P. Cunningham, A. Furtos-Matei, S. Dahan, X. F. Qi, J. P. Crow, N. R. Cashman, L. H. Kondejewski, and A. Chakrabartty. Oxidation-induced misfolding and aggregation of superoxide dismutase and its implications for amyotrophic lateral sclerosis. J.Biol.Chem. 277 (49):47551-47556, 2002.
35. Rodriguez, P. J. Hart, L. J. Hayward, J. S. Valentine, and S. S. Hasnain. The structure of holo and metal-deficient wild-type human Cu, Zn superoxide dismutase and its relevance to familial amyotrophic lateral sclerosis. J.Mol.Biol. 328 (4):877-891, 2003.
36. J. P. Crow, J. B. Sampson, Y. Zhuang, J. A. Thompson, and J. S. Beckman. Decreased zinc affinity of amyotrophic lateral sclerosis-associated superoxide dismutase mutants leads to enhanced catalysis of tyrosine nitration by peroxynitrite. J.Neurochem. 69 (5):1936-1944, 1997.
37. T. J. Lyons, H. Liu, J. J. Goto, A. Nersissian, J. A. Roe, J. A. Graden, C. Cafe, L. M. Ellerby, D. E. Bredesen, E. B. Gralla, and J. S. Valentine. Mutations in copper-zinc superoxide dismutase that cause amyotrophic

lateral sclerosis alter the zinc binding site and the redox behavior of the protein. Proc.Natl.Acad.Sci.U.S.A 93 (22):12240-12244, 1996.

38. Y. Christen. [Proteins and mutations: a new vision (molecular) of neurodegenerative diseases]. J.Soc.Biol. 196 (1):85-94, 2002.
39. J. D. Wood, T. P. Beaujeux, and P. J. Shaw. Protein aggregation in motor neurone disorders. Neuropathol.Appl.Neurobiol. 29 (6):529-545, 2003.
40. A. Okado-Matsumoto and I. Fridovich. Amyotrophic lateral sclerosis: a proposed mechanism. Proc.Natl.Acad.Sci.U.S.A 99 (13):9010-9014, 2002.
41. M. Patel and B. J. Day. Metalloporphyrin class of therapeutic catalytic antioxidants. Trends Pharmacol.Sci. 20 (9):359-364, 1999.
42. A. S. Wu, M. Kiaei, N. Aguirre, J. P. Crow, N. Y. Calingasan, S. E. Browne, and M. F. Beal. Iron porphyrin treatment extends survival in a transgenic animal model of amyotrophic lateral sclerosis. J.Neurochem. 85 (1):142-150, 2003.
43. L. Brasile, R. Buelow, B. M. Stubenitsky, and G. Kootstra. Induction of heme oxygenase-1 in kidneys during ex vivo warm perfusion. Transplantation 76 (8):1145-1149, 2003.
44. C. Jung, Y. Rong, S. Doctrow, M. Baudry, B. Malfroy, and Z. Xu. Synthetic superoxide dismutase/catalase mimetics reduce oxidative stress and prolong survival in a mouse amyotrophic lateral sclerosis model. Neurosci.Lett. 304 (3):157-160, 2001.
45. Texaphyrins: Synthesis and Development of a Novel Class of Therapeutic Agents: Mody, T. D.; Fu, L.; Sessler, J. L. In Progress in Inorg. Chem. Karlin, K. D. (Ed.), Chichester: John Wiley & Sons, Ltd. 2001, 551-598.
46. H. Peluffo, J. J. Shacka, K. Ricart, C. G. Bisig, L. Martinez-Palma, O. Pritsch, A. Kamaid, J. P. Eiserich, J. P. Crow, L. Barbeito, and A. G. Estevez. Induction of motor neuron apoptosis by free 3-nitro-L-tyrosine. *J.Neurochem.* 89 (3):602-612, 2004.

Chapter 18

Salen Manganese Complexes: Multifunctional Catalytic Antioxidants Protective in Models for Neurodegenerative Diseases of Aging

Susan R. Doctrow[1], Michel Baudry[2], Karl Huffman[1], Bernard Malfroy[1], and Simon Melov[3]

[1]Eukarion, Inc., 6F Alfred Circle, Bedford, MA 01730
[2]Neuroscience Program, University of Southern California, Los Angeles, CA 90095
[3]Buck Institute for Age Research, 8001 Redwood Boulevard, Novato, CA 94945

Abstract

Salen Mn complexes are SOD and catalase mimetics which also have activity against reactive nitrogen species. These properties make them well suited as potential therapeutic agents for complex diseases involving oxidative stress. Structure-activity relationship research has yielded compounds with widely varying catalase and cytoprotective activities, but no differences in SOD activity. Further chemical modification has addressed properties such as lipophilicity and stability. Much evidence implicates mitochondrial dysfunction and oxidative stress in neurodegenerative diseases and aging. Salen Mn complexes are effective as "mito-protective" agents, suppressing brain mitochondrial oxidative injury and alleviating neurological impairments in animal models for neurodegeneration and age-associated decline.

Reactive oxygen species (ROS) are potentially toxic oxygen metabolites formed as a normal byproduct of oxidative metabolism and accumulating in many pathological conditions (reviewed in *1*). The term "oxidative stress" (*2*), describes a situation where ROS levels in a tissue exceed the neutralization capacity of endogenous antioxidant defense systems. A vast literature supports the deleterious roles of ROS in a broad array of diseases, including autoimmune, inflammatory, and ischemic diseases as well as organ failure from sepsis or trauma. Furthermore, there is substantial evidence that mitochondrial dysfunction, and the oxidative stress that results, plays a key role in such neurodegenerative diseases as Alzheimer's Disease, Parkinson's Disease, and amyotrophic lateral sclerosis. Therapy of such diseases is an unmet medical need, and will become more pressing as the population ages. Thus, therapeutic approaches aimed at reducing oxidative stress, particularly those effective in the mitochondria, would be a welcome addition to the current arsenal used against various forms of neurodegeneration. In this monograph, we describe a class of synthetic inorganic compounds that have catalytic ROS scavenging activity, focusing on their potential benefit for neurodegenerative diseases.

Search for Synthetic Superoxide Dismutase Mimetics

Superoxide dismutases (SODs), with active-site Mn or Cu and Zn, are key cellular antioxidant enzymes. Because of their presumed role in endogenous antioxidant defense, there has been much effort to use SOD, primarily CuZnSOD from bovine or recombinant sources, in therapeutic applications. But proteins have potential limitations as drugs, including difficulties in production and formulation, potential immunogenicity, and poor bioavailability. Such concerns, as well as the lack of clinical success with protein SODs, have fueled efforts to develop synthetic metal coordination complexes as SOD mimetics. While some Cu complexes had been considered (*3*), investigators from Fridovich's laboratory used Mn complexes, in part because of the lesser toxicity of this metal. They demonstrated this concept by describing the SOD activity of Mn complexed to desferrioxamine (*4*). Two of us, Malfroy and Baudry, founded a biotechnology company known as Eukarion, envisioning broad therapeutic applications for synthetic antioxidant enzymes. In collaboration with the Fridovich laboratory, Baudry's laboratory demonstrated that the desferrioxamine Mn complex had neuroprotective activity in the rat (*5*). However, instability of desferrioxamine Mn complexes was a hindrance (*6*, *7*). In search of more pharmaceutically useful Mn complexes, Malfroy learned of salen Mn complexes, which had been synthesized by chemists over many decades. For example, certain chiral salen Mn complexes were described by Jacobsen and colleagues (*8*) as industrial catalysts for epoxidation. A series of these compounds were tested and found to have SOD activity (*6*). Of these, one was selected for further pharmacological investigation, primarily because, while not the most active SOD of the series, it was nonetheless quite active and, in addition, non-chiral and water-soluble. This unsubstituted salen Mn complex (C7, (*6*)). re-designated as "EUK-8" by Eukarion, became the template for new complexes optimized for pharmaceutical properties, including formulation, stability, catalytic activities, and compatibility with biological systems. This structure-activity relationship research will be further described below. Meanwhile, as work on salen Mn complexes as potential therapeutics was underway, other researchers independently described the SOD activities of macrocyclic Mn complexes (*9*) and Mn containing metalloporphyrins (*10*).

Early Studies Testing EUK-8 in Biological Models for Disease

Shortly after the SOD activity of EUK-8 was reported, a number of biological studies examined whether this prototype could protect tissues against oxidative stress. The first used *ex vivo* biological models simulating tissue injury in acute ischemic or chronic neurodegenerative diseases. For example, Musleh *et al.* (*11*) showed that EUK-8 prevented synaptic injury and lipid peroxidation in rat hippocampal slices subjected to anoxia followed by reoxygenation. Another *ex vivo* study showed that iron-overloaded rat hearts perfused with EUK-8 were resistant to functional and morphological damage induced by global ischemia followed by reperfusion (*12*). In a finding of potential relevance to Alzheimer's disease, Bruce *et al.* (*13*). demonstrated that EUK-8 reduced oxidative stress and promoted neuronal survival in organotypic rat hippocampal slice cultures exposed to neurotoxic beta-amyloid peptides. Next followed a variety of studies testing the efficacy of EUK-8 *in vivo*. EUK-8, administered by intraperitoneal (ip) injection, prevented the development of paralysis in murine EAE, a well-recognized model for multiple sclerosis (*14*). Infused intravenously, the compound preserved lung function and reduced tissue edema and oxidative injury in a highly stringent pig model for the Adult Respiratory Distress Syndrome (*15*).

Salen Mn Complexes Scavenge Multiple Reactive Species

Although, salen Mn complexes for therapeutic use were originally conceived as SOD mimetics, it soon became clear that EUK-8 also exhibited catalase activity, the ability to metabolize hydrogen peroxide (*15*). The catalase activity of EUK-8 was not unexpected, since Mn porphyrins had been studied as catalase models by the Meunier laboratory (*16*) and, like the porphyrins, salen ligands form stable complexes with Mn(III) (*6*). As described previously (*17*), similar to that of mammalian heme-iron based catalases (*18*), the catalase activity of salen Mn complexes is not saturable with respect to hydrogen peroxide. As has been reported for protein catalases (*18*), salen Mn complexes exhibit peroxidase activity, in the presence of an electron donor substrate, as an alternative to a catalatic pathway. This supports the analogy between the behavior of these mimetics and that of catalase enzymes, and is consistent with the following mechanistic scheme (*16, 17*):

$$\text{(ligand)Mn(III)} + H_2O_2 \rightarrow \text{(ligand)Mn(V)=O} + H_2O \quad (1)$$

$$(\text{ligand})\text{Mn(V)=O} + H_2O_2 \rightarrow (\text{ligand})\text{Mn(III)} + O_2 + H_2O \quad (2)$$

$$(\text{ligand})\text{Mn(V)=O} + AH_2 \rightarrow (\text{ligand})\text{Mn(III)} + A + H_2O \quad (3)$$

Catalase reaction (1 + 2):
salen-ligandMn(III) + $2H_2O_2 \rightarrow$ salen-ligandMn(III) + O_2 + $2H_2O$

Peroxidase reaction (1 + 3):
salen-ligandMn(III) + H_2O_2 + AH_2 $\rightarrow$ salen-ligandMn(III) + A + $2H_2O$

In this scheme, (ligand)Mn(V)=O is functionally analogous to "Compound I", the intermediate first described for the heme catalases (*18*). When salen Mn complexes are incubated with hydrogen peroxide, they undergo a spectral change consistent with formation of an oxidized intermediate and an analogous oxomanganese species has been proposed as the active intermediate in oxidative reactions by porphyrin Mn (*16*) and salen Mn (*19*) complexes.

A marked difference between the peroxidase activity of salen Mn complexes and protein catalases is the broader substrate-specificity of the synthetic complexes, a difference likely due to steric factors. Bovine tetrameric catalase recognizes primarily short-chain alcohols, while the dissociated subunits are able to oxidize larger substrates (*20, 18*). For convenience, the peroxidase activity of salen Mn complexes can be monitored with colorimetric substrates not recognized by protein catalases (*21, 17*).

Reactive nitrogen species (RNS), in particular peroxynitrite, have been implicated in many of the same disease processes as ROS, including inflammatory and neurodegenerative pathologies. Nitric oxide is produced in such circumstances through inducible enzymes that are regulated via classic proinflammatory/stress-induced pathways. Nitric oxide and superoxide combine to form peroxynitrite, a cytotoxic, tissue-damaging agent that oxidizes protein residues, in particular, forming 3-nitrotyrosine derivatives that have been detected, for example, in aged and diseased brain tissues (*22*). Metalloporphyrins are known to scavenge peroxynitrite (*23,24,25*) as are another class of Mn complexes bound to expanded porphyrin ligands known as texaphyrins (*26*). Even mammalian catalase has been shown to scavenge nitric oxide in a hydrogen-peroxide dependent fashion (*27*). Indirect evidence from cytoprotection and other *in vitro* studies indicated that salen Mn complexes can do so as well (*28*). Sharpe et al. (*29*) confirmed this with direct demonstration that salen Mn complexes scavenge peroxynitrite and other RNS *in vitro*, through mechanisms mediated by an oxomanganese intermediate, entirely analogous to the catalase/peroxidase schemes presented above.

Thus, salen Mn complexes, as well as Mn porphyrins, are able to target multiple potentially injurious reactive agents, including superoxide, hydrogen peroxide, and peroxynitrite. Unlike the Mn(III) complexes, the macrocyclic Mn(II) complexes, as exemplified by compounds such as M40401 and M40403(*30*), are reported to be highly selective for superoxide (*30*). The relative advantages of a selective SOD mimetic and a multi-functional scavenger for therapeutic use have been debated, with one cited advantage being that a selective SOD mimetic could be used to probe the role of superoxide in disease (*30*). However, there is good reason to believe that added activity against other reactive species, in particular hydrogen peroxide and peroxynitrite, would confer significant therapeutic advantage. Much evidence exists to implicate these diffusible and highly cytotoxic agents in diseases of inflammation and oxidative stress, for example in neurodegenerative diseases. SOD activity alone can decrease peroxynitrite through removing one of its components, superoxide. However, the SOD reaction produces hydrogen peroxide, implying that in some circumstances SOD alone might enhance toxicity. For example, whereas EUK-8, as cited above (*11*), protected brain slices from anoxia/reoxygenation, bovine SOD potentiated the injury (*21*). Similarly, EUK-8 protected an intestinal cell model against oxidative injury during lactic acidosis, while bovine SOD exacerbated injury unless administered in combination with bovine liver catalase (*31*, *21*). EUK-8 was protective in a rodent stroke model, but was significantly less effective than EUK-134, a 3'3-methoxy analog with greater catalase, but equivalent SOD, activity (*21*).

Such observations suggest that compounds exhibiting both SOD and catalase activities, along with the ability to scavenge RNS, should have distinct, qualitative advantages over those having only SOD activity. These include not only the potential to inactivate numerous damaging species but also the reduced likelihood of exacerbating injury through generation of hydrogen peroxide. It should also be considered, however, that superoxide, hydrogen peroxide, and nitric oxide have all been implicated in cellular signaling. Thus, the potential exists, as well, that any of these agents could interrupt normal biological processes. Furthermore, it is possible that the protection offered by these agents in disease models occurs via activities other than those for which they were originally screened, or by intervention at multiple "layers" of pathological regulation. In support of the latter concept, treatment with a salen Mn complex suppresses activation of the proinflammatory transcriptional factors AP-1 and NF kappa B in the brain of rats subjected to an excitotoxic insult (*33*). These "stress-activated" transcriptional factors are known to be regulated by cellular redox state (*32, 33*). This implies that, by interrupting stress-associated events at the level of transcription, synthetic catalytic antioxidants can have profound effects on disease-associated proteomics. Overall, with the enormously complex array of biochemical processes that participate in normal and abnormal physiology, it is clear that multiple screening criteria must be employed, and that, ultimately, empirical demonstrations of efficacy and toxicity will dictate the usefulness of a given metal coordination complex in medicine. Also, it is quite likely that different disease indications will have requirements that favor different types of mimetics. With this level of complexity in mind, we now review structure-activity (SAR) research conducted with salen Mn complexes.

Structure-Activity Relationship (SAR) Research Guided by Multiple Parameters

Beginning with EUK-8 as a template, many salen-Mn complexes were synthesized and evaluated for their catalytic and other pharmacologically relevant properties. Of several metal substitutions evaluated with the EUK-8 ligand, only the Mn complexes had catalase activity, while Co and VO complexes also showed some superoxide scavenging properties. Axial ligand substitutions affected solubility, but not catalytic activities (*17*). Some of the most interesting effects occurred with substitutions on the salen rings (Table I). In particular, alkoxy substituents at the 3,3' and 5, 5' positions increased catalase activity, while those at the 4,4' and 6,6' decreased it. As a result of these and other findings, EUK-134, a 3',3'-methoxy- substituted analog, was advanced as a new prototype for study in biological models (*34, 35*). EUK-134 was not only a more active catalase than EUK-8, but was greatly more effective at protecting human fibroblasts from toxicity by extracellular hydrogen peroxide (Fig 1A). In contrast, both compounds were fully effective at protecting cells from toxicity by the peroxynitrite generator sin-1 (Fig 1B). Modification of the "bridge" region of the salen ligand, by replacing ethylenediamine moiety with various other structures (Fig 2), had enormous impact (Table II), with aromatic bridges conferring substantially greater catalase activity. Despite their broad range of effects on catalase activity, these structural modifications of the EUK-8 prototype, with one exception, had no effect on SOD activity (*17*).

Table I. Structures and Catalase Activities of EUK-8 Analogs

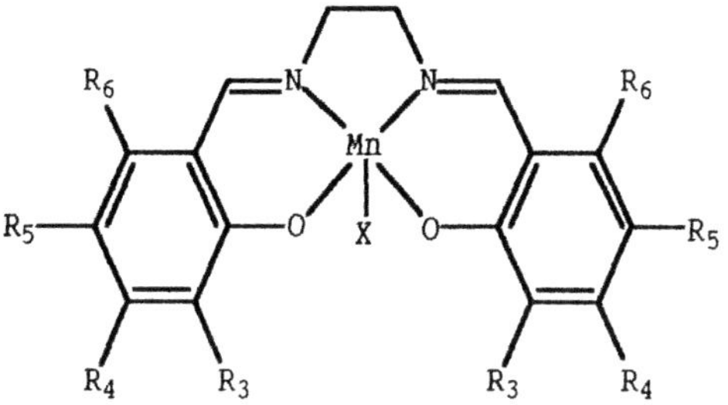

Compound	*Substituents (R3-R6)*	*X*	*Catalase activity Rate*	*Endpoint*
EUK-8	none	Cl	148 ± 27	27 ± 6
EUK-108	none	OAc	149 ± 21	25 ± 4
EUK-122	3-F	Cl	201 ± 39	28 ± 2
EUK-121	3-F	OAc	197 ± 27	28 ± 1
EUK-15	5-OMe	Cl	186 ± 2	84 ± 1
EUK-123	5-OMe	OAc	112 ± 14	55 ± 4
EUK-134	3-OMe	Cl	243 ± 18	81 ± 2
EUK-113	3-OMe	OAc	260 ± 45	79 ± 9
EUK-114	4-OMe	Cl	70 ± 8	11 ± 3
EUK-115	4-OMe	OAc	81 ± 14	12 ± 0.2
EUK-118	4,6-OMe	OAc	35 ± 6	9 ± 4
EUK-160	3,5-OMe	OAc	235 ± 47	57 ± 9
EUK-189	3-OEt	OAc	180 ± 7	80 ± 3

Where a substituent is not named, it is H. Abbreviations are OMe, methoxy; OEt, ethoxy; OAc, acetoxy. All substitutions are symmetrical. Catalase rate (μM O_2/min) and endpoints (maximal μM O_2) were measured as described (17). Data reprinted from reference 17 (copyright, American Chemical Society).

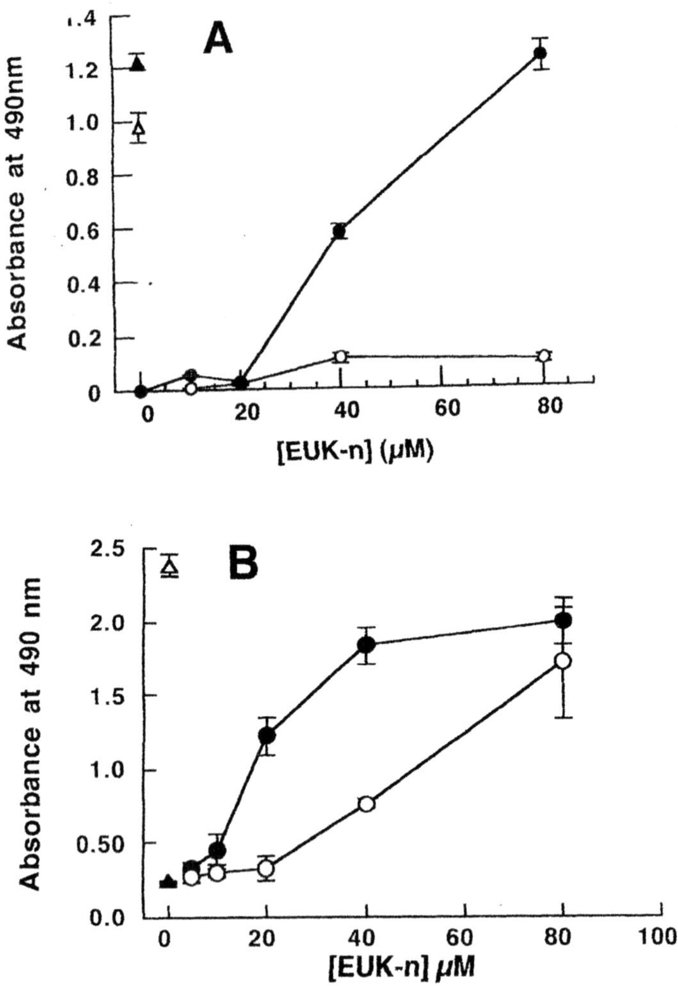

Fig. 1. Cytoprotective activities of EUK-8 and EUK-134. Cytoprotection assays were conducted as described previously (17, 28), with living cells detected by absorbance of the viability dye XTT. (A) Human fibroblasts exposed to hydrogen peroxide (generated by glucose and glucose oxidase, GO). All cells received GO except for the Control (open triangle). Other additions: bovine liver catalase (solid triangle), EUK-134 (solid circles), or EUK-8 (open circles). (B) PC12 cells exposed to Sin-1, with Control EUK-8, and EUK-134 as indicated for Fig 1A. Reprinted from reference 28, copyright (2003), with permission from World Scientific Publishing Co. Fig. 1A is reprinted, as well, from reference 34, copyright (1998), with permission from the American Society of Pharmacology and Experimental Therapeutics.

Table II. Catalase Activities of Bridge-Modified Analogs (Fig 3)

Compound	*Catalase rate ($\mu M\ O_2/min$)*	*Enpoint (maximal $\mu M\ O_2$)*
EUK-177	552 ± 131	119 ± 13
EUK-178	814 ± 155	230 ± 7
EUK-161	890 ± 144	206 ± 19
EUK-172	1073 ± 174	300 ± 14
EUK-159	251 ± 29	91 ± 5
EUK-163	36 ± 6	5 ± 1

Compounds are shown in Fig. 3 and assays conducted as for Table I. Data reprinted from reference 17 (copyright, American Chemical Society).

Because of the prominence of catalase activity in this SAR research, all compounds were evaluated for cytoprotection based on their ability to protect human cells from hydrogen peroxide toxicity, as described for Fig 1A. The most effective cytoprotection was observed with 3,3'- and 5,5' alkoxy- analogs and their potency correlated well with their *in vitro* catalase activity (as did the potency of bovine liver catalase used as positive control). EUK-161 and EUK-177, showed poor cytoprotective activity, despite their high catalase activities. Thus, both catalase activity and alkoxy substitution at the salen rings is required for cytoprotection in this model. Neither alone is sufficient, as evidenced, for example, by poor cytoprotective activity of EUK-163 and EUK-118. Toxicity, stability, or lipophilicity could not explain the advantages of the alkoxy-substituted analogs. Interestingly, when 9 analogs were compared for their effectiveness in a rodent stroke model, compounds with 3,3-alkoxy groups and higher catalase activity were, again, the most active (*17*). However, neither EUK-178 nor EUK-172 was more effective *in vivo* than compounds such as EUK-134. EUK-178 showed neurotoxicity at higher doses, consistent with its instability and formation of cytotoxic breakdown products in solution (*17*).

Altogether, such data illustrate that for these, as for any catalytic ROS scavenger, SAR is highly dependent on the nature of the model for assessing activity, including factors such as type of ROS (or RNS) generated, intra- or extracellular site of ROS insults, and specific cell types, including their available endogenous defense mechanisms. This is not at all surprising and, in fact, quite interesting. It demonstrates that the predictive value of any single *in vitro* model is limited and must be interpreted carefully in relationship to cellular and *in vivo* activity.

Pharmacokinetic and tissue distribution issues have considerable influence on in vivo activity, although they are not necessarily predicted by *in*

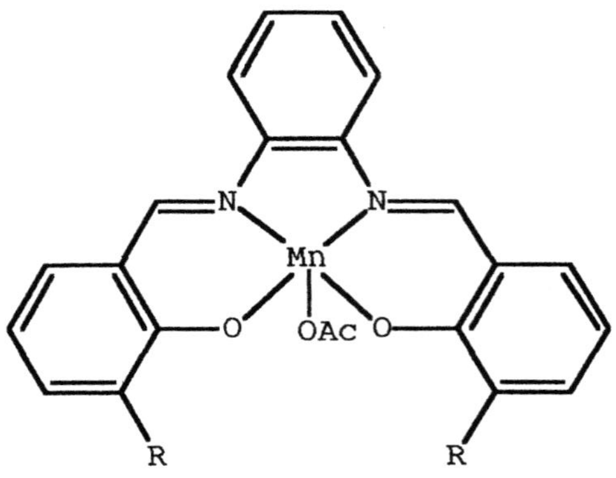

N
N
Mn
OAc
O
O
R
R
EUK-177: R = H
EUK-178: R = OMe 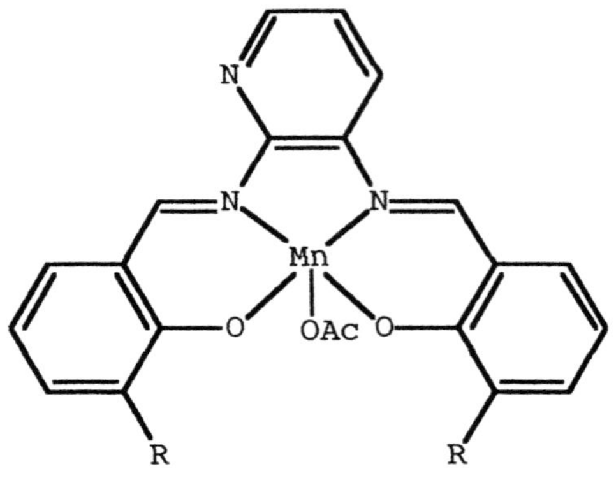
N
N
N
Mn
OAc
O
O
R
R
EUK-161: R = H
EUK-172: R = OMe

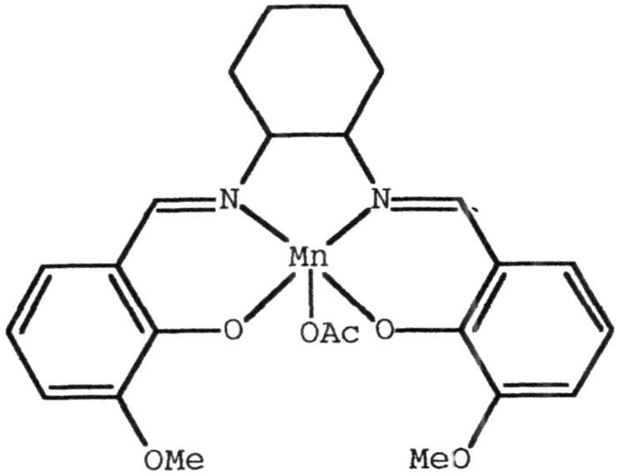

EUK-159

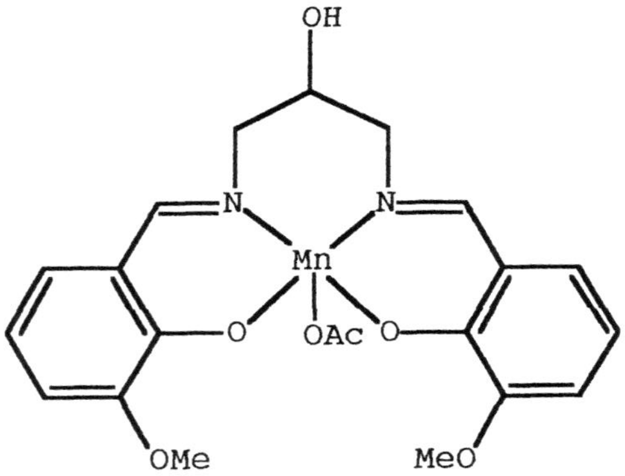

EUK-163

Fig. 2. Structures of bridge-modified salen Mn complexes. Reprinted from reference 17, copyright (2002), American Chemical Society.

vitro properties. Tracer "biodistribution" studies using radiolabelled compounds have indicated that salen Mn complexes injected into rodents distribute widely into tissues, including the brain. One SAR strategy we've taken is to make the compounds slightly more lipophilic, in an attempt to improve their intracellular and brain accessibility, without sacrificing their aqueous solubility properties. EUK-189 (Table I) is a 3, 3'-ethoxy substituted analog with catalytic properties equivalent to those of EUK-134 but about twice the lipophilicity, by solvent partitioning (*36*). The two compounds are equally effective in the extracellular hydrogen peroxide cytotoxity model discussed above. However, EUK-189 is about two orders of magnitude more potent than EUK-134 at preventing staurosporine-induced apoptosis in neurons (*37*), a cellular model involving mitochondrial injury and, hence, requiring intracellular access. As will be discussed further below, EUK-189 is also more effective than EUK-134 at rescuing, *in vivo*, a neurological syndrome resulting from mitochondrial oxidative stress. More lipophilic analogs, having longer-chain 3,3'-alkoxy substitutions, show reduced peroxidase activity but maintain catalase activity (*17*). Steric hindrance most likely explains this phenomenon, but the pharmacological consequences have not yet been studied.

The salen ligands employed in these studies form tightly bound, very stable complexes with Mn (*6, 17*). Nonetheless, structural modifications have been made to try to further enhance stability, with a goal of producing even more potent agents with improved pharmacokinetic properties. Among the most promising approaches to increasing the stability of these complexes has been addition of a lower bridge, essentially cyclizing the ligand. Research examining various lower bridge structures, manipulating length as well as composition, will not be reviewed here. The lead structure, exemplified by the analog EUK-207 (Fig 3), contains an 8-membered poly-ether ring. This compound retains compatibility with biological systems, as well as the catalytic and cytoprotective advantages of the 3,3'-alkoxy substitution. As compared to EUK-134 or EUK-189, it has a slightly higher catalase activity (rate = 330 ± 10 uM O_2/min), equivalent SOD activity, and is less lipophilic (*38*). It has been suggested that stability in the presence of bovine serum albumin or chelators such as EDTA are useful criteria for predicting the relative stability of metal coordination complexes in a biological system (*7, 6*). Salen Mn complexes are very resistant to these agents. Nonetheless, in the presence of a high molar excess of EDTA, some breakdown does occur over a period of hours to days, making this criterion useful to compare various analogs to one another. Table III compares the stability of 4 salen Mn complexes in the presence of ~100-fold molar excess of EDTA.

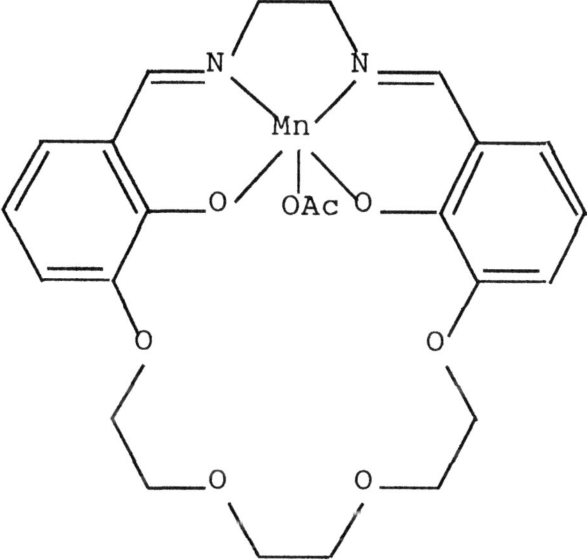

Fig. 3. Structure of EUK-207. Reprinted from reference 38, copyright (2003), with permission from the National Academy of Sciences, USA

Table III. Stability of Salen Mn Complexes in EDTA

Time (hr)	*% Compound remaining*			
	EUK-113	*EUK-189*	*EUK-178*	*EUK-207*
1.7	99.5	99.5	85.9	99.0
3.2	98.2	95.8	72.4	99.2
70	49.3	38.1	0.6	82.7

Compounds (220 uM) were incubated at room temperature with 25 mM EDTA at pH 7.4 and aliquots were withdrawn to measure the intact complex by HPLC-uv. At 70 hr, there was no detectable degradation in control solutions without EDTA for any compound except EUK-178 (33% remaining).

These data demonstrate the high stability of the original salen Mn complexes, EUK-189 and EUK-113, but also show that EUK-207 represents an improvement. In addition, the poor stability of the aromatic bridged analog EUK-178, discussed above, is apparent. Whether EUK-207, or other cyclized salen Mn compounds, will have substantial advantages over EUK-189 *in vivo* is under investigation. Ongoing pharmacokinetic studies indicate that it has a longer plasma half-life than EUK-189, but, with its lower lipophilicity, it may not distribute into tissues as well. The next section will include discussion of its protective activity in two animal models.

Salen Mn Complexes are "Mito-Protective" Agents to Treat Neurodegenerative Diseases and Aging

Approximately 0.4 to 4% of oxygen consumed during normal respiration is converted to superoxide within the mitochondria (*18, 39, 40*). The brain, which accounts for about 20% of the total oxygen consumption in humans while comprising only ~2% of body weight, is particularly vulnerable to oxidative stress (*41*). (Another particularly vulnerable organ, due to its high oxygen consumption, is the heart). Mitochondrial dysfunction, resulting from genetic modification or damage, can lead to accelerated ROS production, exacerbating cellular oxidative stress. There is much evidence implicating oxidative stress and mitochondrial dysfunction in Alzheimer's and Parkinson's diseases, amyotrophic lateral sclerosis (ALS), Friedreich Ataxia, prion disease and, indeed, in aging and age-associated decline (*42,43,44,45,46,47,48,49,50*).

The mitochondrion, as a primary source of ROS, is thus, itself, quite vulnerable to oxidative damage. Mn-SOD (sod2) is a mitochondrial form of the enzyme believed to play a key role in protecting the mitochondria from oxidative injury. CuZnSOD (sod1), by far the more abundant form of SOD in the cell, exists primarily in the cytosol, although it now has been detected, as well, in mitochondria (*51*). The importance of sod2 to the mitochondria and the severe implications of mitochondrial oxidative injury are illustrated dramatically by the effects of genetic deletions of the two SODs in mice. Mice deleted of sod1 exhibit a generally normal phenotype and lifespan, with some increased susceptibility to oxidative challenges and mild neuronal abnormalities (*52*, *53*). However, mice deleted of sod2 (sod2-/- mice) exhibit a profoundly severe phenotype. Mice from a strain developed by Li *et al.* (*54*) die within the first week of life from dilated cardiomyopathy, with hepatic lipid accumulation, mitochondrial enzymatic defects, DNA oxidative damage, and organic aciduria.

Studies conducted with these mice assessed whether treatment with synthetic catalytic antioxidants would alleviate these oxidative pathologies. In the first such study, mice were treated with a Mn-porphyrin catalytic antioxidant, MnTBAP (*55*). MnTBAP treatment approximately doubled the lifespan of the sod2-/- mice and ameliorated dilated cardiomyopathy and hepatic lipid accumulation. However, MnTBAP is believed not to cross the intact blood– brain barrier. In extending the life of sod2-/-mice beyond 2 weeks of age through protection of peripheral tissues, MnTBAP treatment uncovered a severe neurological disorder, attributable to the inability of MnTBAP to protect against ROS-induced mitochondrial injury within the brain. The neurological phenotype seen in the MnTBAP-treated sod2-/- mice is characterized by a severe disturbance in motor control, and the underlying neuropathology is that of spongiform change predominantly within the frontal cortex and focally in brainstem nuclei (*55*). An analogous neurological phenotype occurs in the small percentage of longer-lived"outlier" sod2-/- mice (Melov, unpublished) as well as in sod2-/- mice on certain other genetic backgrounds (*56*).

The next studies investigated whether salen Mn complexes could treat the neurological syndrome in sod2-/- mice. As compared to MnTBAP, EUK-8, EUK-134, and EUK-189 further increased survival, alleviated the severe ataxia and eliminated the spongiform pathology (Fig 4) in the brain (*36*). The cause of death in these treated mice appeared to be yet another unmasked neurological syndrome, one not characterized by spongiform histopathology. Importantly, biochemical studies on brain mitochondria isolated from sod2-/- and control mice demonstrate that salen Mn complexes significantly rescued mitochondrial enzymes, including cis-aconitase (*36*) and complexes I, II, III, and IV (*57*) from oxidative damage. This is consistent with their acting as brain permeable "mito-protective" agents.

Table IV summarizes survival data from studies conducted with sod2-/- mice. The 4 salen Mn complexes were effective at prolonging survival and alleviating the spongiform neurodegeneration. Mice treated with EUK-189 (30

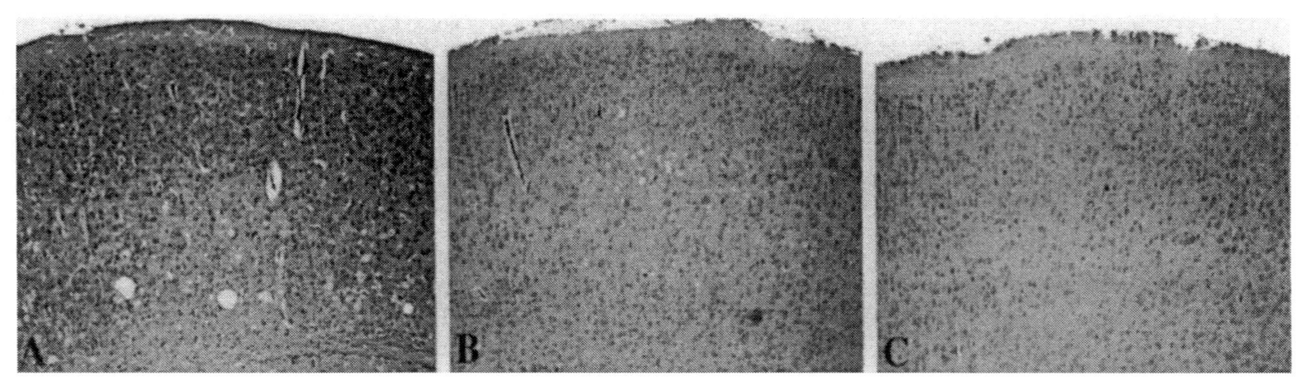

Fig. 4. Spongiform neural pathology in sod2-/- mice treated with catalytic antioxidants. The figure shows the cerebral cortex of 18 day old mice treated with A, MnTBAP (5 mg/kg); B, EUK-8 (1 mg/kg), and C, EUK-8 (30 mg/kg). Reprinted from reference 28, copyright (2003), with permission from World Scientific Publishing Co.

mg/kg) survived significantly longer than those treated with EUK-134 (*36*). Since, as discussed above, the two compounds have the same SOD and catalase activities, it is hypothesized that the greater lipophilicity of EUK-189 enhances its brain and mitochondrial accessibility in this model. EUK-207, the cyclized analog with improved stability properties, is at least as effective as EUK-189. In contrast to findings in the rodent stroke model (*34*), EUK-134 was not more effective than EUK-8 in sod2-/- mice (*36*). Because of the nature of the mutation, SOD activity might be expected to be more important than catalase for protection in this model. Nonetheless, as Table IV shows, the selective SOD mimetic M40403 was ineffective in these mice. Similarly, treatment with acetyl-L-carnitine (ALCAR) and lipoic acid (LA), mitochondrial metabolites reported to be neuroprotective in certain models (*58*), did not enhance the lifespan of sod2-/- mice. The factors required for neuroprotection in these mice, including brain and mitochondrial accessibility and suitable activity against ROS and/or RNS, remain to be fully defined. However, with the recognized importance of mitochondrial dysfunction and oxidative stress in neurodegenerative disease, sod2-/- mice are likely to have predictive value for identifying promising therapies for such disorders.

Table IV: Survival data for sod2-/- mice treated with synthetic catalytic antioxidants

Treatment (mg/kg)	*n*	*Median lifespan (days)*
Untreated	150	8
MnTBAP (5)	145	18**
EUK-8 (1)	109	19**
EUK-8 (30)	136	26**
EUK-134 (1)	8	21*
EUK-134 (30)	24	24**
EUK-189 (1)	79	18**
EUK-189 (30)	80	30**
EUK-207 (30)	4	30**
M40403 (10)	22	10 (NS)
ALCAR (400) + LA (150)	6	6*(-)

*These data summarize the results of Kaplan-Meier survival analysis as described (31). Statistical difference from untreated group: * ($p < 0.001$), ** ($p < 0.0001$). All mice were from a reduced litter size (36, 55). All received ip injections at the indicated dose, except for LA which was administered subcutaneously. Compared to the untreated group, (-) indicates reduction of lifespan and NS indicates no significant difference. This table represents an update that includes previously published data. ALCAR, acetyl-L-carnitine, LA, lipoic acid (58).*

Indeed, the ability of salen Mn complexes to reduce oxidative injury to mitochondria in sod2-/- mice and, thereby, to alleviate severe neurological consequences, implies that they should have beneficial effects in a number of neurodegenerative disease models and, potentially, in models for aging. Several studies have investigated this hypothesis, one of the earliest using the "MPTP model" in mice. Parkinson's disease (PD) is characterized by progressive loss of dopaminergic (DA) neurons in the striatum. The toxin MPTP, which metabolizes to the mitochondrially active agent 1-methyl-4-phenylpyridinium (MPP+), has long been known to cause selective DA neuronal loss and PD-like symptoms in humans and in rodent and primate experimental models. In a mouse MPTP model, EUK-134 (0.02 and 0.2 mg/kg-day, ip) prevented loss of DA neurons, while not affecting metabolism of MPTP to MPP^+ (*28*). A later study showed that, in DA neuronal cultures, EUK-134 (0.5 uM) prevented neuronal toxicity by MPP^+ or 6-hydroxy-dopamine and, in addition, inhibited tyrosine-nitration of tyrosine hydroxylase, a key enzyme in DA neurons (*59*).

Another neurodegenerative disease in which mitochondrial dysfunction has been implicated is amyotrophic lateral sclerosis (ALS) (*50*). The existing experimental model for the disease uses mice expressing certain sod1 mutations identified as causative in some forms of inherited ALS. As a "gain of function" mutation, the mutated sod1, by mechanisms yet to be elucidated, exerts neurotoxic effects leading to spinal cord oxidative stress and progressive loss of motoneurons. The ALS mice exhibit symptoms analogous to those of ALS patients, motor weakness progressing to total paralysis and death. (See the chapter by Hart and Valentine for more detailed discussion of these ALS mutations.) Jung et al. (*60*) treated an ALS mouse line, G93A "low expressors", with EUK-8 and EUK-134 (30 mg/kg, ip) from the age of 60 days until their death. Mean survival of the G93A mice was increased by 17 and 23 days in groups treated with EUK-8 and EUK-134, respectively. Survival after disease onset, as detected by an abrupt loss of muscle strength, was increased by 22% and 68%, respectively. In spinal cord samples, collected at death, indicators of oxidative stress, malonyldialdehyde and protein carbonyl content, were decreased in treated groups. In addition, treated groups had decreased staining for protein tyrosine-nitration in the motorneurons (Fig 5).

A recent study (*61*) examined the effects of salen Mn complexes on a mouse model for ataxia-telangiectasia (AT). AT is a genetic disease affecting young children, characterized by cerebellar degeneration, immunodeficiency, sterility, increased risk of cancer, and genomic instability. It is untreatable and patients survive about 20 years. The disease is caused by mutations in the ATM

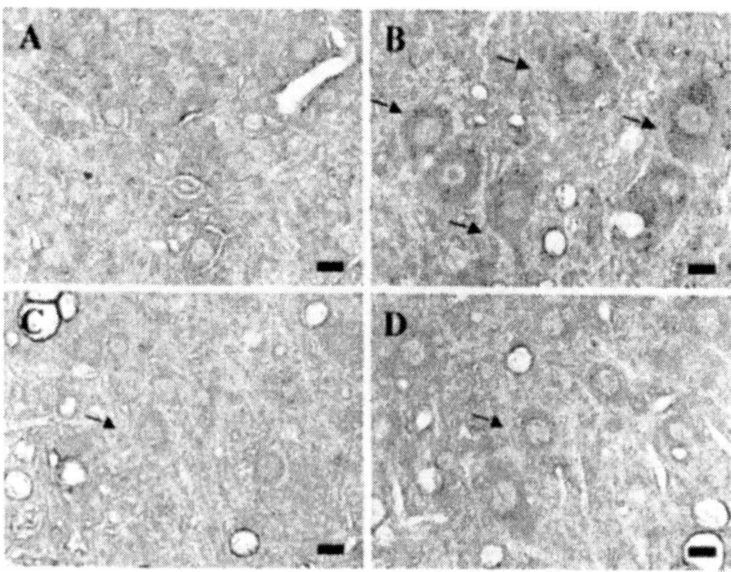

Fig. 5 Protein tyrosine-nitration in motorneurons from ALS mice. This experiment is described in detail by Jung et al. (ref). Spinal cord samples were stained with an antibody specific for nitrotyrosine. The spinal cord samples were obtained from: (A) wildtype mice receiving no treatment; or ALS mice (G93A) treated with (B) vehicle; (C) EUK-134; or D) EUK-8. Three animals from each group were analyzed with similar results. Motorneurons (arrows) show cytoplasmic staining indicative of protein nitration in ALS mice. Reprinted from reference 60, copyright (2001), with permission from Elsevier.

(ataxia-telangiectasia mutated) gene, a protein kinase "early-responder" to double strand DNA breaks. Mice with an inactivated ATM gene (Atm -/-) serve as a model that recapitulates most symptoms of the disease (*62*). These mice do not show overt neuronal degeneration, but they do develop cerebellar dysfunction, as demonstrated by their ataxic gait and by functional tests such as the rotarod test, where their ability to remain on a rotating rod is assessed. Failure to respond to DNA damage explains most of the symptoms in mice and humans, such as the propensity for cancer, but does not explain the neurological dysfunction. Through mechanisms yet to be determined, target organs of Atm -/- mice show increased oxidative stress (*63*). The existence of a link between this localized oxidative stress and cerebellar neuronal injury is now under active investigation. To test this hypothesis, and perhaps provide some insight into new treatments, Atm-/- mice were treated chronically with EUK-189. For efficient chronic delivery, EUK-189 was administered by continuous infusion via a subcutaneous pump (~1.2 mg/kg-day). EUK-189 fully corrected the neurobehavioral defect attributed to cerebellar dysfunction, as indicated by normalized performance on the rotarod test (Fig 6). Brains from the EUK-189-treated group had, in addition, significantly lower ($p=0.01$) levels of F_4-neurofurans and a marginal decrease ($p = 0.06$) in F_4-neuroprostanes, both products of nonenzymatic fatty acid oxidation. In the second of two experiments, mice were followed longer, to 5 months of age. While the study was not structured to test an increase in lifespan, the data indicated that EUK-189 did, indeed, extend lifespan in Atm-/- mice and warrants more rigorous testing of this hypothesis (Fig 7).

In many respects, the ultimate therapeutic challenge is aging itself. Oxidative stress has long been hypothesized as being a causative factor in age-related decline. Evidence indicates that factors such as antioxidant defenses, mitochondrial function, and stress-resistance become impaired in aging organisms. Thus, it has been tempting to speculate that efficient antioxidant therapies could enhance lifespan in addition to treating disease. Whether or not this is possible is still, for the most part, an open question. Nonetheless, salen Mn complexes are beneficial in certain models for aging. In one study (*64*), EUK-8 and EUK-134, added to the liquid culture medium of the nematode C Elegans, increased the lifespan of these worms by an average of about 50%. The compounds also increased survival of a C Elegans mutant strain, mev-1, having a mitochondrial defect producing increased oxidative stress. Another team of investigators (*65*) reported that EUK-8 did not increase C Elegans lifespan, using a slightly different experimental protocol. In a study conducted with houseflies, EUK-8 did not extend lifespan and the higher dose seemed instead to shorten lifespan (*66*). But, it was noted that the flies consumed less liquid food when it contained EUK-8 and, also, exhibited no decrease in tissue oxidative stress, consistent with lack of delivery of active compound. It is

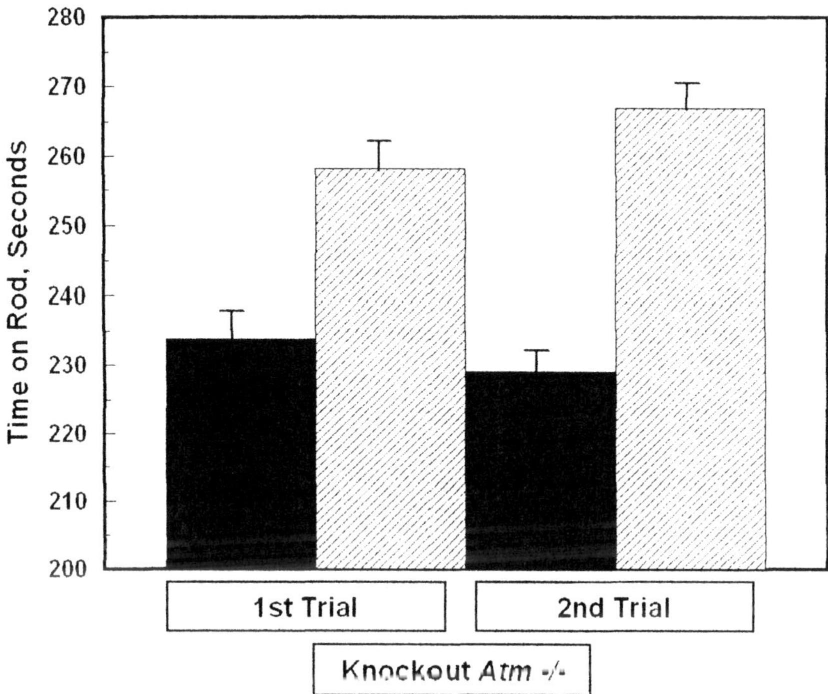

Fig. 6. Rotarod performance of Atm-/- mice. The length of time a mouse is able to stay on the rotating rod decreases as a consequence of cerebellar dysfunction in this model. Atm-/- mice were treated with vehicle (solid bars) or EUK-189 at ~1.2 mg/kg-day (hatched bars) continuously infused from 40 days of age for 56 (trial 1) or 84 (trial 2) days. Mice were tested twice weekly for rotarod performance and results were averaged, excluding the learning phase. The corresponding value for untreated wildtype mice (Atm +/+) was about 260 sec. Reprinted from reference 61, copyright (2004), with permission from Elsevier.

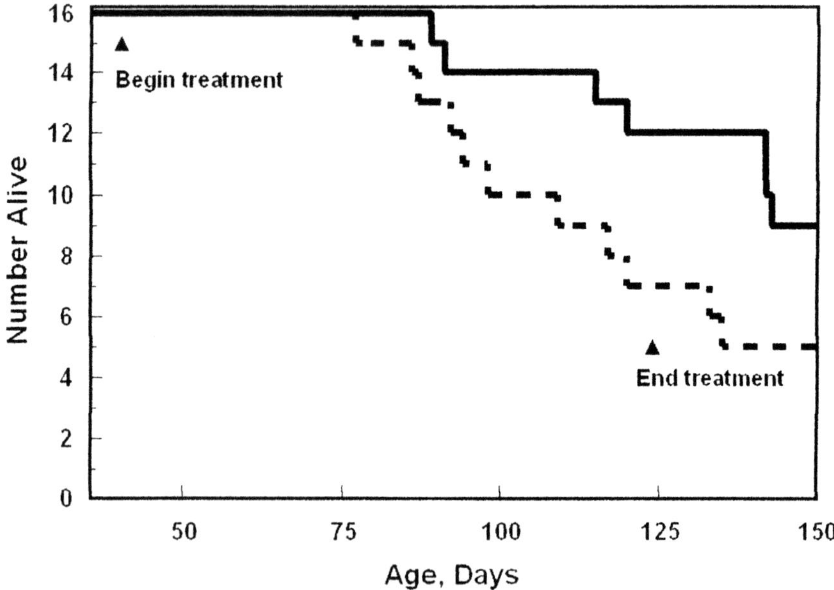

Fig. 7 . Life table analysis in Atm-/- mice. Treatment was as described for 7 (trial 2). Sold line, mice receiving EUK-189; broken line, mice receiving vehicle. By plan, the study was stopped at 150 days of age. All animals were sacrificed to confirm the presence of thymoma. The Cox-Mantel log rank test yielded p = 0.08. Reprinted from reference 61, copyright (2004), with permission from Elsevier.

apparent that pharmacological aging studies in these small organisms can be difficult to conduct and interpret, and are confounded by uncertainties in drug bioavailability. Thus, the most informative pharmacology studies are likely to be those in mammals, where delivery of salen Mn by injection or infusion has been well-established and is relatively controllable, as well as more relevant to potential methods of treating patients. These compounds are well-tolerated in mammals, enabling effective chronic administration. For example, no deleterious effects were noted in normal mice treated with EUK-8 (ip injection, ~ 25 mg/kg, 3 times/week) for about 7 months, or with EUK-189 (~2 mg/kg-day continuous infusion by subcutaneous pump) for 8 months. Formal toxicity studies conducted with EUK-189 further support a favorable profile. An orally-active formulation, of considerable advantage for chronic delivery, has not been established for these compounds and will require further study. This is an important issue that needs to be addressed for all metal coordination complexes intended as drugs.

Mammalian studies to address the effects of chronically administered salen Mn complexes on aging are underway, including lifespan and genomics studies in mice (Melov et al., in progress). Kregel's laboratory has found that aged rats, after treatment for 28 days with continuously-infused EUK-189 (~0.15 mg/kg-day), are resistant to severe oxidative liver injury induced by exposure to a mild heat-stress (*67, 68*). Liu et al. (*38*) recently reported that EUK-189 and EUK-207 were effective at reversing age-related learning deficits in mice. In this study, 8 month old mice were treated by chronic infusion with salen Mn complexes for 3 months. For each compound, two different doses were administered, equivalent to about 1.8 or 0.18 mg/kg-day. Compared to 8-month old mice, the 11-month old control mice showed a decline in cognitive function, as measured by two behavioral tests involving fear conditioning. However, mice treated with salen Mn complexes for 3 months showed significantly improved behavior in these assays, indicating prevention of age-associated mild cognitive impairment. Of the 4 treatments, the most effective was the lower dose of EUK-207 (Fig. 8). All 4 treatment groups showed a significant decrease in brain levels of malonyldialdehyde and protein carbonyls (*38*). EUK-207 treatment decreased levels of oxidized nucleic acids in the brain (Fig 9).

Conclusions

Salen Mn complexes are SOD and catalase mimetics, scavenging multiple reactive species including superoxide, hydrogen peroxide and reactive nitrogen species. Multi-faceted SAR research has led to enhancements of catalytic and cytoprotective activities and other key properties. Salen Mn complexes are protective in many models for disease and, most notably, are

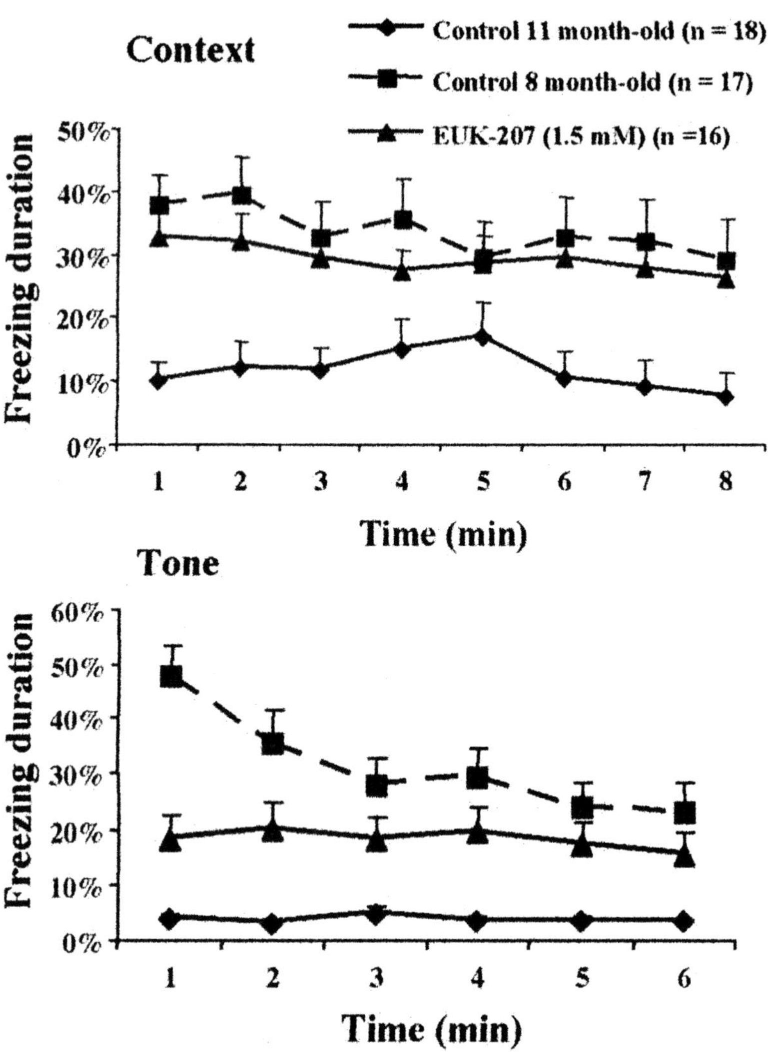

Fig. 8. Effect of EUK-207 in tests related to cognitive function. As described in the text and in reference 38, mice were treated continuously with EUK-207 (0.18 mg/kg-day) or vehicle from 8 months to 11 months of age. Context and tone conditioning tests were performed as described (38). Performance is compared to that of 8 month old untreated mice. Reprinted from reference 38, copyright (2003), with permission from the National Academy of Sciences, USA.

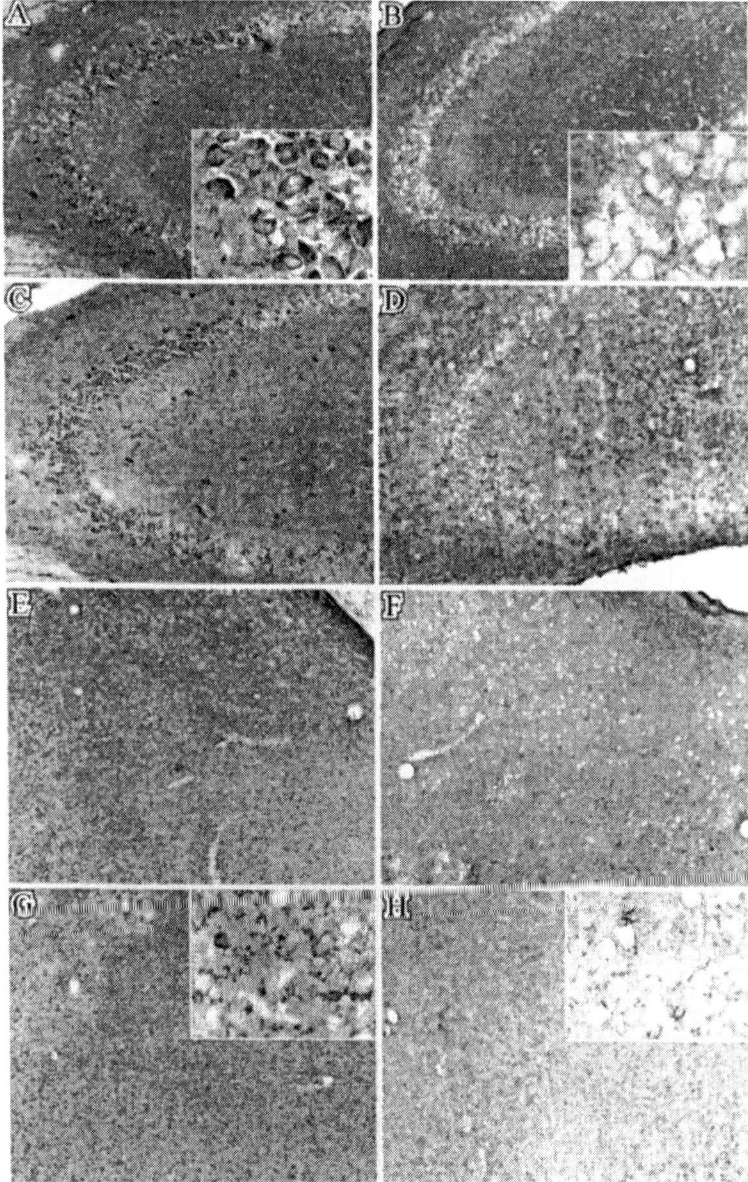

Fig. 9. Effect of EUK-207 treatment on DNA oxidation in the brain. Mice were treated until 11 mos of age as described for Fig. 8. Brains were then stained for oxidized nucleic acids with an antibody against 8-oxoG/8-oxo-dG. Brain regions: hippocampus (A-D) and amygdala (E-H); Treatment: Vehicle (A,C,E,G); EUK-207 (B,D,F,H). Reprinted from reference 38, copyright (2003), with permission from the National Academy of Sciences, USA.

substantially more effective than similarly intended agents at protecting the brain against the consequences of mitochondrial dysfunction and oxidative stress. This "mito-protective" property, in particular, makes these compounds well suited as potential treatments for neurodegenerative diseases of aging. In support of this, two lead compounds developed through SAR research, when delivered chronically, are highly effective against functional, morphological and biochemical forms of injury in models for neurodegeneration and aging.

References

1. *Free Radicals in Biology and Medicine*; Halliwell, B.; Gutteridge, J.M.C.; Second ed. 1989, Oxford: Clarendon Press. 543.
2. Sies, H. *Angewandte Chemie* **1986**, 25, 1058-1071.
3. Kensler, T.W.; Bush, D.M.; Kozumbo, W.J. *Science* **1983**, 221, 75-77.
4. Darr, D.; Zarilla, K.A.; Fridovich, I. *Arch. Biochem. Biophys.* **1987**, 258, 351-355.
5. Bruce, A.J.; Najm, I.; Malfroy, B.; Baudry, M. *Neurodegeneration* **1992**, 1, 265-271.
6. Baudry, M.; Etienne, S.; Bruce, A.; Palucki, M.; Jacobsen, E.; Malfroy, B. *Biochem. Biophys. Res. Comm.* **1993**, 192, 964-968.
7. Faulkner, K.M.; Stevens, R.D.; Fridovich, I. *Arch. Biochem. Biophys.* **1994**, 310, 341-346.
8. Zhang, W.; Loeback, J.L.; Wilson, S.R.; Jacobsen, E.N. *J. Am. Chem. Soc.* **1990**, 112, 2801-2803.
9. Riley, D.P.; Weiss, R.H. *J. Amer. Chem. Soc.* **1994**, 116, 387-388.
10. Faulkner, K.M.; Liochev, S.I.; Fridovich, I. *J. Biol. Chem.* **1994**, 269, 23471-23476.
11. Musleh, W.; Bruce, A.; Malfroy, B.; Baudry, M. *Neuropharmacology* **1994**, 33, 929-934.
12. Pucheu, S.; Boucher, F.; Sulpice, T.; Tresallet, N.; Bonhomme, Y.; Malfroy, B.; deLeiris, J. *Cardiovasc. Drugs Therap.* **1996**, 10, 331-339.
13. Bruce, A.J.; Malfroy, B.; Baudry, M. *Proc. Natl. Acad. Sci. (USA)* **1996**, 93, 2312-2316.
14. Malfroy, B.; Doctrow, S.R.; Orr, P.L.; Tocco, G.; Fedoseyeva, E.V.; Benichou, G. *Cellular Immunology* **1997**, 177, 62-68.
15. Gonzalez, P.K.; Zhuang, J.; Doctrow, S.R.; Malfroy, B.; Benson, P.F.; Menconi, M.J.; Fink, M.P. *J. Pharmacol. Exper. Therapeutics* **1995**, 275, 798-806.
16. Robert, A.; Loock, B.; Momenteau, M.; Meunier, B. *Inorg. Chem.* **1991**, 30, 706-711.

17. Doctrow, S.R.; Huffman, K.; Marcus, C.B.; Tocco, G.; Malfroy, E.; Adinolfi, C.A.; Kruk, H.; Baker, K.; Lazarowych, N.; Mascarenhas, J.; Malfroy, B. *J. Med. Chem.* **2002**, 45, 4549-4558.
18. Chance, B.; Sies, H.; Boveris, A. *Physiological Rev.* **1979**, 59, 527-605.
19. Srinivasan, K.; Michaud, P.; Kochi, J.K. *J. Am. Chem. Soc.* **1986**, 108, 2309-2320.
20. Oshino, N.; Oshino, R.; Chance, B. *Biochem. J.* **1973**, 131, 555-567.
21. Doctrow, S.R.; Huffman, K.; Marcus, C.B.; Musleh, W.; Bruce, A.; Baudry, M.; Malfroy, B., in *Antioxidants in Disease Mechanisms and Therapeutic Strategies*, H. Sies, Editor. 1997, Academic Press: New York. p. 247-270.
22. Beal, M.F. *Free Radic Biol Med* **2002**, 32, 797-803.
23. Crow, J.P. *Free Rad. Biol. Med.* **2000**, 28, 1487-94.
24. Patel, M.; Day, B.J. *Trends Pharmacol. Sci.* **1999**, 20, 359-364.
25. Shimanovich R, G.J. *ARch. Biochem. Biophys.*, 387, 307-317.
26. Shimanovich, R.; Hannah, S.; Lynch, V.; Gerasimchuk, N.; Mody, T.D.; Magda, D.; Sessler, J.; Groves, J.T. *J Am Chem Soc.* **2001**, 123, 3613-3614.
27. Poduslo, J.F.; Whelan, S.L.; Curran, G.L.; Wengenack, T.M. *Ann. Neurol.* **2000**, 48, 943-947.
28. Doctrow, S.R., Adinolfi, C., Baudry, M., Huffman, K., Malfroy, B., Marcus, C.B., Melov, S., Pong, K., Rong, Y., Smart, J., and Tocco, G., in *Oxidative Stress and Aging: Advances in Basic Science, Diagnostics, and Intervention*, H. Rodriguez and R. Cutler, Editors. 2003, World Scientific Publishing Company: Singapore, London, NJ. p. 1324-1342.
29. Sharpe, M.A.; Ollosson, R.; Stewart, V.C.; Clark, J.B. *Biochem. Journal Immediate Publication* **2002**.
30. Salvemini, D.; Wang, Z.Q.; Zweier, J.L.; Samouilov, A.; Macarther, H.; Misko, T.P.; Currie, M.G.; Cuzzocrea, S.; Sikorski, J.A.; Riley, D.P. *Science* **1999**, 286, 304-306.
31. Gonzalez, P.K.; Doctrow, S.R.; Fink, M.P. *Shock* **1997**, 8, 108-114.
32. Karin M, T.T., Kapahi P, Delhase M, Chen Y, Makris C, Rothwarf D, Baud V, Natoli G, Guido F, Li N. *Biofactors* **2001**, 15, 87-89.
33. Gius, D.; Botero, A.; Shah, S.; Curry, H.A. *Toxicol. Lett.* **1999**, 106, 93-106.
34. Baker, K.; Bucay-Marcus, C.; Huffman, C.; Kruk, H.; Malfroy, B.; Doctrow, S.R. *J. Pharmacol. Exp. Therapeutics* **1998**, 284, 215-221.
35. Rong, Y.; Doctrow, S.R.; Tocco, G.; Baudry, M. *Proc. Natl. Acad. Sci. (USA)* **1999**, 96, 9897-9902.
36. Melov, S.; Doctrow, S.R.; Schneider, J.A.; Haberson, J.; Patel, M.; Coskun, P.E.; Huffman, K.; Wallace, D.C.; Malfroy, B. *J. Neuroscience* **2001**, 21, 8348-8353.
37. Pong, K.; Doctrow, S.R.; Huffman, K.; Adinolfi, C.A.; Baudry, M. *Experimental Neurology* **2001**, 171, 84-97.
38. Liu, R.; Liu, I.Y.; Bi, X.; Thompson, R.F.; Doctrow, S.R.; Malfroy, B.; Baudry, M. *Proc Natl Acad Sci U S A* **2003**, 100, 8526-8531.

39. Boveris, A. *Methods Enzymol.* **1984**, 105, 429-435.
40. Turrens, J.F.; Alexandre, A.; Lehninger, A.L. *Arch. Biochem. Biophys.* **1985**, 237, 408-414.
41. *Mitochondria and Free Radicals in neurodegenerative diseases*; Beal, M.F.; Howell, N.; Bodis-Wollner, I.; 1st ed. 1997, New York: Wiley-Liss.
42. Beal, M.F. *Curr. Opin. Neurobiol.* **1996**, 6, 661-666
43. Rotig, A.; deLonlay, P.; Chretien, D.; F.Foury; Koenig, M.; Sidi, D.; Munnich, A.; Rustin, P. *Nature Genetics* **1997**, 17, 215-217.
44. Beckman, K.B.; Ames, B.N. *Physiol. Rev.* **1998**, 78, 547-581.
45. Choi, S.I.; Ju, W.K.; Choi, E.K.; Kim, J.; Lea, H.Z.; Carp, R.I.; Wisniewski, H.M.; Kim, Y.S. *Acta Neuropath. (Berl.)* **1998**, 96, 279-286.
46. Schapira, A.H.V. *Biochim Biophys. Acta* **1998**, 1366, 225-233.
47. Smith, M.A.; Sayre, L.M.; Anderson, V.E.; Harris, P.L.; Beal, M.F.; Kowall, N.; Perry, G. *J. Histochem. Cytochem.* **1998**, 46, 731-735.
48. Behl, C. *Prog. Neurobiol.* **1999**, 57, 301-323.
49. White, A.R.; Huang, X.; Jobling, M.F.; Barrow, C.J.; Beyreuther, K.; Masters, C.L.; Bush, A.I.; Cappai, R. *J. Neurochem.* **2001**, 76, 1509-1520
50. Jung, C.; Higgins, C.M.; Xu, Z. *J. Neurochem.* **2002**, 83, 535-545.
51. Sturtz LA, D.K., Jensen LT, Lill R, Culotta VC. *J. Biol. Chem.*, 276, 38084-38089.
52. Fujimura, M.; Morita-Fujimura, Y.; Copin, J.; Yoshimoto, T.; Chan, P.H. *Brain Res.* **2001**, 889, 208-213.
53. Shefner, J.M.; Reaume, A.G.; Flood, D.G.; Scott, R.W.; Kowall, N.W.; Ferrante, R.J.; Siwek, D.R.; Upton-Rice, M.; Jr, R.H.B. *Neurology* **1999**, 53, 1239-1246.
54. Li, Y.; Huang, T.T.; Carlson, E.J.; Melov, S.; Ursell, P.C.; Olson, J.L.; Noble, L.J.; Yoshirmura, M.P.; Berger, C.; Chan, P.H.; Wallace, D.C.; Epstein, C.J. *Nature Genetics* **1995**, 11, 376-381.
55. Melov, S.; Schneider, J.A.; Day, B.J.; Hinerfeld, D.; Coskun, P.; Mirra, S.S.; Crapo, J.D.; Wallace, D.C. *Nature Genetics* **1998**, 18, 159-163
56. Huang, T.T.; Carlson, E.J.; Raineri, I.; Gillespie, A.M.; Kozy, H.; Epstein, C.J. **1999**, 893, 95-112.
57. Hinerfeld, D., Train, M.D., Weinberger, R.P., Cochran, B., Doctrow, S.R., Harry, J., and Melov, S. *J. Neurochem.* **2004**, 88, 657-667.
58. Liu, J.; Atamna, H.; Kuratsune, H.; Ames, B.N. *Ann. NY Acad. Sci.* **2002**, 959, 133-66.
59. Pong, K.; Doctrow, S.R.; Baudry, M. *Brain Res.* **2000**, 881, 182-189.
60. Jung, C.; Rong, Y.; Doctrow, S.; Baudry, M.; Malfroy, B.; Xu, Z. *Neuroscience Letters* **2001**, 304, 157-160.
61. Brown, S.E.; II, L.J.R.; Dennery, P.A.; Doctrow, S.R.; Beal, M.F.; Barlow, C.; Levine, R.L. *Free Radic Biol Med* **2004**, 36, 938-942.

62. Barlow, C.; Hirotsune, S.; Paylor, R.; Liyanage, M.; Eckhaus, M.; Collins, F.; Shiloh, Y.; Crawley, J.N.; Ried, T.; Tagle, D.; Wynshaw-Boris, A. *Cell* **1996**, 86, 159-171.
63. Barlow, C.; Dennery, P.A.; Shigenaga, M.K.; Smith, M.A.; Morrow, J.D.; Roberts, L.J.; Wynshaw-Boris, A.; Levine, R.L. *Proc Natl Acad Sci U S A* **1999**, 96, 9915-9919.
64. Melov, S.; Ravenscroft, J.; Malik, S.; Gill, M.S.; Walker, D.S.; Clayton, P.E.; Wallace, D.C.; Malfroy, B.; Doctrow, S.; Lithgow, G.J. *Science* **2000**, 289, 1567-1569.
65. Keaney, M.; Gems, D. *Free Radic Biol Med* **2003**, 34, 277-282.
66. Bayne, A.C.; Sohal, R.S. *Free Radic Biol Med* **2002**, 32, 1229-1234.
67. Zhang, H.; Doctrow, S.R.; Xu, L.; Oberley, L.W.; Beecher, B.; Morrison, J.,: Oberley, T.D.; Kregel, K.C. *FASEB J.* **2004,** in press.
68. Zhang, H.J.; Drake, V.J.; Xu, L.; Xie, L.; Oberley, L.W.; Kregel , K.C. *FASEB J.* **2003**, 17, 2293-2295.

Acknowledgements

Studies described here were supported in part by grants from the NIH (GM57770, CA83575, AG19531, and AG018679) and by the AT Children's Project. We thank Dr. Rodney Levine (Laboratory of Biochemistry, NHLBI) for his insightful comments on this manuscript. ALCAR and LA were kindly supplied by Viatris GmbH & Co. KG, Germany, formerly ASTA. M40403 was kindly provided by Dr. Daniella Salvemini of MetaPhore Pharmaceuticals, Inc.

Chapter 19

Metal-Deficient Copper–Zinc Superoxide Dismutase and Familial Amyotrophic Lateral Sclerosis

Pathogenic SOD1 Oligomerization through Non-Native Protein–Protein Interactions

P. John Hart[1] and Joan Selverstone Valentine[2]

[1]Department of Biochemistry and the X-ray Crystallography Core Lab, The University of Texas Health Science Center, San Antonio, TX 78229
[2]Department of Chemistry and Biochemistry, University of California, Los Angeles, CA 90095

Since the link between mutations in copper-zinc superoxide dismutase (SOD1) and familial ALS (FALS) was first described approximately 10 years ago, laboratories worldwide have sought to understand how the mutations render the SOD1 protein toxic to motor neurons. Evidence is accumulating that this toxic property comes from the ability of the mutant SOD1 proteins to assemble into higher order structures, either soluble oligomers or insoluble aggregates, that somehow interfere with the neuronal cellular machinery. We observe that five different metal-deficient pathogenic SOD1 proteins can form higher order filamentous arrays reminiscent of the types of seen in other neurodegenerative disorders such as Alzheimer's and Parkinson's diseases.

The Connection of ALS to SOD

The hallmark of amyotrophic lateral sclerosis (ALS, Lou Gerhrig's Disease, Motor Neuron Disease) is the loss of large motor neurons from the spinal cord and brain. This leads to progressive paralysis, resulting in death within two to five years of the appearance of symptoms (*1*). About 10% of cases are familial, and approximately 20% of familial ALS (FALS) cases are linked to dominantly inherited mutations in SOD1, a 32 kDa homodimeric copper- and zinc-containing antioxidant enzyme. Because superoxide is a ubiquitous reactive oxygen species produced in all aerobic cells, SOD1 is needed to protect against oxidative damage by catalyzing its disproportionation to hydrogen peroxide and dioxygen (*2, 3*). Although eleven different families each with a different point mutation were initially identified (*2*), the number has grown to about 100 (more information can be found at www.alsod.org). Plate 1A shows that these mutations are widely scattered throughout the 3-D scaffold of the protein. Indeeed, one of the most puzzling aspects of SOD1-linked FALS is how such a structurally diverse collection of single amino acid substitutions can all lead to motor neuron death.

Molecular Mechanisms of FALS SOD1 Mutant Proteins

Although it is logical to assume that SOD1-linked FALS is a loss of function disorder coming from increased oxidative damage due to diminished SOD1 activity (*2*), SOD1Δ mice do not develop motor neuron disease (*4*). Transgenic mice expressing human FALS SOD1 mutants become paralyzed, however, despite the fact that they possess normal (*5, 6*) or elevated (*7, 8*) levels of SOD1 activity. Together, these observations suggest that pathogenic SOD1 molecules act through the gain of a cytotoxic property and not a loss of function. Explanatory hypotheses regarding this acquired property range from aberrant copper-mediated catalysis (*9-11*) to mutant SOD1 misfolding and aggregation (*6, 12, 13*). These hypotheses are known as the "oxidative" and the "aggregation" hypotheses, respectively (*14*). Brief descriptions of each are given below.

The Oxidative Hypothesis: Reactions with Hydrogen Peroxide and Peroxynitrite

The oxidative hypothesis states that FALS mutant SOD1 proteins catalyze reactions with hydrogen peroxide or peroxynitrite that damage cellular components that are critical for motor neuron viability (*9*). For these chemistries to occur, copper must be bound in the active site of SOD1 to promote the oxidative reactions.

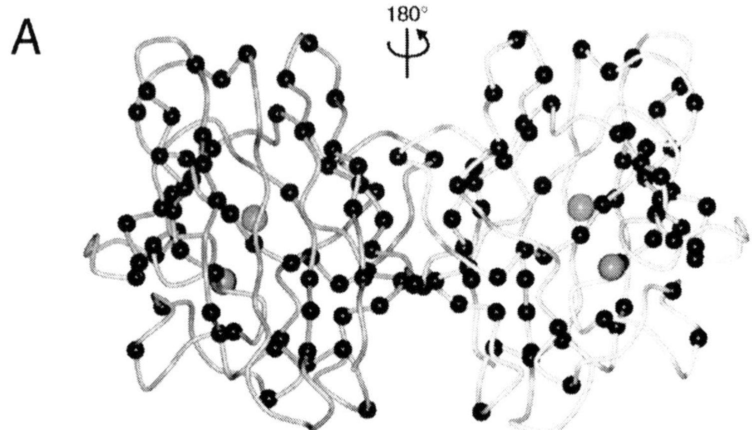

Plate 1. A. Human SOD1 showing the locations of the FALS mutations. The two subunits of the molecule are in orange and cyan, respectively. The relationship between the two monomers is indicated. The copper and zinc ions are shown as green and magenta spheres, respectively. The sites of single amino acid substitutions that lead to FALS are represented as black spheres. B. Human SOD1 showing its negative design and the locations of the wild type like (WTL) and metal binding region (MBR) FALS mutations. In the leftmost monomer, the gain-of-interaction (GOI) interface is boxed. In the wild type protein, the zinc loop projects from the plane of the paper toward the viewer, preventing SOD1-SOD1 protein-protein interactions from occurring at the edge strands. The copper and zinc ions are shown as green and magenta spheres, respectively. In the rightmost monomer, the positions of the MBR mutations are shown as yellow spheres and the WTL mutations (not shown for clarity) are scattered throughout the β-barrel depicted in gold. Metal binding stabilizes the conformation of the zinc and electrostatic loop elements and thus plays an intimate role in the negative design of the molecule.

(See page 6 of color insert.)

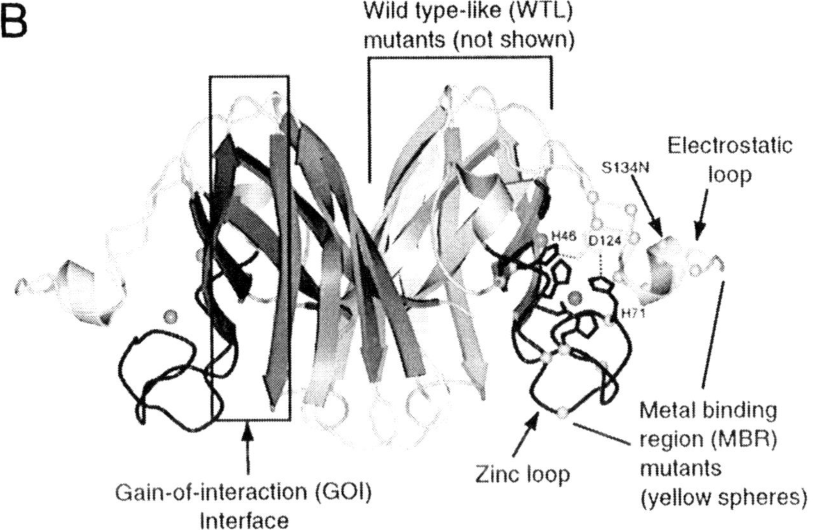
B
Wild type-like (WTL)
mutants (not shown)
S134N
Electrostatic
loop
H46
D124
H71
Metal binding
region (MBR)
mutants
(yellow spheres)
Zinc loop
Gain-of-interaction (GOI)
Interface

FALS mutant SOD1 proteins have been reported to function as peroxidases, using hydrogen peroxide to generate hydroxyl radical (•OH). This reactivity was discovered by Fridovich (*15*) and later by Yim and Stadtman who found that bovine wild type SOD1 could catalyze the oxidation of a model substrate, spin trap DMPO (5,5'-dimethyl-1-pyrroline N-oxide) by hydrogen peroxide (*16, 17*). Importantly, studies of human FALS mutants A4V, G93A, and L38V demonstrate that the peroxidative reaction occurs at significantly higher rates for these pathogenic SOD1 proteins than for wild type human SOD1 (*9, 10, 18*).

The mechanism of peroxidation involves the reduction of Cu(II)-SOD1 by hydrogen peroxide (reaction 1) followed by the reaction with another molecule of peroxide to produce a species capable of transferring the elements of a hydroxyl radical, •OH, to an exogenous substrate (reaction 2).

$$H_2O_2 + \text{SOD-Cu(II)} \rightarrow O_2^- + 2H^+ + \text{SOD-Cu(I)} \quad (1)$$
$$H_2O_2 + \text{DMPO} + \text{SOD-Cu(I)} \rightarrow \text{DMPO-OH} + OH^- + \text{SOD-Cu(II)} \quad (2)$$

In vivo, reducing agents such as ascorbate may take the place of hydrogen peroxide in reaction 1 (*19*). In addition, the presence of bicarbonate anion plays an intimate mechanistic role in the peroxidase chemistry catalyzed by pathogenic SOD1. Based on combination of 3-D structural and chemical data, we recently proposed that bicarbonate anion enhances the peroxidase activity of SOD1 through an enzyme-bound peroxycarbonate intermediate (*20*).

An increasing body of evidence points to oxidative damage of tissues in ALS. Elevated levels of markers of oxidative damage to nucleic acids, proteins, and lipids are reported in tissue samples obtained from both sporadic ALS (SALS) and SOD1-associated FALS patients (*21*). In spinal cord extracts of G93A-expressing transgenic mice, increased oxidation of the spin trap azulenyl nitrone is observed when compared with those of nontransgenic animals or transgenic mice expressing wild type human SOD1 (*22, 23*). It remains unclear, however, whether the observed increased levels of oxidation are the cause of toxicity in motor neurons or whether they represent a downstream effect of neuronal cell misfunction, particularly since mitochondrial abnormalities are among the earliest signs of pathology in the ALS transgenic mice (*8*).

A related oxidative hypothesis implicates the molecules nitric oxide, superoxide, and peroxynitrite in the modification of proteins in cells (*24, 25*). Nitric oxide (NO), produced by interneurons surrounding motor nuclei, reacts with superoxide (O_2^-) approximately three times faster than superoxide does with native SOD1, to form the strong oxidant peroxynitrite ($ONOO^-$). It has been proposed that peroxynitrite can react with SOD1 to form a nitronium-like intermediate which can in turn nitrate tyrosine residues (reactions 3 and 4).

$$\text{SOD-Cu(II)} + {}^-\text{OONO} \rightarrow \text{SOD-CuO-NO}_2^+ \quad (3)$$

$$SOD\text{-}CuO\text{--}NO_2^+ + H\text{-}Tyr \rightarrow SOD\text{-}Cu(II) + HO^- + NO_2\text{-}Tyr \qquad (4)$$

Nitration of critical cellular targets such as tyrosine kinase receptors or neurofilament proteins could injure motor neurons, and since motor neurons cannot regenerate, this damage may accumulate and eventually cause the cell to die. Increased 3-nitrotyrosine immunoreactivity is observed in motor neurons of both sporadic and familial ALS patients (*21*). As with the markers of oxidative damage mentioned above, however, it is unknown whether the observed increased levels of nitrotyrosine are the cause or a downstream effect due to neuronal dysfunction.

The Aggregation Hypothesis: Insoluble and Soluble Oligomers

The aggregation hypothesis states that pathogenic SOD1 proteins become misfolded and oligomerize into higher molecular-weight species that eventually are incorporated into proteinaceous inclusions. The assumption is that aggregated SOD1 proteins are selectively toxic to motor neurons. Recent reports suggest that high molecular weight insoluble protein aggregates, composed in part of FALS SOD1, may play a role in pathogenesis either by sequestering heat shock proteins (*26, 27*), and/or interfering with the neuronal axonal transport (*28, 29*) and protein degradation (*30*) machineries. Support for the aggregation hypothesis comes from the fact that proteinaceous inclusions rich in mutant SOD1 have been found in tissues from ALS patients (*31-33*) and ALS-SOD1 transgenic mice (*6, 12*).

While the offending oligomeric species of pathogenic SOD1 could indeed be the insoluble SOD1-containing aggregates themselves, an alternative notion is that they are actually somewhat benign, resulting from a cellular defense mechanism that occurs when the levels of misfolded or damaged proteins exceed the capacity of the protein degradation machinery to eliminate them (*34*). Accumulating evidence suggests that instead of these insoluble aggregates, it may be their soluble oligomeric precursors termed “protofibrils” that exert toxic effects, possibly by forming annular “pore-like” structures that can inhibit the proteasome and/or damage cell or mitochondrial membranes (see below) (*35, 36*).

The possibility that a molecular species other than the insoluble protein aggregate could be pathogenic became evident when soluble oligomers were found to be discrete intermediates in the fibrillization of β-amyloid (Aβ) (Alzheimer's disease) and of α-synuclein (Parkinson's disease) *in vitro* (*37, 38*). An intermediate protofibril might therefore be pathogenic and become "detoxified" by conversion to an insoluble fibril. This is supported by the observation that there is no significant correlation between the quantity of fibrillar deposits at autopsy and the clinical severity of Alzheimer's or

Parkinson's disease. In addition, transgenic mouse models of these conditions have disease-like phenotypes before fibrillar deposits can be detected (*13, 39*). Similarly in ALS transgenic mice, defects in axonal transport can be detected approximately 30 days after birth, well before any symptoms or insoluble proteinaceous deposits are manifest, suggesting that soluble oligomers of pathogenic SOD1 could be the interfering entity (*40*). It is important to note, however, that whether or not the toxic oligomeric SOD1 species is soluble or insoluble, it is clear that these pathogenic SOD1 proteins must have *some* feature distinct from the wild type protein that facilitates their self-association (see below), and it may be this feature that underlies the toxic gain of function in these proteins.

"Metal-binding Region" Mutants versus "Wild Type-like" Mutants

Based on their positions on the 3-D structure, the FALS mutant SOD1 proteins are divided into two groups. The first we term "wild type-like" (WTL) mutants because these proteins are isolated from their expression systems with copper and zinc levels similar to those found for the wild type protein expressed under the same conditions. The SOD1 activities and the spectroscopic characteristics of this class of FALS mutant SOD1 proteins are virtually identical to those of wild type SOD1 (*41, 42*). The second class of FALS mutant SOD1 proteins we term "metal-binding region" (MBR) mutants. These include mutations in the metal-binding ligands themselves and/or in the electrostatic and zinc loop elements that are heavily involved with metal binding. The rightmost monomer of SOD1 in Plate 1B shows the positions of the MBR mutations, which are represented as yellow spheres (the WTL mutations are not shown, but are scattered throughout the β-barrel of the protein shown in gold). The electrostatic (cyan) and zinc (black) loop elements are in close contact, and the conformation of one of these loops influences the conformation of the other. Particularly important is the side chain of Asp 124 that directly links the electrostatic and zinc loops and contributes to the stabilization of both the copper and zinc binding sites by forming hydrogen bonds simultaneously to the non-liganding imidazole nitrogen atoms of copper ligand His 46 and zinc ligand His 71. Due to this cross-talk between the electrostatic and zinc loop elements, it is not surprising that the MBR mutants are isolated from their expression systems with very low zinc and copper (*41, 43*).

Negative Design of Wild Type SOD1

Plate 1B shows that although each subunit of SOD1 possesses an eight-stranded Greek Key β-barrel fold, not all of the strands forming the β-barrel are continuously hydrogen-bonded to each other. β-strands 5 and 6 (boxed in the leftmost monomer of Plate 1B) can be considered edge strands. Edge strands are perilous for proteins because they provide a hydrogen-bonding surface that can accommodate interaction with other edge strands, which in turn can lead to the formation of higher-order structures such as those found in amyloid fibers. Proteins containing β-sheets have evolved a wide range of strategies to protect their edge strands from nonproductive associative interactions by incorporating "negative design" into their architecture (reviewed in (*44*)). The negative design of SOD1 comes from the well ordered electrostatic and zinc loop elements that prevent SOD1-SOD1 protein-protein interactions from occurring at these edge strands. As mentioned previously, the ligands for the copper and zinc ions are located in these loops, and thus, their conformations are critically dependent upon the presence of metal ions (see below).

Structures of WTL SOD1 Mutants Provide Little Insight into Toxicity

Given the negative design of SOD1 described above, it is difficult to envision how members of the WTL class, when fully metallated, might be prone to participate in nonproductive self-association. We determined the X-ray crystal structures of wild type human SOD1 (*45*) and WTL mutants A4V (*46*), G37R (*47*), G93A (unpublished results), and I113T (*46*). In each case, the WTL pathogenic SOD1 proteins are fully metallated and have an overall 3-D architecture that is nearly indistinguishable from the wild type protein. Thus, the negative design inherent in the wild type appears to be intact in these WTL mutant proteins. Taken alone, these structures shed little insight into how mutant SOD1 oligomerization or aggregation might occur.

Structural Studies of MBR Mutants Reveal Filamentous Assemblies

We recently determined the X-ray crystal structures of human MBR pathogenic SOD1 mutants S134N and apo-H46R to 1.3 Å and 2.5 Å resolution, respectively (*48*). The S134N substitution falls in the small helix that is part of the electrostatic loop (see rightmost monomer in Plate 1B). The H46R mutation occurs in the copper binding site. As with others of the MBR mutant class, the S134N and H46R proteins are metal deficient when isolated from their expression systems, and their low levels of metallation are likely similar to their states *in vivo*. Plate 2A shows that while the eight-stranded Greek key β-barrel components of both of these metal-deficient, pathogenic structures (green) are preserved when compared to the wild type protein (gray), they demonstrate disorder and conformational changes in their zinc and electrostatic loop elements. In both structures, the conformational mobility of these loops deprotects the edges of β-strands 5 and 6, which form a cleft between the two β-sheets of the SOD1 β-barrel (red). This depression serves as the molecular interface for non-native SOD1-SOD1 protein-protein contacts, where it interacts with part of the electrostatic loop of a neighboring molecule in the crystal lattice (blue in Plate 2A). We term these new contacts 'gain-of-interaction' (GOI) contacts to distinguish them from the interfaces between subunits of the native SOD1 dimer. When the GOI contacts are combined with the native dimeric interactions, the pathogenic SOD1 dimers assemble into linear filamentous arrays (shown in Plate 2B).

S134N and apo-H46R linear amyloid-like filaments

The S134N and apo-H46R filaments are nearly identical and their GOI interfaces (boxed) are symmetrical and run parallel to the symmetry axes of the natural dimers. Plate 2B*i* shows that these GOI interfaces bury nearly the same amount of solvent accessible surface area per polypeptide as the highly stable native SOD1 homodimeric interface. The arrangement of pathogenic SOD1 dimers is such that the β-sheets of the molecules run parallel to the long axis of the fiber, while their constituent β-strands run perpendicular to this axis (Plate 2B*iii*). Importantly, superposition of metal-loaded, native SOD1 dimers onto mutant dimers that comprise the linear filaments results in steric and electrostatic repulsion through the zinc loops at the mutant GOI interface (Plate 2C). Well-ordered SOD1 electrostatic and zinc loops therefore appear critical to the negative design of SOD1 and to prevent the observed self-association. Preliminary analyses of newly obtained crystals of metal-deficient pathogenic SOD1 mutants H80R, G85R, and E133Δ reveal that they too form filamentous arrays nearly identical to those formed by the S134N and H46R proteins described above. The fact that five different mutants form the *same* higher order filamentous structure is possibly suggestive of what might be occurring *in vivo*.

Zn-H46R "Pore-like" helical filament

Alzheimer's, Parkinson's and prion diseases are characterized by neuronal loss and protein aggregates that may or may not be fibrillar. However, the exact identity of the neurotoxic species and the mechanism by which it kills neurons are unknown. Biophysical studies support the emerging notion that a prefibrillar oligomer (protofibril) may be responsible for cell death and that the fibrillar form that is typically observed post mortem may actually be neuroprotective. The laboratory of Peter Lansbury suggests that a subpopulation of the soluble protofibrils may function as pathogenic pores that might have the ability to permeabilize cell or mitochondrial membranes (*35*). Annular, pore-like structures are observed in familial mutants of α-synuclein (Parkinson's disease) and Alzheimer's precursor protein (Alzheimer's disease) as shown in Plate 3A (*36*).

Intriguingly, we also observe an annular, pore-like entity in the X-ray crystal structure of zinc-loaded MBR mutant H46R (*48*). Plate 3C shows that a GOI interface between Zn-H46R dimers results in the assembly of helical filaments consisting of four Zn-H46R dimers per turn. A helical filament rather than a linear filament arises because the intermolecular contact surface is shifted several amino acids relative to that of the linear filaments, and residues of the zinc loop rather than the electrostatic loop participate (Plate 3B). Zinc loop residues 78-81 hydrogen bond to the deprotected edge of β-strand 6, extending the β-sheet formed by β-strands 1, 2, 3, and 6 by one strand. Repetition of this GOI in the Zn-H46R crystal results in hollow 'nanotubes' with an overall diameter of approximately 95 Å and an inner, water-filled cavity with a diameter of approximately 30 Å. These dimensions are intriguingly similar to those observed for the "pores" formed by α-synuclein and Alzheimer's precursor protein.

Conversion of "wild type-like" mutants to metal-deficient proteins

As described above, the metal-deficient MBR mutant class of pathogenic SOD1 proteins appear to be primed and ready to self-associate. How might the WTL mutant class fit into the "aggregation hypothesis?" The answer may lie in the peroxidase activity of SOD1 described above. Once formed, the strong oxidant at the copper center can inactivate SOD1 by damaging nearby active site histidine copper ligands, resulting in copper loss (*49-52*). The following reaction scheme has been proposed:

$$\text{SOD-Cu(I)} + H_2O_2 \rightarrow \text{SOD-Cu(II)(*OH)} + OH^- \quad (5)$$

$$\text{SOD-Cu(II)(*OH)} + \text{XH} \rightarrow \text{SOD-Cu(II)} + H_2O + \text{X*} \quad (6)$$

where XH represents amino acids at the active site (*53, 54*). Figure 1A (next page) shows that in the presence of H_2O_2, wild type human SOD1 self-inactivates as a function of time. Intriguingly, the presence of bicarbonate anion enhances the rates of self-inactivation 2-3 fold (*20*). The potential relevance of this activity is underscored by the significant concentration of HCO_3^- found *in vivo* (~25 mM). Figure 1B represents a schematic version of reaction 6, where the copper-bound oxidant (hydroxyl radical) can attack one of the histidine copper ligands to form 2-oxo-histidine. As described above, this chemistry is known to be significantly enhanced in the WTL mutant class for FALS SOD1 proteins (*9, 10, 18*). Over time, oxidative modification of metal ligands could compromise the ability of these mutants to bind metal ions and may convert them to forms that are much more prone to aggregation. In this way, the WTL and the MBR classes of pathogenic SOD1 mutants can become fused into a single class of molecules that can oligomerize and lead to motor neuron disfunction (*14*).

A

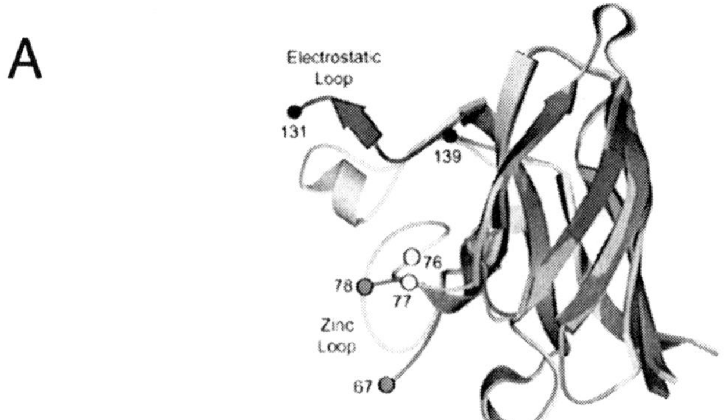

Plate 2. A. S134N and apo-H46R SOD1 monomers (green) superimposed human wild type SOD1 (gray). In both structures, elements of the zinc and electrostatic loops are disordered. Residues N- and C-terminal to these disordered regions are indicated by red and black spheres for the zinc and electrostatic loop elements, respectively. This disorder deprotects a cleft on the β-barrel (red). The gain-of-interaction (GOI) interface between S134N and apo-H46R SOD1 molecules is formed by reciprocal interactions between residues of the electrostatic loop (blue) with the exposed cleft (red) of adjacent molecules in the crystal lattice (see Plate 2B). Asp 76 and Glu 77 of the zinc loop in the wild type protein are represented by yellow spheres. These residues repel the approach of the same residues on other SOD1 molecules and prevent SOD1-SOD1 interactions from occurring at the cleft (see Plate 2C). B. Gain-of-interaction (GOI) in pathogenic SOD1 gives rise to cross-β fibrils in two different crystal systems. The filaments are represented by three dimers shown from top to bottom in green, gold, and blue, respectively. The GOI interface is boxed. In image *iii*, which is rotated 90° relative to images *i* and *ii*, β-strands 1, 2, 3, and 6, comprising one-half of each SOD1 β-barrel, are shown in red. The long axes of the β-strands run perpendicular to the long axis of the filament, an architecture similar to the "cross-β" structure observed in amyloid fibrils (shown schematically in *iv*). C. Negative design of SOD1 prevents formation of the GOI interface. When metal bound, the zinc loop elements are well-ordered and prevent self-association into the linear filaments through electrostatic and steric considerations (see text and Plate 2A).

(Reproduced with permission from reference 48. Copyright 2003.)
(See page 7 of color insert.)

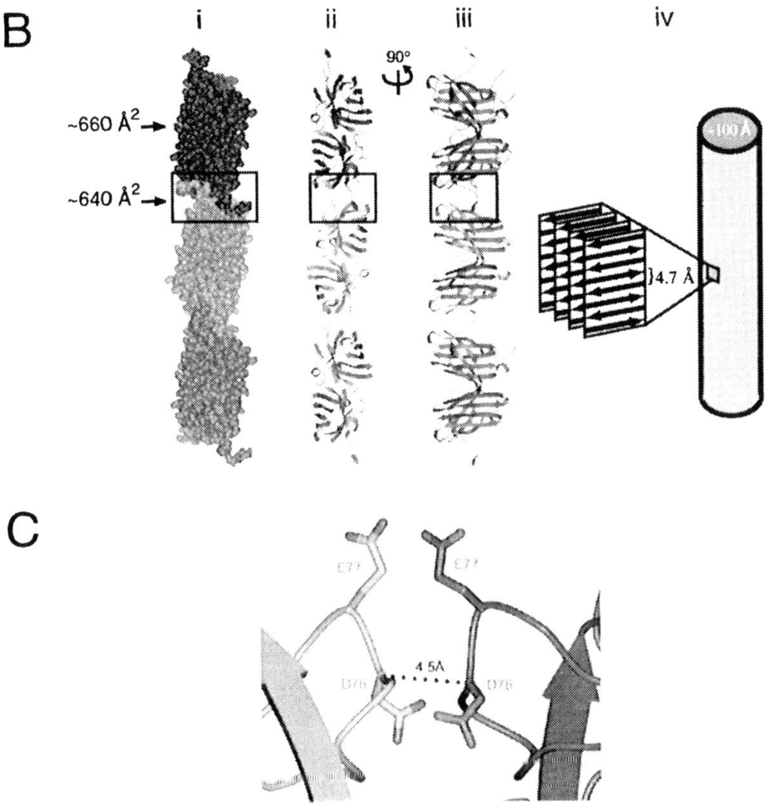

Plate 2. Continued.

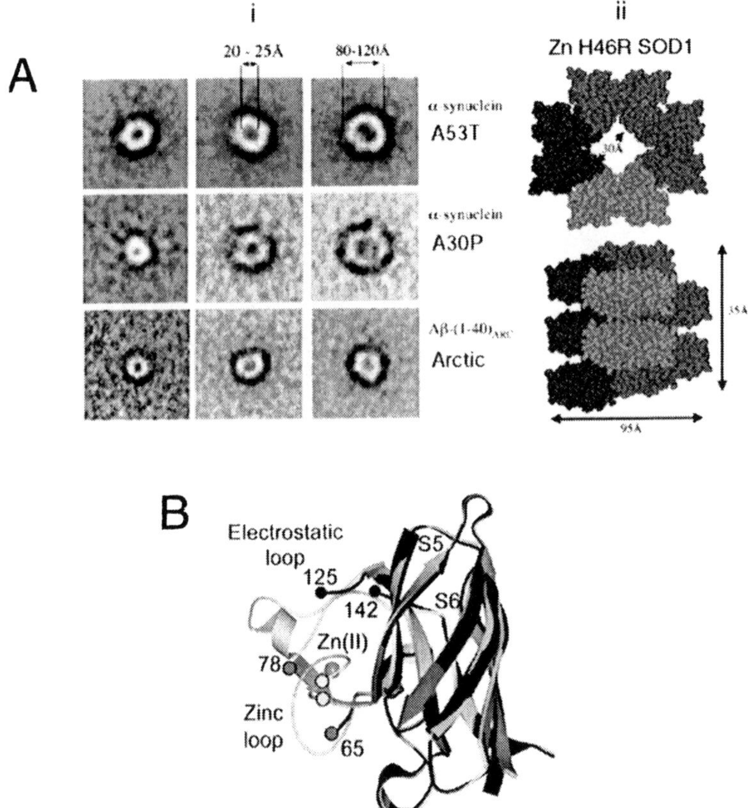

Plate 3. A. Annular, pore-like structures observed in proteins causing Parkinson's and Alzheimer's disease (left) are similar to the pore-like structure observed for Zn-loaded H46R SOD1 (right). The inner and outer dimensions of the pores are given. B. Zn-H46R SOD1 monomer (black) superimposed on human wild type SOD1 (gray). Elements of the zinc and electrostatic loops are disordered. Residues N- and C-terminal to these disordered regions are indicated by red and black spheres for the zinc and electrostatic loop elements, respectively. This disorder deprotects a cleft on the β-barrel (red). The GOI interface between Zn-H46R SOD1 molecules is formed by reciprocal interactions between residues of the zinc loop (blue) with the uncovered edges of β-strands 5 and 6 (red) of adjacent molecules in the crystal lattice (see Plate 3C). The zinc ion is shown as a green sphere. C. Gain-of-interaction in Zn-H46R leading to hollow, pore-like structures. β-strands 1, 2, 3, and 6, that form one-half of each SOD1 β-barrel, are shown in red. Residues of the zinc loop undergo a conformational change and form a short β-strand (shown in blue) that reciprocally adds to this β-sheet in neighboring Zn-H46R dimers, stabilizing the GOI interface. Zinc ions are shown as magenta spheres.

(Plate 3A is reproduced with permission from reference 36. Copyright 2002.)
(See page 8 of color insert.)

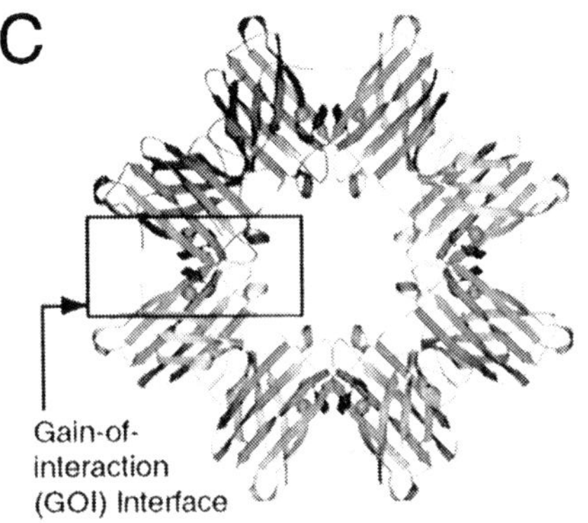

Plate 3. Continued.

Concluding Remarks

Our X-ray crystal structures suggest a model for pathogenic SOD1 aggregation that relies on the dimeric form of the protein. However, the possibility that a monomer of pathogenic SOD1 is the aggregating species *in vivo* cannot currently be ruled out. Metal binding and the presence of the intrasubunit disulfide bond in each subunit of SOD1 have considerable stabilizing effects on the dimeric form of the enzyme. More studies are underway to address this possibility.

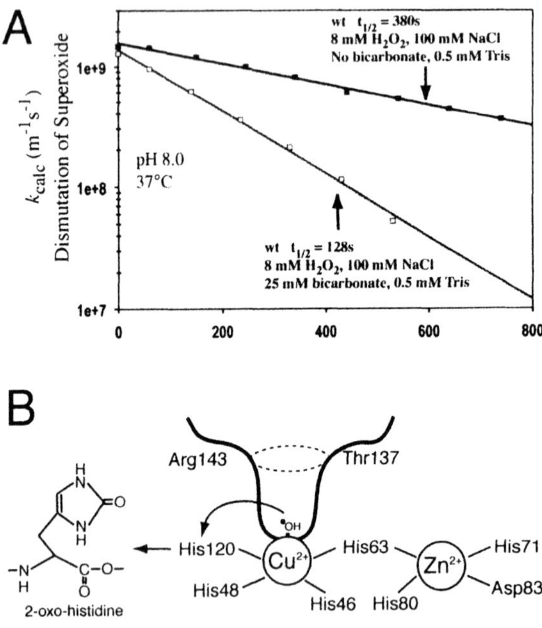

Figure 1. A) Effect of bicarbonate on hydrogen peroxide-mediated SOD1 self-inactivation. B) The attack of an oxygen radical species on one of the histidine copper ligands in SOD1 leads to the formation of a 2-oxo histidine adduct, leading to cofactor loss and enzyme inactivation (see text and (*20*)).

Acknowledgements

This work was supported by grants from the NIH NS39112 (PJH), NIH GM28222 (JSV), the Robert A. Welch Foundation (PJH), and the ALS Association (PJH, JSV). Collaborative efforts on the Zn-H46R structure by R.

Strange, S. Antonyuk, and S.S. Hasnain (Daresbury Lab, U.K.) are also gratefully acknowledged.

Literature Cited

1. Haverkamp, L. J., Appel, V., and Appel, S. H. (1995) *Brain 118*, 707-19.
2. Deng, H. X., Hentati, A., Tainer, J. A., Iqbal, Z., Cayabyab, A., Hung, W. Y., Getzoff, E. D., Hu, P., Herzfeldt, B., Roos, R. P., and et al. (1993) *Science 261*, 1047-51.
3. Rosen, D. R., Siddique, T., Patterson, D., Figlewicz, D. A., Sapp, P., Hentati, A., Donaldson, D., Goto, J., O'Regan, J. P., Deng, H. X., and et al. (1993) *Nature 362*, 59-62.
4. Reaume, A. G., Elliott, J. L., Hoffman, E. K., Kowall, N. W., Ferrante, R. J., Siwek, D. F., Wilcox, H. M., Flood, D. G., Beal, M. F., Brown, R. H., Jr., Scott, R. W., and Snider, W. D. (1996) *Nat Genet 13*, 43-7.
5. Ripps, M. E., Huntley, G. W., Hof, P. R., Morrison, J. H., and Gordon, J. W. (1995) *Proc Natl Acad Sci U S A 92*, 689-93.
6. Bruijn, L. I., Becher, M. W., Lee, M. K., Anderson, K. L., Jenkins, N. A., Copeland, N. G., Sisodia, S. S., Rothstein, J. D., Borchelt, D. R., Price, D. L., and Cleveland, D. W. (1997) *Neuron 18*, 327-38.
7. Gurney, M. E., Pu, H., Chiu, A. Y., Dal Canto, M. C., Polchow, C. Y., Alexander, D. D., Caliendo, J., Hentati, A., Kwon, Y. W., Deng, H. X., and et al. (1994) *Science 264*, 1772-5.
8. Wong, P. C., Pardo, C. A., Borchelt, D. R., Lee, M. K., Copeland, N. G., Jenkins, N. A., Sisodia, S. S., Cleveland, D. W., and Price, D. L. (1995) *Neuron 14*, 1105-16.
9. Wiedau-Pazos, M., Goto, J. J., Rabizadeh, S., Gralla, E. B., Roe, J. A., Lee, M. K., Valentine, J. S., and Bredesen, D. E. (1996) *Science 271*, 515-8.
10. Yim, M. B., Kang, J. H., Yim, H. S., Kwak, H. S., Chock, P. B., and Stadtman, E. R. (1996) *Proc Natl Acad Sci U S A 93*, 5709-14.
11. Beckman, J. S., Chen, J., Ischiropoulos, H., and Crow, J. P. (1994) *Methods Enzymol 233*, 229-40.
12. Bruijn, L. I., Houseweart, M. K., Kato, S., Anderson, K. L., Anderson, S. D., Ohama, E., Reaume, A. G., Scott, R. W., and Cleveland, D. W. (1998) *Science 281*, 1851-4.
13. Wang, J., Xu, G., Gonzales, V., Coonfield, M., Fromholt, D., Copeland, N. G., Jenkins, N. A., and Borchelt, D. R. (2002) *Neurobiol Dis 10*, 128-38.
14. Valentine, J. S., and Hart, P. J. (2003) *Proc Natl Acad Sci U.S.A 100,* 3617-22.
15. Hodgson, E. K., and Fridovich, I. (1975) *Biochemistry 14*, 5299-303.
16. Yim, M. B., Chock, P. B., and Stadtman, E. R. (1990) *Proc Natl Acad Sci U S A 87*, 5006-10.

17. Yim, M. B., Chock, P. B., and Stadtman, E. R. (1993) *J Biol Chem 268*, 4099-105.
18. Yim, H. S., Kang, J. H., Chock, P. B., Stadtman, E. R., and Yim, M. B. (1997) *J Biol Chem 272*, 8861-3.
19. Lyons, T. J., Liu, H., Goto, J. J., Nersissian, A., Roe, J. A., Graden, J. A., Cafe, C., Ellerby, L. M., Bredesen, D. E., Gralla, E. B., and Valentine, J. S. (1996) *Proc Natl Acad Sci U S A 93*, 12240-4.
20. Elam, J. S., Malek, K., Rodriguez, J. A., Doucette, P. A., Taylor, A. B., Hayward, L. J., Cabelli, D. E., Valentine, J. S., and Hart, P. J. (2003) *J Biol Chem 278*, 21032-21039.
21. Ferrante, R. J., Browne, S. E., Matthews, R. T., Kowall, N. W., and Brown, R. H., Jr. (1997) *Annals of Neurology 42*, 644-54.
22. Gurney, M. E., Liu, R., Althaus, J. S., Hall, E. D., and Becker, D. A. (1998) *J Inherit Metab Dis 21*, 587-97.
23. Liu, R., Althaus, J. S., Ellerbrock, B. R., Becker, D. A., and Gurney, M. E. (1998) *Ann Neurol 44*, 763-70.
24. Beckman, J. S., and Crow, J. P. (1993) *Biochem Soc Trans 21*, 330-4.
25. Beckman, J. S., Chen, J., Crow, J. P., and Ye, Y. Z. (1994) *Prog Brain Res 103*, 371-80.
26. Bruening, W., Roy, J., Giasson, B., Figlewicz, D. A., Mushynski, W. E., and Durham, H. D. (1999) *J Neurochem 72*, 693-9.
27. Okado-Matsumoto, A., and Fridovich, I. (2002) *Proc Natl Acad Sci U S A 99*, 9010-4.
28. Borchelt, D. R., Wong, P. C., Becher, M. W., Pardo, C. A., Lee, M. K., Xu, Z. S., Thinakaran, G., Jenkins, N. A., Copeland, N. G., Sisodia, S. S., Cleveland, D. W., Price, D. L., and Hoffman, P. N. (1998) *Neurobiol Dis 5*, 27-35.
29. Williamson, T. L., and Cleveland, D. W. (1999) *Nat Neurosci 2*, 50-6.
30. Johnston, J. A., Dalton, M. J., Gurney, M. E., and Kopito, R. R. (2000) *Proc Natl Acad Sci U S A 97*, 12571-6.
31. Hirano, A., Kobayashi, M., Sasaki, S., Kato, T., Matsumoto, S., Shiozawa, Z., Komori, T., Ikemoto, A., and Umahara, T. (1994) *Neuroscience Letters 179*, 149-52.
32. Udaka, F., Kameyama, M., Kusaka, H., Ito, H., Imai, T., and Matsumoto, S. (1996) *Clinical Neuropathology 15*, 209-14.
33. Asayama, K., Hirano, A., Kobayashi, M., and Shibata, N. (1996) *Developmental Neuroscience 18*, 492-8.
34. Sherman, M. Y., and Goldberg, A. L. (2001) *Neuron 29*, 15-32.
35. Caughey, B., and Lansbury, P. T. (2003) *Annu Rev Neurosci 26*, 267-298.
36. Lashuel, H. A., Hartley, D., Petre, B. M., Walz, T., and Lansbury, P. T., Jr. (2002) *Nature 418*, 291.
37. Lansbury, P. T., Jr. (1999) *Proc Natl Acad Sci U S A 96*, 3342-4.

38. Goldberg, M. S., and Lansbury, P. T., Jr. (2000) *Nature Cell Biology 2*, E115-9.
39. Wang, J., Xu, G., and Borchelt, D. R. (2002) *Neurobiol Dis 9*, 139-148.
40. Shah, J. V., Cleveland, D. W., and Williamson, T. L. (2002) *Current Opinion in Cell Biology 14*, 58-62.
41. Hayward, L. J., Rodriguez, J. A., Kim, J. W., Tiwari, A., Goto, J. J., Cabelli, D. E., Valentine, J. S., and Brown, R. H., Jr. (2002) *J Biol Chem 277*, 15923-31.
42. Goto, J. J., Zhu, H., Sanchez, R. J., Nersissian, A., Gralla, E. B., Valentine, J. S., and Cabelli, D. E. (2000) *J Biol Chem 275*, 1007-14.
43. Rodriguez, J. A., Valentine, J. S., Eggers, D. K., Roe, J. A., Tiwari, A., Brown, R. H., Jr., and Hayward, L. J. (2002) *J Biol Chem 277*, 15932-15937.
44. Richardson, J. S., and Richardson, D. C. (2002) *Proc Natl Acad Sci U S A 99*, 2754-9.
45. Strange, R. W., Antonyuk, S., Hough, M. A., Doucette, P. A., Rodriguez, J. A., Hart, P. J., Hayward, L. J., Valentine, J. S., and Hasnain, S. S. (2003) *J Mol Biol 328,* 877-91.
46. Hough, M. A., Grossmann, J. G., Antonyuk, S., Strange, R. W., Doucette, P. A., Rodriguez, J. A., Whitson, L. J., Hart, P. J., Hayward, L. J., Valentine, J. S., and Hasnain, S. S. (2003) *Proc Natl Acad Sci U.S.A. 101,* 5976-81.
47. Hart, P. J., Liu, H., Pellegrini, M., Nersissian, A. M., Gralla, E. B., Valentine, J. S., and Eisenberg, D. (1998) *Protein Sci 7*, 545-55.
48. Elam, J. S., Taylor, A. B., Strange, R., Antonyuk, A., Doucette, P. A., Rodriguez, J. A., Hasnain, S. S., Hayward, L. J., Valentine, J. S., Yeates, T. O., and Hart, P. J. (2003) *Nat Struct Biol 10*, 461-467.
49. Kurahashi, T., Miyazaki, A., Suwan, S., and Isobe, M. (2001) *J Am Chem Soc 123*, 9268-78.
50. Bray, R. C., Cockle, S. A., Fielden, E. M., Roberts, P. B., Rotilio, G., and Calabrese, L. (1974) *Biochem J 139*, 43-8.
51. Blech, D. M., and Borders, C. L., Jr. (1983) *Arch Biochem Biophys 224*, 579-86.
52. Uchida, K., and Kawakishi, S. (1994) *J Biol Chem 269*, 2405-10.
53. Hodgson, E. K., and Fridovich, I. (1975) *Biochemistry 14*, 5294-9.
54. Liochev, S. I., and Fridovich, I. (1999) *Free Radic Biol Med 27*, 1444-7.

Chapter 20

Impact of the Lipophilicity of Desferrithiocin Analogues on Iron Clearance

Raymond J. Bergeron[1,*], Jan Wiegand[1], James S. McManis[1], William R. Weimar[1,4], Jeong-Hyun Park[1], Eileen Eiler-McManis[1], Jennifer Bergeron[2], and Gary M. Brittenham[3]

[1]Department of Medicinal Chemistry, University of Florida, Gainesville, FL 32610–0485
[2]Department of Educational Psychology, University of Florida, Gainesville, FL 32611–7047
[3]Departments of Pediatrics and Medicine, Columbia University College of Physicians and Surgeons, New York, NY 10032
[4]Current address: Department of Pharmaceutical Sciences, South University School of Pharmacy, Savannah, GA 31406

Increasing chelator lipophilicity can substantially augment iron clearing efficiency in *Cebus apella* primates. The impact of altering the lipophilicity, i.e., octanol—water partition properties (log *P*), of analogues of desazadesferrithiocin, (*S*)-4,5-dihydro-2-(2-hydroxyphenyl)-4-methyl-4-thiazolecarboxylic acid, on the ligands' iron clearing properties is described. The complications of iron overload are often associated with the metal's interaction with hydrogen peroxide, generating hydroxyl radicals (Fenton chemistry) and, ultimately, other related deleterious species. In fact, some iron chelators actually promote this chemistry. A number of the title compounds synthesized and tested in our laboratory are shown to be both inhibitors of the iron-mediated oxidation of ascorbate as well as effective radical scavengers.

Essentially all prokaryotes and eukaryotes have a strict requirement for iron. Although this element comprises 5% of the earth's crust, living systems have great difficulty in accessing and managing this vital micronutrient. The low solubility of Fe(III) hydroxide ($K_{sp} = 1 \times 10^{-39}$) (*1*), the predominant form of the metal in the biosphere, has led to the development of sophisticated iron storage and transport systems in nature. Microorganisms utilize low molecular weight, virtually ferric iron-specific, ligands, siderophores; higher eukaryotes tend to employ proteins to transport (e.g., transferrin) and store (e.g., ferritin) iron (*2, 3, 4*).

Abnormally high iron levels in particular biological compartments can have a profound effect on a number of disease processes. Thus, iron-mediated diseases can be separated into two fundamental categories, systemic and focal iron overload disorders. The systemic disorders are typified by hemochromatosis and transfusional iron overload associated with either hemolytic anemias, such as thalassemia and sickle cell anemia, or bone marrow deficiency, for example, myelodysplasia. Disorders with focal iron sequestration include ischemia-reperfusion injury (*5, 6*), Parkinson's disease, Alzheimer's disease, and Friedreich's ataxia, to name a few (*7, 8, 9, 10, 11, 12*).

In primates, iron metabolism is highly efficient (*13, 14, 15, 16, 17*): Little of the metal is absorbed; no specific mechanism exists for its elimination. Because it cannot be effectively cleared, the introduction of "excess iron" (*18, 19, 20*) into this closed metabolic loop leads to chronic overload and ultimately to peroxidative tissue damage. For example, patients with severe refractory anemia, such as Cooley's anemia (β-thalassemia major), require chronic transfusions, which increase their body iron by 200–250 mg with each unit of blood administered. Unless these individuals receive chelation therapy, they frequently die in their second or third decade from complications associated with iron overload.

The hydroxamate bacterial siderophore desferrioxamine B (DFO, Figure 1), produced by large-scale fermentation of a strain of *Streptomyces pilosus* (*21*), has been the drug of choice for the treatment of transfusional iron overload for more than three decades. Although the ligand is very specific for iron, treatment with DFO is nevertheless expensive, unwieldy, inefficient, and unpleasant. This chelator is poorly absorbed from the gastrointestinal tract and rapidly eliminated from the circulation. Therefore, prolonged parenteral [subcutaneous (sc) or intravenous (iv)] infusion, usually for 8–12 h daily, is needed (*22, 23, 24, 25*), and patient compliance with this regimen is problematic. Because of these difficulties, considerable interest has focused on a search for either a parenteral deferration agent with superior iron clearing efficacy or, ideally, an iron chelator that is orally active.

Cooley's anemia (β-thalassemia major), with chronic transfusional iron overload, probably best illustrates the hurdles which must be overcome in chelator design. The lifelong treatment necessary in this disorder dictates that both the long-term toxicity of the drug and the efficiency with which the ligand

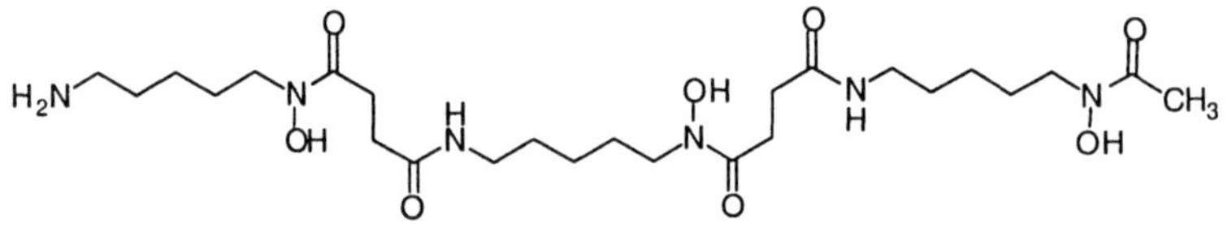

DFO

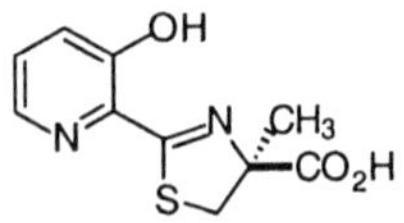

DFT

Figure 1. Structures of desferrioxamine B (DFO) and desferrithiocin (DFT).

removes iron are crucial. Because the efficiency after oral administration is negligible, desferrioxamine must be given parenterally. In addition, because the drug is rapidly cleared from the circulation, prolonged parenteral infusion is needed to remove enough iron to keep pace with the amounts supplied within the transfused blood. Both of these features of desferrioxamine adversely affect patient compliance. The parameters set by the nature of a particular iron-mediated disease on the toxicity profile and modes of administration of the ligands can vary significantly. For example, in reperfusion injury, more flexibility exists regarding toxicity profiles and modes of administration, inasmuch as short-term exposure to the drug is foreseen. Thus, ligands that are unacceptable for the treatment of systemic iron overload may be completely appropriate and useful in other maladies in which the iron overload is more localized. Despite these differences, some considerations are common to the design of chelators for treatment of any iron overload disorder, in addition to their ability to remove the metal. These considerations focus on the capacity of the chelators to prevent the promotion of the production of free radicals by iron and to scavenge already existing radicals.

The toxicity associated with excess iron, whether a systemic or a focal problem, derives from its interaction with reactive oxygen species, for instance, hydrogen peroxide (H_2O_2). This species is derived either from the spontaneous dismutation of superoxide anion or from the reaction catalyzed by superoxide dismutase. In the presence of Fe(II), H_2O_2 is reduced to the hydroxyl radical ($HO^\bullet$) and HO^-, often referred to as the Fenton reaction (Equation 1):

$$\mathrm{Fe(II)} + H_2O_2 \rightleftharpoons \mathrm{Fe(III)} + HO^\bullet + HO^- \quad (1)$$

Either the hydroxyl radical, the resulting lower-energy radicals that it produces in turn, or the radical chain reactions that it initiates can cause damage to cell components and promote the formation of carcinogens (*26, 27, 28*). Unfortunately, this is not a single event; there are any number of physiological reductants that can reduce Fe(III) back to Fe(II), such as ascorbate (Equation 2):

$$\text{Fe(III)} + \text{ascorbate} \rightleftharpoons \text{Fe(II)} + \text{ascorbyl radical} \tag{2}$$

Certain Fe(III) chelators, such as ethylenediaminetetraacetic acid (EDTA), (*29*) nitrilotriacetic acid (NTA) (*30*), 5-aminosalicylic acid (5-ASA) (*31, 32*), and 1,2-dimethyl-3-hydroxypyridin-4-one (L1) (*29, 33*) actually promote the reduction of Fe(III) to Fe(II). Not surprisingly, one of the major issues in the design of therapeutic iron chelators revolves around the impact of a given ligand on Fenton chemistry. Hexacoordinate Fe(III) ligands are generally more effective at preventing Fenton chemistry than are ligands of lower denticity (*26, 29, 34*). Hexacoordinate chelators surround the metal, preventing its reduction to Fe(II) and thus the Fe(II)-mediated reduction of H_2O_2. Although a ligand that prevents the conversion of Fe(III) to Fe(II) altogether is the most desirable, a chelator that does not actively promote the reduction, but still facilitates iron excretion from the animal, would be clinically useful. In the absence of iron removal, Fenton chemistry would continue.

The capacity of a chelator to trap free radicals is another notable characteristic. A ligand that can trap these reactive species might well render moot some of the problems mediated by radical species. In fact, such radical trapping properties have been observed with a variety of hexacoordinate ligands, including the naturally occurring hydroxamate DFO (*35*) and the synthetic polycarboxylate *N,N'*-bis(2-hydroxybenzyl)ethylenediamine-*N,N'*-diacetic acid (HBED) (*36*).

The Desferrithiocin Pharmacophore

Iron clearing efficiency is a measure of how much iron excretion is promoted by a chelator. It is a comparison of actual iron excretion versus theoretical iron clearance; this number is given as a percentage. For example, because DFO forms a very tight 1:1 complex with Fe(III), one millimole of DFO (656 mg as its mesylate salt) given sc to a patient should cause the excretion of one milligram-atom of Fe(III) (56 mg). Unfortunately, the efficiency of DFO in clinical use is only about 5 to 7% (*25, 37*).

Desferrithiocin (DFT, Figure 1) was one of the first iron chelators that was shown to be orally active (*38*). This agent performed well in both the bile duct-cannulated rodent model (iron clearing efficiency, 5.5%) (*39*) and in the iron-overloaded *Cebus apella* primate (iron clearing efficiency, 16%) (*40, 41*). Although this is three times as efficient as sc DFO in the primates, the compound was severely nephrotoxic (*41*). Nevertheless, the outstanding oral activity spurred us to launch a structure–activity study from the DFT pharmacophore in an effort to identify an orally active DFT analogue that has diminished toxicity. Our first goal was to identify the minimal structural platform compatible with iron clearance upon oral administration to the iron-overloaded primates.

The initial stage of the study entailed simplifying the platform. The thiazoline methyl of DFT was deleted to produce (*S*)-4,5-dihydro-2-(3-hydroxy-2-pyridinyl)-4-thiazolecarboxylic acid [(*S*)-DMDFT]; this derivative was less than one-third as efficient (4.8%) as the parent in the primate model (*41*). Removal of DFT's aromatic nitrogen left (*S*)-4,5-dihydro-2-(2-hydroxyphenyl)-4-methyl-4-thiazolecarboxylic acid [(*S*)-DADFT] (Figure 2) and only moderately diminished the compound's activity, an efficiency of 13% in *C. apella* (*42*). Taking away the thiazoline methyl from (*S*)-DADFT generated (*S*)-4,5-dihydro-2-(2-hydroxyphenyl)-4-thiazolecarboxylic acid [(*S*)-DADMDFT] (Figure 2); this modification had little additional effect on efficacy, 12% versus 13% efficiency (*41, 42*). These observations are consistent with the idea that more lipophilic chelators are more active.

Thus, a simple, synthetically easily accessible framework, (*S*)-DADMDFT, was available for subsequent structure–activity studies. The outcome of this second phase of the program revealed that very few further structural changes could be made to (*S*)-DADMDFT without loss of activity in the *C. apella* model. The C-4 carbon must remain in the (*S*)-configuration. (*42, 43, 44*) Both the thiazoline ring and the distances between the ligating centers must remain unaltered (*45, 46*). Benz-fusion of the aromatic rings was also counterproductive (*43, 46*).

The third phase of the study then focused on structural alterations that could reduce the toxicity of (*S*)-DADMDFT. Both desazadesferrithiocin analogues, (*S*)-DADMDFT and (*S*)-DADFT, were still quite toxic (*41, 42, 46*). This structure–activity analysis focused on the effect of changing the partition properties and potential metabolic profiles of the compounds on their toxicity parameters by introducing various substituents into the aromatic ring. We found that addition of aromatic hydroxyl groups, as in the systems (*S*)-2-(2,4-dihy-

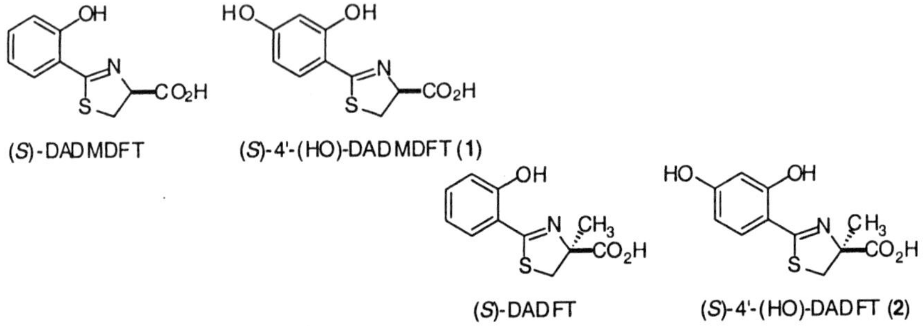

Figure 2. Structures of (S)*-4,5-dihydro-2-(2-hydroxyphenyl)-4-thiazolecarboxylic acid [(*S*)-DADMDFT] and its 4'-hydroxylated analogue [(*S*)-4'-(HO)-DADMDFT; **1**] (upper left) and (*S*)-4,5-dihydro-2-(2-hydroxyphenyl)-4-methyl-4-thiazolecarboxylic acid [(*S*)-DADFT] and its 4'-hydroxylated analogue [(*S*)-4'-(HO)-DADFT; **2**] (lower right).*

droxyphenyl)-4,5-dihydro-4-thiazolecarboxylic acid [(*S*)-4'-(HO)-DADMDFT, **1**] and (*S*)-2-(2,4-dihydroxyphenyl)-4,5-dihydro-4-methyl-4-thiazolecarboxylic acid [(*S*)-4'-(HO)-DADFT, **2**] (Figure 2), strikingly diminished ligand toxicity. In both instances, the hydroxylated compounds are less active than their corresponding parent drugs, 5.3% efficiency for **1** [vs 12.4% for (*S*)-DADMDFT when administered orally (po) at a dose of 300 μmol/kg] and 17% efficiency for **2** [vs 21% for (*S*)-DADFT when administered po at a dose of 75 μmol/kg] (*42, 46*). Again, the fact that introduction of a 4'-hydroxyl reduced the ligand's activity is consistent with the importance of lipophilicity in the chelator's iron clearing efficiency.

It was thus not unexpected that hydroxylation changes the partition between octanol and water, favoring the aqueous phase (*46*). This compelled us to consider methoxylation as an alternative to hydroxylation at the 3'- or 4'-position of these DFT derivatives. Methoxylation could potentially increase lipophilicity, altering the octanol–water partition values, and improve iron clearing efficiency. The methoxy group would still maintain the electron-feeding properties of the 4'-hydroxyl. Accordingly, the current work focuses on the impact of lipophilicity within two desferrithiocin analogue structural families both on the efficiencies of iron clearance and on the mode of iron excretion, that is, biliary versus renal metal clearance. Our studies also assess the ability of these systems to prevent ascorbate-mediated reduction of Fe(III) to Fe(II) and to act as radical scavengers.

Design Concept and Synthesis

Design Concept

Initial considerations (*46*) suggested that the correlation between chelator lipophilicity, that is, octanol–water partition coefficients (log *P*) and iron clearing efficiencies was poor. For example, although the octanol–water partition coefficients of DFT and (*S*)-DMDFT are very similar (–1.77 and –1.87, respectively), their iron clearing efficiencies in primates were quite different, 16.1 ± 8.5% and 4.8 ± 2.7%, respectively. However, similar considerations within the same structural group were somewhat more revealing. For example, a comparison of the partition properties of DADMDFT and (*S*)-4'-(HO)-DADMDFT (**1**) (–0.93 and –1.2, respectively) and their respective iron clearing efficiencies (12.4 ± 7.6% vs. 5.3 ± 1.7%) does imply a relationship between lipophilicity and iron clearing efficiency. Accordingly, we have assembled a group of structurally similar desferrithiocins that have graded variations in their lipophilic behavior. Based on the substitutions in the aromatic ring, two different core systems were considered, 2,4-dihydroxyphenyl (**1**–**5**) and 2,3-dihydroxyphenyl (**6**–**9**) compounds (Scheme 1, Table I). The 4'-substituted systems included (*S*)-4'-(HO)-DADMDFT (**1**), (*S*)-4'-(HO)-DADFT (**2**), (*S*)-4,5-

dihydro-2-(2-hydroxy-4-methoxyphenyl)-4-thiazolecarboxylic acid [(*S*)-4'-(CH_3O)-DADMDFT, **3**], (*S*)-4,5-dihydro-2-(2-hydroxy-4-methoxyphenyl)-4-methyl-4-thiazolecarboxylic acid [(*S*)-4'-(CH_3O)-DADFT, **4**], and (*S*)-2-(2,4-dihydroxyphenyl)-4,5-dihydro-5,5-dimethyl-4-thiazolecarboxylic acid (**5**). The 3'-substituted systems encompassed (*S*)-2-(2,3-dihydroxyphenyl)-4,5-dihydro-4-thiazolecarboxylic acid [(*S*)-3'-(HO)-DADMDFT, **6**], (*S*)-2-(2,3-dihydroxyphenyl)-4,5-dihydro-4-methyl-4-thiazolecarboxylic acid [(*S*)-3'-(HO)-DADFT, **7**], (*S*)-4,5-dihydro-2-(2-hydroxy-3-methoxyphenyl)-4-thiazolecarboxylic acid [(*S*)-3'-(CH_3O)-DADMDFT, **8**], and (*S*)-4,5-dihydro-2-(2-hydroxy-3-methoxyphenyl)-4-methyl-4-thiazolecarboxylic acid [(*S*)-3'-(CH_3O)-DADFT, **9**].

Synthetic Methods

Two different cyclizations were employed in the assembly of the ligands (Scheme 1). Condensation of the appropriate nitrile (generic structure **10**) with D-cysteine (R_3, R_4, and R_5 = H) produced compounds **1**, **3**, **6**, and **8**; reaction with D-α-methyl cysteine ($R_3 = CH_3$; $R_4 = R_5$ = H) provided derivatives **2**, **4**, **7**,

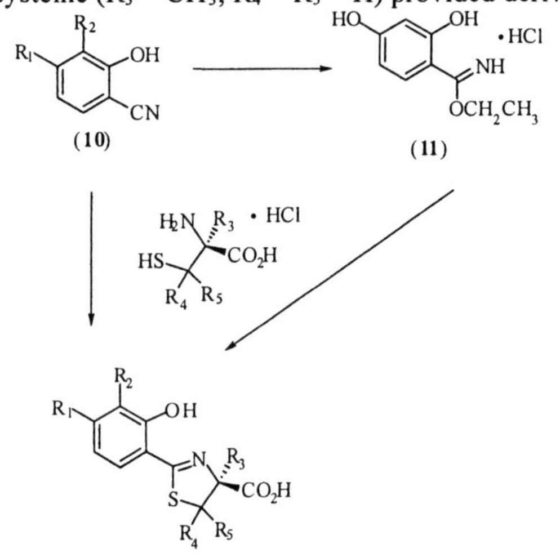

(**10**) (**11**) •HCl

(**1**) R_1 = OH; $R_2 = R_3 = R_4 = R_5$ = H 66%
(**2**) R_1 = OH; R_2 = H; $R_3 = CH_3$; $R_4 = R_5$ = H 57%
(**3**) $R_1 = OCH_3$; $R_2 = R_3 = R_4 = R_5$ = H 71%
(**4**) $R_1 = OCH_3$; R_2 = H; $R_3 = CH_3$; $R_4 = R_5$ = H 46%
(**5**) R_1 = OH; $R_2 = R_3$ = H; $R_4 = R_5 = CH_3$ 43%
(**6**) R_1 = H; R_2 = OH; $R_3 = R_4 = R_5$ = H 19%
(**7**) R_1 = H; R_2 = OH; $R_3 = CH_3$; $R_4 = R_5$ = H 15%
(**8**) R_1 = H; $R_2 = OCH_3$; $R_3 = R_4 = R_5$ = H 90%
(**9**) R_1 = H; $R_2 = OCH_3$; $R_3 = CH_3$; $R_4 = R_5$ = H 39%

*Scheme 1. Synthesis of desferrithiocin analogues **1–9**; the yields (%) for cyclization are given. Generation of derivatives **1** and **8** was presented in reference 46; synthesis of analogue **2** was described in reference 42. The details of the production of compounds **3–5** are given in reference 47.*

and 9. These cyclizations were carried out in heated phosphate-buffered methanol at pH 6. The nitriles for **1–4** (R_1 = OH or OCH_3; R_2 = H) were prepared by treatment of the corresponding 2,4-disubstituted benzaldehyde with nitroethane and sodium acetate in acetic acid at reflux. The nitrile for ligands **8** and **9** (R_1 = H; R_2 = OCH_3) resulted from heating 2-hydroxy-3-methoxybenzaldehyde, hydroxylamine hydrochloride, and sodium formate in aqueous formic acid. 2,3-Dihydroxybenzonitrile (**10**, R_1 = H and R_2 = OH), the precursor to chelators **6** and **7**, was obtained by exposure of 2,3-dimethoxybenzonitrile to BBr_3 in methylene chloride.

In the case of **5**, condensation of ethyl 2,4-dihydroxybenzimidate (**11**) with D-penicillamine (R_3 = H; R_4 = R_5 = CH_3) in refluxing aqueous ethanol in the presence of sodium bicarbonate afforded the target compound. Utilization of the imidate was necessary, owing to the low yields realized when the condensation was attempted with the nitrile. Imidate **11** was accessed by reacting 2,4-dihydroxybenzonitrile (**10**, R_1 = OH and R_2 = H) with hydrogen chloride gas in ethanol.

Iron Clearance

The animal model selected for this series of experiments was the iron-overloaded *C. apella* monkey. In this model, we are able to measure both total iron excretion, expressed as iron clearing efficiency, and the modes of iron clearance, that is, biliary and renal. Data are presented (Table I) as the mean ± standard error of the mean. For comparison of the means between two compounds, a two-sample *t*-test (without the assumption of equality of variances) was performed. All tests were two-tailed, and a significance level of $P < 0.05$ was used.

In the primate model, (*S*)-4'-(HO)-DADMDFT (**1**) performs moderately well when given orally (*42*, *46*). At a po dose of 150 μmol/kg, the efficiency was 4.2 ± 1.4%; 70% of the iron excretion was fecal. Introduction of a methyl group into the thiazoline ring of **1** [(*S*)-4'-(HO)-DADFT, **2**] led to a substantial increase in iron clearing efficiency, to 13.4 ± 5.8% ($P < 0.05$ vs **1**); the stool/urine ratio was 86:14 (*42*). Methylation of the 4'-hydroxyl of **1** to afford **3** more than tripled the deferration efficacy to 16.2 ± 3.2% ($P < 0.005$ vs **1**); 81% of the iron excretion was fecal, and 19% was urinary (*47*). The same modification to **2**, affording (*S*)-4'-(CH_3O)-DADFT (**4**), further augmented the iron clearing efficiency, to 24.4 ± 10.8%. This effectiveness is significantly greater than that of **1** ($P < 0.005$), although the differences in efficiency between **4** and either **2** or **3** were not statistically significant. In the case of analogue **4**, slightly more of the iron excretion, 91%, was via the stool; only 9% was via the urine (*47*).

The situation was similar with the 2',3'-dihydroxy derivatives. When a methyl group was introduced into the 4-position of the thiazoline ring of (*S*)-3'-(HO)-DADMDFT (**6**) to afford (*S*)-3'-(HO)-DADFT (**7**), both the iron clearing efficiency and the modes of excretion changed. The efficiency of compound **6**

Table I. Desferrithiocin Analogues' Iron Clearing Activity when Administered Orally to *Cebus apella* Primates and the Partition Coefficients of the Compounds

4'-Substituted Compounds			*3'-Substituted Compounds*		
Desferrithiocin Analogue (compd. no.)	*Iron Clearing Efficiency (%)*[a]	*log* P[b]	*Desferrithiocin Analogue (compd. no.)*	*Iron Clearing Efficiency (%)*[a]	*log* P[b]
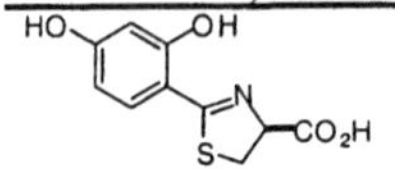 **1**	4.2 ± 1.4[c] [70/30]	−1.33	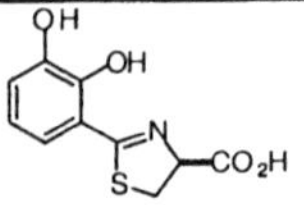 **6**	5.8 ± 3.4 [91/9]	−1.67
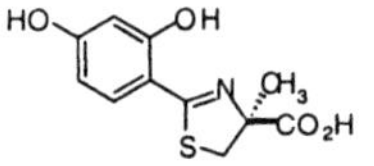 **2**	13.4 ± 5.8[d] [86/14]	−1.05	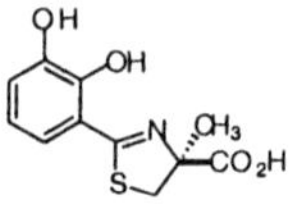 **7**	23.1 ± 5.9 [83/17]	−1.17
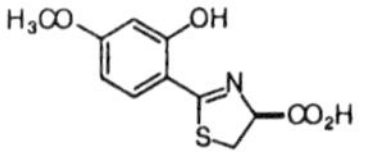 **3**	16.2 ± 3.2[e] [81/19]	−0.89	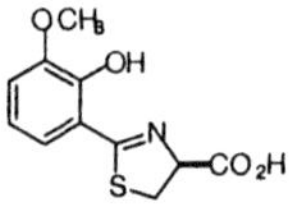 **8**	15.5 ± 7.3 [87/13]	−1.52
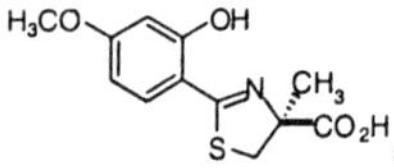 **4**	24.4 ± 10.8[e] [91/9]	−0.70	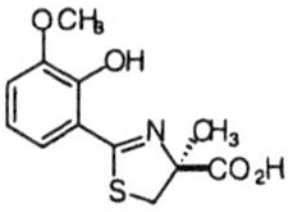 **9**	22.5 ± 7.1 [91/9]	−1.12
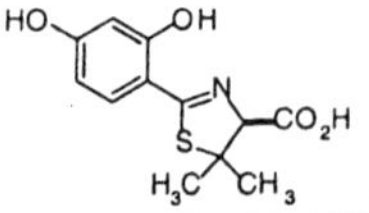 **5**	12.3 ± 2.7[e] [64/36]	−0.91			

SOURCE: Adapted from reference 47. Copyright 2003 American Chemical Society.

[a] In the monkeys [n = 4 (**1**, **2**, **3**, **5**, **7**), 7 (**4**), 5 (**9**), 6 (**8**), or 8 (**6**)], the dose was 150 μmol/kg. The efficiency of each compound was calculated by averaging the iron output for 4 days before the administration of the drug, subtracting these numbers from the 2-day iron clearance after the administration of the drug, and then dividing by the theoretical output; the result is expressed as a percent. The relative percentages of the iron excreted in the stool and urine are in brackets.

[b] Data are expressed as the log of the fraction in the octanol layer; measurements were done in TRIS buffer, pH 7.4, using a "shake flask" direct method. The values obtained for compounds **1–5** are from reference 47.

[c] Data are from reference 46.

[d] Data are from reference 42.

[e] Data are from reference 47.

was 5.8 ± 3.4%; the stool:urine ratio was 91:9. Upon methylation of the thiazoline ring to produce (**7**), the efficiency increased to 23.1 ± 5.9% ($P < 0.01$ vs. **6**). Eighty-three percent of the iron was in the stool, and 17% was in the urine. An increase in iron clearing efficiency was observed in the (*S*)-3'-(CH_3O)-DADMDFT series (analogues **8** and **9**, Table I), although the difference between these two compounds was not statistically significant. Whereas the efficiency of derivative **8** was 15.5 ± 7.3%, that of 4-methyl compound **9** was 22.5 ± 7.1% ($P > 0.05$). 4-Methylation also modestly elevated the proportion of the iron in the stool, 91% for **9** versus 87% for **8**. Again, although methylation of the 3'-hydroxyl also increased the iron clearing efficiency (e.g., **8** vs **6**, $P = 0.02$), methylation of the thiazoline ring exerted a more profound effect.

Finally, addition of two methyl groups to the 5-position of the thiazoline ring of **1** (DM, **5**) had an obvious impact on iron clearing efficacy; the efficiency was 12.3 ± 2.7% ($P < 0.005$ vs **1**). When **5** is compared with **3** and **4**, considerably less of the iron was excreted in the feces, 64%, and 36% of the iron was in the urine (*47*). In general, whereas methylation did alter the modes of iron excretion, the effect on overall iron clearing efficiency was much more striking.

Relationship between Partition Coefficients and Iron Clearing Efficacy and Modes of Excretion

The partition coefficients of ligands **1**–**9** were evaluated in buffered octanol–water. The log *P* values (Table I) ranged from –0.70 for **4** to –1.67 for **6**. Attaching alkyl groups to the ligands increases a compound's lipophilicity; furthermore, when the iron clearing efficiency is plotted versus partition coefficient (Figure 3A), it is obvious that increasing the lipophilicity also augments the iron clearing efficiency within a particular family. A similar relationship can be observed when fecal iron clearance is plotted versus log *P* (Figure 3B). To assess the significance of the apparent linear trends consistent with these observations, weighted regression analyses of the 2',4'-disubstituted analogues were carried out both with and without ligand **5** (*47*). The results indicated that there was sufficient evidence to conclude a significant ($P < 0.001$) relationship between efficiency and log *P* regardless of whether **5** was included, although the strength of the slope was greater when **5** was excluded. In the case of the 2',4'-dihydroxy compounds, these findings indicated that methylation of the 4'-hydroxyl of the parent **1**, as in **3**, seems to have a more dramatic effect on lipophilicity, iron clearing efficiency, and fecal iron clearance than does appending a methyl to the 4-position of the thiazoline ring (**2**), or attaching two methyls to the 5-position of the thiazoline ring (**5**).

In the case of the 2',3'-dihydroxy compounds, there was also an apparent relationship between log *P* and iron clearing efficiency. However, what is most

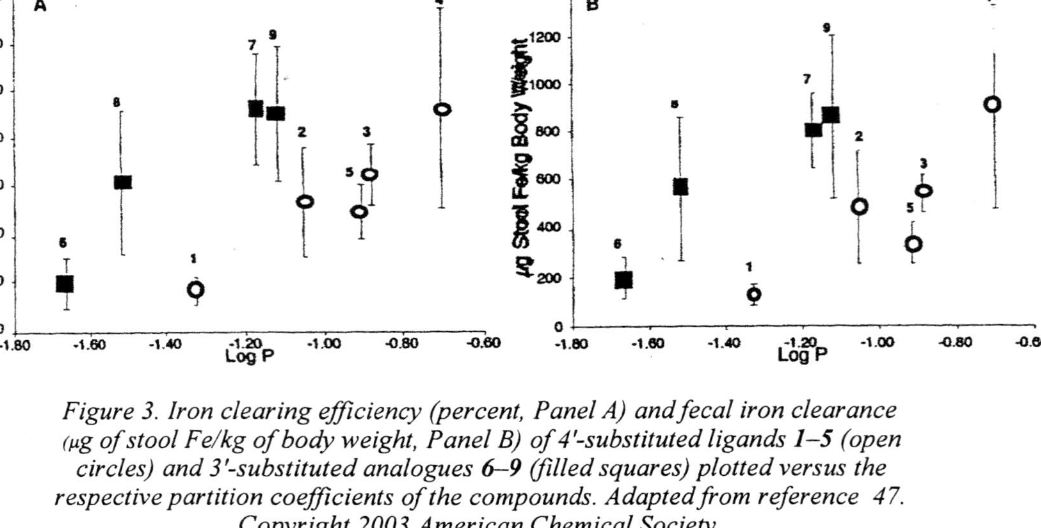

*Figure 3. Iron clearing efficiency (percent, Panel A) and fecal iron clearance (µg of stool Fe/kg of body weight, Panel B) of 4'-substituted ligands **1–5** (open circles) and 3'-substituted analogues **6–9** (filled squares) plotted versus the respective partition coefficients of the compounds. Adapted from reference 47. Copyright 2003 American Chemical Society.*

interesting is the difference in the impact that changing log *P* has on the efficiency of the 2',3'-substituted derivatives vs the effect on that of the 2',4'-substituted analogues. Linear regression analyses were conducted to find whether the relationship between log *P* and iron clearing efficiency was the same for both series of compounds. The data consisited of the log *P* for all of the drug groups within both families. A preliminary inspection of the data indicated that ordinary least squares estimation could be applied to test for the relationship between iron clearing efficiency and log *P*. Although the data do not appear to be perfectly linear (Figure 3), any curvature was negligible so that ordinary least squares could still be applied. A group * log *P* interaction was tested to determine whether the slopes of the lines for both families were equal (i.e., H_o: group * log $P = 0$), that the relationship between iron clearing efficiency and log *P* was the same for both groups. The results indicated that the interaction was not significant, suggesting that the slopes for the two groups were not significantly different from each other ($P = 0.815$). The interaction term (group * log *P*) was then dropped from the model. The effect for group was tested to indicate for which group iron clearing efficacy was the greater (i.e., H_o: groups = 0). These results clearly show a significant group effect with the 2',3'-disubstituted set of analogues having a greater overall efficiency (Figure 3).

Antioxidant and Free Radical Scavenging Activity

Effect of Methoxylation on the Ability of Desferrithiocin Analogues to Prevent Fenton Chemistry

The principal reason for removing excess iron from tissue is to prevent redox cell damage. Thus, one of the issues of major concern is related to the effect of any potential therapeutic ligand on the reduction of Fe(III) to Fe(II). In the +2 oxidation state, iron drives Fenton chemistry (Equation 1). Thus, a situation in which the reduction of Fe(III) to Fe(II) is promoted by a chelator would worsen any Fenton chemistry-mediated cell damage. Because some iron chelators are known to catalyze this reduction (*29, 30*), a determination of whether a ligand promotes, prevents, or has no effect on Fe(III) reduction is critical in designimg therapeutic chelators for clinical use. Of the many physiological reductants that can reduce Fe(III) to Fe(II), ascorbate is a frequently used model. Ascorbate oxidation is easy to follow qualitatively by its disappearance spectrophotometrically (*29, 33*). The data in Figure 4 are plotted as the change in ascorbate concentration versus the ligand/metal ratio. Diminished Fe(III)-mediated ascorbate oxidation is indicated by a value less than 100% of control. Because the equilibria unfolding during the reaction are quite complicated, these results should not be overinterpreted. The question addressed is, simply, in the presence of a test ligand, is the rate of ascorbate oxidation/iron reduction increased, decreased, or the same relative to controls containing iron(III) and ascorbate without the compound? Desferrioxamine functions as a positive control, as it decreases ascorbate loss (*29*); 1,2-dimethyl-3-hydroxypyridin-4-one (L1), which increases ascorbate disappearance at ratios less than 3:1, serves as a negative control. Recall that the desferrithiocin analogues form 2:1 complexes with Fe(III); accordingly, the ligand-to-metal ratios from 0.5 to 2 are measured for ligands **1–9** (Figure 4). The 4'-substituted derivatives (**1–5**) (*47*), 3'-hydroxy (**6** and **7**), and 3'-methoxy (**8** and **9**) analogues slow Fe(III) reduction considerably, even at a 0.5:1 ratio, as did the parent compound DFT (data not shown) (*48, 49*). The results with the parent ligand are consistent with those reported in this same assay (*29, 33*) and in a cultured cell system (*33*). These results indicate that an enhancement of Fenton chemistry is unlikely to account for the toxicity of desferrithiocin and some of its derivatives.

Effect of Methoxylation on the Ability of Desferrithiocin Analogues to Scavenge Free Radicals

Radical traps can help to attenuate Fenton chemistry-induced, free radical-mediated damage; thus, the radical scavenging properties of a particular ligand are of importance. Two hexacoordinate iron chelators, *N,N'*-bis(2-

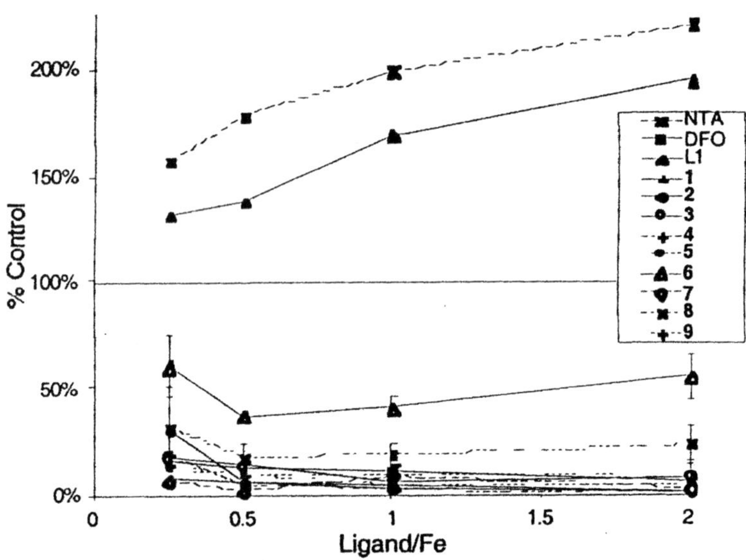

Figure 4. Effect of various chelators on the iron-mediated oxidation of ascorbate (Fenton chemistry) (percent of control, y*-axis): 1,2-dimethyl-3-hydroxypyridin-4-one (L1), desferrioxamine (DFO), and DFT analogues 1–9 at several ligand:metal ratios (*x*-axis). Typically, each assay contained 3 controls with only ascorbate and* $FeCl_3$*; these varied less than 5% (± s.d.). Each assay also usually included a "negative control" containing ascorbate,* $FeCl_3$*, and L1 at a ligand:iron ratio of 2:1; this value was 174 ± 27%..*

hydroxybenzyl)ethylenediamine-*N*,*N'*-diacetic acid (HBED) (*36*) and DFO (*35, 48*), are excellent radical traps. The issue of Fenton chemistry and chelator design has a dimension beyond the prevention of the reduction of Fe(III). The liberated $HO^{\bullet}$ molecules are very short-lived, reacting with most surrounding molecules at a diffusion-controlled rate. Ultimately, less active, more selective radicals, which can initiate a radical chain process, are produced and can cause significant cell damage (*50*).

We have evaluated the 4'-methoxylated analogues (**3** and **4**), their parent molecules (**1** and **2**), and the 5,5-dimethyl analogue (**5**) in a free radical scavenging assay (*47*); the 3'-hydroxylated compounds (**6** and **7**) and their respective O-methylated derivatives (**8** and **9**) have also been assessed. The capacity of each of these ligands to function as a one-electron donor to the preformed radical monocation of 2,2'-azinobis(3-ethylbenzothiazoline-6-sulfonic acid) ($ABTS^{\bullet +}$) was compared with that of Trolox, an analogue of vitamin E (*51*). Table II shows the calculated slopes of the decrease in A_{734} versus ligand concentration line for each compound; a more negative slope indicates a more effective radical scavenger. As anticipated, the 3'- and 4'-methoxylated compounds were less effective radical scavengers than were the corresponding 3'- and 4'-hydroxylated molecules; nevertheless, **3**, **4**, **8**, and **9** were as effective as Trolox at trapping free radicals and much better than the parent DFT. Analogue **5**, on the other hand, was about twice as active as Trolox. Compounds **6** and **7** scavenged the model radical cation at a level comparable to that of analogue **5** and considerably less than those of the corresponding 4'-hydroxy analogues **1** and **2**. Although it is uncertain what effect the iron complexes of these analogues would have on free radicals, it is very likely that in any given dosing situation, considering that the efficiencies of these compounds are under 25%, most of the ligand would remain uncomplexed and thus be available for free radical scavenging.

Conclusion

Overall, the octanol–water partition properties (log *P*) of the desferrithiocin analogues are clearly related to both their iron clearing efficiency and modes of excretion. Nonetheless, this relationship holds more strictly within a given structural group, for example, 2',3'-dihydroxy or 2',4'-dihydroxy compounds. Methylation increases both iron clearing efficiency and the fraction of biliary iron excretion (Figure 3). Interestingly, the effects are roughly the same in both the 2',3'- and 2',4'-disubstituted chelators.

Although methylation of the 3'- or 4'-hydroxyl did not alter the ability of the ligand to prevent ascorbate reduction of Fe(III), it did decrease the ligands' free radical scavenging capacity relative to their respective parent non-methylated systems. Nevertheless, these compounds were still at least as active as the positive control, the vitamin E analogue Trolox.

Table II. ABTS Radical Cation Quenching Activity of Desferrithiocin Derivatives

Compound	*Slope × 10³ OD units/μM*[a]	*Compound*	*Slope × 10³ OD units/μM*[a]
DFT	–0.9[b]	**7**	–58
3	–33[c]	**6**	–62
8	–35	**5**	–70[c]
4	–36[c]	**1**	–102[b]
Trolox	–37[b]	**2**	–106[b]
9	–39		

SOURCE: Adapted from reference 49. Copyright 2003 American Chemical Society.

[a] The slope was derived from A_{734} versus concentration data after a 6-min reaction period between the chelator of interest and the 2,2'-azinobis(3-ethylbenzothiazoline-6-sulfonic acid) radical cation ($ABTS^{\bullet +}$), which was formed from the reaction between ABTS and persulfate. A negative slope represents a decrease in the amount of highly colored radical cation over the time interval from an initial A_{734} of about 0.900. Trolox, an analogue of vitamin E, served as a positive control.

[b] Data are from reference 48.

[c] Data are from reference 47.

Acknowledgments

The U.S. National Institutes of Health (Grant No. R01-DK49108) is gratefully acknowledged for its support of the research from the Bergeron laboratory. We appreciate the technical assistance of Harold Snellen, Tim Vinson, and Samuel E. Algee.

References

1. Raymond, K. N.; Carrano, C. J. *Acc. Chem. Res.* **1979,** *12,* 183-190.
2. Bergeron, R. J. *Trends Biochem. Sci.* **1986,** *11,* 133-136.
3. Theil, E. C.; Huynh, B. H. *J. Inorg. Biochem.* **1997,** *67,* 30.
4. Ponka, P.; Beaumont, C.; Richardson, D. R. *Semin. Hematol.* **1998,** *35,* 35-54.
5. Glozman, S.; Yavin, E. *Neurochem. Res.* **1997,** *22,* 201-208.
6. Palmer, C.; Roberts, R. L.; Bero, C. *Stroke* **1994,** *25,* 1039-1045.
7. Double, K. L.; Riederer, P.; Gerlach, M. *Drug News Perspect.* **1999,** *12,* 333-340.
8. Loeffler, D. A.; Connor, J. R.; Juneau, P. L.; Snyder, B. S.; Kanaley, L.; DeMaggio, A. J.; Nguyen, H.; Brickman, C. M.; LeWitt, P. A. *J. Neurochem.* **1995,** *65,* 710-716.
9. Ben-Shachar, D.; Eshel, G.; Riederer, P.; Youdim, M. B. *Ann. Neurol.* **1992,** *32,* S105-S110.
10. Riederer, P.; Sofic, E.; Rausch, W. D.; Schmidt, B.; Reynolds, G. P.; Jellinger, K.; Youdim, M. B. *J. Neurochem.* **1989,** *52,* 515-520.
11. Becker, E.; Richardson, D. R. *Int. J. Biochem. Cell Biol.* **2001,** *33,* 1-10.
12. Puccio, H.; Simon, D.; Cossée, M.; Criqui-Filipe, P.; Tiziano, F.; Melki, J.; Hindelang, C.; Matyas, R.; Rustin, P.; Koenig, M. *Nat. Genet.* **2001,** *27,* 181-186.
13. Finch, C. A.; Deubelbeiss, K.; Cook, J. D.; Eschbach, J. W.; Harker, L. A.; Funk, D. D.; Marsaglia, G.; Hillman, R. S.; Slichter, S.; Adamson, J. W.; Ganzoni, A.; Giblett, E. R. *Medicine (Baltimore)* **1970,** *49,* 17-53.
14. Finch, C. A.; Huebers, H. A. *N. Engl. J. Med.* **1982,** *306,* 1520-1528.
15. Finch, C. A.; Huebers, H. A. *Clin. Physiol. Biochem.* **1986,** *4,* 5-10.
16. Hallberg, L. *Annu. Rev. Nutr.* **1981,** *1,* 123-147.
17. Conrad, M. E.; Umbreit, J. N.; Moore, E. G. *Am. J. Med. Sci.* **1999,** *318,* 213-229.
18. O'Connell, M. J.; Ward, R. J.; Baum, H.; Peters, T. J. *Biochem. J.* **1985,** *229,* 135-139.
19. Seligman, P. A.; Klausner, R. D.; Huebers, H. A. In: *The Molecular Basis of Blood Diseases,* G. Stamatoyannopoulos; A. W. Nienhuis; P. Leder; P. W. Majeris, Eds.; W. B. Saunders: Philadelphia, PA, 1987; p. 219.
20. Thomas, C. E.; Morehouse, L. A.; Aust, S. D. *J. Biol. Chem.* **1985,** *260,* 3275-3280.
21. Bickel, H.; Hall, G. E.; Keller-Schierlein, W.; Prelog, V.; Vischer, E.; Wettstein, A. *Helv. Chim. Acta* **1960,** *43,* 2129-2138.

22. Porter, J. B. *Semin. Hematol.* **2001,** *38,* 63-68.
23. Lee, P.; Mohammed, N.; Marshall, L.; Abeysinghe, R. D.; Hider, R. C.; Porter, J. B.; Singh, S. *Drug Metab. Dispos.* **1993,** *21,* 640-644.
24. Pippard, M. J.; Callender, S. T.; Finch, C. A. *Blood* **1982,** *60,* 288-294.
25. Pippard, M. J. *Bailliere's Clin. Haematol.* **1989,** *2,* 323-343.
26. Babbs, C. F. *Free Radic. Biol. Med.* **1992,** *13,* 169-181.
27. Halliwell, B. *Nutr. Rev.* **1994,** *52,* 253-265.
28. Hazen, S. L.; d'Avignon, A.; Anderson, M. M.; Hsu, F. F.; Heinecke, J. W. *J. Biol. Chem.* **1998,** *273,* 4997-5005.
29. Dean, R. T.; Nicholson, P. *Free Radic. Res.* **1994,** *20,* 83-101.
30. Graf, E.; Mahoney, J. R.; Bryant, R. G.; Eaton, J. W. *J. Biol. Chem.* **1984,** *259,* 3620-3624.
31. Grisham, M. B. *Biochem. Pharmacol.* **1990,** *39,* 2060-2063.
32. Grisham, M. B.; Ware, K.; Marshall, S.; Yamada, T.; Sandhu, I. *Dig. Dis. Sci.* **1992,** *37,* 1383-1389.
33. Cragg, L.; Hebbel, R. P.; Miller, W.; Solovey, A.; Selby, S.; Enright, H. *Blood* **1998,** *92,* 632-638.
34. Hiraishi, H.; Terano, A.; Ota, S.; Mutoh, H.; Razandi, M.; Sugimoto, T.; Ivey, K. J. *Am. J. Physiol.* **1991,** *260,* G556-G563.
35. Denicola, A.; Souza, J. M.; Gatti, R. M.; Augusto, O.; Radi, R. *Free Radic. Biol. Med.* **1995,** *19,* 11-19.
36. Samuni, A. M.; Afeworki, M.; Stein, W.; Yordanov, A. T.; DeGraff, W.; Krishna, M. C.; Mitchell, J. B.; Brechbiel, M. W. *Free Radic. Biol. Med.* **2001,** *30,* 170-177.
37. Pippard, M. J.; Jackson, M. J.; Hoffman, K.; Petrou, M.; Modell, C. B. *Scand. J. Haematol.* **1986,** *36,* 466-472.
38. Wolfe, L. C.; Nicolosi, R. J.; Renaud, M. M.; Finger, J.; Hegsted, M.; Peter, H.; Nathan, D. G. *Br. J. Haematol.* **1989,** *72,* 456-461.
39. Bergeron, R. J.; Wiegand, J.; Dionis, J. B.; Egli-Karmakka, M.; Frei, J.; Huxley-Tencer, A.; Peter, H. H. *J. Med. Chem.* **1991,** *34,* 2072-2078.
40. Bergeron, R. J.; Streiff, R. R.; Wiegand, J.; Vinson, J. R. T.; Luchetta, G.; Evans, K. M.; Peter, H.; Jenny, H.-B. *Ann. N. Y. Acad. Sci.* **1990,** *612,* 378-393.
41. Bergeron, R. J.; Streiff, R. R.; Creary, E. A.; Daniels, R. D., Jr.; King, W.; Luchetta, G.; Wiegand, J.; Moerker, T.; Peter, H. H. *Blood* **1993,** *81,* 2166-2173.
42. Bergeron, R. J.; Wiegand, J.; McManis, J. S.; McCosar, B. H.; Weimar, W. R.; Brittenham, G. M.; Smith, R. E. *J. Med. Chem.* **1999,** *42,* 2432-2440.
43. Bergeron, R. J.; Wiegand, J.; Wollenweber, M.; McManis, J. S.; Algee, S. E.; Ratliff-Thompson, K. *J. Med. Chem.* **1996,** *39,* 1575-1581.
44. Bergeron, R. J.; Wiegand, J.; Weimar, W. R.; McManis, J. S.; Smith, R. E.; Abboud, K. A. *Chirality* **2003,** *15,* 593–599.
45. Bergeron, R. J.; Liu, C. Z.; McManis, J. S.; Xia, M. X. B.; Algee, S. E.; Wiegand, J. *J. Med. Chem.* **1994,** *37,* 1411-1417.
46. Bergeron, R. J.; Wiegand, J.; Weimar, W. R.; Vinson, J. R. T.; Bussenius, J.; Yao, G. W.; McManis, J. S. *J. Med. Chem.* **1999,** *42,* 95-108.

47. Bergeron, R. J.; Wiegand, J.; McManis, J. S.; Bussenius, J.; Smith, R. E.; Weimar, W. R. *J. Med. Chem.* **2003,** *46,* 1470-1477.
48. Bergeron, R. J.; Wiegand, J.; Weimar, W. R.; Nguyen, J. N.; Sninsky, C. A. *Dig. Dis. Sci.* **2003,** *48,* 399-407.
49. Bergeron, R. J.; Huang, G.; Weimar, W. R.; Smith, R. E.; Wiegand, J.; McManis, J. S. *J. Med. Chem.* **2003,** *46,* 16-24.
50. Grisham, M. B.; Granger, D. N. *Dig. Dis. Sci.* **1988,** *33,* 6S-15S.
51. Re, R.; Pellegrini, N.; Proteggente, A.; Pannala, A.; Yang, M.; Rice-Evans, C. *Free Radic. Biol. Med.* **1999,** *26,* 1231–1237.

Chapter 21

In Vivo Coordination Chemistry and Biolocalization of Bis(ligand)oxovanadium(IV) Complexes for Diabetes Treatment

Katherine H. Thompson[1,*], Barry D. Liboiron[1,3], Graeme R. Hanson[2], and Chris Orvig[1,*]

[1]Medicinal Inorganic Chemistry Group, Department of Chemistry, University of British Columbia, Vancouver, British Columbia, Canada
[2]Centre for Magnetic Resonance, University of Queensland, St. Lucia, Queensland, Australia
[3]Current address: Department of Chemistry, Stanford University, Stanford, CA 94305–5080

Many vanadyl complexes of the form VOL_2, where L is a monoprotic bidentate ligand, have been investigated for the treatment of diabetes mellitus. Not all such compounds are effective in counteracting diabetic symptomatology. An emerging picture of the metabolic fate of successful vanadium complexes is one of rapid de-complexation following oral administration, calling into question the role of the ligand in determining relative potency. A model of VOL_2 uptake, distribution and excretion is proposed that takes into account coordination, biolocalization and *in vivo* speciation results from recent studies.

Vanadium, as sodium vanadate, was first recognized as having orally available insulin enhancing potential in the mid-1980's (1). Although earlier anecdotal reports of anti-diabetic efficacy dated to the late 1800's (reported in (2)), no formal experimental testing was carried out until after *in vitro* testing had demonstrated a potential biochemical role for vanadium (3,4). The first well-controlled *in vivo* trials were reported in 1985 (1). There, and in many subsequent trials (5), it has been shown that vanadium compounds (of both V(V) and V(IV)) tend to 'mimic' insulin in their diverse biochemical actions, stimulating or inhibiting many of the same metabolic pathways *in vitro* (6,7) and in experimental animals (for review see (8), and references therein).

Despite a burgeoning interest in the anti-diabetic effects of vanadium species, neither the specific mechanism of action nor the metabolic fate of vanadium compounds is well understood. The term 'insulin enhancement' is used frequently as a reminder that vanadium never entirely substitutes for insulin, and, in fact, cannot function *in vivo* in the absence of insulin (9). Supplementation with vanadium reduces the requirement for exogenous insulin, especially in models of type 2 diabetes (9), and in that sense, 'enhances' existing insulin stores. With that understanding, we will use the term insulin mimetic agent (IMA) for the remainder of this article.

Vanadium compounds are potentially unique among metal-containing compounds in their ability to alleviate diabetic symptomatology. Candidate pro-drugs containing a variety of other metal ions have been tried as alternates to vanadium compounds as insulin mimetic agents (IMA's), with modest success (10,11). However, in a head-to-head comparison between vanadium- and other metallo-maltol complexes (12), in the same animal model, at the same dose, with the same ligand, and by the same method of administration, only the vanadium compound was effective in normalizing blood glucose levels in streptozotocin (STZ) diabetic rats.

Not all chelated vanadium compounds are equally effective; in fact the range of efficacies observed for a wide variety of compounds includes negative rankings (13). The most intensively studied, and the most commonly used as 'benchmark' compounds, the maltol and ethylmaltol complexes of vanadyl ($[VO]^{2+}$) ions, are now frequently incorporated in over-the-counter pharmaceutical preparations. Both bis(maltolato)oxovanadium(IV) (BMOV), and bis(ethylmaltolato)oxovanadium(IV), BEOV, are several times more potent as IMA's in diabetic rats than is the inorganic congener, $VOSO_4$ (14). Both BEOV and $VOSO_4$ have completed phase I human clinical trials.

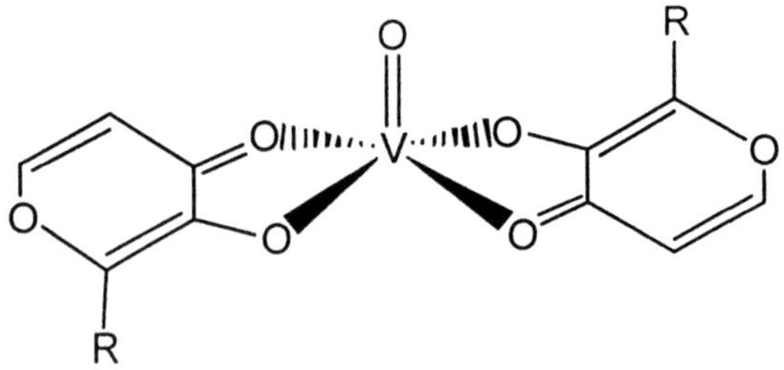

R= CH_3, Bis(maltolato)oxovanadium(IV), BMOV
R= C_2H_5, Bis(ethylmaltolato)oxovanadium(IV), BEOV

Recent evidence has shown that BEOV is mostly, if not entirely, dissociated prior to tissue uptake (15). This begs the question: how is the ligand important in determining relative therapeutic efficacy? Is the intact compound relevant to the purported mechanism of action? Proposed models of vanadium metabolism, based primarily on studies of inorganic vanadium compounds, assume interconversion between V(V) and V(IV) species *in vivo* (16,17). Are there distinct changes in this pathway with chelated complexes? These, and other related, questions, will be considered in this overview, with emphasis on very recent experimental findings.

Coordination Compounds of Vanadium for Insulin Enhancement

Coordination complexes of $[V^{IV}O]^{2+}$ with a variety of ligands constitute the majority of vanadium compounds proposed as IMA's (13,14,18,19). A select few complexes have demonstrated significantly increased activity over inorganic vanadium sources (e.g. $VOSO_4$ or $[VO_4]^{3-}$) through both *in vivo* (20) and/or *in vitro* (19,21,22) assays of potential biological effectiveness.

Biologically effective VOL_2 complexes have been synthesized with various ligands, many of which do not appear to have much in common chemically (Scheme 1). Maltol and ethylmaltol have the advantage of being approved food additives in the U.S.A., U.K. and Canada (11,23). Picolinic acid is a natural metabolite of tryptophan (24,25). Isomaltol is a structural isomer of maltol that occurs naturally in roasted coffee (26). Metformin is used worldwide as an oral hypoglycemic agent for treatment of type 2 diabetes, alone, or in combination with sulfonylureas (27). Thiazolidinediones are also commonly prescribed antidiabetic compounds (27); conjugation of kojic acid to the active portion of a thiazolidinedione molecule was used to create the bifunctional ligand, 5-[4-(5-

hydroxy-4-oxo-4H-pyran-2ylmethoxy)benzylidene]-thiazolidene-2,4-dione (Scheme 1) (28). All of these ligands, when complexed in a 2:1 ratio with $[V^{IV}O]^{2+}$, formed effective IMA's with remarkably similar glucose-lowering responses when tested in a standard acute testing protocol (13,20,24,28,29). In other words, at the ED_{50} dose for BMOV (0.12 mmol kg^{-1} i.p. or 0.55 mmol kg^{-1} by oral gavage), VOL_2 compounds, where L is one of the ligands shown in Scheme 1, were equivalently effective at 12, 24 and 48 h after administration in normalizing blood glucose in STZ-diabetic rats.

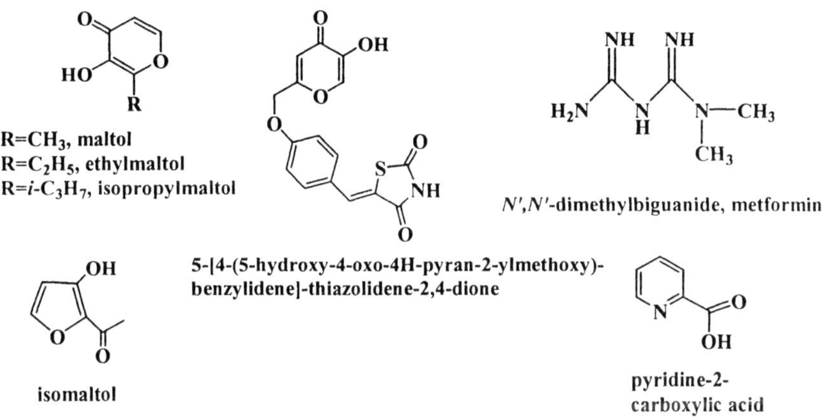

Scheme 1. Ligands that have been incorporated into bis(ligand)oxovanadium(IV) complexes (VOL_2) showing similar effectiveness as IMA's in the same in vivo acute testing protocol in STZ-diabetic rats.

V(III) complexes of some of these same ligands have also been synthesized and characterized (30,31). On average, $V^{III}L_3$ complexes were less efficacious in counteracting the hyperglycemia of STZ-induced diabetes in rats (13). Complicated oxidation processes to V(IV) and V(V) occur rapidly under physiological conditions (30).

V(V)-containing complexes have also been proposed as IMA's (19,32). The coordination chemistry of vanadium(V) and vanadium(IV) usually involves oxygen-rich ligands, but nitrogen- and sulfur-bonding are also well-represented

(19,33). Indeed, vanadium(V) has a particularly varied coordination chemistry. Non-rigid stereochemical requirements for V(V) permit coordination geometries from tetrahedral and octahedral to trigonal- and pentagonal-bipyramidal. The higher çoordinative flexibility of vanadate compared to phosphate has, in fact, been used to advantage for structural characterization of phosphatase enzymes (34). Unlike phosphate, vanadate is readily reduced *in vivo* (to vanadyl). The vanadyl ion does not serve as a phosphate analog, but instead competes preferentially with other divalent cations (35,36).

Coordination and Speciation *in Vivo*

In aqueous solution, under physiologically relevant conditions, VOL_2 may be subject to oxidation, dissociation and ligand substitution effects. The second order rate constant for the oxidation of BMOV by O_2 (pH = 7.25, T= 25°C) is 0.21 M^{-1} sec^{-1} (37), with a known temperature dependence suggesting that at 37°C, the rate would be severalfold faster. The oxidation product, $[VO_2(ma)_2]^-$, can readily interact with ascorbate, and secondarily, with glutathione (38).

The biological milieu typically includes a large number of potentially competing ligands (39). Speciation studies of inorganic vanadium compounds (40) suggest rapid binding in the bloodstream to transferrin primarily, and to albumin, citrate, and smaller organic ligands to a lesser degree. Kiss *et al.* (41), studying ligand substitution effects on BMOV, concluded that "in the absence of specific kinetic effects maltol can readily be replaced by a stronger ligand and VO^{2+} can rapidly form thermodynamically [more] stable complexes in ligand rich environments such as biological fluids." A modeling study (39) of chelated vanadium binding to both low molecular mass (oxalic acid, citric acid, lactic acid and phosphate) and high molecular mass (albumin and transferrin) binders predicted that ~70% of V supplied as VOL_2, especially as BMOV, or bis(picolinato)oxovanadium(IV) ($VO(pic)_2$), would be bound to transferrin in the circulation and the remaining ~30% bound to low molecular mass binders, especially citrate. This result was subsequently confirmed empirically in speciation studies by pH potentiometry (42). An initial EPR spectroscopic investigation of BMOV binding to transferrin confirmed that complexation would readily take place at ambient temperature (43), and recently, a thorough variable temperature EPR study of BMOV binding to transferrin and other serum proteins showed unequivocally that BMOV binds to transferrin in a manner indistinguishable from that of $VOSO_4$ (44) (Figure 1).

Intact Compound vs. Dissociated Metal Ion: Which Is It?

The evidence for dissociation of VOL_2 *in vivo* includes several seminal studies. As noted above, coordination of vanadyl to human serum transferrin was shown to be the same, regardless of whether $VOSO_4$ or BMOV is combined with apo-transferrin (Fig. 1) (44,45). Vanadium, whether from chelated or unchelated sources, appears to accumulate preferentially in bone (46-48). Coordination binding of VO^{2+} derived from BEOV given in the drinking water was identical to that seen for $VOSO_4$-fed rats, in both kidney and bone (49). Detailed ESEEM analysis of vanadyl binding to a triphosphate model system (50) demonstrated the feasibility of vanadyl coordination to phosphates on the bone mineral surface, rather than incorporation into the apatitic lattice.

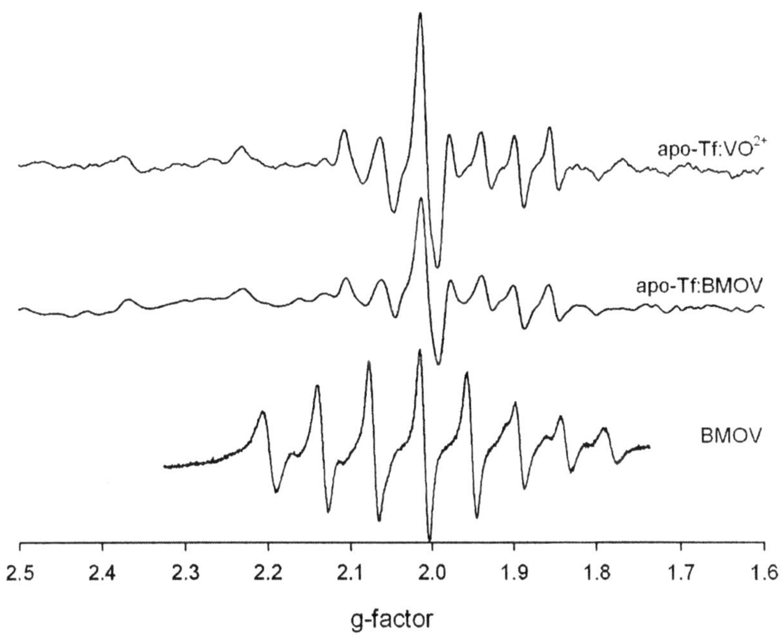

Figure 1. EPR spectra of (top) apo-transferrin, 0.29 mM, vanadyl sulfate, 1.11 mM; (middle) apo-transferrin, 0.13 mM, BMOV, 0.11 mM; and (bottom) BMOV, 1.10 mM (T = 298K, pH 7.4, 0.16 M NaCl).

In the kidney, vanadyl ions appear equatorially coordinated to amine nitrogen atoms, whether the vanadyl originated from $VOSO_4$ (51) or from BEOV (49), resulting in EPR spectra that are entirely congruent. By contrast, VO^{2+} from bis(picolinato)oxovanadium(IV), $VO(pic)_2$ (52,53) in kidney and liver was seen to bind principally to amine nitrogens, but secondarily to imine (pyridyl) nitrogens, suggesting a possible residual fraction of the dose as undissociated $VO(pic)_2$ (52); thus, the likelihood of residual intact $VO(pic)_2$ in kidney tissue was considerably higher than for residual intact BEOV. However, the principal reason for this discrepancy is an experimental limitation: O^{16}-coordination (which would be anticipated from BEOV) does not result in an observable ESEEM pattern. The appearance of both amine and imine binding following i.p. chelated vanadyl demonstrates that at least a major portion of the chelated vanadyl compound undergoes ligand dissociation and substitution even in the absence of absorptive and/or digestive processes (44).

At the cellular level, vanadium from BMOV may be incorporated into phosphotyrosine phosphatases as $[VO_4]^{3-}$ (Figure 2) (54). Insofar as the mechanism of action of vanadium IMA's purports to involve phosphotyrosine phosphatase inhibition, the appearance of vanadate alone in the active site of PTP1B incubated *in vitro* with BMOV (Figure 2) further supports the mounting evidence that vanadium must be released from the compound in order for it to be pharmacologically active (54).

Perhaps most convincing are the pharmacokinetics of disappearance of vanadium, which differed substantially from those of ^{14}C from an oral dose of [ethyl-1-^{14}C]BEOV (Figure 3); the tissue uptake was entirely different for the radiolabelled ligand and for vanadium (15). Instead of the ligand being present at roughly twice the concentration of vanadium in the blood (as would be anticipated from VOL_2), it is present at a lower concentration compared to vanadium from the 1-hour timepoint onwards. Modeling predictions of pharmacokinetic parameters, such as half-life and time of maximal concentration, were also entirely dissimilar for a broad range of tissues studied (liver, kidney, bone, small intestine and lung) (15).

Thus, a number of recent studies have demonstrated fairly conclusively that vanadium IMA's dissociate rapidly once ingested or injected. Elucidation of the coordination environment (43,44,47) and probable speciation (38,41,42) of vanadium compounds *in vivo* suggests an eventual metabolic fate for chelated vanadium complexes that differs little, if any, from that of non-complexed vanadium compounds, when the compounds are administered orally.

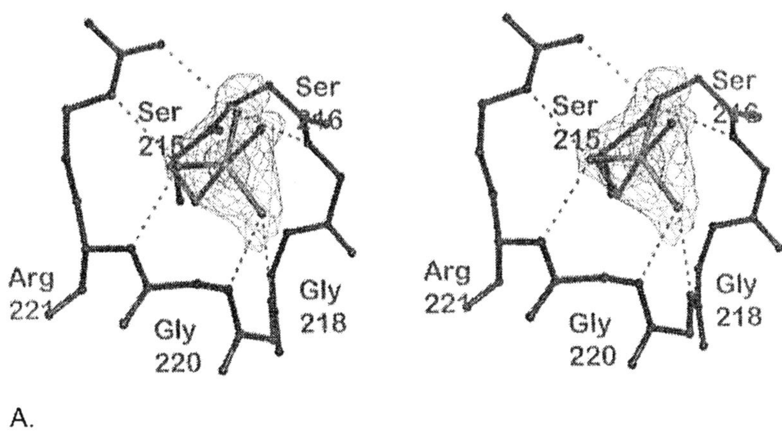

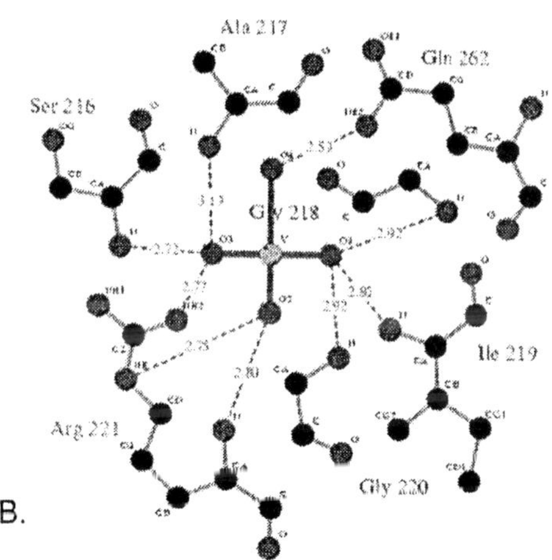

Figure 2. Crystal structure of PTP1B (C215S mutant) soaked with BMOV reveals $[VO_4]^{3-}$ coordinated to serine 215 in a trigonal bipyramidal geometry. (A) Stereoview of the difference electron density in the active site of PTP1B, shown together with the final model of the oxovanadate ligand. Most protein atoms are omitted for clarity. Axial vanadium–oxygen bonds are shown as thick broken lines, select hydrogen bonds between ligand and enzyme are shown as thin broken lines. (B) Detailed representation of the H bonding network of $[VO_4]^{3-}$ with backbone atoms in the PTP1B (C215S mutant) active site. (54). Reprinted from J. Inorg. Biochem. ***96****, Peters et al., "Mechanism of insulin sensitization by BMOV (bis maltolato vanadium); unliganded vanadium (VO_4) as the active component," pp. 321-330, 2003, with permission from Elsevier.*

Oral Bioavailability

If VOL_2 indeed dissociates rapidly after ingestion, then the differential efficacy between $VOSO_4$ and VOL_2 must be due to very early effects on bioavailability, either at the level of absorption, or shortly thereafter. Bioavailability of a metallopharmaceutical is defined as "the amount of a dose that is functionally usable by an organism" (55). Relative efficacy of vanadium-containing therapeutic agents, analogous to other small molecule metal-containing therapeutics, may not be due solely, or even mostly, to the presence of intact compound in the circulation.

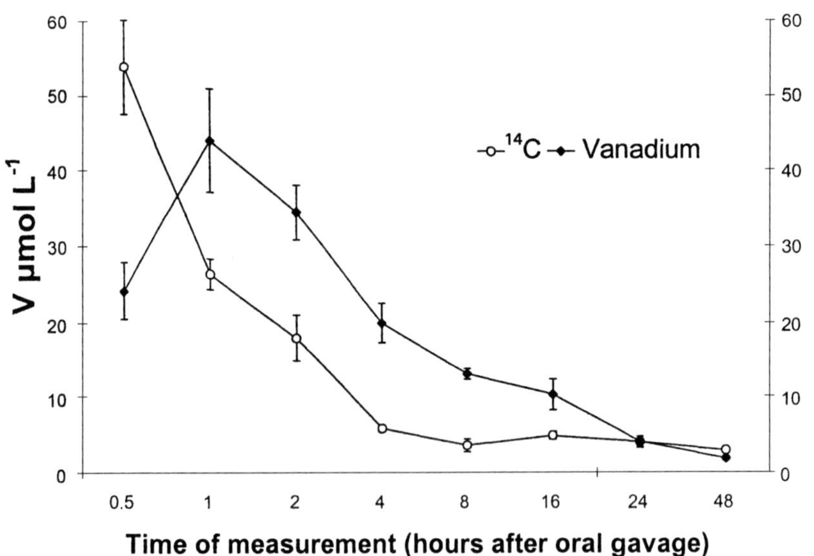

Figure 3. Blood [V] and ^{14}C concentrations after a single bolus dose of [ethyl-1-^{14}C]BEOV (50 mg kg^{-1} in 2 mL 1% carboxymethylcellulose, n = 4 rats for each time point). Data are expressed as means ± SEM (15).

An emerging picture of VOL_2 metabolism (Scheme 2) is thus not substantially different from that of inorganic vanadium compounds. For BEOV (and presumably also BMOV), dissociation of at least a major portion of the dose occurs within an hour of ingestion. VOL_2, including BMOV, $VO(pic)_2$ and bis(iodopicolinato)oxovanadium(IV), $VO(IPA)_2$, are cleared rapidly (less than 0.5 h) from the circulation (46,56,57).

Once out of the systemic circulation, vanadium is clearly taken up by many different tissues (46,58,59). Vanadium, whether administered as a vanadium(IV) or vanadium (V) salt, e.g. $VOSO_4$ or $[VO_4]^{3-}$, respectively, accumulates predominately in bone, liver and kidney tissue (59-62). (Because vanadate ($[VO_4]^{3-}$) has various protonation states under physiological conditions, we use V_i to denote all protonation states, $V_i = [VO_4]^{3-} / [HVO_4]^{2-} / [H_2VO_4]^{-}$.) Binding of V(IV) and/or V(V) to fiber in the lower GI tract and to ferritin in the liver, and conjugation prior to urinary excretion, are unchanged from models appropriate to inorganic vanadium compounds (16,17). Miscellaneous losses in hair, skin and nails (63) are presumed to account for significant decreases in overall body burden over time (64).

Distributions from chelated vanadium sources, e.g. BMOV (46) and vanadyl picolinate (48) show a similar profile. Both the amount and relative vanadium concentrations in these organs vary with the particular compound and dose, but the percentage of dose taken up (whether from an oral or an i.p. dose) is generally significantly higher from chelated sources (46,48) compared to that from inorganic vanadyl compounds (56,59,61,65). Thus, chelation improves bioavailability, reducing the dose required for therapeutic effect (55).

A dose adjustment is necessary when administering VOL_2 by i.p. injection, as opposed to VOL_2 by oral ingestion (whether in the drinking water or by oral gavage) (Scheme 2). Dose adjustment was empirically determined to be roughly a factor of five lower for i.p. administration (9), implying a gastrointestinal absorption of 20%. This result is consistent with compartmental modeling predictions (46).

Improved bioavailability of some VOL_2 compared to that of $VOSO_4$ may involve not only improved uptake into the circulation, but also delayed release from the bloodstream (and consequent tissue uptake). In a recent study comparing V(IV) clearance from the bloodstream following several different VOL_2 administered i.p., delayed clearance of the compound from the systemic circulation appears to correlate with improved pharmacological potency (53,66). This mechanism for increased potency would fit well with the proposed formation of a ternary complex between some VOL_2 (e.g. BMOV and bis(acetylacetonato)oxovanadium(IV) ($VO(acac)_2$ (67)) and human serum albumin, observed by EPR spectroscopy (43,44). Albumin binding to metallopharmaceuticals has been shown previously to effectively delay circulatory clearance, facilitating appropriate tissue targetting (68).

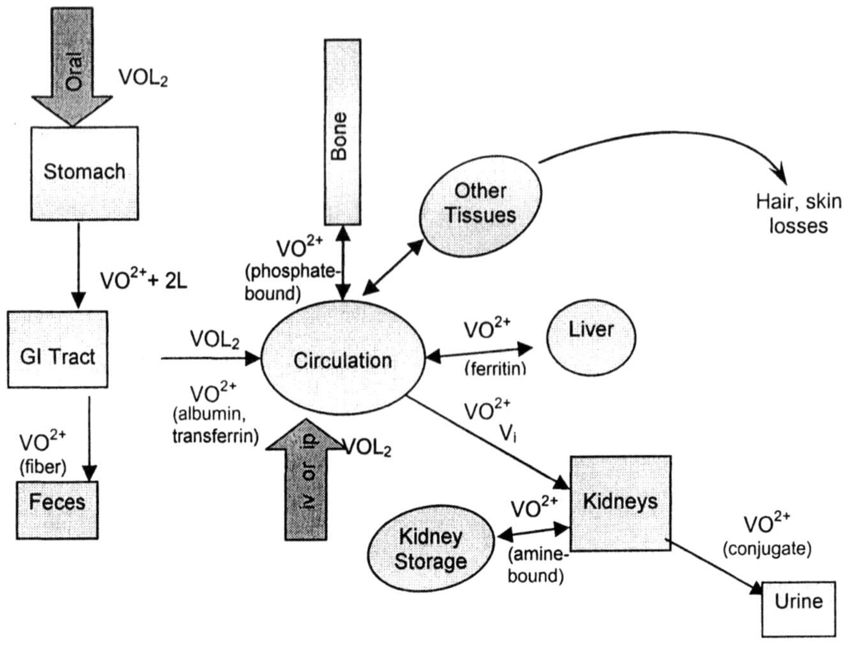

Scheme 2. Schematic outline of VOL_2 metabolism (adapted from 16,64).

The lack of efficacy of some VOL_2 could then be due to either high lability, in which case the dissociated compound behaves exactly like $VOSO_4$ in the biological milieu, or to tight coordination, in which case VO^{2+} is never released, and the compound may be absorbed even more poorly than is $VOSO_4$. Ideally, the ligand remains bound to the metal ion long enough to delay oxidation and/or favor binding to albumin, with consequent ternary complex formation (43,44) and prevention of gastrointestinal irritation (69).

The proposed vanadium delivery to tissues by VOL_2 (Scheme 2) is not unlike that of a number of other biologically relevant metal ions, e.g. copper and manganese, which are both essential and toxic (55,70). A well-studied example is the tissue uptake of ferric maltol, developed specifically to lessen the gastrointestinal irritation associated with ferrous compounds in treating iron deficiency (23). Rather than diffusing across the intestinal wall, dissociating immediately, and allowing Fe^{3+} binding to transferrin, the ferric ion appears to be first reduced, then donated to an endogenous uptake carrier, and thus remains under regulatory control. Saturable uptake kinetics for radiolabelled iron entry from ferric maltol are evidence of this improved pathway for delivery of iron to intestinal cell surface carriers (23).

A similar process of uptake, distribution and metabolism may be operative for chromium picolinate, ($Cr(pic)_3$). In a study in which ^{51}Cr or ^{3}H-labelled complex was administered intravenously (i.v.), metabolic fates of the metal ion and the ligand were compared (71). $Cr(pic)_3$ was clearly dissociated prior to urinary elimination of ^{51}Cr; size exclusion chromatography indicated that the chromium-containing fraction had a molecular weight of ~1500. Blood chromium concentrations were similar at both 4 and 24 h after consuming a diet containing either $CrCl_3$ or $Cr(pic)_3$; tissue concentrations were, on the other hand, four to five times higher following $Cr(pic)_3$ oral administration (71). Although $CrCl_3$ is readily taken up by apo-transferrin (72), $Cr(pic)_3$ may not be so readily bound, and hence may be bound preferentially by albumin, with consequent slower clearance from the bloodstream. Decomposition of $Cr(pic)_3$, as well as Cr(III) propionate, has been studied under physiologically relevant conditions (neutral aqueous solutions, and in artificial gastric juice, pH ~2) (73) by UV-Visible spectroscopy, electrospray mass spectrometry and X-ray absorption spectroscopy. Although ligand exchange for Cr(III) complexes is relatively slow compared to that of vanadyl complexes (on the order of several hours) (74), the dissociation process is nonetheless biologically significant.

Whether administered as V(III), V(IV) or V(V), vanadium *in vivo* speciation is dominated by interconversion between V(IV) and V(V) (38), which would presumably underlie the biolocalization of any of the chelated vanadium complexes. Amine- (and possibly also imine) binding in kidneys, phosphate-binding and adsorption on the bone, and potential ternary complex formation in the circulation are all specific to particular VOL_2, according to evidence gathered so far, but may have more general application to chelated vanadium compounds.

Summary and Conclusions

Ensuring delivery of the vanadium ion to the site of incorporation appears to be the main role of the ligand in VOL_2. Prior to this incorporation, the metal ion must be prevented from incurring toxicity through gastrointestinal irritation, then chaperoned through the systemic circulation such that overly rapid clearance is avoided, and finally sequestered in tissues well-equipped to maintain the vanadium in storage for release and utilization as needed. A major role of the complexation of vanadium ions may, in fact, be in slowing down both the uptake into and release of vanadium from the circulation.

Acknowledgement

The authors gratefully acknowledge past co-workers (cited below) and financial support from the Canadian Institutes of Health Research (operating grants), the Natural Sciences and Engineering Research Council (fellowships and operating grants), the Science Council of British Columbia (B.D.L., GREAT program), and an Australian Research Council International Research Exchange (IREX) program grant to G.R.H. and C.O. We thank Prof. John McNeill and his group, as well as Kinetek Pharmaceuticals, Inc., for long-term, fruitful collaboration.

Literature Cited

1)Heyliger, C. E.; Tahiliani, A. G.; McNeill, J. H. *Science* **1985**, *227*, 1474-1477.

2)Henquin, J. C.; Brichard, S. M. *La Presse Medicale* **1992**, *21*, 1100-1101.

3)Tolman, E. L.; Barris, E.; Burns, M.; Pansisni, A.; Partridge, R. *Life Sci.* **1979**, *25*, 1159-1164.

4)Cantley, L. C.; Resh, M. D.; Guidotti, G. *Nature (London)* **1978**, *272*, 552-554.

5)Orvig, C.; Thompson, K. H.; Battell, M.; McNeill, J. H. *Metal Ions Biol. Syst.* **1995**, *31*, 575-594.

6)Shechter, Y. *Diabetes* **1990**, *39*, 1-5.

7)Tsiani, E., Fantus, I.G. *Trends Endocrinol. Metab.* **1997**, *8*, 51 - 58.

8)Thompson, K. H.; McNeill, J. H.; Orvig, C. *Chem. Rev.* **1999**, *99*, 2561-2571.

9)Yuen, V. G.; Pederson, R. A.; Dai, S.; Orvig, C.; McNeill, J. H. *Can. J. Physiol. Pharmacol.* **1996**, *74*, 1001-1009.

10)Sakurai, H.; Kojima, Y.; Yoshikawa, Y.; Kawabe, K.; Yasui, H. *Coord. Chem. Rev.* **2002**, *226*, 187-198.

11)Lord, S. J.; Epstein, N. A.; Paddock, R. L.; Vogels, C. M.; Hennigar, T. L.; Zaworotko, M. J.; Taylor, N. J.; Driedzic, W. R.; Broderick, T. L.; Westcott, S. A. *Can. J. Chem.* **1999**, *77*, 1249-1261.

12)Thompson, K. H.; Chiles, J.; Yuen, V. G.; Tse, J.; McNeill, J. H.; Orvig, C. *J. Inorg. Biochem.* **2004**, *98*, 683-690.

13)Thompson, K. H.; Orvig, C. *Metal Ions Biol. Syst.* **2004**, *41*, 221-252.

14)McNeill, J. H.; Yuen, V. G.; Hoveyda, H. R.; Orvig, C. *J. Med. Chem.* **1992**, *35*, 1489-1491.

15)Thompson, K. H.; Liboiron, B. D.; Sun, Y.; Bellman, K. D. D.; Setyawati, I. A.; Patrick, B. O.; Karunaratne, V.; Rawji, G.; Wheeler, J.; Sutton, K.; Bhanot, S.; Cassidy, C.; McNeill, J. H.; Yuen, V. G.; Orvig, C. *J. Biol. Inorg. Chem.* **2003**, *8*, 66-74.

16)Chasteen, N. D.; Lord, E. M.; Thompson, H. J. in *Frontiers in Bioinorganic Chemistry*; Xavier, A., Ed.; VCH Publishing: Weinheim, Federal Republic of Germany, 1986, pp 133-145.

17)Baran, E. J. *Biol. Soc. Chil. Quim.* **1997**, *42*, 247-256.

18)Sakurai, H.; Tsuchiya, K.; Nukatsuka, M.; Kawada, J.; Ishikawa, S.; Yoshida, H.; Komatsu, M. *J. Clin. Biochem. Nutrit.* **1990**, *8*, 193-200.

19)Rehder, D.; Costa Pessoa, J.; Geraldes, C. F.; Castro, M. C.; Kabanos, T.; Kiss, T.; Meier, B.; Micera, G.; Pettersson, L.; Rangel, M.; Salifoglou, A.; Turel, I.; Wang, D. *J. Biol. Inorg. Chem.* **2002**, *7*, 384-396.

20)Yuen, V. G.; Orvig, C.; McNeill, J. H. *Can. J. Physiol. Pharmacol.* **1993**, *71*, 263-269.

21)Crans, D. C.; Mahroof-Tahir, M.; Johnson, M. D.; Wilkins, P. C.; Yang, L.; Robbins, K.; Johnson, A.; Alfano, J. A.; Godzala, I., Michael E.; Austin, L. T.; Willsky, G. R. *Inorg. Chim. Acta* **2003**, 365-378.

22)Sakurai, H.; Fujii, K.; Watanabe, H.; Tamura, H. *Biochem. Biophys. Res. Commun.* **1995**, *214*, 1095-1101.

23)Barrand, M. A.; Callingham, B. A. *Br. J. Pharmacol.* **1991**, *102*, 408-414.

24)Melchior, M.; Thompson, K. H.; Jong, J. M.; Rettig, S. J.; Shuter, E.; Yuen, V. G.; Zhou, Y.; McNeill, J. H.; Orvig, C. *Inorg. Chem.* **1999**, *38*, 2288-2293.

25)Rehder, D. *Inorg. Chem. Commun.* **2003**, *6*, 604-617.

26)Tressl, R.; Bahri, D.; Koppler, H.; Jensen, A. *Z. Lebensm.Unters Forsch.*. **1978**, *167*, 111-114.

27)Inzucchi, S. E. *J. Am. Med. Assoc.* **2002**, *287*, 360-372.

28)Storr, T.; Mitchell, D.; Buglyo, P.; Thompson, K. H.; Yuen, V. G.; McNeill, J. H.; Orvig, C. *Bioconjugate Chem.* **2003**, *14*, 212-221.

29)Woo, L. C. Y.; Yuen, V. G.; Thompson, K. H.; McNeill, J. H.; Orvig, C. *J. Inorg. Biochem.* **1999**, *7*, 251-257.

30)Melchior, M.; Rettig, S. J.; Liboiron, B. D.; Thompson, K. H.; Yuen, V. G.; McNeill, J. H.; Orvig, C. *Inorg. Chem.* **2001**, *40*, 4686-4690.

31)Thompson, K. H.; Orvig, C. *Coord. Chem. Rev.* **2001**, *219-221*, 1033-1053.
32)Crans, D. C.; Yang, L.; Jakusch, T.; Kiss, T. *Inorg. Chem.* **2000**, *39*, 4409-4416.
33)Butler, A. in *Vanadium in Biological Systems*; Chasteen, N. D., Ed.; Kluwer Academic: Dordrecht, 1990, pp 25-50.
34)Plass, W. *Angew. Chem.* **1999**, *38*, 909-912.
35)Cantley, L. C. Jr; Aisen, P. *J. Biol. Chem.* **1979**, *254*, 1781-1784.
36)Thompson, K. H.; Tsukada, Y.; Xu, Z.; Battell, M.; McNeill, J. H.; Orvig, C. *Biol. Trace Elem. Res.* **2002**, *86*, 31-44.
37)Sun, Y.; James, B. R.; Rettig, S. J.; Orvig, C. *Inorg. Chem.* **1996**, *35*, 1667-1673.
38)Song, B.; Aebischer, N.; Orvig, C. *Inorg. Chem.* **2002**, *41*, 1357-1364.
39)Kiss, T.; Kiss, E.; Garribba, E.; Sakurai, H. *J. Inorg. Biochem.* **2000**, *80*, 65-73.
40)Chasteen, N. D.; Grady, J. K.; Holloway, C. E. *Inorg. Chem.* **1986**, *25*, 2754-2760.
41)Kiss, E.; Fabian, I.; Kiss, T. *Inorg. Chim. Acta* **2002**, *340*, 114-118.
42)Buglyo, P., Kiss, T.; Sanna, D.; Micera, G. *J. Chem. Soc. Dalton Trans.*, **2002**, 2275-2282.
43)Willsky, G. R.; Goldfine, A. B.; Kostyniak, P. J.; McNeill, J. H.; Yang, L. Q.; Khan, H. R.; Crans, D. C. *J. Inorg. Biochem.* **2001**, *85*, 33-42.
44)Liboiron, B. D.; Thompson, K. H.; Lam, E.; Aebischer, N.; Orvig, C.; Hanson, G. R. Submitted for publication.
45)Chasteen, N. D.; Lord, E. M.; Thompson, H. J.; Grady, J. K. *Biochim. Biophys. Acta* **1986**, *884*, 84-92.
46)Setyawati, I. A.; Thompson, K.H.; Sun, Y.; Lyster, D.M.; Vo., C.; Yuen, V.G.; Battell, M.; McNeill, J.H.; Ruth, T.J.; Zeisler, S.; Orvig, C. *J. Appl. Physiol.* **1998**, *84*, 569-575.
47)Dikanov, S. A.; Liboiron, B. D.; Thompson, K. H.; Vera, E.; Yuen, V. G.; McNeill, J. H.; Orvig, C. *J. Am. Chem. Soc.* **1999**, *121*, 11004-11005.
48)Fujimoto, S.; Fujii, K.; Yasui, H.; Matsushita, R.; Takada, J.; Sakurai, H. *J. Clin. Biochem. Nutr.* **1997**, *23*, 113-129.
49)Dikanov, S. A.; Liboiron, B. D.; Thompson, K. H.; Vera, E.; Yuen, V. G.; McNeill, J. H.; Orvig, C. *Bioinorg. Chem. Appl.* **2003**, *1*, 70-83.
50)Dikanov, S. A.; Liboiron, B. D.; Orvig, C. *J. Am. Chem. Soc.* **2002**, *124*, 2969-2978.
51)Fukui, K.; Ohya-Nishiguchi, H.; Nakai, M.; Sakurai, H.; Kamada, H. *FEBS Lett.* **1995**, *368*, 31-35.
52)Fukui, K.; Fujisawa, Y.; Ohya-Nishiguchi, H.; Kamada, H.; Sakurai, H. *J. Inorg. Biochem.* **1999**, *77*, 215-224.
53)Yasui, H.; Tamura, A.; Takino, T.; Sakurai, H. *J. Inorg. Biochem.* **2002**, *91*, 327-338.

54)Peters, K. G.; Davis, M. G.; Howard, B. W.; Pokross, M.; Rastogi, V.; Diven, C.; Greis, K. D.; Eby-Wilkens, E.; Maier, M.; Evdokimov, A.; Soper, S.; Genbauffe, F. *J. Inorg. Biochem.* **2003**, *96*, 321-330.
55)Thompson, K. H.; Orvig, C. *Science* **2003**, *300*, 936-939.
56)Yasui, H.; Takechi, K.; Sakurai, H. *J. Inorg. Biochem.* **2000**, *78*, 185-196.
57)Takino, T.; Yasui, H.; Yoshitake, A.; Hamajima, Y.; Matsushita, R.; Takada, J.; Sakurai, H. *J. Biol. Inorg. Chem.* **2001**, *6*, 133-142.
58)Wiegmann, T. B.; Day, H.D.; Patak, R.V. *J. Toxicol. Environ. Health* **1982**, *10*, 233-245.
59)Al-Bayati, M.; Raabe, O. G.; Giri, S. N.; Knaak, J. B. *J. Amer. Coll. Toxicol.* **1991**, *10*, 233-241.
60)Hopkins, L. L., Jr.; Tilton, B.E. *Am. J. Physiol.* **1966**, *211*, 169-172.
61)Bogden, J. D.; Higashino, H.; Lavenhar, M..A.; Bauman, J. W., Jr.; Kemp, F.W.; Aviv, A. *J. Nutr.* **1982**, *112*, 2279-2285.
62)Dai, S.; Vera, E.; McNeill, J. H. *Pharmacol. Toxicol.* **1995**, *76*, 263-268.
63)Kucera, J.; Lener, J.; Mnukova, J.; Bayerova, E. *Vanadium in the Environment. Part 2. Health Effects*; Nriago, J. O., Ed.; John Wiley & Sons, Inc.: New York, 1998, pp 55-73.
64)Sandstrom, B.; Fairweather-Tait, S.; Hurrell, R.; Van Dokkum, W. *Nutr. Res. Rev.* **1993**, *6*, 71-95.
65)Conklin, A. W.; Skinner, C.S.; Felten, T.L.; Sanders, C.L. *Toxicol. Letts.* **1982**, *11*, 199-203.
66)Fugono, J.; Yasui, H.; Sakurai, H. *J. Pharm. Pharmacol.* **2001**, *53*, 1247-1255.
67)Makinen, M. W.; Brady, M. J. *J. Biol. Chem.* **2002**, *277*, 12215-12220.
68)Caravan, P.; Cloutier, N. J.; Greenfield, M. T.; McDermid, S. A.; Dunham, S. U.; Bulte, J. W. M.; Amedio, J., J. C.; Lobby, R. J.; Supkowski, R. M.; Horrocks, J., W. D.; McMurry, T. J.; Lauffer, R. B. *J. Am. Chem. Soc.* **2002**, *124*, 3152-3162.
69)Thompson, K. H.; Battell, M.; McNeill, J.H. *Vanadium in the Environment. Part 2. Health Effects*; Nriagu, J. O., Ed.; John Wiley & Sons, Inc.: Ann Arbor, 1998; Vol. I, pp 21-37.
70)Luk, E.; Jensen, L. T.; Culotta, V. C. *J. Biol. Inorg. Chem.* **2003**, *8*, 803-809.
71)Hepburn, D. D. D.; Vincent, J. B. *Chem. Res. Toxicol.* **2002**, *15*, 93-100.
72)Hopkins, L. L., Jr.; Schwartz, K. *Biochim. Biophys. Acta* **1964**, *90*, 484-491.
73)Mulyani, I.; Levina, A.; Lay, P. A. *J. Inorg. Biochem.* **2003**, *96*, 196-196.
74)Levina, A.; Cood, R.; Dillon, C. T.; Lay, P. A. *Progr. Inorg. Chem.* **2003**, *51*, 145-250.

Chapter 22

Targeting Melanoma via Metal-Based Stress

Patrick J. Farmer[1], Daniel Brayton[1], Christina Moore[1], Donny Williams[1], Babbak Shahandeh[2], Dazhi Cen[2], and Frank Meyskens, Jr.[2]

Departments of [1]Chemistry and [2]Medicine, University of California, Irvine, CA 92687–2025

Melanoma have an unusual susceptibility to certain lipophilic metal complexes, which is likely due to the pro-oxidant effect of metal ions on their endogenous catecholic pigment, melanin. We have investigated the speciation, metal-binding affinity and redox reactivity of the polymeric pigment using synthetic melanins derived from dihydroxyindole, DHI. The binding of certain metal ions to DHI-melanin enhances redox cycling in air and the generation of reactive oxygen species. Our hypothesis is that the toxicity of metal ions to melanoma is via this pro-oxidant reactivity of the pigment. A screening methodology utilizing DHI-melanin reactivity has been developed to identify new candidates for melanoma chemotherapy. Lipophilic metallodithiocarbamate complexes also display selective toxicity against melanoma in culture, which is increased under high O_2 concentrations. Disulfiram, a dithiocarbamate disulfide, has strong activity which is shown to be Cu-dependent. The ability of dithiocarbamates to promote metal uptake and mobility within organisms is discussed.

Melanin, metals and melanoma

Melanomas are among the deadliest forms of cancer as they have a high recurrence but as yet no effective chemotherapy.(*1, 2*) The drug resistance of melanoma has been attributed to the presence of melanin, a redox-active polymeric pigment formed from the oxidation of tyrosine within cells.(*3*) The formation of melanin itself depends on fine control of oxidative chemistry: a peroxide-dependent enzyme, tyrosinase, catalyzes two successive reactions, the hydroxylation of tyrosine and the oxidation of the product L-dopa, Scheme 1.(*4*) The product of dopa oxidation cyclizes to a 5,6-dihydroxyindole (**DHI**) intermediate, which is highly reactive and gives rise to black eumelanin polymers by a pathway dependent on further oxidation by oxygen.(*5*)

Scheme 1

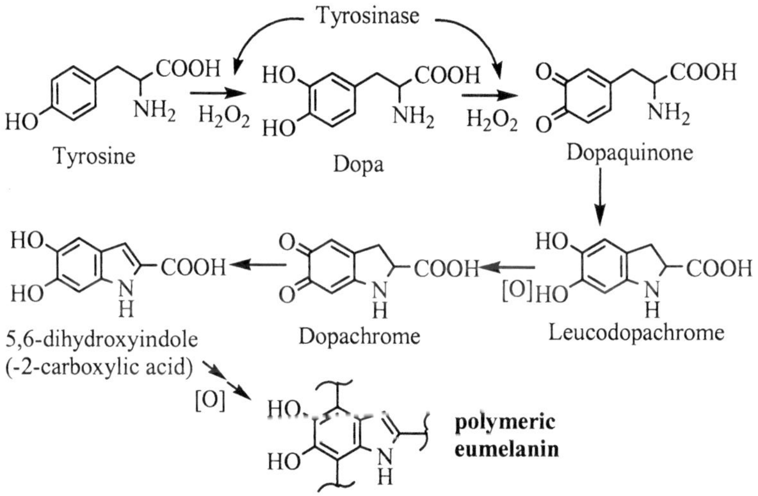

Tyrosinase

Tyrosine $\xrightarrow{H_2O_2}$ Dopa $\xrightarrow{H_2O_2}$ Dopaquinone

Dopaquinone → Leucodopachrome $\xrightarrow{[O]}$ Dopachrome → 5,6-dihydroxyindole (-2-carboxylic acid) $\xrightarrow{[O]}$ **polymeric eumelanin**

In normal melanocytes, the black melanin particles are generated and contained within suborganelles called melanosomes, but in melanoma cells melanosomes are poorly compartmentalized with malformed or twinned membranes, occlusions within the melanin, and structural disorganization evident.(*6, 7*) These structural differences are significant, as melanosomal compartmentalization helps protect the cell from the highly reactive, small-molecule catechols that are generated during melanogenisis. Such protection is likely lost within melanomas, and the complex redox reactivity of these melanin DHI catechols may provide a unique target for chemotherapy.(*8*)

One of melanin's presumed biological roles is to act as buffer against different forms of chemical stress.(*9*) In aggregate form, their dark coloring protects against photo-induced damage, their catecholic subunits protect against oxidative stress by neutralizing oxidants and by sequestering reactive metal ions.(*10*) But melanins may also act as pro-oxidants; e.g., melanins generate peroxide during slow bleaching in air, the rate of bleaching is increased by the presence of different metal ions.(*11-13*)

Several recent studies suggest that melanomas themselves are predisposed to internal oxidative stress,(*14, 15*) and have a weakened ability to mediate extra-cellular oxidative stress.(*16, 17*) Melanomas have unusually high uptake of Cu and Fe ions, even higher than normal melanocytes which themselves accumulate metal ions.(*18-20*) And importantly, melanomas have been shown to be much more sensitive to increases of certain divalent metal salts; for instance, Zn and Cd, which induce melanoma death at concentrations several orders of magnitude lower than that which affects melanocytes.(*21, 22*)

Scheme 2

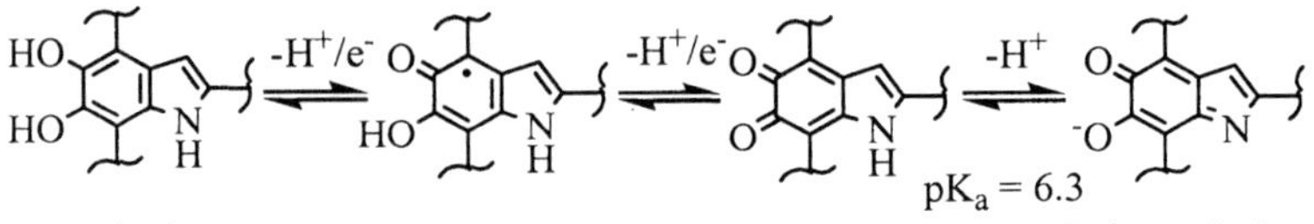

Using synthetic DHI-melanins, we have demonstrated that the binding of metal ions to the catecholic subunits of polymeric melanin can dramatically affect its redox properties. For example, Zn(II) saturation of a melanin film shifts the melanin-based oxidation by up to +100 mV,(*23*) and increases its semiquinone concentration six-fold.(*24*) We attribute the unusual effect of metal ion binding to a tautomerization of the quinone form of DHI to a quinone-imine, Scheme 2, thus providing an excellent chelator for metal ions at neutral pH. Potentiometric titrations suggest the quinone imine has a high affinity for metal ions, e.g., a K_a ca. 10^{10} for Zn(II) at pH 7, as compared with that of the quinole at K_a ca. 10^2.(*24*) This implies that melanin's metal-binding ability increases upon partial oxidation, making them more susceptible to further oxidation. Therefore, the presence of metal ions promotes melanin oxidation, and suggests that this pro-oxidant response is the souce of melanoma's sensitivity to metal ions.

Zn-binding also increases the auto-oxidation under air of the synthetic DHI-melanin, termed redox cycling, which produces a flux of reactive oxygen species (**ROS**) such as superoxide, peroxide and hydroxyl radicals.(*25*) As illustrated in

Scheme 3, we postulate that the reactivity with dioxygen is via interaction with the low concentration of radical semiquinone within the polymer. Most significant is the observed increase in hydroxyl radicals generated from melanin redox-cycling after treatment with divalent Cu or Zn salts.

Scheme 3

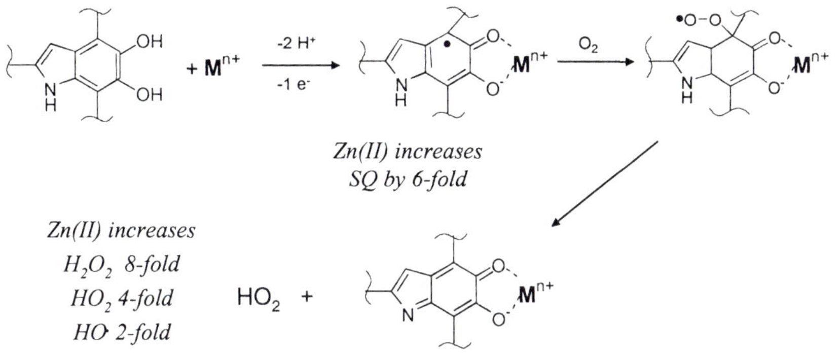

+ M^{n+} $\xrightarrow[-1\ e^-]{-2\ H^+}$ $\xrightarrow{O_2}$

Zn(II) increases SQ by 6-fold

Zn(II) increases
H_2O_2 8-fold
HO_2 4-fold
HO· 2-fold

HO_2 +

The formation of ROS was quantified using trapping experiments in which synthetic melanin films are treated with various metal salt solutions. After equilibration and extensive washing, the samples were allowed to auto-oxidize under air in the presence of the radical trap 5,5-dimethyl-1-pyrroline-N-oxide (**DMPO**). Electron paramagnetic resonance characterization of the trapped ROS showed an distinctive increase in melanin's auto-oxidation upon metal-ion binding.(*25*) The increase in HO radical trapped by DMPO after treating the melanin with solutions of Zn(II) and Cu(II) salts correlated with the increase in quinone imine concentrations predicted.

To confirm the generation of hydroxyl radicals, we exposed the melanin to supercoiled plasmid DNA under similar conditions and observed extensive clipping, clearly indicative of hydroxyl radical damage. The quantity of DNA clipping qualitatively follows the intensity of the observed DMPO-OH signal under the same conditions. Previous studies had shown that binding of redox-active Cu(II) (11) or Fe(III) (*13*) induced a pro-oxidant effect from melanins. Our results showed that non-redox metals such as Zn also induce such an effect. This is quite important, as it suggests that production of ROS by melanin/oxygen may be increased by binding typically non-toxic metal ions. The ability of melanins to induce DNA damage in vitro led to the suggestion that exposed melanin within melanoma cells may be related to the observed high rates of DNA mutations. (*26, 27*)

Melanoma response to oxidative stress

Several previous reports suggest that melanomas in culture are more susceptible to oxidative stress than other cells.(*14*) In one study, normal melanocytes, melanoma and other cancer cells lines were exposed to extracellular peroxide stress generated by glucose/glucose oxidase.(*17*) Normal cells neutralized the stress within a short time, but melanoma cell lines could not and a continued buildup of ROS over time was seen. Addition of catalase (**CAT**) decreased the ROS signal as might be expected, but addition of superoxide dismutase (**SOD**) produced a several-fold increase in extracellular ROS signal, implying that a high flux of superoxide. The differential response suggested an essential change had occurred within transformed melanoma cells that render them susceptible to oxidative stress.

Using the same experimental conditions, we measured hydroxyl radical formation, using radical trap DMPO, for both DHI-melanin and melanoma cells in culture.(*25*) The comparison was telling: for both, addition of CAT decreased the signal, while SOD caused a large increase in the hydroxyl-radical adduct. Our hypothesis is that the pro-oxidant response in both chemical and cell culture studies are due to the pro-oxidant reactivity of melanin.

If the melanins within melanoma are as reactive as our synthetic melanins, then oxygen itself should be toxic to them. Indeed, melanoma cultures grown under high O_2 atmosphere show a distinct loss in viability that is not seen for normal melanocytes.(*25*) This toxicity is enhanced by drugs which induce metal-uptake by the cells, such as dithiocarbamate derivatives that will be described more fully below. Melanocyte cultures treated with low doses of

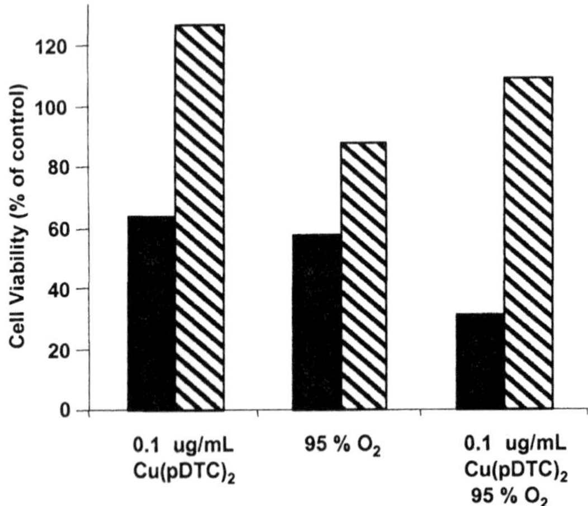

Figure 1. Effect of $Cu(pDTC)_2$ and oxygen on 81-61a melanoma cell line (solid) and melanocyte (striped) viability after 72 hrs. Data from reference 25.

bis(pyrollidinedithiocarbamate)copper, **Cu(pDTC)$_2$**, show enhancements in viability even under high O_2 conditions, perhaps due to antioxidant abilities of the dithiocarbamate. As seen in Figure 1, the effect of the drug itself on melanoma under ambient oxygen is evident. Under higher O_2 concentrations, the difference between melanoma and melanocyte viability is particularly striking, and suggests a chemotherapeutic strategy which could selectively target melanoma.

Melanoma response to metal-ion stress

Melanoma cells have a demonstrated susceptibility to metal-induced stress, which we believe to be a consequence of the pro-oxidant response of the constitutive melanin pigments. Melanomas have long been known for an unusually high concentrations of Cu and other metals, even higher than normal melanocytes which themselves accumulate metal ions.(*18*) Most importantly, melanomas are uniquely sensitive to certain metal salts; for instance, divalent Zn and Cd are toxic to melanoma at concentrations several orders of magnitude lower than that which affects melanocytes.(*21, 22*) We suggest that the cytosolic melanin fragments within melanoma promote high metal ion accumulation, which then increases internal oxidative stress by the production of ROS. In such a scenario, an auto-catalytic redox-cycling may generate enough stress to overcome the melanoma's defenses. Therefore, melanomas should be hyper-sensitive to drugs which increase metal-uptake.

Using the hypothetical mechanism of action given in Scheme 3, i.e., that increased semiquinone subunit concentrations in bulk melanins leads to increased redox cycling and ROS formation, we have begun screening metal salts for their effect on synthetic melanins as measured by the melanin-based oxidation potential, EPR quantification of the semiquinone radical, and by plasmid DNA clipping by hydroxyl radicals produced during redox cycling. Metal ions with apparent high reactivity are then tested in cell viability studies of melanoma cell cultures, in combination with lipophilic chelators that assist transport into the cells.

An initial success in this screening was the discovery a high anti-melanoma activity for In(III)/maltol combinations. Sealy first demonstrated that binding of In(III) to melanin increases its semiquinone content by ca. 7- fold.(*28*) In the plasmid DNA assay, $InCl_3$ effected clipping comparable to that of Cu(II). Most tellingly, initial viability studies showed that it had good anti-melanoma activity in the presence of excess maltol, Figure 2. Maltol is a non-toxic chelating agent widely used to promote passive uptake of metal ions into cells.(*29, 30*)

The pro-oxidant effect of metal ions on melanins may also play a role in the carcinogenesis of melanoma. Although widely attributed to childhood ultraviolet light exposure, this accounts for only half of the attributable risk of melanoma.(*31*) Many epidemiological studies have revealed an increased risk of melanoma in the electronic and chemical industries,(*32*), and we have proposed that exposure to redox active metals or other compounds that bind to melanin may play a co-carcinogenic role in melanoma pathogenesis.(*33*)

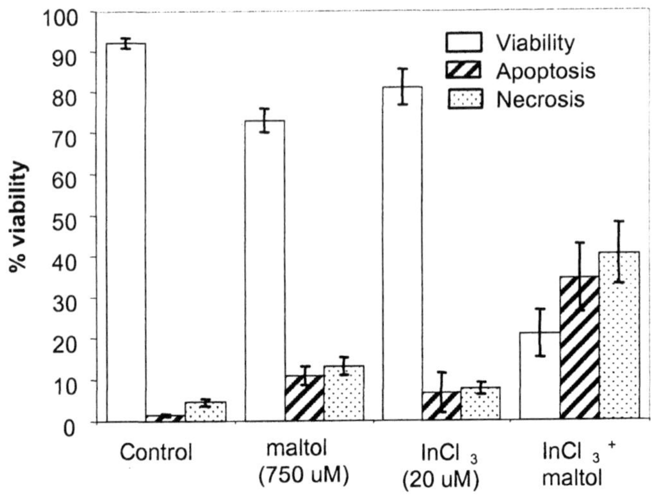

Figure 2. The effect of $InCl_3$ and maltol on melanoma cell line A375 over 72 hrs, by FACs analysis. Bars as percent total cell count: white, percent viable; striped, apoptotic; dotted, necrotic.

Dithiocarbamate-based drugs

Dithiocarbamates (DTCs) are chelators derived from CS_2, commonly used as anti-oxidants in cell-studies. However, Slater and coworkers have recently shown that DTCs may also act as pro-oxidants by increasing Cu uptake into cells.(*34, 35*) DTCs form strong lipophilic complexes with Cu and Zn, which can pass through cell membranes. Inside the cell, Slater postulates the complexes are oxidatively decomposed, releasing free Cu which may then induce oxidative stress and the generation of hydroxyl radicals.

We screened the cytotoxicity of metal-dithiocarbamate complexes against melanoma cell cultures and found that Cu, Zn and Fe DTC complexes decreases melanoma viability, but that Ni complex enhance it. As the first complexes had low solubility in water, we then investigated several water-soluble DTC Cu complexes, such as the N-methyl glutamate derivative $Cu(mgDTC)_2$, but all were

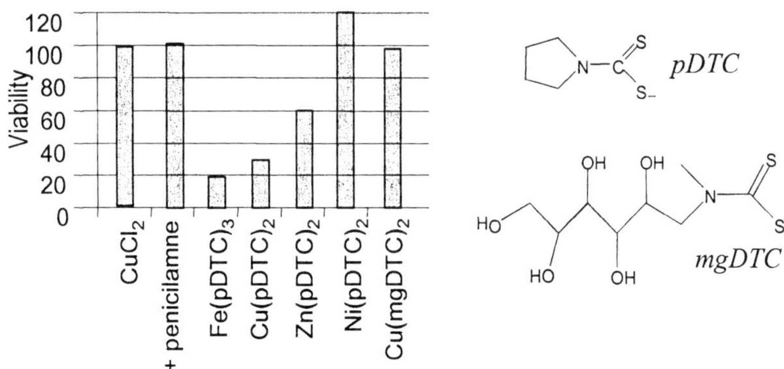

Figure 3. Cell viability as % of control in human metastatic melanoma cell line c81-61 following 72 hr incubation with 1 ug/mL of various DTC complexes; viability of cells treated with $CuCl_2$ and penicilamine are shown as controls.

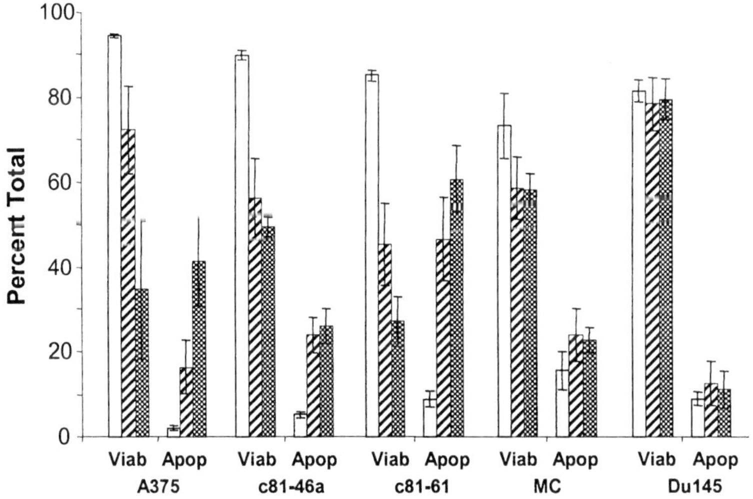

Figure 4. The effect of DSF on melanoma (A375, c81-46a, c81-61), melanocytes (MC), and prostate cancer cell lines (Du145). Bars as percent total cell count: white control; striped 25 ng/mL DSF (84 nM); dotted, 50 ng/mL DSF (168 nM). Data from reference 33.

completely inactive. Thus both metal-specific and fat-solubility requirements are evident for anti-melanoma activity. Subsequent studies centered on pDTC complexes of Cu and Zn, as in Figure 3, using fluorescence-based cell cytometric assays (FACS). We find that both compounds significantly increase apoptosis, and both have increased activity under high oxygen conditions

Importantly, disulfiram (**DSF**) which is an oxidized form of diethyldithiocarbamate (**deDTC**), **Scheme 4**, was found to induce dramatic increases in apoptosis in a number of melanoma cell lines.(*36*) As shown in Figure 4, melanoma cells in culture are much more sensitive to DSF than those of normal melanocytes (MC) or prostate cancer (Du145). This is significant as DSF has been used for fifty years as the main component of alcohol-aversion therapy approved by the FDA.(*37*) The use of DSF in clinical treatment of melanoma is currently being investigated in a phase I trial.(*38*)

Scheme 4

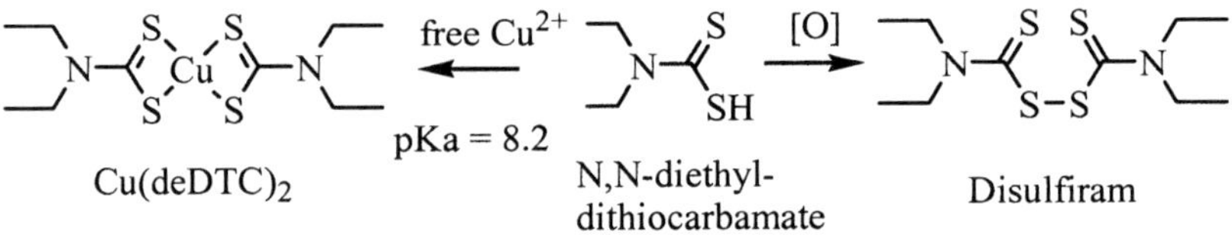

free Cu^{2+} ← ; pKa = 8.2 ; [O] →

$Cu(deDTC)_2$ — N,N-diethyl-dithiocarbamate — Disulfiram

We have recently found that the anti-melanoma activity of DSF is Cu dependent,(*39*) and several lines of evidence suggest that the active species is $Cu(deDTC)_2$. First of all, the anti-melanoma activity of DSF is enhanced by the addition of $CuCl_2$ and attenuated in the presence of large excess of chelators, such as bathocuproine disulfonic acid, BCPD. Likewise, DSF treatment leads to significant increases in intracellular Cu concentrations, which can also be inhibited by BCPD. Chemical studies have shown that in aqueous media, Cu(II) salts react with DSF to directly form $Cu(deDTC)_2$ in almost stoichiometric yields, via oxidative decomposition of small amounts of disulfiram.

Conclusions and considerations

There is a growing interest in metal-based chemotherapeutic drugs and radiopharmaceuticals, as well as in metal-chelators that may control metal-ion trafficking or inhibit specific metalloenzymes.(*40-44*) A number of metal-complexes have been reported to have activity against melanoma,(*45-50*) thus the proven metal-sensitivity of melanoma suggests it may be a likely target of other lipophilic metal complexes and chelators.

While our work demonstrates high anti-melanoma activity of various metal complexes, it does not address the mechanism of action that results in the dominant apoptotic response. For the most active compound, $Cu(deDTC)_2$, this response is closely tied to the uptake and accumulation of Cu. Previous investigations of the activity of both DSF and other DTC derivatives have focused on bio-transformations of the DTC ligand itself, but little consideration was given to what effect metal-uptake might induce.(*51-54*) DTC cytotoxicity has been attributed to disruption of critical glutathione-redox status within the cell,(*53*) but many recent publications demonstrate that the cytotoxicity of DTCs depends on the presence metal ions such as Cu and Zn.(*55-57*) For example, pyrrolidine dithiocarbamate (pDTC) inhibits activation of NF-kB by increasing intracellular Cu levels, and that this effect can be reversed by high concentrations of non-cell permeable chelators which compete for the available extracellular Cu.(*58*)

Dithiocarbamates are widely used in industry and agriculture,(*59*) as well as medicine.(37) The U.S. EPA's Office of Pesticide Programs recently published a Scientific Advisory Panel review on the cumulative risks assignable to DTC pesticides, but little mention was given to their metal-transport properties.(*60*) There is ample evidence of DTCs ability to transport metals in biological systems. Cu and Zn uptake and accumulation has been demonstrated in phytoplankton exposed to DTC pesticides.(*61*) Likewise, DTCs increase the uptake and vary the distribution of Pb,(*62*, *63*) Cd,(*64*) and Hg(*65*) in rats. The growing problem of toxic metals in technological waste and widespread usage of DTCs as pesticides gives rise to a possibility of heavy metal/DTC combinations as environmental problems. For instance, a recent study showed that Cu chelation inhibits acid-catalyzed hydrolysis of mDTC in agricultural waste waters reaching San Franscisco Bay, dramatically lengthening their average lifetimes.(*66*) Thus, further study of the chemical and biological activities of DTC derivatives will be of wide ranging importance.

Experimental Section

Abbreviations. Disulfiram (DSF), tetraethythiuram disulfide; pDTC, pyrollidine- dithiocarbamate; deDTC, diethyldithiocarbamate; maltol, 2-methyl-3-hydroxy-4-pyrone; BCPD, bathocuproine disulfonic acid.

General Procedures. Sodium diethyldithiocarbamate trihydrate, sodium diethyl carbamate, tetraethythiuram disulfide, and fetal bovine serum were purchased from Aldrich; cupric chloride was purchased from Kodak and Aldrich; disodium bathocuproinedisulfonic acid (BCPD) was purchased from Sigma; Dulbecco's modified eagle medium was purchased from Invitrogen. Synthetic melanin films for assays were prepared as in reference 23. Where anaerobic techniques were required, a dry glove-box and standard Schlenk techniques were used. ^{1}H and ^{13}C NMR spectra were recorded on either a Bruker DR 400 or an Omega 500 spectrometers at ambient temperatures with chemical shifts in ppm relative to the specified solvent.

Cell culture. Metastatic melanoma cells from the A375 cell line were cultured in correspondent media DMEM medium (Invitrogen, Grand Island, NY), c81-46a and c81-61 were cultured in HAM'S F10 (CellGrowth) with 5% fetal calf serum, 5% newborn calf serum, penicillin (100 U/ml) and streptomycin (0.1 mg/ml). Cells were split 1 to 2 days prior to drug treatment, with a density of 4-6 x 10^4/ml (confluence level of about 75%). Du145 cells were cultured in the same medium as the A375 cells. Melanocytes were processed from pooled neonatal foreskins and cultured in MCDB 153 (Sigma, St Louis, MO) medium containing 2% fetal calf serum, 0.3% bovine pituitary extract (Clonetics, San Diego, CA), 10ng/ml phorbol myristate-13-acetate, 2mM calcium chloride, 5μg/ml insulin, and 0.1mM 3-isobutyl-methylzanthine (Sigma).

Drug administration. Drugs were dissolved in the culture medium or in dimethyl sulfoxide and added to the culture directly. Control samples were added with the same amount of medium/or DMSO. The DMSO in the cell media was <0.2%, and previous experiments have shown that DMSO at concentrations below 1% did not affect cell viability. To control the oxygen concentration during viability studies, confluent cell cultures were incubated in a Billups-Rothenburg MIC-101 Hypoxia/Tissure Culture Incubator. A mixture of 95% oxygen and 5% carbon dioxide was pumped into the chamber and sealed in. The chamber was then placed in and incubator at 37°C. Every 24 hours the chamber was opened to accommodate another cell culture for different timepoints. Every time a new culture was added, more oxygen was re-pumped into the chamber. Control cultures were grown along side the chamber in the same incubator. Doubling the media volume had no effect on viability.

Apoptosis Assay. Apoptosis assay was performed according to the manufacturer's protocol (Pharmingen). Briefly, cells were trypsinized, washed twice in PBS and re-suspended in binding buffer at a concentration of 1 x 10^6 cells/ml, of which 100 μl was incubated with 5 μl of Annexin V (AV) conjugated to FITC (Molecular Probes, Eugene, OR) and 10 mM propidium iodide (PI) for 15 minutes at room temperature. Cells were then analyzed by flow cytometry using a Becton-Dickinson FACScan. The proportion of apoptotic cells was estimated by the percentage of cells that stained positive for AV while remaining impermeable to PI (AV+/PI-); necrosis was defined as positive stain with both AV and PI (AV+/PI+); and viability was defined as AV-/PI-. This method was based on the previous published literature that in the early stages of apoptosis, phosphatidylserine (PS) is translocated from the inner to the outer leaflet of the plasma membrane at the cell surface. Annexin V has a high affinity for PS binding on the cell surface. In late stage of cell death, membrane integrity is lost, and propidium iodide uptake, a non-membrane permeable fluorophores, is assayed for such necrotic cells. Total cell count is the sum of assigned amounts viable, apoptotic and necrotic cells.

Acknowledgements. This research was supported by the American Cancer Society Research Scholar grant (PJF, RSG-03-251-01) and in part by the Waltmar Foundation (DC, P30CA62203) and the Chao Family Comprehensive Cancer Center (FM).

References

1. Helmbach, H.; Rossmann, E.; Kern, M.A.; Schadendorf, D. *Int. J. Cancer* **2001**, *93*, 617-622.
2. Kleeberg, U.R. *Melanoma Res.* **1997,** 7, S143.
3. Larsson, B.S. *Pigment Cell Res.* **1993**, 127-133.
4. Crippa, R.; Horak, V.; Prota, G.; Sworonos, P.; Wolfram, L. In *Alkaloids*, Brossi, A., Ed.; Academic Press: New York, NY, 1989; pp. 253-323.
5. Montagna, W.; Prota, G.; Kenney Jr., J.A. In *Black Skin, Structure and Function;* Academic Press: San Diego, CA., 1993; pp. 117-130.
6. Borovansky, J.; Mirejovsky, P.: Riley, P.A. *Neoplasma* **1991**, *38*, 4-10
7. Curran R.C.; McCann B.G. *J. Pathology.* **1976**, *119*, 135-46.
8. Jimbow, K.; Iwashina, T.; Alena, F.; Yamada, K.; Pankovich, J.; Umemura, T. *J. Invest. Derma.* **1993**, *100*, 231S.
9. Bustamante, J.; Bredeston, L.; Malanga, G.; Mordoh, J. *Pigment Cell Res.* **1993**, *6*, 348-353.
10. Riley, P.A. *Int. J. Biochem. Cell Biol.* **1997**, *29*, 1235-1239.
11. Korytowski, W.; Sarna, T. *J. Biol. Chem.* **1990**, *285*, 12410.
12. Sarna, T.; Duleba, A.; Korytowski, W.; Swartz, H. *Arch. Biochem. Biophys.* **1980**, *200*, 140-148.
13. Sotomatsu, A.; Tanaka, M.; Hirai, S. *FEBS Lett.* **1994**, *342*, 105,
14. Meyskens, F.L.; Farmer, P.J.; Fruehauf, J. *Pigment Cell Res.* **2001**, *14*, 148-154.
15. Meyskens, F.L.; Buckmeier, J.A.; McNulty, S.E.; Tohidian, N.B. *Clin. Canc. Res.* **1999**, *5*, 1197-1202.
16. Applegate, L.A.; Scaletta, C.; Labid, F.; Vile, G.F.; Frenk, E. *Int. J. Cancer* **1996,** *67*, 430.
17. Meyskens F.L.; Van Chau, H.; Tohidian, N.; Buckmeier, J. *Pigment Cell Res.* **1997**, *10*, 184-189.
18. Bedrick, A.E.; Ramaswamy, G.; Tchertkoff, V. *Am. J. Clin. Pathol.* **1986**, *86*, 637-640.

19. Sarzanini, C.; Mentasti, E.; Abollino, O.; Fasano, M. *Marine Chem.* **1992,** *39*, 243.
20. Enochs, W.S.; Petherick, P.; Bogdanova, A.; Mohr, U.; Weissleder, R. *Radiology* **1997**, *204*, 417,.
21. Borovansky, J.; Mirejovsky, P.: Riley, P.A. *Eur. J. Cancer Clin. Oncol.* **1983**, *19*, 91-99.
22. Borovansky, J.; Blasko, M.; Siracky, J.; Schothorst, A.A.; Smit, N.P.M.; Pavel, S. *Melanoma Res.* **1997**, *7*, 449-453.
23. Gidanian, S.; Farmer, P.J. *J. Inorg. Biochem.* **2002**, *89*, 54-60.
24. Szpoganicz, B.; Kong, P.; Farmer, P.J. *J. Inorg. Biochem.* **2002**, *89*, 45-53.
25. Farmer, P.J.; Gidanian, S.; Shahandeh, B.; Di Bilio, A.J.; Tohidian, N.; Meyskens, F.L. *Pigment Cell Res.* **2002**, *15*, 33.
26. Hubbard-Smith, K.; Hill, H.Z.; Hill, G.J. *Radiation. Res.* **1992**, *130*, 160-165.
27. Husain, S.; Hadi, S.M. *Mutat. Res.* **1998**, *397*, 161-168.
28. Felix, C.C.; Hyde, J.S.; Sarna, T.; Sealy, R.C. *J. Am. Chem. Soc.* **1978,** *100*, 3922-3926.
29. McNeill, J.H.; Yuen, V.G.; Hoveyda, H.R.; Orvig, C. *J. Med. Chem.* **1992,** *35*, 1489-91.
30. Hider, R.C.; Ejim, L.; Taylor, P.D.; Gale, R.; Huehns, E.; Porter, J.B. *Biochem. Pharm.* **1990**, *39*, 1005-1012.
31. MacKie, R.M.; Marks, R.; Green, A. *Br. Med. Journal* **1996**, *312*, 1362-1363.
32. Nelemans, P.J.; Scholte, R.; Groendendal, H. *Br. J. Ind. Med.* **1993,** *50*, 642-646.
33. Meyskens, F.L.; Farmer, P.J.; Anton-Culver, H. *Clin. Canc. Res.* **2004,** *10,* 2581-2583.
34. Nobel, C.S.I.; Kimland, M.; Lind, B.; Orrenius, S.; Slater, A.F.G. *J. Biol. Chem.* **1995**, *270*, 26202-26208.
35. Orrenius, S.; Nobel, C.S.I.; Bandendobbelsteen, D.J.; Burkitt, M.J.; Slater, A.F.G. *Biochem. Soc. Trans.* **1996**, *24*, 1032-1038.
36. Cen, D.; Gonzalez, R.I.; Buckmeier, J.A.; Kahlon, R.S.; Tohidian, N.B.; Meyskens, F.L. *Mol. Canc. Ther.* **2002**, *1*, 197-204.
37. Brewer, C. *Alcohol and Alcoholism*, **1993**, *28*, 383-395.
38. Cen, D., Phase I Clinical Trial Protocol, UCI 01-01, **2003**.
39. Cen, D.; Brayton, D.; Shahandeh, B.; Meyskens, F.L. Jr, Farmer, P.J.; *submitted.*
40. Zhang, C.X.; Lippard, S.J. *Curr. Opin. Chem. Biol.* **2003**, *7*, 481-489.
41. Thompson, K.H.; Orvig, C. *Science* **2003**, *300*, 936-939.
42. Paschke, R.; Paetz, C.; Mueller, T.; Schmoll, H.J.; Mueller, H.; Sorkau, E.; Sinn, E. *Curr. Med. Chem.* **2003**, *10*, 2033-2044.
43. Clarke, M.J. *Coord. Chem. Rev.* **2002**, *232*, 69-93.

44. Thompson, K.H.; Orvig, C. *Coord. Chem. Rev.* **2001**, *219*, 1033-1053.
45. Collins, M.; Ewing, D.; Mackenzie, G.; Sinn, E.; Sandbhor, U.; Padhye, S.; Padhye, S. *Inorg. Chem. Comm.* **2000**, *3*, 453-457.
46. Arbiser, J.L.; Kraeft, S.-K.; Van Leeuwen, R.; Hurwitz, S.J.; Selig, M.; Dickersin, G.R.; Flint, A.; Byers, H.R.; Chen, L.B. *Molec. Med.* **1998**, *4*, 665-670.
47. Giblin, M.F.; Wang, N.; Hoffman, T.J.; Jurisson, S.S.; Quinn, T.P. *PNAS* **1998**, *95*, 12814-12818.
48. Garcia-Tojal, J.; Garcia-Orad, A.; Alvarez Diaz, A.; Serra, J. L.; Urtiaga, M. K.; Arriortua, M. I.; Rojo, T. *J. Inorg. Biochem.* **2001**, *84*, 271-278.
49. Lukevics, E.; Arsenyan, P.; Shestakova, I.; Domracheva, I.; Nesterova, A.; Pudova, O. *Euro. J. Med. Chem.* **2001**, *36*, 507-515.
50. Miyamoto, D.; Endo, N.; Oku, N.; Arima, Y.; Suzuki, T.; Suzuki, Y. *Biol. Pharm. Bull.* **1998**, *21*, 1258-1262.
51. Hart, B.W.; Faiman, M.D. *Biochem. Pharm.* **1992**, *43*, 403-406.
52. Madan, A.; Faiman, M.D. *Drug. Met. Dispos.* **1994**, 324-330.
53. Burkitt, M.J.; Bishop, H.S.; Milne, L.; Tsang, S.Y.; Provan, G.J.; Nobel, C.S.I.; Orrenius, S.; Slater, A.F.G. *Arch. Biochem. Biophys.* **1998**, *353*, 73-84.
54. Stefan, C.; Burgess, D.H.; Zhivotovsky, B.; Burkitt, M.J.; Orrenius, S.; Slater, A.F.G. *Chem. Res. Tox.* **1997**, *10*, 636-643.
55. Erl, W.; Weber, C.; Hansson, G.K. *Amer. Journ. Phys. Cell Phys.* **2000**, *278*, C1116-25.
56. Chung, K.C.; Park, J.H.; Kim, C.H.; Lee, H.W.; Sato, N.; Uchiyama, Y.; Ahn, Y.S. *J. Neurosci. Res.* **2000**, *59*, 117-25.
57. Kim, C.H.; Kim, J.H.; Xu, J.; Hsu, C.Y.; Ahn, Y.S. *J. Neurochem.* **1999**, *72*, 1586-92.
58. Iseki, A.; Kambe, F.; Okumura, K.; Niwata, S.; Yamamopto, R.; Hayakawa, T.; Seo, H. Biochem. *Biophys. Res. Comm.* **2000**, *276*, 88-92.
59. Reid, E.E. *Organic Chemistry of Bivalent Sulfur*; Chemical Publishing Co.: New York, NY, **1962**; Chp 3.
60. Mulkey, M.E.; as Director of Office of Pesticide Programs, *EPA MEMORANDUM*, **2001**, available on EPA website.
61. Phinney, J.T.; Bruland, K.W. *Envir. Toxicol. Chem.* **1997**, *16*, 2046-53.
62. Adamis, Z.; Tatrai, E.; Honma, K.; Ungvary, G. *J. Appl. Toxicol.* **1999**, *19*, 347-350.
63. Tatrai, E.; Naray, M.; Brozik, M.; Adamis, Z.; Ungvary, G. *J. Appl. Toxicol.* **1998**, *18*, 33-37.
64. Funakoshi, T.; Ohta, O.; Shimada, H.; Kojima, S. *Toxicol. Lett.* **1995**, *78*, 183-8.
65. Aaseth, J.; Alexander, J.; Wannag, A. *Arch. Toxicol.* **1981**, *48*, 29-39.
66. Weissmahr, K.W.; Sedlak, D.L. *Environ. Tox. Chem.* **2000**, *19*, 820-826.

Chapter 23

Medicinal Applications of Metal Complexes of N-Heterocyclic Carbenes

Wiley J. Youngs*, Claire A. Tessier, Jered C. Garrison, Carol A. Quezada, Abdulkareem Melaiye, Semih Durmus, Matthew J. Panzner, and Aysegul Kascatan-Nebioglu

Department of Chemistry, The University of Akron, 190 East Buchtel Commons, Akron, OH 44325–3601
***Corresponding author: youngs@uakron.edu**

The focus of this chapter is the synthesis and medicinal applications of metal complexes of N-heterocyclic carbenes (NHCs). Silver NHC complexes will be examined for their potential uses as antimicrobials and ^{111}Ag based radiopharmaceuticals. Rhodium NHC complexes are discussed in regards to their potential use as radiopharmaceuticals based on ^{105}Rh.

Introduction

NHCs have novel chelating properties and have been shown to bind very strongly to a variety of metals to produce very stable complexes. In 1968 Öfele (*1*) and Wanzlick (*2*) used imidazolium salt precursors to synthesize the first transition metal N-heterocyclic carbene (NHC) complexes of type **1**. In 1982 Lappert synthesized several complexes in which NHCs were bound to transition metals (*3*) and in 1991 Arduengo (*4*) reported the synthesis of a free isolable NHC. NHCs and their binding to transition metals are the subjects of several review articles (*5*, *6*) and are of considerable interest in the field of catalysis.

The pπ- pπ electron donation from the two nitrogen atoms to the carbene carbon atom gives a stabilization energy of approximately 70 kcal/mol, and is the major contributor to the stability of NHCs (*5*). The aromaticity of **1** contributes about 25 kcal/mol to the stabilization energy. Figure 1 illustrates the different resonance forms of NHCs. The multiple bond character of the nitrogen to carbene carbon bond is represented by structures B and C and summarized by D, while the overall aromatic stabilized structure can be represented by E.

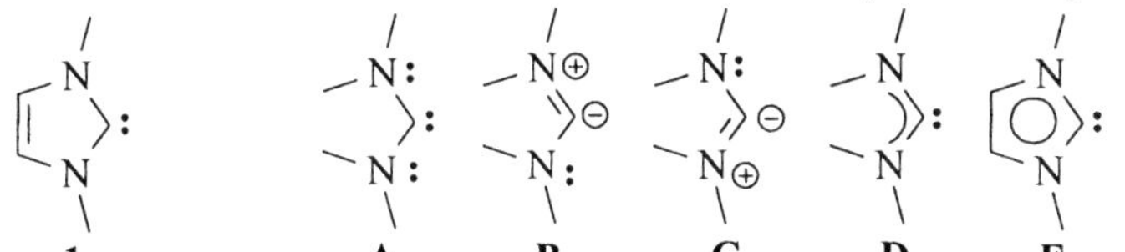

1 **A** **B** **C** **D** **E**

Figure 1. *Resonance structures for NHCs*

The stability of the metal complexes used as therapeutic radiopharmaceuticals is extremely important. A major problem in this area is dissociation of the metal from the metal complex that is administered to the patient. This can result in the accumulation of radioactive elements in healthy tissue of the patient. N-heterocyclic carbenes bind very strongly to transition metals and current data indicate that N-heterocyclic carbenes will bind more strongly to metal centers than do the other chelating and monodentate ligands used in the above applications. The metals to be focused on will be silver and rhodium because ^{111}Ag and ^{105}Rh show promise for use in radiation therapy as a β particle emitting radionuclides. Both ^{105}Rh and ^{111}Ag have convenient half-lives of 1.5 days and 7.47 days, respectively. Both of these radioisotopes emit relatively low levels of imageable γ-radiation, with significantly less γ-radiation than ^{131}I which is used in the treatment of thyroid cancer. The high level of γ-radiation of ^{131}I has been considered one of its drawbacks.

NHC Precursors and Metal Complexes

The synthetic chemistry of the N-heterocyclic carbenes has been developed using non-radioactive naturally abundant $^{107,109}Ag$ and ^{103}Rh in place of the β particle emitting isotopes ^{111}Ag and ^{105}Rh. Our group has synthesized a number of macrocyclic N-heterocyclic carbene complexes of silver and their imidazolium precursors. The results relevant to the synthesis of stable silver complexes are summarized here.

Pyridine Based NHC Precursors and Metal Complexes

The combination of 2,6-bis(imidazolemethyl)pyridine **2** with 2,6-bis(bromomethyl)pyridine **3** in CH_2Cl_2 gives the bisimidazolium-pyridine cyclophane **4** in high yield (*7*). The combination of the PF_6^- salt of **4** with an equimolar amount of Ag_2O gives the silver biscarbene dimer **5** in nearly quantitative yield. The bromide salt of **5** (X=Br) is synthesized in water using 4 equivalents of Ag_2O. Compound **5** (X=Br) is very soluble and stable in water. Under analogous reaction conditions, the combination of **4** (X=PF_6) with 4 equivalents of Ag_2O gives the tetranuclear biscarbene dimer **6** (*8*).

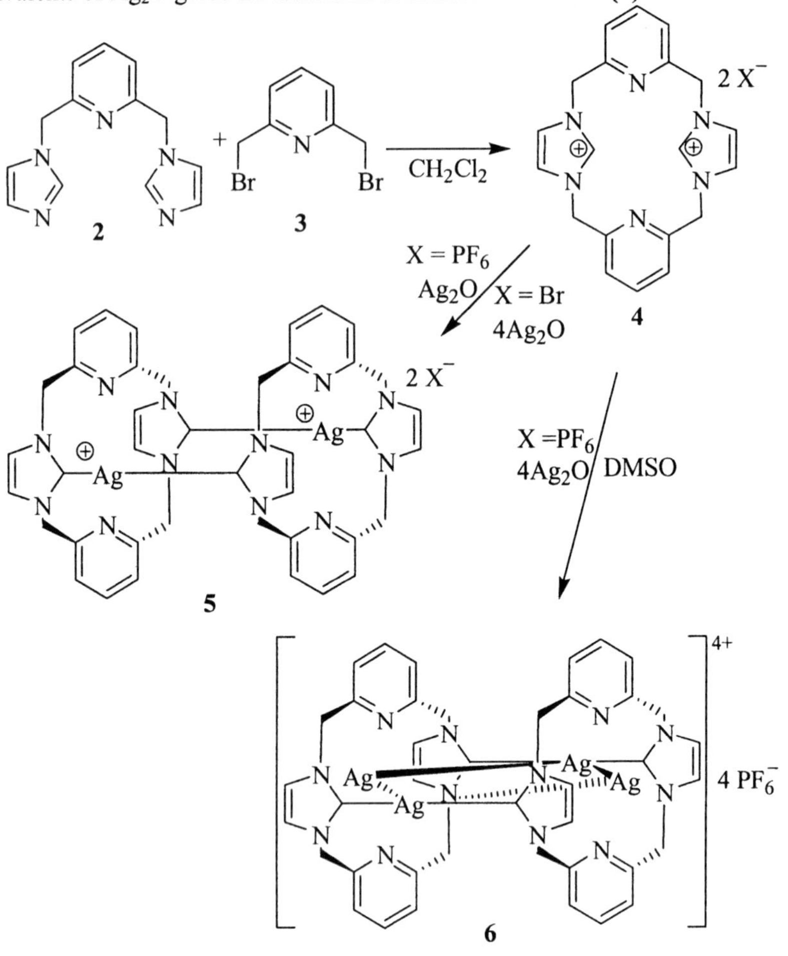

The hexafluorophosphate salts of **5** and **6** were crystallographically characterized and are depicted in Figures 2 and 3, respectively. The silver atoms of **5** are bonded to the NHCs from each of the two cyclophanes. The bonding of the carbenes to the silvers is essentially linear. The tetranuclear silver complex, **6**, is similar to **5** but has Ag atoms (Ag2 and Ag2A) inserted into the pocket of the cyclophane. The Ag atoms form a planar cluster that has Ag-Ag bonding distances (Ag1-Ag2) similar to that of metallic silver. The pocket silvers are stabilized by the pyridines of the cyclophanes as well as the N-heterocyclic carbenes. Preliminary studies show that the bromide and chloride salts of **5** are excellent antimicrobials.

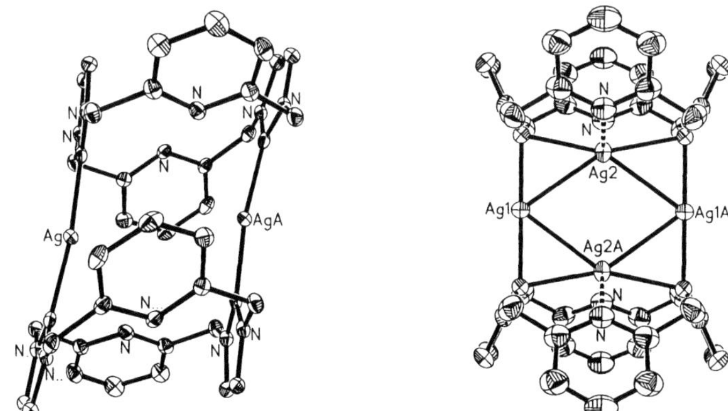

Figure 2. Dicationic portion of ***5***   *Figure 3. Tetracationic portion of* ***6***

The 3 + 1 condensation of **7** with **2** gives the iodide salt **8**. Anion exchange on **8** (X=I) with $NH_4^+PF_6^-$ gives **8** as the hexafluorophosphate salt. The combination of **8** ($X=PF_6$) with four equivalents of Ag_2O gives the tetranuclear biscarbene dimer **9** (*9*). Compound **9** is a dimer composed of a silver cluster that is stabilized by two cyclophanes. The structural bonding of **9** is similar to **6** except that each pocket silver is stabilized by only one pyridine ring.

$2\,I^-$

7 + **2** $\xrightarrow{CH_2Cl_2}$ **8** ($X = I, PF_6$; $2\,X^-$)

8 $\xrightarrow{4\,Ag_2O,\ DMSO}$ **9** ($[\ ]^{4+}$ $4\,PF_6^-$)

A general synthesis of pincer N-heterocyclic carbenes with pyridines as the bridging unit is presented below. The combination of **2** with 2-iodoethanol or 3-bromopropanol gave **10a** and **10b**, respectively (*10*). The combination of the halide salt of **10a** or **10b** with an equimolar amount of Ag_2O gives the silver biscarbene polymers **11a** and **11b**, respectively. Compound **11a** has been crystallographically characterized. The hydroxide salts **11a** and **11b** are very soluble and slowly decompose in water. The decomposition leads to a slow release of silver atoms. The thermal ellipsoid plot (atoms shown isotropically) of **11a** is shown in Figure 4.

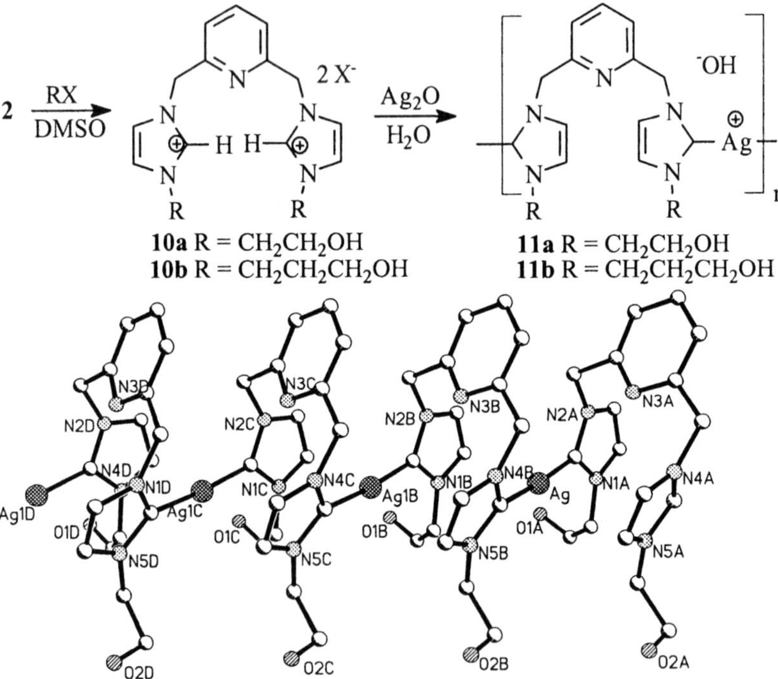

*Figure 4. Thermal Ellipsoid Plot showing the polymeric **11a***

The antimicrobial activity of water soluble silver (I) N-heterocyclic carbene **11a** was investigated (*10*) on four types of fungi (*Candida albicans*, *Aspergillus niger*, *Mucor*, *Saccharomyce cerevisiae*) using the LB broth dilutions technique, and bacteria (*E. coli, S. aureus, P. aeruginosa*) of clinical importance. The silver NHC complex **11a** was effective on bacteria and fungi at about the same concentrations as $AgNO_3$. However, the silver NHC complex exhibited a longer period of silver activity than silver nitrate over the 7 day time course of the experiment. Toxicity studies with rats have shown that ligand **10a**, the precursor to **11a** and the product that forms on degradation of **11a**, is of low toxicity and clears within two days through the kidneys as determined by mass spectroscopic study of the urine.

The combination of **2** with 1,3-dichloroacetone in acetonitrile at 70°C for 2hrs gave the imidazolium salt **12**. The silver (I) carbene complex **13** was obtained by combining Ag_2O with **12** in methanol and stirring for 1 hr. The x-ray crystal structure of **13** is shown in Figure 5. The silver (I) carbene compound is air and light stable, soluble and stable in ethanol solution for more than 24 hrs, and slightly soluble in water. These properties suggest that **13** is an excellent candidate for embedding into fibers using the electrospin coating techniques to form silver impregnated bandages with antimicrobial properties. The TEM images of the spun mat of **13** embedded in Tecophilic® (polyurethane) polymer showed uniform dispersion (*11*). Silver particles were visible when the spun silver-polymer mat was soaked in water. The antimicrobial activity of the embedded silver-polymer mat was impressive. No growth of the organism was observed within the area of the mat tested over a lawn of organism.

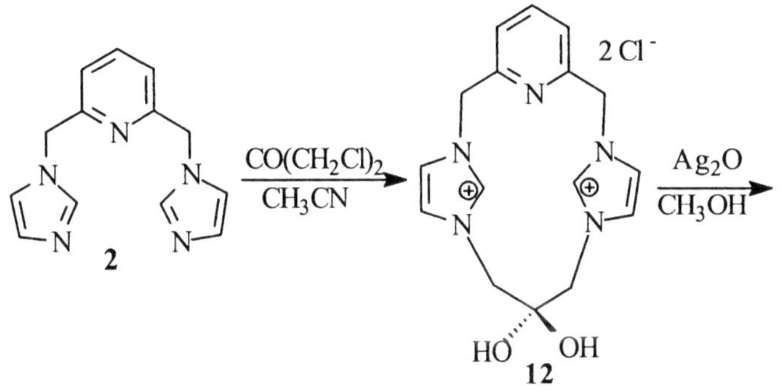

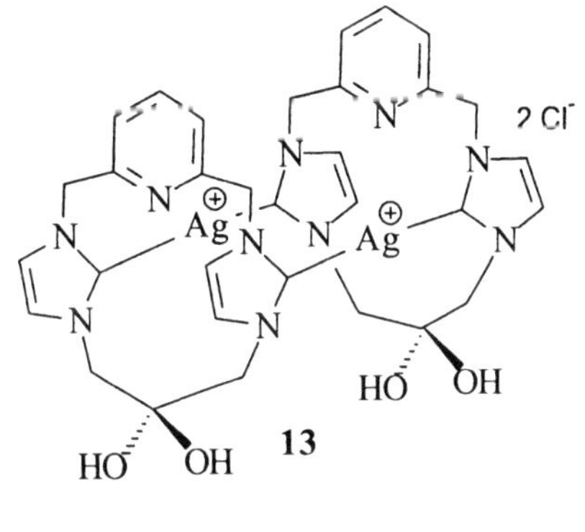

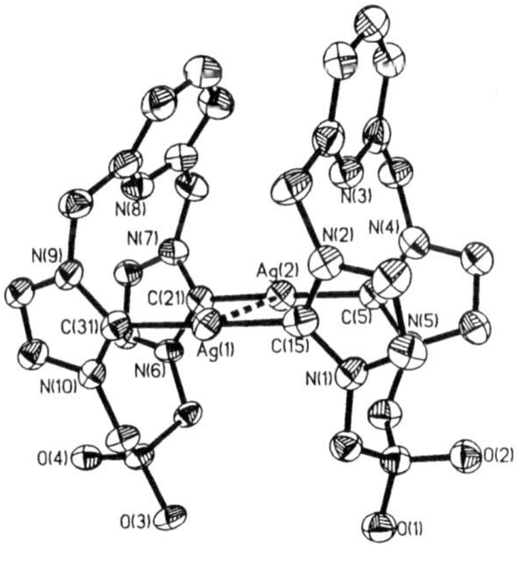

$CO(CH_2Cl)_2$, CH_3CN; Ag_2O, CH_3OH

Figure 5 TEP of **13**

Corrole and sapphyrin analogs are examples of how the pocket sizes of imidazolium cyclophanes can be altered to accommodate different sized metals into the pockets. The corrole analog **16** would be best suited for first row transition metals because of its decreased size. The pocket cavity of the sapphyrin analog **17** is larger and would be potentially better suited for coordinating both second and third row transition metals. The synthesis of **16** and **17** is outlined below (*12*). Thermal ellipsoid plots of compounds **16** and **17** are shown in Figure 6 and 7, respectively.

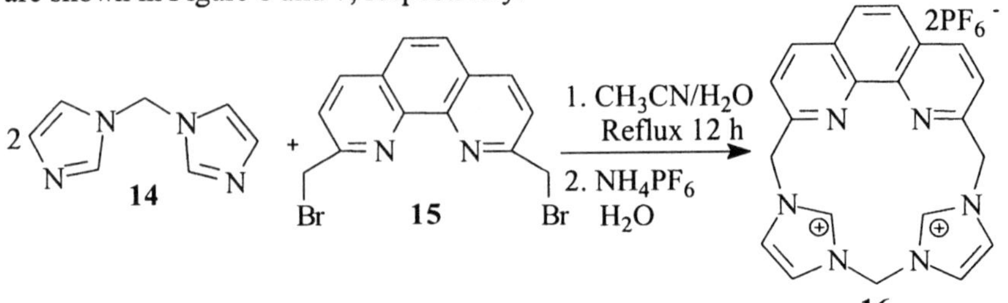

2 **14** + **15** → 1. CH_3CN/H_2O Reflux 12 h; 2. NH_4PF_6 H_2O → **16** $2PF_6^-$

2 + **15** → CH_2Cl_2 → **17** $2\ X^-$

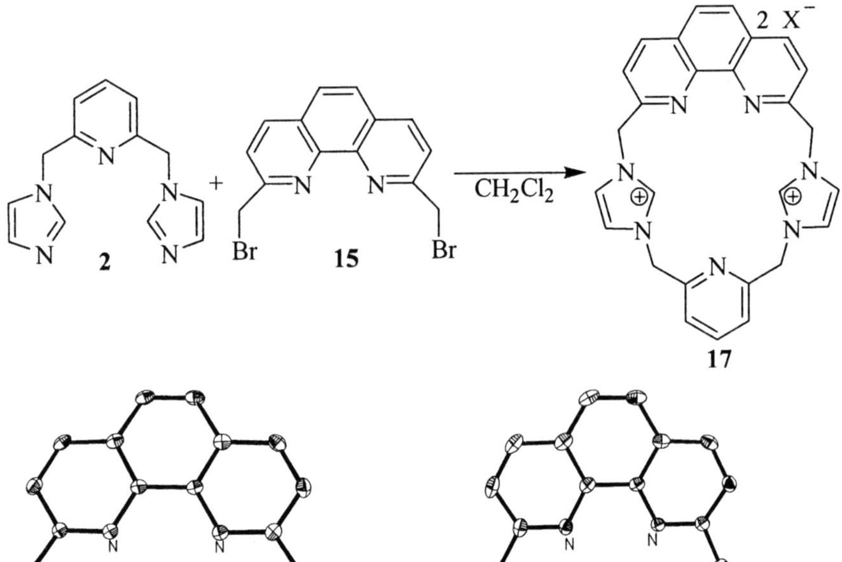

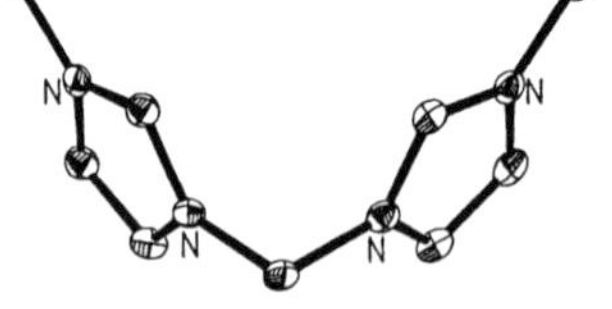

Figure 6. TEP of ***16***

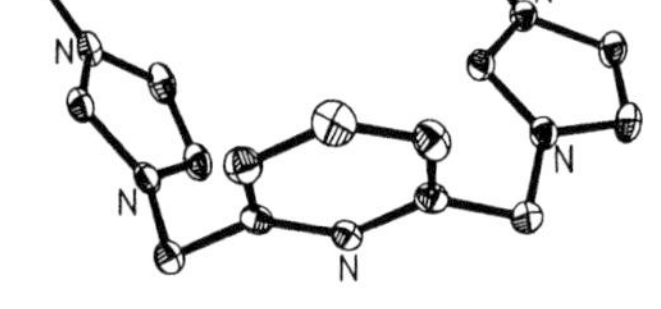

Figure 7. TEP of ***17***

NHC Precursors Functionalized for Targeting

Histamine and histidine can be used to construct cyclophanes with side chains attached for incorporation of targeting moieties. The reaction of histamine dihydrochloride **18** with carbonyldiimidazole (*13*) in DMF resulted in 5,6,7,8-tetrahydro-5-oxoimidazo [1,5-c]pyrimidine **19**. The combination of two equivalents of **19** with 2,6-bis(bromomethyl)pyridine **3** in acetonitrile resulted in the formation of **20** in very high yield. Treatment of **20** with diisopropylethylamine and methanol resulted in the formation of **21**. The combination of 2,6-bis(bromomethyl)pyridine **3** with **21** gives the macrocyclic ring **22** in high yield. Compounds **19-22** have been crystallographically characterized (*14*). The thermal ellipsoid plot of **22** is shown in Figure 8. The protecting group on the amine portion of **22** can be readily removed. This should allow for the attachment of targeting groups such as peptides and 2-nitroimidazole ultimately leading to the formation of radioimmunoconjugates.

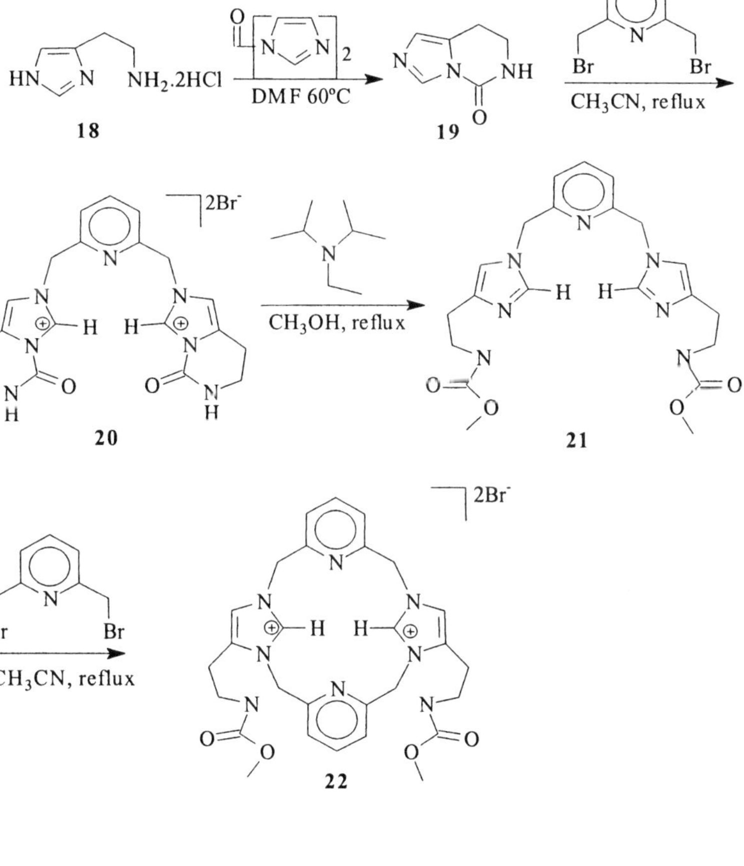

HN N NH$_2$.2HCl **18** → (DMF 60°C) N N NH O **19** → (Br Br, CH$_3$CN, reflux)

20 (2Br$^-$) → (CH$_3$OH, reflux) **21**

→ (Br Br, CH$_3$CN, reflux) **22** (2Br$^-$)

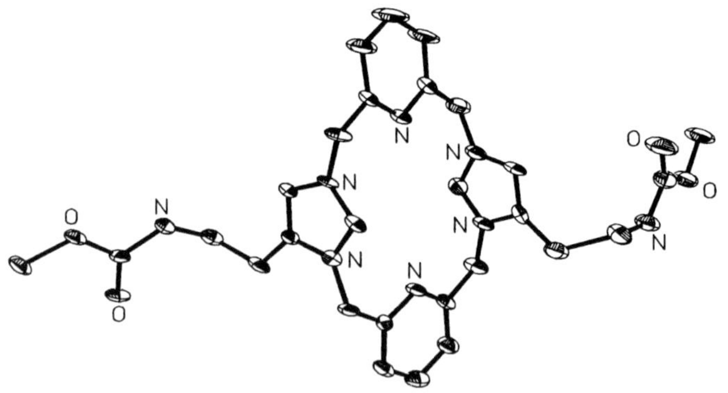
N
N
N
N
N
N
N
N
O
O
O
O

Figure 8. Dication of ***22****.*

The 3 and 3' positions of 1,1'-methylene bis(imidazole) **14** can be readily substituted with different bromoester **23** compounds to form ester functionalized bisimidazolium salts **24**. The carboxylic functionality of these salts can be potentially used for condensations with small peptides known to have binding affinities to overexpressed receptor sites of certain tumor cells. Other targeting groups such as 2-nitroimidazole could also be incorporated in a similar manner. The hexafluorophosphate salt of the ethyl ester compound reacts with Ag_2O in DMSO to form the silver carbene complex **25**.

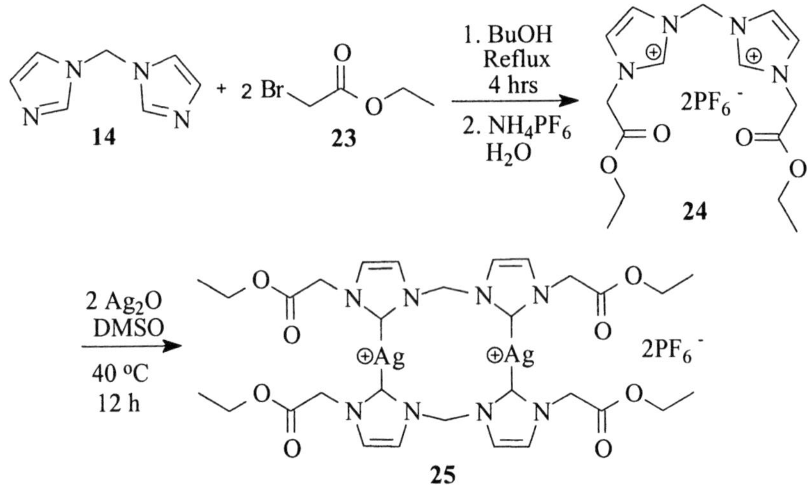
14
+ 2 Br
23
1. BuOH
Reflux
4 hrs
2. NH4PF6
H2O
2PF6-
24
2 Ag2O
DMSO
40 °C
12 h
⊕Ag
⊕Ag
2PF6-
25

NHC Rhodium Complexes

A goal of this research is to evaluate the possibility of using rhodium N-heterocyclic carbene (Rh NHC) complexes as ^{105}Rh radiopharmaceuticals. The fact that ^{105}Rh emits beta particles in the ratio of 70% 0.560 MeV and 30% 0.250 MeV that are suitable for radiation therapy as well as a 319 keV gamma ray that is suitable for imaging makes using ^{105}Rh radiopharmaceuticals an excellent choice for tumor therapy. The ^{105}Rh half-life of 35.5 hours is advantageous in that it provides sufficient time to kill tumor cell lines when appropriately targeted, but it is also short enough to avoid prolonged accumulation of radioactive material in the body (*15*). Because the radioactive ^{105}Rh is available only as $RhCl_3.xH_2O$, the challenge exists of making a robust, chelating Rh NHC complex starting with $RhCl_3.xH_2O$ well within the 35.5 hour half life with relatively simple chemistry. It could be envisioned to do this in two ways: (a) by rapidly transforming the $RhCl_3.xH_2O$ into a form that had been previously used to form rhodium carbene complexes (*16 - 20*) or (b) by the direct synthesis of Rh NHC complexes from $RhCl_3.xH_2O$. Both approaches are shown here.

In the first approach $RhCl_3.xH_2O$ is converted to $(Rh(OAc)_2)_2.2MeOH$ (*21*). The combination of $(Rh(OAc)_2)_2.2MeOH$, **26a** (*22*), sodium acetate, and KI in acetonitrile at reflux for 12 hours gave the Rh NHC **27** in good yield. The major problems with this approach are: (a) the synthesis is performed in acetonitrile, a toxic solvent which is not suitable for medicinal use, and (b) if applied to ^{105}Rh, it would require several synthetic steps with radioactive materials, making this approach impractical.

An alternative approach is to synthesize a Rh NHC directly from $^{105}RhCl_3.xH_2O$. This synthesis has been developed and bypasses the previously mentioned problems, in that it is performed in H_2O and DMSO, DMSO is a biologically benign solvent (*23*, *24*), and the Rh is introduced at the end, thereby avoiding using radioactive materials in every step of a multi-step synthesis. The synthesis of the Rh complexes, **29a** and **29b**, is outlined below. The Ag complexes, **28a** and **28b**, are formed by stirring a solution of the bisimidazolium salt (*25*), **26a** or **26b**, with Ag_2O at RT for 15 minutes and 2 hours, respectively. After filtration an aqueous solution of the Ag NHC complex **28a** or **28b** is combined with $RhCl_3.3H_2O$ in air at RT. The resultant red powder is isolated by vacuum filtration and heated in DMSO at 100°C for 1 hour in air. The rhodium complexes **29a** and **29b** are stable in sodium chloride solution for months, and they are extremely stable *in vitro* with decomposition points of 208-210°C and 237-238°C, respectively (*26*). See Figures 9 and 10 for thermal ellipsoid plots of **27** and **29b**.

The research has demonstrated that extremely stable Rh NHC complexes can be synthesized directly from $RhCl_3 \cdot 3H_2O$. This suggests that ^{105}Rh complexes of bisimidazole ligands containing appropriate targeting substituents such as peptides may be suitable complexes for radiation therapy. Work continues to improve the synthesis, ideally completing it solely in water.

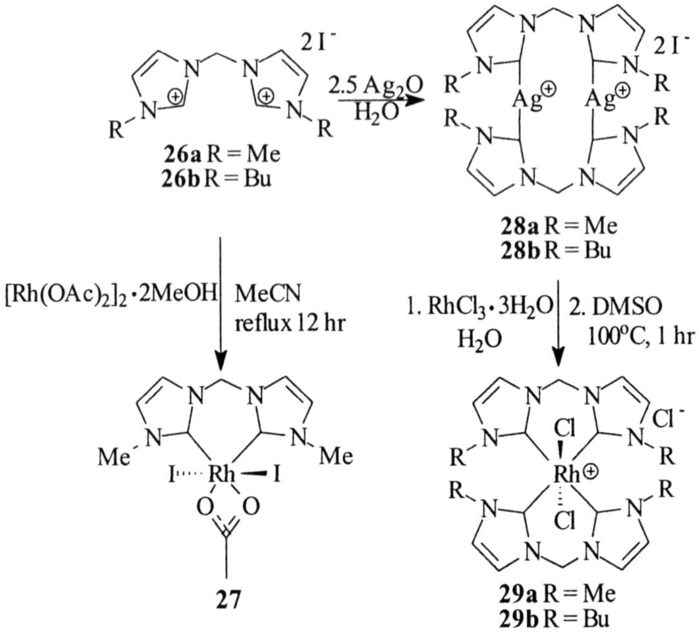

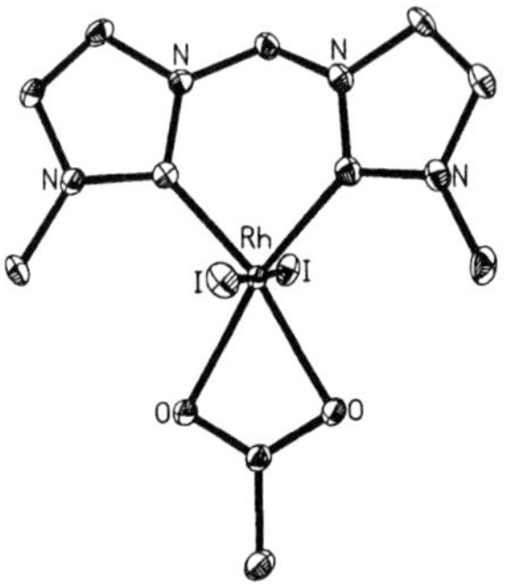

2.5 Ag_2O, H_2O

26a R = Me
26b R = Bu

28a R = Me
28b R = Bu

$[Rh(OAc)_2]_2 \cdot 2MeOH$, MeCN, reflux 12 hr

1. $RhCl_3 \cdot 3H_2O$, H_2O 2. DMSO, 100°C, 1 hr

27

29a R = Me
29b R = Bu

Figure 9. TEP of 27

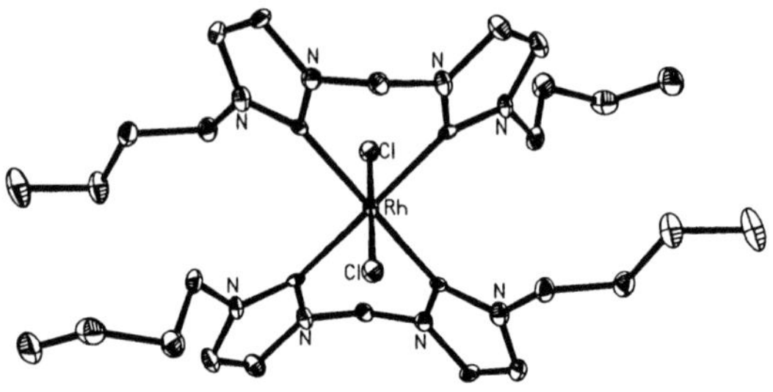

*Figure 10. TEP of **29b***

Methylated caffeine, 1,3,7,9-tetramethylxanthinium **30a**, is synthesized by the reaction of caffeine with dimethyl sulfate and converted to **30b** by anion exchange. Water soluble ligand **30a** reacts with Ag_2O in water to form **31a** (*27*).

Similarly, ligand **30b** readily reacts with Ag_2O in DMSO at 60°C to yield the silver NHC complex **31b** in high yield confirmed by X-ray crystallography shown in Figure 11. This reaction of methylated caffeine with Ag_2O was first reported in 1926 (*28*), however, the authors believed they had synthesized the hydroxide salt of **30**. Complex **31a** is stable in water and the absence of light for five days. Complex **31b** is stable in air and light up to its melting point and only soluble in DMSO. It is stable in wet DMSO for months in the light. The reaction of equimolar amounts of complex **31a** and $[Rh(COD)Cl]_2$ in DMSO gives the corresponding Rh(I) complex **32**. Complex **32** is stable in air up to its melting point. Compound **32** has been crystallographically characterized and is depicted in Figure 12.

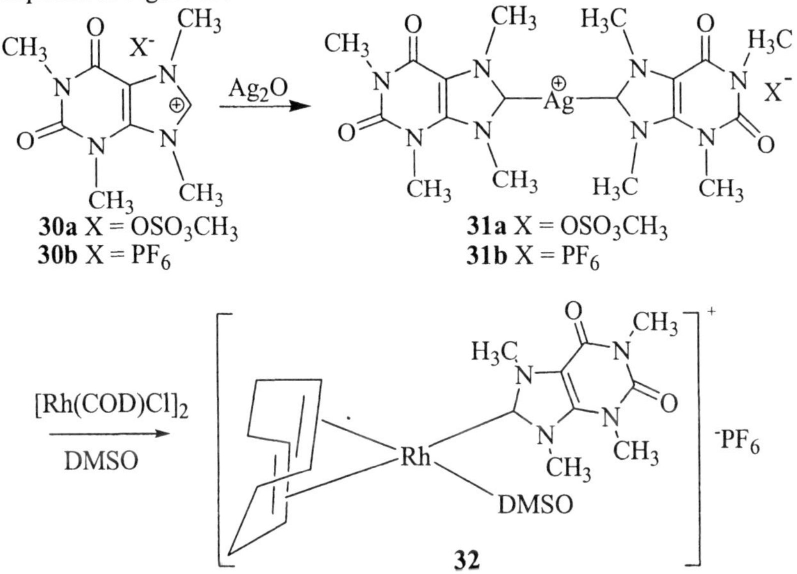

Ag_2O

30a X = OSO_3CH_3
30b X = PF_6

31a X = OSO_3CH_3
31b X = PF_6

$[Rh(COD)Cl]_2$
DMSO

32

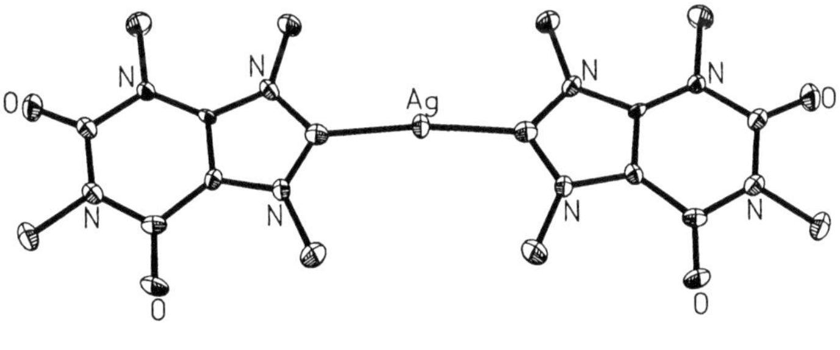

*Figure 11. TEP of **31b***

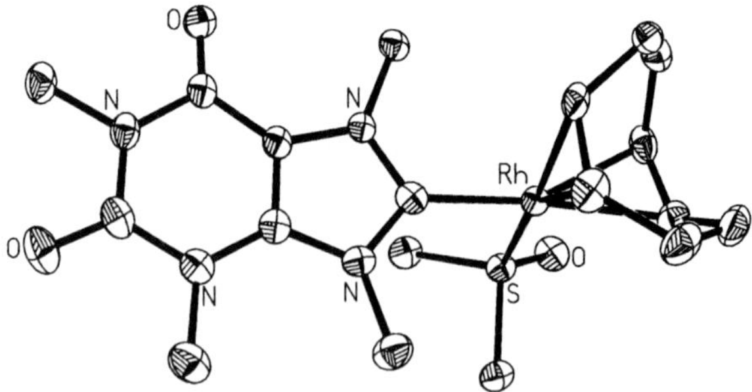

Figure 12. TEP of ***32***

Conclusions

It has been demonstrated that N-Heterocyclic carbenes are promising new ligands for medicinal use. Our preliminary studies show silver NHC complexes have the desired properties to make them good antimicrobial agents. Further studies are being performed to ascertain the effectiveness of current Ag NHC complexes as antimicrobial agents and to develop Ag NHC complexes, that are stable to biological conditions, for use as ^{111}Ag radioimmunoconjugates. Although the Ag NHC complexes are easily generated in water, and other benign solvents, currently the synthetic route to the formation of Rh NHC complexes in water has not been achieved. Our recent results do show promise for the formation of stable Rh complexes in benign solvents such as DMSO while using $RhCl_3 \cdot 3H_2O$, which is currently the available form of ^{105}Rh for synthesis. Further investigation into developing a faster and easier synthetic route to Rh NHC complexes is underway.

References:

[1] Öfele, K. *J. Organomet. Chem.* **1968**, *12*, P42-P43.

[2] Wanzlick, H.-W.; Schönherr, H.-J. *Angew. Chem. Int. Ed. Engl.* **1968**, *7,* 141-142.

[3] Hitchcock, P. B.; Lappert, M. F.; Terreros, P. *J. Organomet. Chem.* **1982**, *239*, C26-C30.

[4] Arduengo, A. J. III; Harlow, R. L.; Kline, M. *J. Am. Chem.Soc.* **1991**, *113*, 361-363.

[5] Bourissou, D.; Guerret, O.; Gabbaie, F. P.; Bertrand, G. *Chem. Rev.* **2000**, *100*, 39-91.

[6] Herrmann, W. A.; Köcher, C. *Angew. Chem. Int. Ed. Engl.* **1997**, *36,* 2162-2187.

[7] Garrison, J. C.; Simons, R. S.; Talley, J. M.; Wesdemiotis, C.; Tessier, C. A.; Youngs, W. J. *Organometallics*, **2001**, *20*, 1276-1278.

[8] Garrison, J. C.; Simons, R. S.; Tessier, C. A.; Youngs, W. J. *J. Organomet. Chem.*, **2003**, *673*, 1-4.

[9] Garrison, J. C.; Simons, R. S.; Kofron, W. G.; Tessier, C. A.; Youngs, W. J. *Chem. Commun.*, **2001**, 1780-1781.

[10] Melaiye, A.; Simons, R. S.; Milsted, A.; Pingitore, F.; Wesdemiotis, C.; Tessier, C. A.; Youngs, W. J. *J. Med.Chem.* **2004**, *47*, 973-977.

[11] Melaiye, A.; Sun, Z.; Reneker, D. H.; Milsted, A.; Tessier, C. A.; Youngs, W. J. *J. Med. Chem.*, **2004**, In Press.

[12] Garrison, J. C.; Panzner, M. J.; Tessier, C. A.; Youngs, W. J. *Synlett.* Submitted.

[13] Jain, R.; Cohen, L. A., *Tetrahedron*, **1996**, 52, 5363-5370.

[14] Durmus, S.; Garrison, J. C.; Panzner M. J.; Tessier, C. A.; Youngs, W. J., *Tetrahedron*, Submitted.

[15] Jurisson, S. S.; Ketring, A. R.; Volkert, W. A., *Transition Met. Chem.*, **1997**, 22, 315-317.

[16] Poyatos, M.; Mas-Marza, E.; Mata, J. A.; Sanau, M.; Peris, E. *Eur. J. Inorg. Chem.* **2003**, 1215-1221.

[17] Poyatos, M.; Sanau, M.; Peris, E. *Inorg. Chem.* **2003**, *42*, 2572-2576.

[18] Poyatos, M.; Uriz, P.; Mata, J. A.; Claver, C.; Fernandez, E.; Peris, E. *Organometallics*, **2003**, *22*, 440-444.

[19] Albrecht, M.; Crabtree, R.H.; Mata, J. A.; Peris, E. *Chem. Commun.*, **2002**, 32-33.

[20] Simons, R. S.; Custer, P.; Tessier, C. A.; Youngs, W. J. *Organometallics*, **2003**, 22, 1979-1982.

[21] Rempel, G. A.; Legzdins, P.; Smith, H.; Wilkinson, G. In Inorganic Synthesis, Cotton, F. A. Ed; McGraw-Hill Book Company; 1972, Vol. 13; pp 90-91.

[22] Herrmann, W. A.; Goossen, L. J.; Kocher, C.; Artus, G. R. J. *Angew. Chem. Int. Ed. Engl.* **1996**, *35*, 2805-2807.

[23] Gavrilin, M. V.; Karpenya, L. I.; Ushakova, L. S.; Senchukoa, G. V.; Kompantseva, E. V. *Pharmaceutical Chem. J.*, **2001**, *35*, 284-285.

[24] Yu, L. *J. Agric. Food Chem.*, **2003**, *51*, 2344-2347.

[25] Quezada, C. A.; Garrison, J. C.; Tessier, C. A.; Youngs, W. J. *J. Organomet. Chem.*, **2003**, *671*, 183-186.

[26] Quezada, C. A.; Garrison, J. C.; Panzner, M. J.; Tessier, C. A.; Youngs, W. J., *Organometallics*, **2004**, In Press.

[27] Kascatan-Nebioglu, A.; Garrison, J. C.; Panzner, M. J.; Tessier, C. A.; Youngs, W. J. *Organometallics*, **2004**, *23*, 1928-1931.

[28] Paderi, C. *Archivio di Farmacologia Sperimentale e Scienze Affini*, **1926**, 41, 92-101.

Indexes

Author Index

Subject Index

B

C

D

E

F

G

H

I

M

N

O

P

Q

R

T

U

V

W

X

Z